# STUDENT'S SOLUTIONS MANUAL

TONI GARCIA

*Arizona State University*

# ELEMENTARY STATISTICS

NINTH EDITION

## Neil A. Weiss

*School of Mathematical and
Statistical Sciences
Arizona State University*

**PEARSON**

Boston  Columbus  Indianapolis  New York  San Francisco
Amsterdam  Cape Town  Dubai  London  Madrid  Milan  Munich  Paris  Montreal  Toronto
Delhi  Mexico City  São Paulo  Sydney  Hong Kong  Seoul  Singapore  Taipei  Tokyo

The author and publisher of this book have used their best efforts in preparing this book. These efforts include the development, research, and testing of the theories and programs to determine their effectiveness. The author and publisher make no warranty of any kind, expressed or implied, with regard to these programs or the documentation contained in this book. The author and publisher shall not be liable in any event for incidental or consequential damages in connection with, or arising out of, the furnishing, performance, or use of these programs.

Reproduced by Pearson from electronic files supplied by the author.

Copyright © 2016, 2012, 2008, 2005 Pearson Education, Inc.
Publishing as Pearson, 501 Boylston Street, Boston, MA 02116.

All rights reserved. No part of this publication may be reproduced, stored in a retrieval system, or transmitted, in any form or by any means, electronic, mechanical, photocopying, recording, or otherwise, without the prior written permission of the publisher. Printed in the United States of America.

ISBN-13: 978-0-321-98965-9
ISBN-10: 0-321-98965-1

www.pearsonhighered.com

# Contents

| | | |
|---|---|---|
| Chapter 1 | The Nature of Statistics | 1 |
| Chapter 2 | Organizing Data | 13 |
| Chapter 3 | Descriptive Measures | 69 |
| Chapter 4 | Descriptive Methods in Regression and Correlation | 117 |
| Chapter 5 | Probability and Random Variables | 157 |
| Chapter 6 | The Normal Distribution | 191 |
| Chapter 7 | The Sampling Distribution of the Sample Mean | 221 |
| Chapter 8 | Confidence Intervals for One Population Mean | 257 |
| Chapter 9 | Hypothesis Tests for One Population Mean | 289 |
| Chapter 10 | Inferences for Two Population Means | 315 |
| Chapter 11 | Inferences for Population Proportions | 343 |
| Chapter 12 | Chi-Square Procedures | 361 |
| Chapter 13 | Anaylsis of Variance (ANOVA) | 389 |
| Chapter 14 | Inferential Methods in Regression and Correlation | 419 |

**1.45** (a)

| E,M,P,L | E,M,L,B | E,P,A,B | M,P,A,B |
|---|---|---|---|
| E,M,P,A | E,M,A,B | E,L,A,B | M,L,A,B |
| E,M,P,B | E,P,L,A | M,P,L,A | P,L,A,B |
| E,M,L,A | E,P,L,B | M,P,L,B | |

(b) One procedure for taking a random sample of four representatives from the six is to write the initials of the representatives on six separate pieces of paper, place the six slips of paper into a box, and then, while blindfolded, pick four of the slips of paper. Or, number the representatives 1-6, and use a table of random numbers or a random-number generator to select four different numbers between 1 and 6.

(c) 1/15; 1/15

**1.47** (a)

| F,T | F,G | F,H | F,L | F,B | F,A |
|---|---|---|---|---|---|
| T,G | T,H | T,L | T,B | T,A | G,H |
| G,L | G,B | G,A | H,L | H,B | H,A |
| L,B | L,A | B,A | | | |

(b) 1/21; 1/21

**1.49** (a) I am using Table I to obtain a list of 10 random numbers between 1 and 500 as follows.

I start at the three digit number in line number 14 and column numbers 10-12, which is the number 452.

I now go down the table and record the three-digit numbers appearing directly beneath 452. Since I want numbers between 1 and 500 only, I throw out numbers between 501 and 999, inclusive. I also discard the number 000.

After 452, I skip 667, 964, 593, 534, and record 016.

Now that I've reached the bottom of the table, I move directly rightward to the adjacent column of three-digit numbers and go up.

I record 343, 242, skip 748, 755, record 428, skip 852, 794, 596, record 378, skip 890, record 163, skip 892, 847, 815, 729, 911, 745, record 182, 293, and 422.

I've finished recording the 10 random numbers. In summary, these are:

452   016   343   242   428
378   163   182   293   422

(b) We can use Minitab to generate random numbers. Following the instructions in The Technology Center, our results are 489, 451, 61, 114, 389, 381, 364, 166, 221, and 437. Your result may be different from ours.

**1.51** (a) First re-assign the elements 93 though 118 as elements 01 to 26.

Select a random starting point in Table I of Appendix A and read in a pre-selected direction until you have encountered 8 different elements. For example, if we start at the top of the column 10 and read two digit numbers down and then up in the following columns, we encounter the elements 04, 01, 03, 08, 11, 18, 22, and 15. This corresponds to a sample of the elements Cm, Np, Am, Fm, Lr, Ds, Fl, and Bh. Your answer may be different from this one.

(b) We can use Minitab to generate random numbers. Following the instructions in The Technology Center, our results are 8, 2, 9, 20, 24, 19, 21, and 13. Thus our sample of 8 elements is Fm, Pu, Md, Cn, Lv, Rg, Uut, and Db. Your result may be different from ours.

**4  Chapter 1**

**1.53**  (a)  One of the dangers of nonresponse is that the individuals who do not respond may have a different observed value than the individuals that do respond causing a nonresponse bias in the estimate. Nonresponse bias may make the measured value too small or too large.

   (b)  The lower the response rate, the more likely there is a nonresponse bias in the estimate. Therefore the estimate will either under or over estimate the generalized results to the entire population.

**Exercises 1.3**

**1.55**  Systematic random sampling is easier to execute than simple random sampling and usually provides comparable results. The exception is the presence of some kind of cyclical pattern in the listing of the members of the population.

**1.57**  Ideally, in stratified sampling, the members of each stratum should be homogeneous relative to the characteristic under consideration.

**1.59**  (a)  Answers will vary, but here is the procedure: (1) Divide the population size, 372, by the sample size, 5, and round down to the nearest whole number if necessary; this gives 74. Use a table of random numbers (or a similar device) to select a number between 1 and 74, call it $k$. (3) List every $74^{th}$ number, starting with $k$, until 5 numbers are obtained; thus, the first number of the required list of 5 numbers is $k$, the second is $k + 74$, the third is $k + 148$, and so forth.

   (b)  Following part (a) with $k = 10$, the first number of the sample is 10, the second is $10 + 74 = 84$. The remaining three numbers in the sample would be 158, 232, and 306. Thus, the sample of 5 would be 10, 84, 158, 232, and 306.

**1.61**  (a)  Answers will vary, but here is the procedure: (1) The population of size 50 is already divided into five clusters of size 10. (2) Since the required sample size is 20, we will need to take a SRS of 2 clusters. Use a table of random numbers (or a similar device) to select two numbers between 1 and 5. These are the two clusters that are selected. (3) Use all the members of each cluster selected in part (2) as the sample.

   (b)  Following part (a) with clusters #1 and #3 selected, we would select all the members in cluster 1, which are 1 – 10, and all the members in cluster 3, which are 21 – 30.

**1.63**  (a)  From each strata, we need to obtain a SRS of a size proportional to the size of the stratum. Therefore, since strata #1 is 30% of the population, a SRS equal to 30% of 20, or 6, should be sampled from strata #1. Since strata #2 is 20% of the population, a SRS equal to 20% of 20, or 4, should be sampled from strata #2. Similarly, a SRS of size 8 should be sampled from strata #3 and a SRS of size 2 should be sampled from strata #4. The sample sizes from stratum #1 through #4 are 6, 4, 8, and 2 respectively.

   (b)  Answers will vary following the procedure in part (a).

**1.65**  Stratified Sampling. The entire population is naturally divided into subpopulations, one from each lake, and random sampling is done from each lake. The stratified sampling is not with proportional allocation since that would require knowing how many fish were in each lake.

**1.67**  Systematic Random Sampling. Kennedy selected his sample using the fixed periodic interval of every $50^{th}$ letter, which is the similar to the method presented in procedure 1.1.

**1.69**  Cluster Sampling. The clusters of this sampling design are the 46 schools. A random sample of 10 clusters was selected and then all of the parents of the nonimmunized children at the 10 selected schools were sent a questionnaire.

**1.71** (a) Answers will vary, but here is the procedure: (1) Divide the population size, 500, by the sample size, 10, and round down to the nearest whole number if necessary; this gives 50. (2) Use a table of random numbers (or a similar device) to select a number between 1 and 50, call it $k$. (3) List every 50th, starting with $k$, until 10 numbers are obtained; thus, the first number on the required list of 10 numbers is $k$, the second is $k+50$, the third is $k+100$, and so forth (e.g., if $k=6$, then the numbers on the list are 6, 56, 106, ...).

(b) Systematic random sampling is easier.

(c) The answer depends on the purpose of the sampling. If the purpose of sampling is not related to the size of the sales outside the U.S., systematic sampling will work. However, since the listing is a ranking by amount of sales, if k is low (say 2), then the sample will contain firms that, on the average, have higher sales outside the U.S. than the population as a whole. If the k is high, (say 49) then the sample will contain firms that, on the average, have lower sales than the population as a whole. In either of those cases, the sample would not be representative of the population in regard to the amount of sales outside the U.S.

**1.73** (a) Number the suites from 1 to 48, use a table of random numbers to randomly select three of the 48 suites, and take as the sample the 24 dormitory residents living in the three suites obtained.

(b) Probably not, since friends are more likely to have similar opinions than are strangers.

(c) There are 384 students in total. Freshmen make up 1/3 of them. Sophomores make up 7/24 of them, Juniors 1/4, and Seniors 1/8. Multiplying each of these fractions by 24 yields the proportional allocation, which dictates that the number of freshmen, sophomores, juniors, and seniors selected should be, respectively, 8, 7, 6, and 3. Thus a stratified sample of 24 dormitory residents can be obtained as follows: Number the freshmen dormitory residents from 1 to 128 and use a table of random numbers to randomly select 8 of the 128 freshman dormitory residents; number the sophomore dormitory residents from 1 to 112 and use a table of random numbers to randomly select 7 of the 112 sophomore dormitory residents; and so forth.

**1.75** (a) Answers will vary, but here is the procedure: (1) Divide the population size, 435, by the sample size, 15, and round down to the nearest whole number if necessary; this gives 29. Use a table of random numbers (or a similar device) to select a number between 1 and 29, call it $k$. (3) List every 29th number, starting with $k$, until 15 numbers are obtained; thus, the first number of the required list of 15 numbers is $k$, the second is $k + 29$, the third is $k + 58$, and so forth.

(b) Following part (a) with $k = 12$, the first number of the sample is 12, the second is $12 + 29 = 41$. The third number selected is $12 + 58 = 70$. The remaining twelve numbers are similarly selected. Thus, the sample of 15 would be 12, 41, 70, 99, 128, 157, 186, 215, 244, 273, 302, 331, 360, 389, and 418.

**1.77** (a) This is a poll taken by calling randomly selected U.S. adults. Thus, the sampling design appears to be simple random sampling, although it is possible that a more complex design was used to ensure that various political, religious, educational, or other types of groups were proportionately represented in the sample.

(b) The sample size for the second question was 78% of 1010 or 788.

(c) The sample size for the third question was 28% of 788 or 221.

**1.79** (a) It is also true for systematic random sampling if the population size divided by the sample size results in an <u>integer</u> for $m$. The chance for each member to be selected is then still equal to the sample size divided by the population size.  For example, suppose the population size is N=10 and the sample size is n=2.  The chance that each member in simple random sampling to be selected is 2/10 = 1/5.  In systematic random sampling for the same example, m=5.  The possible samples of size two are 1 and 6, 2 and 7, 3 and 8, 4 and 9, and 5 and 10.  Therefore, the chance that a member is selected is equal to the chance of one of those five samples being selected, which is the same as simple random sampling of 1/5.

(b) It is not true for systematic random sampling if the population size divided by the sample size <u>does not result in an integer</u> for $m$.  For example, suppose the population size is N=15 and the sample size is n=2.  After dividing the population size by the sample size and rounding down to the nearest whole number, we get $m$=7.  You would select every 7th member after a random starting place $k$, between 1 and 7, is determined.  If $k$=1, you would select the first and eighth member.  If $k$=7, you would select the seventh and fourteenth member.  In this situation, the last member (fifteenth) can never be selected.  Therefore, the last member of the sample does not have the same chance of being selected as any other member in the population.

### Exercises 1.4

**1.81** (a) Experimental units are the individuals or items on which the experiment is performed.

(b) When the experimental units are humans, we call them subjects.

**1.83** (a) The response variable is the characteristic of the experimental outcome that is to be measured or observed.

(b) A factor is a variable whose effect on the response variable is of interest in the experiment.

(c) The levels are the possible values of the factor.

(d) For a one-factor experiment, the treatments are the levels of the factor.  For multifactor experiments, the treatments are the combinations of levels of the factors.

**1.85** In a one-factor experiment, the number of treatments is equal to the number of levels of the factor.  Therefore, there are four treatments.

**1.87** (a)

|   |   | B |   |   |   |
|---|---|---|---|---|---|
|   |   | $b_1$ | $b_2$ | $b_3$ | $b_4$ |
| A | $a_1$ | $a_1b_1$ | $a_1b_2$ | $a_1b_3$ | $a_1b_4$ |
|   | $a_2$ | $a_2b_1$ | $a_2b_2$ | $a_2b_3$ | $a_2b_4$ |
|   | $a_3$ | $a_3b_1$ | $a_3b_2$ | $a_3b_3$ | $a_3b_4$ |

(b) There are twelve combinations of the levels of the factors.  Therefore, there are twelve treatments.

(c) Yes, you could have multiplied the number of levels in each factor.  There are three levels of factor A and four levels of factor B.  Therefore, there are (3)(4) = 12 treatments.

**1.89** You can multiply the number of levels in each factor.  There are $m$ levels in the first factor and $n$ levels in the second factor.  Therefore, there are $(m)(n) = m \times n$ treatments.

**1.91** (a) There were three treatments.

(b) The first group, the one receiving only the pharmacologic therapy, would be considered the control group.

(c) There were three treatment groups. The first received only pharmacologic therapy, the second received pharmacologic therapy plus a pacemaker, and the third received pharmacologic therapy plus a pacemaker-defibrillator combination.

(d) The first group (control) contained 1/5 of the 1520 patients or 304. The other two groups each contained 2/5 of the 1520 patients or 608.

(e) Each patient could be randomly assigned a number from 1 to 1520. Any patient assigned a number between 1 and 304 would be assigned to the control group; any patient assigned to the next 608 numbers (305 to 912) would be assigned to receive the pharmacologic therapy plus a pacemaker; and any patient assigned a number between 913 and 1520 would receive pharmacologic therapy plus a pacemaker-defibrillator combination. Each random number would be used only once to ensure that the resulting treatment groups were of the intended sizes.

**1.93** (a) Experimental units: the drivers

(b) Response variable: the detection distance, in feet

(c) Factors: two factors - sign size and sign material

(d) Levels of each factor: three levels of sign size (small, medium, and large) and three levels of sign material (1, 2, and 3)

(e) Treatments: the nine different combinations of sign size and sign material resulting from testing each of the three sign sizes with each of the three sign materials

**1.95** (a) Experimental units: female lions

(b) Response variable: whether or not the female lion approached the male lion dummy

(c) Factors: length and color of the mane on the male lion dummy

(d) Levels of each factor: two different mane lengths and two different mane colors

(e) Treatments: the four combinations of mane length and color

**1.97** (a) Experimental units: the children

(b) Response variable: IQ score

(c) Factor: Whether they were given dexamethasone (control or dexamethasone group)

(d) Levels of each factor: two levels of the single factor (control or dexamethasone group)

(e) Treatments: the two levels of the single factor

**1.99** (a) This is a randomized block design. The experiment first blocked by gender. All the experimental units are not randomly assigned among all the treatments.

(b) The blocks are the two genders (male and female).

**1.101** (a) Simple random sampling corresponds to completely randomized designs since selection is randomly made from the entire population.

(b) Stratified sampling corresponds to randomized block designs since selection is randomly made from within each strata.

# 8   Chapter 1

**Review Problems for Chapter 1**

1. Student exercise.

2. Descriptive statistics are used to display and summarize the data to be used in an inferential study. Preliminary descriptive analysis of a sample often reveals features of the data that lead to the choice or reconsideration of the choice of the appropriate inferential analysis procedure.

3. (a) An *observational study* is a study in which researchers simply observe characteristics and take measurements.

   (b) A *designed experiment* is a study in which researchers impose treatments and controls and *then* observe characteristics and take measurements.

4. A literature search should be made before planning and conducting a study.

5. (a) A representative sample is one that reflects as closely as possible the relevant characteristics of the population under consideration.

   (b) Probability sampling involves the use of a randomizing device such as tossing a coin or die, using a random number table, or using computer software that generates random numbers to determine which members of the population will make up the sample.

   (c) A sample is a simple random sample if all possible samples of a given size are equally likely to be the actual sample selected.

6. (a) This method does not involve probability sampling. No randomizing device is being used and people who do not visit the campus cafeteria have no chance of being included in the sample.

   (b) The dart throwing is a randomizing device that makes all samples of size 20 equally likely. This is probability sampling.

7. (a) Systematic random sampling is done by first dividing the population size by the sample size and rounding the result down to the next integer, say $m$. Then we select one random number, say $k$, between 1 and $m$ inclusive. That number will be the first member of the sample. The remaining members of sample will be those numbered $k+m$, $k+2m$, $k+3m$,... until a sample of size $n$ has been chosen. Systematic sampling will yield results similar to simple random sampling as long as there is nothing systematic about the way the members of the population were assigned their numbers.

   (b) In cluster sampling, clusters of the population (such as blocks, precincts, wards, etc.) are chosen at random from all such possible clusters. Then every member of the population lying within the chosen clusters is sampled. This method of sampling is particularly convenient when members of the population are widely scattered and is most appropriate when the members of each cluster are representative of the entire population. Cluster sampling can save both time and expense in doing the survey, but can yield misleading results if individual clusters are made up of subjects with very similar views on the topic being surveyed.

   In stratified random sampling with proportional allocation, the population is first divided into subpopulations, called strata, and simple random sampling is done within each stratum. Proportional allocation means that the size of the sample from each stratum is proportional to the size of the population in that stratum. This type of sampling may improve the accuracy of the survey by ensuring that those in each stratum are more proportionately represented than would be the case with cluster sampling or even simple random sampling. Ideally, the members of each stratum should be homogeneous relative to the characteristic under consideration. If they are not homogeneous within each stratum, simple random sampling would work just as well.

8. The three basic principles of experimental design are control, randomization, and replication. Control refers to methods for controlling factors other than those of primary interest. Randomization means randomly dividing the subjects into groups in order to avoid unintentional selection bias in constituting the groups. Replication means using enough experimental units or subjects so that groups resemble each other closely and so that there is a good chance of detecting differences among the treatments when such differences actually exist.

9. Descriptive study. It is a summary of the scores of major league baseball games on August 14, 2013.

10. (a) Descriptive study. It is a summary of the responses from those that participated in the poll.

    (b) Inferential statement. It is an implied estimate of the responses of all adults in the U.S.

11. Inferential study. The results of a sample are used to make inferences about the age distribution of all British backpackers in South Africa.

12. (a) Descriptive study. It is a summary of the percentages of Jewish children sampled in Israel and Britain that have peanut allergies.

    (b) Observational study. The researchers simply observed the two groups.

13. This is an observational study. To be a designed experiment, the researchers would have to have the ability to assign some children at random to live in persistent poverty during the first 5 years of life or to not suffer any poverty during that period. Clearly that is not possible.

14. This is a designed experiment since the researcher is imposing a treatment and then observing the results.

15. Because Yale is a very expensive school, incomes of parents of Yale students will not be representative of the incomes of all college students' parents.

16. (a) H,Z,C    H,Z,A    H,Z,J    H,C,A    H,C,J
       H,A,J    Z,C,A    Z,C,J    Z,A,J    C,A,J

    (b) Since each of the 10 samples of size three is equally likely, there is a 1/10 chance that the sample chosen is the first sample in the list, 1/10 chance that it is the second sample in the list, and 1/10 chance that it is the tenth sample in the list.

    (c) (i) Make five slips of paper with each airline on one slip. Draw three slips at random. (ii) Make 10 slips of paper, each having one of the combinations in part (a). Draw one slip at random. (iii) Number the five airlines from 1 to 5. Use a random number table or random number generator to obtain three distinct random numbers between 1 and 5, inclusive.

    (d) Your method and result may differ from ours. We rolled a die (ignoring 6's and duplicates) and got 2, 5, 2, 6, 4. Ignoring duplicates and numbers greater than five, our sample consists of Horizon, Jazz, and Alaska Airlines.

17. (a) Table I can be employed to obtain a sample of 15 random numbers between 1 and 100 as follows. First, I pick a random starting point by closing my eyes and putting my finger down on the table.

    My finger falls on three digits located at the intersection of a line with three columns. (Notice that the first column of digits is labeled "00" rather than "01".) This is my starting point.

    I now go down the table and record all three-digit numbers appearing directly beneath the first three-digit number that are between 001 and

**10    Chapter 1**

100 inclusive. I throw out numbers between 101 and 999, inclusive. I also discard the number 0000. When the bottom of the column is reached, I move over to the next sequence of three digits and work my way back up the table. Continue in this manner. When 10 distinct three-digit numbers have been recorded, the sample is complete.

(b) Starting in row 10, columns 7-9, we skip 484, 797, record 082, skip 586, 653, 452, 552, 155, record 008, skip 765, move to the right and record 016, skip 534, 593, 964, 667, 452, 432, 594, 950, 670, record 001, skip 581, 577, 408, 948, 807, 862, 407, record 047, skip 977, move to the right, skip 422 and all of the rest of the numbers in that column, move to the right, skip 732, 192, record 094, skip 615 and all of the rest of the numbers in that column, move to the right, record 097, skip 673, record 074, skip 469, 822, record 052, skip 397, 468, 741, 566, 470, record 076, 098, skip 883, 378, 154, 102, record 003, skip 802, 841, move to the right, skip 243, 198, 411, record 089, skip 701, 305, 638, 654, record 041, skip 753, 790, record 063.

The final list of numbers is 82, 8, 16, 1, 47, 94, 97, 74, 52, 76, 98, 3, 89, 41, 63.

(c) Using Minitab, our results were the numbers 46, 99, 90, 31, 75, 98, 79, 14, 44, 13, 66, 49, 37, 87, 73, 26, 61, 71, 72, 2. Thus our sample consists of the first 15 numbers 46, 99, 90, 31, 75, 98, 79, 14, 44, 13, 66, 49, 37, 87, 73. Your sample may be different.

**18.** The statement under the vote is a disclaimer as to the validity of the survey. Since the vote reflects only the responses of volunteers who chose to vote, it cannot be regarded as representative of the public in general, some of which do not use the Internet, nor as representative of Internet users since the sample was not chosen at random from either group.

**19.** The data in this study were clearly not collected via a controlled experiment in which some participants were forced to do crossword puzzles, practice musical instruments, play board games, or read while others were not allowed to do any of those activities. Therefore, any data relative to these activities and dementia arose as a result of observing whether or not the subjects in the study carried out any of those activities and whether or no they had some form of dementia. Since this would be an observational study, no statement of cause and effect can rightfully be made.

**20.** The researchers did not impose or manipulate any of the conditions of this study. They didn't decide who had cancer, who didn't have cancer, who had hepatitis B, or who had hepatitis C. This study was an observational study and not a controlled experiment. Observational studies can only reveal an association, not causation. Therefore, the statement in quotes is valid. If the researchers wanted to establish causation, they would need a designed experiment.

**21.** (a) Answers will vary, but here is the procedure: (1) Divide the population size, 100, by the sample size 15, and round down to the nearest whole number; this gives 6. (2) Use a table of random numbers (or a similar device) to select a number between 1 and 6, call it $k$. (3) List every 6th number, starting with $k$, until 15 numbers are obtained; thus the first number on the required list of 15 numbers is $k$, the second is $k+6$, the third is $k+12$, and so forth (e.g., if $k=4$, then the numbers on the list are 4, 10, 16, ...).

(b) Yes, unless for some reason there is some kind of trend or a cyclical pattern in the listing of the athletes.

**22.** (a) Each category of "Distance from Plant" should be represented in the sample in the same proportion that it is present in the population of City of Durham's water distribution system. 1310/11707 = 0.112. Thus, 11.2% of the sample of 80 water samples should be from "Less than 1.5 miles", 27.0% from "1.5 - less than 3.0 miles", 24.1% from "3.0 - less than 4.5 miles", 13.6% from "4.5 – less than 6.0 miles", 11.5% from "6.0 – less than 7.5 miles", and 12.5% from "7.5 miles or greater". Multiplying each of these fractions by 80 gives us the sample sizes from each category. These sample sizes will not necessarily be integers, so we will need to make some minor adjustments of the results. The first category should have (11.2/100)(80) = 8.96. The second should have (27/100)(80) = 21.6. Similarly, the third, fourth, fifth, and sixth categories should have 19.28, 10.88, 9.2, and 10 for their sample sizes. We round the six sample sizes from the categories to 9, 22, 19, 11, 9, and 10 respectively. We would now randomly select water samples from each region.

23. (a) This is a designed experiment.

    (b) The treatment group consists of the 158 patients who took AVONEX. The control group consists of the 143 patients who were given a placebo. The treatments were the AVONEX and the placebo.

24. (a) Experimental units: tomato plants

    (b) Response variable: yield of tomatoes

    (c) Factor(s): tomato variety and density of plants

    (d) Levels of each factor: The four tomato varieties (Harvester, Pusa Early Dwarf, Ife No. 1, and Ibadan Local) would be the levels of variety. The four densities (10,000, 20,000, 30,000, and 40,000 plants/ha) would be the levels of the density.

    (e) Treatments: Each treatment would be one of the combinations of a variety planted at a given plant density.

25. (a) Experimental Units: The children

    (b) Response variable: Whether or not the child was able to open the bottle

    (c) Factors: The container designs

    (d) Levels of each factor: Three (types of containers)

    (e) Treatments: The container designs

26. This is a completely randomized design. All of the experimental units (batches of doughnuts) were assigned at random to the four treatments (four different fats).

27. (a) This is a completely randomized design since the 24 cars were randomly assigned to the 4 brands of gasoline.

    (b) This is a randomized block design. The four different gasoline brands are randomly assigned to the four cars in each of the six car model groups. The blocks are the six groups of four identical cars each.

    (c) If the purpose is to learn about the mileage rating of one particular car model with each of the four gasoline brands, then the completely randomized design is appropriate. But if the purpose is to learn about the performance of the gasoline across a variety of cars (and this seems more reasonable), then the randomized block design is more appropriate and will allow the researcher to determine the effect of car model as well as of gasoline type on the mileage obtained.

# CHAPTER 2 SOLUTIONS

## Exercises 2.1

**2.1** (a) Hair color, model of car, and brand of popcorn are qualitative variables.

(b) Number of eggs in a nest, number of cases of flu, and number of employees are discrete, quantitative variables.

(c) Temperature, weight, and time are quantitative continuous variables.

**2.3** (a) Qualitative data result from observing and recording values of a qualitative variable, such as, color or shape.

(b) Discrete, quantitative data are values of a discrete quantitative variable. Values usually result from counting something.

(c) Continuous, quantitative data are values of a continuous variable. Values are usually the result of measuring something such as temperature that can take on any value in a given interval.

**2.5** Of qualitative and quantitative (discrete and continuous) types of data, only qualitative yields nonnumerical data.

**2.7** (a) The first column consists of *quantitative, continuous* data. This column provides the time that the earthquake occurred.

(b) The second column consists of *quantitative, continuous* data. This column provides the magnitude of each earthquake.

(c) The third column consists of *quantitative, continuous* data. This column provides the depth of each earthquake in kilometers.

(d) The fourth column consists of *quantitative, discrete* data. This column provides the number of stations that reported activity on the earthquake.

(e) The fifth column consists of *qualitative* data since the region of the location of each earthquake is nonnumerical.

**2.9** (a) The first column consists of *quantitative, discrete* data. This column provides the ranks of the deceased celebrities with the top 10 earnings.

(b) The second column consists of *qualitative* data since names are nonnumerical.

(c) The third column consists of *quantitative, discrete* data, the earnings of the celebrities. Since money involves discrete units, such as dollars and cents, the data is discrete, although, for all practical purposes, this data might be considered quantitative continuous data

**2.11** (a) The first column contains types of products. They are *qualitative* data since they are nonnumerical.

(b) The second column contains number of units shipped in the millions. These are whole numbers and are *quantitative, discrete.*

(c) The third column contains money values. Technically, these are *quantitative, discrete* data since there are gaps between possible values at the cent level. For all practical purposes, however, these are *quantitative, continuous* data.

**2.13** The first column contains *quantitative, discrete* data in the form of ranks. These are whole numbers. The second and third columns contain *qualitative* data in the form of names. The last column contains the rating of the program which is *quantitative, continuous*.

**14   Chapter 2**

**2.15**   The first column is *quantitative, discrete* since it is reporting a rank. The second and third columns are *qualitative* since make/model and type are nonnumerical. The last column is *quantitative, continuous* since it is reporting mileage.

**Exercises 2.2**

**2.17**   A frequency distribution of qualitative data is a table that lists the distinct values of data and their frequencies. It is useful to organize the data and make it easier to understand.

**2.19**   (a) True. Having identical frequency distributions implies that the total number of observations and the numbers of observations in each class are identical. Thus, the relative frequencies will also be identical.

(b) False. Having identical relative frequency distributions means that the ratio of the count in each class to the total is the same for both frequency distributions. However, one distribution may have twice (or some other multiple) the total number of observations as the other. For example, two distributions with counts of 5, 4, 1 and 10, 8, 2 would be different, but would have the same relative frequency distribution.

(c) If the two data sets have the same number of observations, either a frequency distribution or a relative-frequency distribution is suitable. If, however, the two data sets have different numbers of observations, using relative-frequency distributions is more appropriate because the total of each set of relative frequencies is 1, putting both distributions on the same basis for comparison.

**2.21**   (a)-(b)

The classes are presented in column 1. The frequency distribution of the classes is presented in column 2. Dividing each frequency by the total number of observations, which is 5, results in each class's relative frequency. The relative frequency distribution is presented in column 3.

| Class | Frequency | Relative Frequency |
|-------|-----------|--------------------|
| A     | 3         | 0.60               |
| B     | 1         | 0.20               |
| C     | 1         | 0.20               |
|       | 5         | 1.00               |

(c) We multiply each of the relative frequencies by 360 degrees to obtain the portion of the pie represented by each class. The result using Minitab is

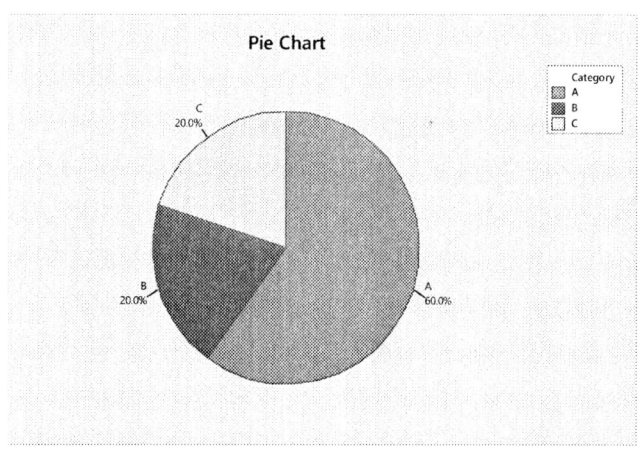

(d) We use the bar chart to show the relative frequency with which each class occurs. The result is

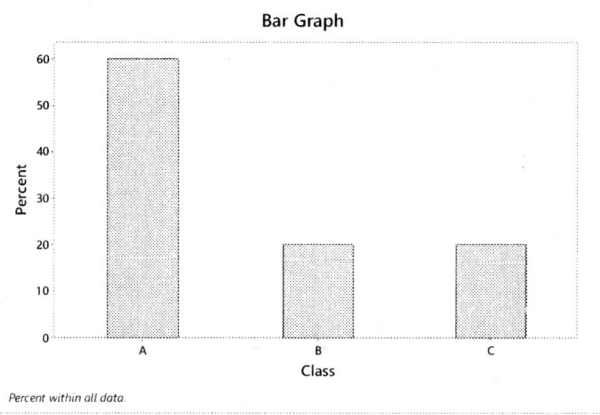

**2.23** (a)-(b)

The classes are presented in column 1. The frequency distribution of the classes is presented in column 2. Dividing each frequency by the total number of observations, which is 10, results in each class's relative frequency. The relative frequency distribution is presented in column 3.

| Class | Frequency | Relative Frequency |
|---|---|---|
| A | 4 | 0.40 |
| B | 3 | 0.30 |
| C | 1 | 0.10 |
| D | 2 | 0.20 |
|   | 10 | 1.00 |

(c) We multiply each of the relative frequencies by 360 degrees to obtain the portion of the pie represented by each class. The result using Minitab is

**16** Chapter 2

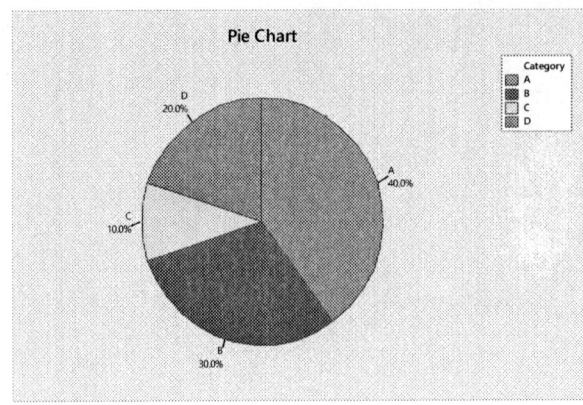

(d) We use the bar chart to show the relative frequency with which each class occurs. The result is

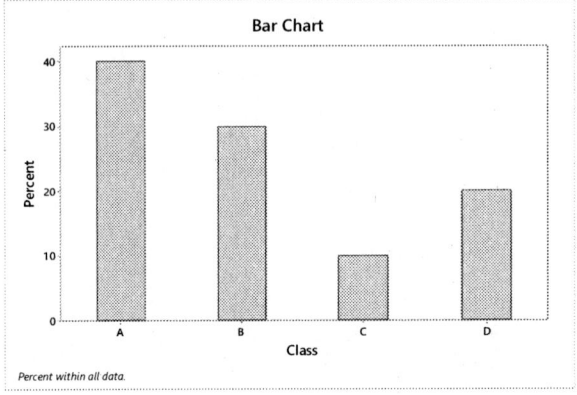

**2.25** (a)-(b)

The classes are presented in column 1. The frequency distribution of the classes is presented in column 2. Dividing each frequency by the total number of observations, which is 20, results in each class's relative frequency. The relative frequency distribution is presented in column 3.

| Class | Frequency | Relative Frequency |
|-------|-----------|--------------------|
| A | 1 | 0.05 |
| B | 3 | 0.15 |
| C | 7 | 0.35 |
| D | 7 | 0.35 |
| E | 2 | 0.10 |
|   | 20 | 1.00 |

(c) We multiply each of the relative frequencies by 360 degrees to obtain the portion of the pie represented by each class. The result using Minitab is

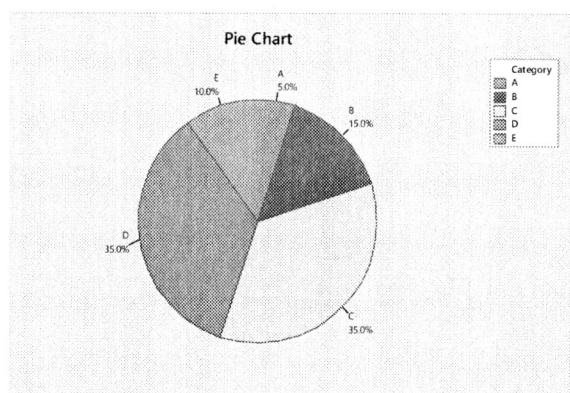

(d) We use the bar chart to show the relative frequency with which each class occurs. The result is

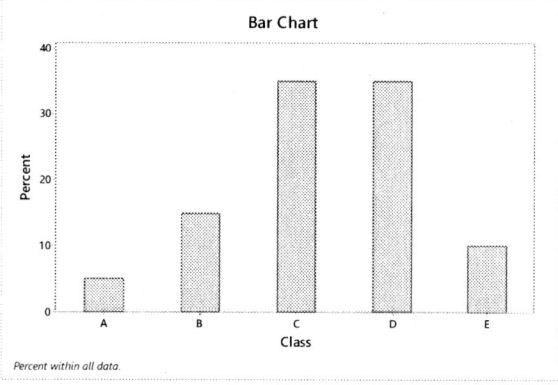

**2.27** (a)-(b)

The classes are the NCAA wrestling champions and are presented in column 1. The frequency distribution of the champions is presented in column 2. Dividing each frequency by the total number of champions, which is 25, results in each class's relative frequency. The relative frequency distribution is presented in column 3.

| Champion | Frequency | Relative Frequency |
|---|---|---|
| Iowa | 12 | 0.48 |
| Penn State | 3 | 0.12 |
| Minnesota | 3 | 0.12 |
| Oklahoma St. | 7 | 0.28 |
| | 25 | 1.00 |

(b) We multiply each of the relative frequencies by 360 degrees to obtain the portion of the pie represented by each team. The result is

18   Chapter 2

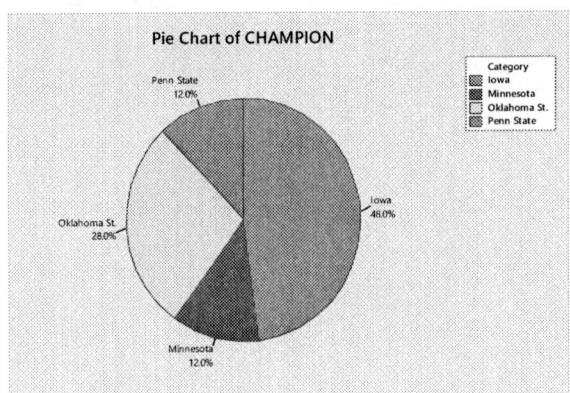

(c) We use the bar chart to show the relative frequency with which each TEAM occurs. The result is

2.29  (a)-(b)

The classes are the class levels and are presented in column 1. The frequency distribution of the class levels is presented in column 2. Dividing each frequency by the total number of students in the introductory statistics class, which is 40, results in each class's relative frequency. The relative frequency distribution is presented in column 3.

| Class Level | Frequency | Relative Frequency |
|---|---|---|
| Fr | 6 | 0.150 |
| So | 15 | 0.375 |
| Jr | 12 | 0.300 |
| Sr | 7 | 0.175 |
|  | 40 | 1.000 |

(c) We multiply each of the relative frequencies by 360 degrees to obtain the portion of the pie represented by each class level. The result is

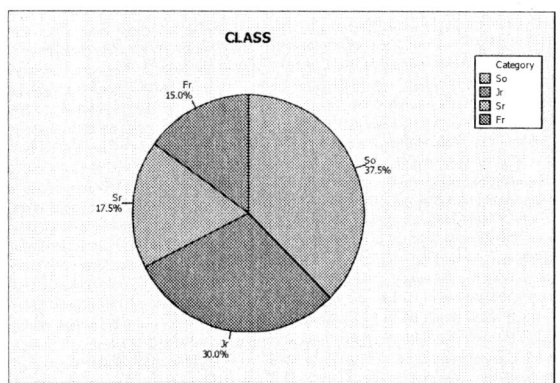

(d) We use the bar chart to show the relative frequency with which each CLASS level occurs. The result is

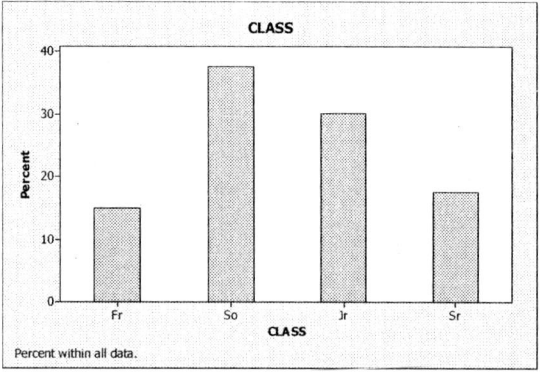

**2.31** (a)-(b)

The classes are the days and are presented in column 1. The frequency distribution of the days is presented in column 2. Dividing each frequency by the total number road rage incidents, which is 69, results in each class's relative frequency. The relative frequency distribution is presented in column 3.

| Class Level | Frequency | Relative Frequency |
|---|---|---|
| Su | 5 | 0.0725 |
| M | 5 | 0.0725 |
| Tu | 11 | 0.1594 |
| W | 12 | 0.1739 |
| Th | 11 | 0.1594 |
| F | 18 | 0.2609 |
| Sa | 7 | 0.1014 |
|  | 69 | 1.0000 |

(c) We multiply each of the relative frequencies by 360 degrees to obtain the portion of the pie represented by each day. The result is

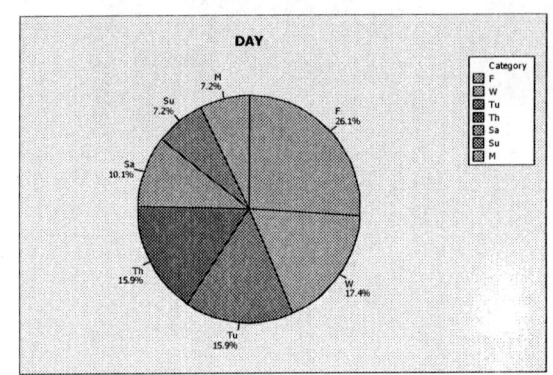

(d) We use the bar chart to show the relative frequency with which each DAY occurs. The result is

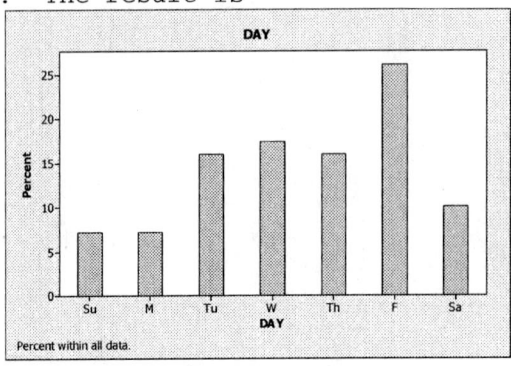

**2.33** (a) We find the relative frequencies by dividing each of the frequencies by the total sample size of 509.

| Color | Frequency | Relative Frequency |
|---|---|---|
| Brown | 152 | 0.2986 |
| Yellow | 114 | 0.2240 |
| Red | 106 | 0.2083 |
| Orange | 51 | 0.1002 |
| Green | 43 | 0.0845 |
| Blue | 43 | 0.0845 |
|  | 509 | 1.0000 |

(b) We multiply each of the relative frequencies by 360 degrees to obtain the portion of the pie represented by each color of M&M. The result is

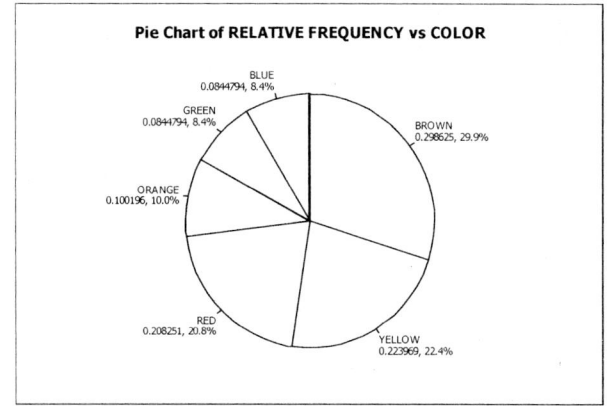

(c) We use the bar chart to show the relative frequency with which each color occurs. The result is

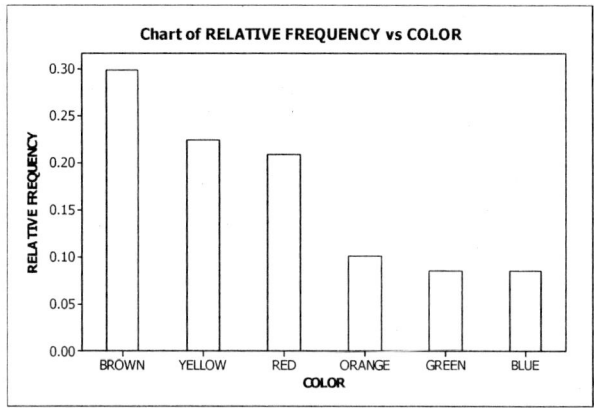

**2.35** (a) We find the relative frequencies by dividing each of the frequencies by the total sample size of 137,925.

| Rank | Frequency | Relative Frequency |
| --- | --- | --- |
| Professor | 32,511 | 0.2357 |
| Associate professor | 28,572 | 0.2072 |
| Assistant professor | 59,277 | 0.4298 |
| Instructor | 14,289 | 0.1036 |
| Other | 3,276 | 0.0238 |
| | 137,925 | 1.0000 |

(b) We multiply each of the relative frequencies by 360 degrees to obtain the portion of the pie represented by each rank. The result is

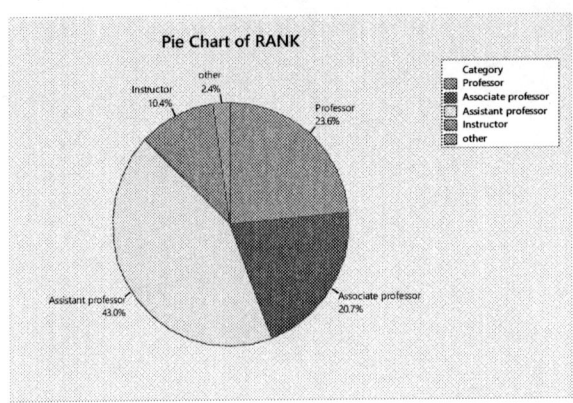

(c) We use the bar chart to show the relative frequency with which each rank occurs. The result is

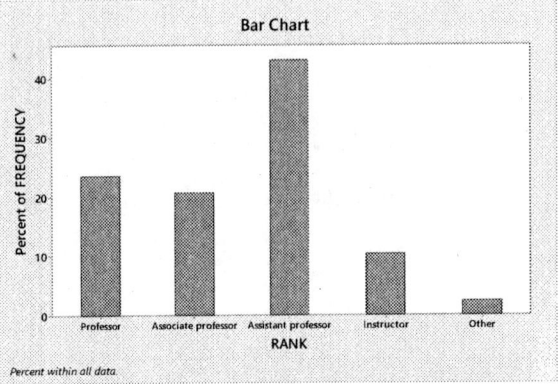

**2.37** (a) We first find the relative frequencies by dividing each of the frequencies by the total sample size of 200.

| Color | Frequency | Relative Frequency |
|---|---|---|
| Red | 88 | 0.44 |
| Black | 102 | 0.51 |
| Green | 10 | 0.05 |
|  | 200 | 1.00 |

(b) We multiply each of the relative frequencies by 360 degrees to obtain the portion of the pie represented by each color. The result is

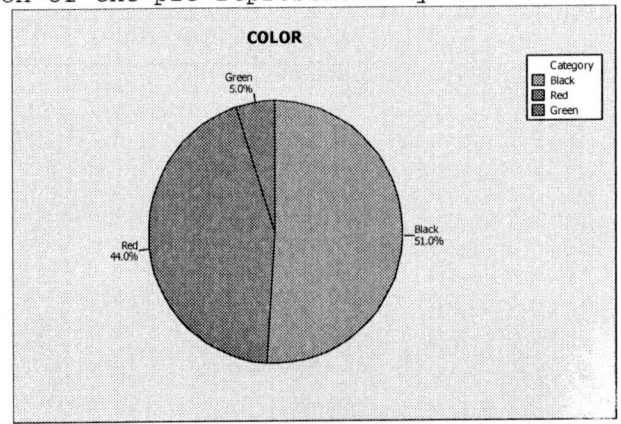

(c) We use the bar chart to show the relative frequency with which each color occurs. The result is

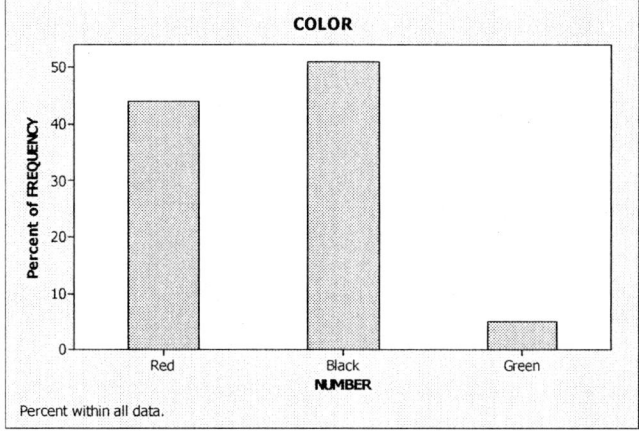

**2.39** (a) Using Minitab, retrieve the data from the WeissStats Resource Site. Column 1 contains the type of the vehicle. From the tool bar, select **Stat ▶ Tables ▶ Tally Individual Variables**, double-click on TYPE in the first box so that TYPE appears in the **Variables** box, put a check mark next to Counts and Percents under Display, and click **OK**. The result is

| TYPE | Count | Percent |
|------|-------|---------|
| Bus | 21 | 2.80 |
| Car | 667 | 88.93 |
| Truck | 62 | 8.27 |
| N= | 750 | |

(b) The relative frequencies were calculated in part(a) by putting a check mark next to Percents. 2.8% of the vehicles were busses, 88.93% of the vehicles were cars, and 8.27% of the vehicles were trucks.

(c) Using Minitab, select **Graph ▶ Pie Chart**, check Chart counts of unique values, double-click on TYPE in the first box so that TYPE appears in the Categorical Variables box. Click Pie Options, check decreasing volume, click OK. Click Labels, enter TYPE in for the title, click Slice Labels, check Category Name, Percent, and Draw a line from label to slice, Click OK twice. The result is

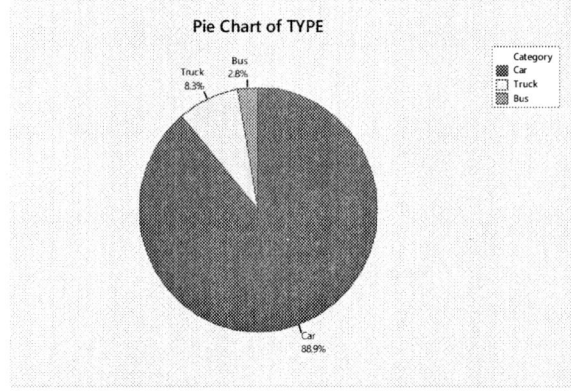

(d) Using Minitab, select **Graph ▶ Bar Chart**, select Counts of unique values, select Simple option, click OK. Double-click on TYPE in the first box so that TYPE appears in the Categorical Variables box.

Select Chart Options, check decreasing Y, check show Y as a percent, click OK. Select Labels, enter in TYPE as the title. Click OK twice. The result is

2.41 (a) Using Minitab, retrieve the data from the WeissStats Resource Site. Column 2 contains the preference for how the members want to receive the ballots and column 3 contains the highest degree obtained by the members. From the tool bar, select **Stat ▶ Tables ▶ Tally Individual Variables**, double-click on PREFERENCE and DEGREE in the first box so that both PREFERENCE and DEGREE appear in the **Variables** box, put a check mark next to Counts and Percents under Display, and click **OK**. The results are

| PREFERENCE | Count | Percent | DEGREE | Count | Percent |
|---|---|---|---|---|---|
| Both | 112 | 19.79 | MA | 167 | 29.51 |
| Email | 239 | 42.23 | Other | 11 | 1.94 |
| Mail | 86 | 15.19 | PhD | 388 | 68.55 |
| N/A | 129 | 22.79 | N= | 566 | |
| N= | 566 | | | | |

(b) The relative frequencies were calculated in part(a) by putting a check mark next to Percents. For the PREFERENCE variable; 19.79% of the members prefer to receive the ballot by both e-mail and mail, 42.23% prefer e-mail, 15.19% prefer mail, and 22.79% didn't list a preference. For the Degree variable; 29.51% obtained a Master's degree, 68.55% obtained a PhD, and 1.94% received a different degree.

(c) Using Minitab, select **Graph ▶ Pie Chart**, check Chart counts of unique values, double-click on PREFERENCE and DEGREE in the first box so that PREFERENCE and DEGREE appear in the Categorical Variables box. Click Pie Options, check decreasing volume, click OK. Click Multiple Graphs, check On the Same Graphs, Click OK. Click Labels, click Slice Labels, check Category Name, Percent, and Draw a line from label to slice, Click OK twice. The results are

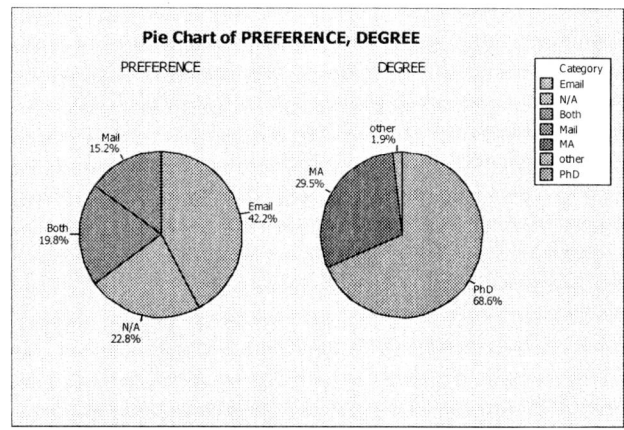

(d) Using Minitab, select **Graph ▶ Bar Chart**, select Counts of unique values, select Simple option, click OK. Double-click on PREFERENCE and DEGREE in the first box so that PREFERENCE and DEGREE appear in the Categorical Variables box. Select Chart Options, check decreasing Y, check show Y as a percent, click OK. Click OK twice. The results are

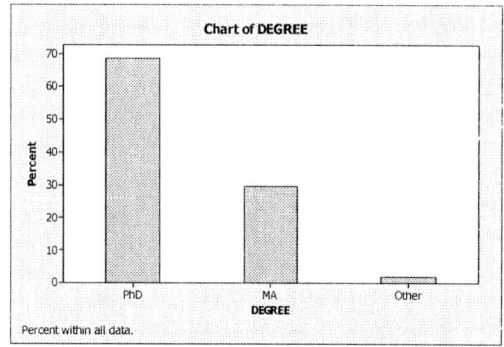

### Exercises 2.3

**2.43** For class limits, marks, cutpoints and midpoints to make sense, data must be numerical. They do not make sense for qualitative data classes because such data are nonnumerical.

**2.45** In the first method for depicting classes called cutpoint grouping, we used the notation **a - under b** to mean values that are greater than or equal to **a** and up to, but not including **b**, such as 30 - under 40 to mean a range of values greater than or equal to 30, but strictly less than 40. In the alternate method called limit grouping, we used the notation **a-b** to indicate a class that extends from **a** to **b**, including both. For example, 30-39 is a class that includes both 30 and 39. The alternate method is especially appropriate when all of the data values are integers. If the data include values like 39.7 or 39.93, the first method is more advantageous since the cutpoints remain integers; whereas, in the alternate method, the upper limits for each class would have to be expressed in decimal form such as 39.9 or 39.99.

**2.47** For limit grouping, we find the class mark, which is the average of the lower and upper class limit. For cutpoint grouping, we find the class midpoint, which is the average of the two cutpoints.

**2.49** An advantage of the frequency histogram over a frequency distribution is that it is possible to get an overall view of the data more easily. A disadvantage of the frequency histogram is that it may not be possible to determine exact frequencies for the classes when the number of observations is large.

**2.51** If the classes consist of single values, stem-and-leaf diagrams and frequency histograms are equally useful. If only one diagram is needed and the classes consist of more than one value, the stem-and-leaf diagram allows one to retrieve all of the original data values whereas the frequency histogram does not. If two or more sets of data of different sizes are to be compared, the relative frequency histogram is advantageous because all of the diagrams to be compared will have the same total relative frequency of 1.00. Finally, stem-and-leaf diagrams are not very useful with very large data sets and may present problems with data having many digits in each number.

**2.53** You can reconstruct the stem-and-leaf diagram using two lines per stem. For example, instead of listing all of the values from 10 to 19 on a '1' stem, you can make two '1' stems. On the first, you record the values from 10 to 14 and on the second, the values from 15 to 19. If there are still two few stems, you can reconstruct the diagram using five lines per stem, recording 10 and 11 on the first line, 12 and 13 on the second, and so on.

**2.55** For the ages of householders, given as a whole number, limit grouping is probably the best because the data are given as whole numbers and there are probably too many distinct observations to list them as single-value grouping.

**2.57** For the number of automobiles per family, single-value grouping is probably the best because the data is discrete with relatively few distinct observations.

**2.59** For carapace length for a sample of giant tarantulas, cutpoint grouping is probably the best because the data is continuous and the data was recorded to the nearest hundredth of a millimeter.

**2.61** (a) Since the data values range from 1 to 4, we construct a table with classes based on a single value. The resulting table follows.

| Single Value Class | Frequency |
|---|---|
| 1 | 2 |
| 2 | 2 |
| 3 | 5 |
| 4 | 1 |
|  | 10 |

(b) To get the relative frequencies, divide each frequency by the sample size of 10.

| Single Value Class | Relative Frequency |
|---|---|
| 1 | 0.20 |
| 2 | 0.20 |
| 3 | 0.50 |
| 4 | 0.10 |
|  | 1.00 |

(c) The frequency histogram in Figure (a) is constructed using the frequency distribution presented in part (a) of this exercise. Column 1 demonstrates that the data are grouped using classes based on a single value. These single values in column 1 are used to label the horizontal axis of the frequency histogram. When classes are based on a single value, the middle of each histogram bar is placed directly over the single numerical value represented by the class. Also, the height of each bar in the frequency histogram matches the respective frequency in column 2.

Figure (a)  Figure (b)

(d) The relative-frequency histogram in Figure (b) is constructed using the relative-frequency distribution presented in part (b) of this exercise. It has the same horizontal axis as the frequency histogram. The middle of each histogram bar is placed directly over the single numerical value represented by the class. Also, the height of each bar in the relative-frequency histogram matches the respective relative frequency in column 2.

**2.63** (a) Since the data values range from 0 to 4, we construct a table with classes based on a single value. The resulting table follows.

| Single Value Class | Frequency |
|---|---|
| 0 | 1 |
| 1 | 1 |
| 2 | 4 |
| 3 | 8 |
| 4 | 6 |
|  | 20 |

(b) To get the relative frequencies, divide each frequency by the sample size of 20.

| Single Value Class | Relative Frequency |
|---|---|
| 0 | 0.05 |
| 1 | 0.05 |
| 2 | 0.20 |
| 3 | 0.40 |
| 4 | 0.30 |
|  | 1.00 |

(c) The frequency histogram in Figure (a) is constructed using the frequency distribution presented in part (a) of this exercise. Column 1 demonstrates that the data are grouped using classes based on a single value. These single values in column 1 are used to label the horizontal axis of the frequency histogram. When classes are based on a single value, the middle of each histogram bar is placed directly over the single numerical value represented by the class. Also, the height of each bar in the frequency histogram matches the respective frequency in column 2.

Figure (a)                                          Figure (b)

 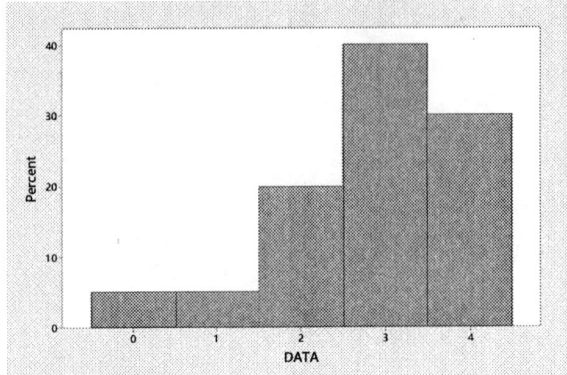

    (d)    The relative-frequency histogram in Figure (b) is constructed using the relative-frequency distribution presented in part (b) of this exercise. It has the same horizontal axis as the frequency histogram. The middle of each histogram bar is placed directly over the single numerical value represented by the class. Also, the height of each bar in the relative-frequency histogram matches the respective relative frequency in column 2.

**2.65**  (a)    The first class to construct is 0-4. The width of all the classes is 5, so the next class would be 5-9. The classes are presented in column 1. The last class to construct is 25-29, since the largest single data value is 26. The tallied results are presented in column 2, which lists the frequencies.

| Limit Grouping Classes | Frequency |
|---|---|
| 0-4 | 5 |
| 5-9 | 5 |
| 10-14 | 1 |
| 15-19 | 4 |
| 20-24 | 2 |
| 25-29 | 3 |
|  | 20 |

    (b)    Dividing each frequency by the total number of observations, which is 20, results in each class's relative frequency. The relative frequencies for all classes are presented in column 2. The resulting table follows.

| Limit Grouping Classes | Relative Frequency |
|---|---|
| 0-4 | 0.25 |
| 5-9 | 0.25 |
| 10-14 | 0.05 |
| 15-19 | 0.20 |
| 20-24 | 0.10 |
| 25-29 | 0.15 |
|  | 1.00 |

    (c)    The frequency histogram in Figure (a) is constructed using the frequency distribution presented in part (a) of this exercise. The lower class limits of column 1 are used to label the horizontal axis of the frequency histogram. The height of each bar in the frequency histogram matches the respective frequency in column 2.

(d) The relative-frequency histogram in Figure (b) is constructed using the relative-frequency distribution presented in part (b) of this exercise. It has the same horizontal axis as the frequency histogram. The height of each bar in the relative-frequency histogram matches the respective relative frequency in column 2.

Figure (a)                                          Figure (b)

 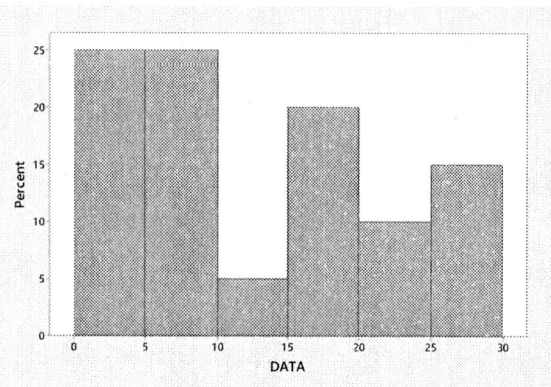

**2.67** (a) The first class to construct is 50-59. The width of all the classes is 10, so the next class would be 60-69. The classes are presented in column 1. The last class to construct is 90-99, since the largest single data value is 98. The tallied results are presented in column 2, which lists the frequencies.

| Limit Grouping Classes | Frequency |
|---|---|
| 50-59 | 6 |
| 60-69 | 3 |
| 70-79 | 6 |
| 80-89 | 7 |
| 90-99 | 3 |
|  | 25 |

(b) Dividing each frequency by the total number of observations, which is 25, results in each class's relative frequency. The relative frequencies for all classes are presented in column 2. The resulting table follows.

| Limit Grouping Classes | Relative Frequency |
|---|---|
| 50-59 | 0.24 |
| 60-69 | 0.12 |
| 70-79 | 0.24 |
| 80-89 | 0.28 |
| 90-99 | 0.12 |
|  | 1.00 |

(c) The frequency histogram in Figure (a) is constructed using the frequency distribution presented in part (a) of this exercise. The lower class limits of column 1 are used to label the horizontal axis of the frequency histogram. The height of each bar in the frequency histogram matches the respective frequency in column 2.

(d) The relative-frequency histogram in Figure (b) is constructed using the relative-frequency distribution presented in part (b) of this exercise. It has the same horizontal axis as the frequency histogram. The height of each bar in the relative-frequency histogram matches the respective relative frequency in column 2.

**30** Chapter 2

Figure (a)     Figure (b)

 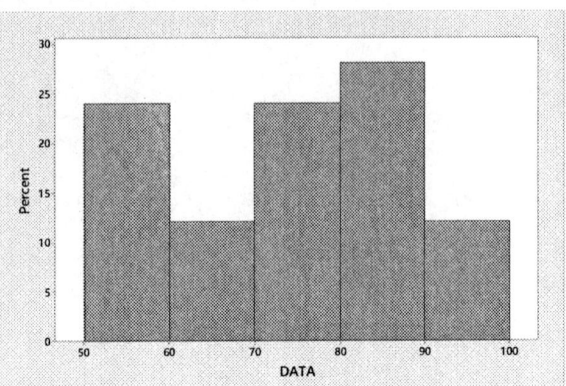

2.69 (a) The first class to construct is 40 – under 46. Since all classes are to be of equal width 6, the second class is 46 – under 52. All of the classes are presented in column 1. The last class to construct is 64 – under 70, since the largest single data value is 65.4. The results of the tallying are presented in column 2, which lists the frequencies.

| Cutpoint Grouping Classes | Frequency |
|---|---|
| 40 – under 46 | 3 |
| 46 – under 52 | 6 |
| 52 – under 58 | 10 |
| 58 – under 64 | 0 |
| 64 – under 70 | 1 |
| | 20 |

(b) Dividing each frequency by the total number of observations, which is 20, results in each class's relative frequency. The relative frequencies for all classes are presented in column 2.

| Cutpoint Grouping | Relative Frequency |
|---|---|
| 40 – under 46 | 0.15 |
| 46 – under 52 | 0.30 |
| 52 – under 58 | 0.50 |
| 58 – under 64 | 0.00 |
| 64 – under 70 | 0.05 |
| | 1.00 |

(c) The frequency histogram in Figure (a) is constructed using the frequency distribution obtained in part (a) of this exercise. The cutpoints are used to label the horizontal axis. Also, the height of each bar in the frequency histogram matches the respective frequency in column 2.

(d) The relative-frequency histogram in Figure (b) is constructed using the relative-frequency distribution obtained in part (b) of this exercise. It has the same horizontal axis as the frequency histogram. The height of each bar in the relative-frequency histogram matches the respective relative frequency in column 2.

Figure (a)                                   Figure (b)

 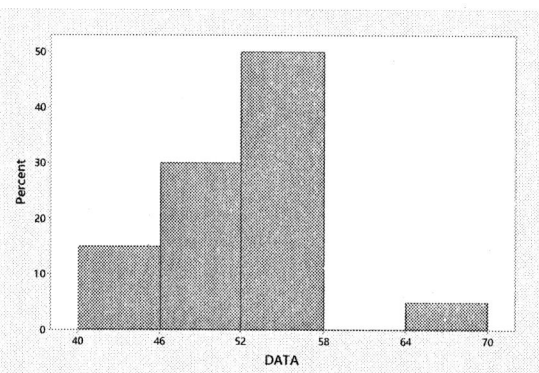

**2.71** (a) The first class' cutpoint is 25 with a width of 3. Therefore, the first class to construct is 25 - under 28. Since all classes are to be of equal width 3, the second class is 28 - under 31. All of the classes are presented in column 1. The last class to construct is 43 - under 46, since the largest single data value is 43.01. The results of the tallying are presented in column 2, which lists the frequencies.

| Cutpoint Grouping Classes | Frequency |
|---|---|
| 25 - under 28 | 1 |
| 28 - under 31 | 2 |
| 31 - under 34 | 5 |
| 34 - under 37 | 7 |
| 37 - under 40 | 5 |
| 40 - under 43 | 4 |
| 43 - under 46 | 1 |
|  | 25 |

(b) Dividing each frequency by the total number of observations, which is 25, results in each class's relative frequency. The relative frequencies for all classes are presented in column 2.

| Cutpoint Grouping Classes | Relative Frequency |
|---|---|
| 25 - under 28 | 0.04 |
| 28 - under 31 | 0.08 |
| 31 - under 34 | 0.20 |
| 34 - under 37 | 0.28 |
| 37 - under 40 | 0.20 |
| 40 - under 43 | 0.16 |
| 43 - under 46 | 0.04 |
|  | 1.00 |

(c) The frequency histogram in Figure (a) is constructed using the frequency distribution obtained in part (a) of this exercise. The cutpoints are used to label the horizontal axis. Also, the height of each bar in the frequency histogram matches the respective frequency in column 2.

(d) The relative-frequency histogram in Figure (b) is constructed using the relative-frequency distribution obtained in part (b) of this exercise. It has the same horizontal axis as the frequency histogram. The height of each bar in the relative-frequency histogram matches the respective relative frequency in column 2.

Figure (a)

Figure (b)

**2.73** The horizontal axis of this dotplot displays a range of possible values.  To complete the dotplot, we go through the data set and record data value by placing a dot over the appropriate value on the horizontal axis.

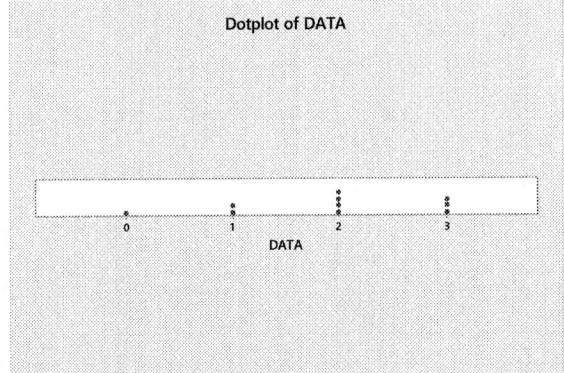

**2.75** The horizontal axis of this dotplot displays a range of possible values.  To complete the dotplot, we go through the data set and record data value by placing a dot over the appropriate value on the horizontal axis.

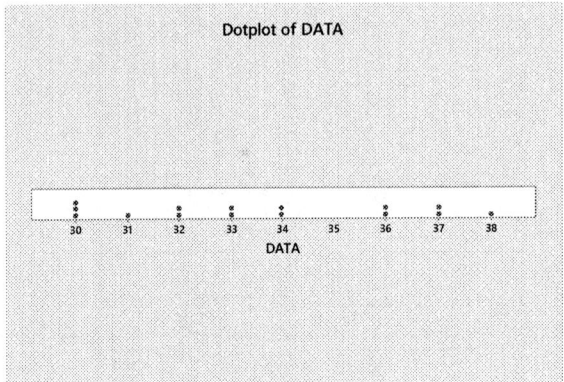

**2.77** Each data value consists of 2 digit numbers ranging from 20 to 62.  The last digit becomes the leaf and the remaining digits are the stems, so we have stems of 2 to 6.  The resulting stem-and-leaf diagram is

```
2| 01
3| 2278
4| 13
5| 5
6| 2
```

**2.79** Each data value consists of 2 digit numbers ranging from 22 to 46. The last digit becomes the leaf and the remaining digits are the stems, so we have stems of 2 to 4. Splitting the stems into two lines per stem, the resulting stem-and-leaf diagram is

```
2 | 224
2 | 5577789
3 | 12234
3 | 67
4 | 0
4 | 56
```

**2.81** (a) Since the data values range from 1 to 7, we construct a table with classes based on a single value. The resulting table follows.

| Number of Persons | Frequency |
|---|---|
| 1 | 7 |
| 2 | 13 |
| 3 | 9 |
| 4 | 5 |
| 5 | 4 |
| 6 | 1 |
| 7 | 1 |
|   | 40 |

(b) To get the relative frequencies, divide each frequency by the sample size of 40.

| Number of Persons | Relative Frequency |
|---|---|
| 1 | 0.175 |
| 2 | 0.325 |
| 3 | 0.225 |
| 4 | 0.125 |
| 5 | 0.100 |
| 6 | 0.025 |
| 7 | 0.025 |
|   | 1.000 |

(c) The frequency histogram in Figure (a) is constructed using the frequency distribution presented in part (a) of this exercise. Column 1 demonstrates that the data are grouped using classes based on a single value. These single values in column 1 are used to label the horizontal axis of the frequency histogram. Suitable candidates for vertical axis units in the frequency histogram are the integers within the range 0 through 13, since these are representative of the magnitude and spread of the frequencies presented in column 2. When classes are based on a single value, the middle of each histogram bar is placed directly over the single numerical value represented by the class. Also, the height of each bar in the frequency histogram matches the respective frequency in column 2.

**34** Chapter 2

Figure (a)

Figure (b)

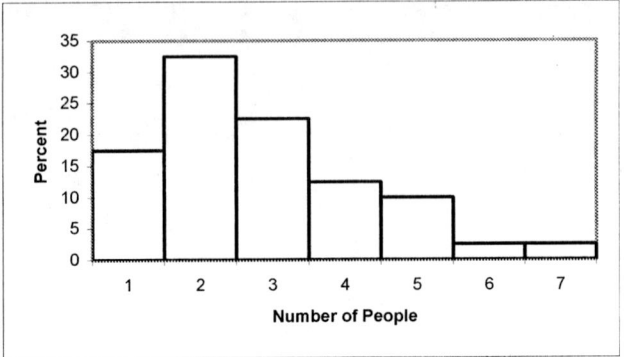

(d) The relative-frequency histogram in Figure (b) is constructed using the relative-frequency distribution presented in part (b) of this exercise. It has the same horizontal axis as the frequency histogram. We notice that the relative frequencies presented in column 2 range in size from 0.025 to 0.325. Thus, suitable candidates for vertical-axis units in the relative-frequency histogram are increments of 0.05 (or 5%), from zero to 0.35 (or 35%). The middle of each histogram bar is placed directly over the single numerical value represented by the class. Also, the height of each bar in the relative-frequency histogram matches the respective relative frequency in column 2.

**2.83** (a) Since the data values range from 0 to 5, we construct a table with classes based on a single value. The resulting table follows.

| Number of Computers | Frequency |
|---|---|
| 0 | 5 |
| 1 | 22 |
| 2 | 11 |
| 3 | 2 |
| 4 | 4 |
| 5 | 1 |
|  | 45 |

(b) To get the relative frequencies, divide each frequency by the sample size of 45. The sum of the relative frequency column is 0.999 due to rounding

| Number of Computers | Relative Frequency |
|---|---|
| 0 | 0.111 |
| 1 | 0.489 |
| 2 | 0.244 |
| 3 | 0.044 |
| 4 | 0.089 |
| 5 | 0.022 |
|  | 0.999 |

(c) The frequency histogram in Figure (a) is constructed using the frequency distribution presented in part (a) of this exercise. Column 1 demonstrates that the data are grouped using classes based on a single value. These single values in column 1 are used to label the horizontal axis of the frequency histogram. When classes are based on a single value, the middle of each histogram bar is placed directly over the single numerical value represented by the class. Also, the height of each bar in the frequency histogram matches the respective frequency in column 2.

Copyright © 2016 Pearson Education, Inc.

Figure (a)                                    Figure (b)

 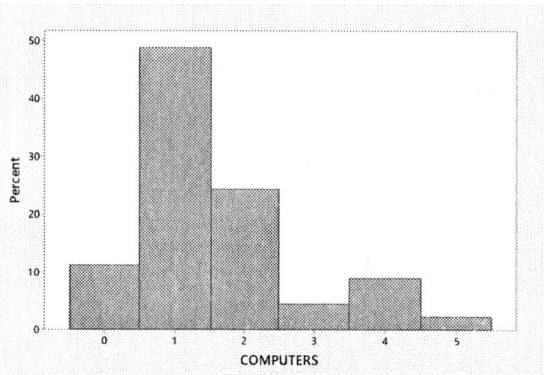

(d) The relative-frequency histogram in Figure (b) is constructed using the relative-frequency distribution presented in part (b) of this exercise. It has the same horizontal axis as the frequency histogram. The middle of each histogram bar is placed directly over the single numerical value represented by the class. Also, the height of each bar in the relative-frequency histogram matches the respective relative frequencies in column 2.

**2.85** (a) The first class to construct is 40-44. Since all classes are to be of equal width, and the second class begins with 45, we know that the width of all classes is 45 - 40 = 5. All of the classes are presented in column 1. The last class to construct is 60-64, since the largest single data value is 61. Having established the classes, we tally the age figures into their respective classes. These results are presented in column 2, which lists the frequencies.

| Age | Frequency |
|---|---|
| 40-44 | 4 |
| 45-49 | 3 |
| 50-54 | 4 |
| 55-59 | 8 |
| 60-64 | 2 |
| | 21 |

(b) Dividing each frequency by the total number of observations, which is 21, results in each class's relative frequency. The relative frequencies for all classes are presented in column 2. The resulting table follows.

| Age | Relative Frequency |
|---|---|
| 40-44 | 0.190 |
| 45-49 | 0.143 |
| 50-54 | 0.190 |
| 55-59 | 0.381 |
| 60-64 | 0.095 |
| | 1.000 |

(c) The frequency histogram in Figure (a) is constructed using the frequency distribution presented in part (a) of this exercise. The lower class limits of column 1 are used to label the horizontal axis of the frequency histogram. Suitable candidates for vertical-axis units in the frequency histogram are the even integers 2 through 8, since these are representative of the magnitude and spread of the frequency presented in column 2. The height of each bar in the frequency histogram matches the respective frequency in column 2.

(d) The relative-frequency histogram in Figure (b) is constructed using the relative-frequency distribution presented in part (b) of this exercise. It has the same horizontal axis as the frequency histogram. We notice that the relative frequencies presented in column 2 vary in size from 0.095 to 0.381. Thus, suitable candidates for vertical axis units in the relative-frequency histogram are increments of 0.10 (or 10%), from zero to 0.40 (or 40%). The height of each bar in the relative-frequency histogram matches the respective relative frequency in column 2.

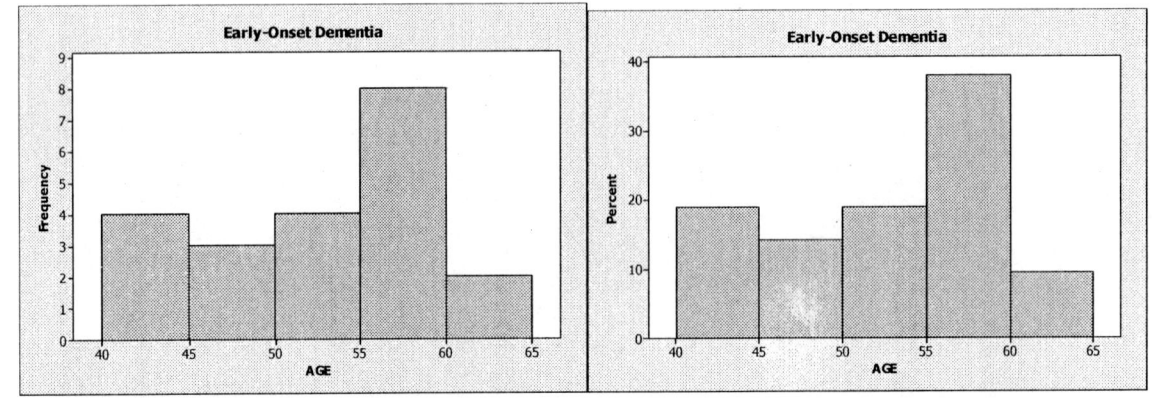

Figure (a)    Figure (b)

**2.87** (a) The first class to construct is 12-17. Since all classes are to be of equal width, and the second class begins with 18, we know that the width of all classes is 18 - 12 = 6. All of the classes are presented in column 1. The last class to construct is 60-65, since the largest single data value is 61. Having established the classes, we tally the anxiety questionnaire score figures into their respective classes. These results are presented in column 2, which lists the frequencies.

| Anxiety | Frequency |
|---------|-----------|
| 12-17   | 2         |
| 18-23   | 3         |
| 24-29   | 6         |
| 30-35   | 5         |
| 36-41   | 10        |
| 42-47   | 4         |
| 48-53   | 0         |
| 54-59   | 0         |
| 60-65   | 1         |
|         | 31        |

(b) Dividing each frequency by the total number of observations, which is 31, results in each class's relative frequency. The relative frequencies for all classes are presented in column 2. The resulting table follows.

| Anxiety | Relative Frequency |
|---------|--------------------|
| 12-17   | 0.065              |
| 18-23   | 0.097              |
| 24-29   | 0.194              |
| 30-35   | 0.161              |
| 36-41   | 0.323              |
| 42-47   | 0.129              |
| 48-53   | 0.000              |
| 54-59   | 0.000              |
| 60-65   | 0.032              |
|         | 1.000              |

(c) The frequency histogram in Figure (a) is constructed using the frequency distribution presented in part (a) of this exercise. The lower class limits of column 1 are used to label the horizontal axis of the frequency histogram. Suitable candidates for vertical-axis units in the frequency histogram are the even integers 0 through 10, since these are representative of the magnitude and spread of the frequency presented in column 2. The height of each bar in the frequency histogram matches the respective frequency in column 2.

(d) The relative-frequency histogram in Figure (b) is constructed using the relative-frequency distribution presented in part (b) of this exercise. It has the same horizontal axis as the frequency histogram. We notice that the relative frequencies presented in column 2 vary in size from 0.000 to 0.323. Thus, suitable candidates for vertical axis units in the relative-frequency histogram are increments of 0.05 (or 5%), from zero to 0.35 (or 35%). The height of each bar in the relative-frequency histogram matches the respective relative frequency in column 2.

Figure (a)                     Figure (b)

 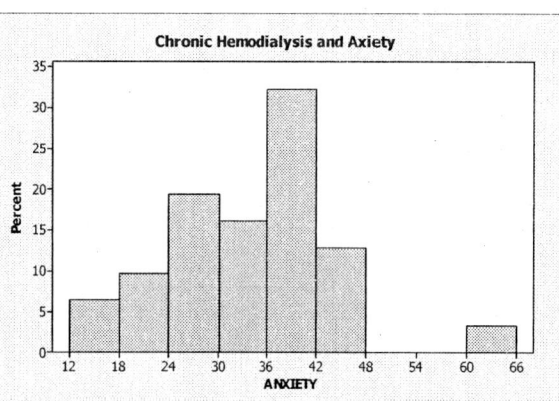

2.89 (a) The first class to construct is 52 - under 54. Since all classes are to be of equal width 2, the second class is 54 - under 56. All of the classes are presented in column 1. The last class to construct is 74 - under 76, since the largest single data value is 75.3. Having established the classes, we tally the cheetah speeds into their respective classes. These results are presented in column 2, which lists the frequencies.

| Speed | Frequency |
|---|---|
| 52 – under 54 | 2 |
| 54 – under 56 | 5 |
| 56 – under 58 | 6 |
| 58 – under 60 | 8 |
| 60 – under 62 | 7 |
| 62 – under 64 | 3 |
| 64 – under 66 | 2 |
| 66 – under 68 | 1 |
| 68 – under 70 | 0 |
| 70 – under 72 | 0 |
| 72 – under 74 | 0 |
| 74 – under 76 | 1 |
|  | 35 |

(b) Dividing each frequency by the total number of observations, which is 35, results in each class's relative frequency. The relative frequencies for all classes are presented in column 2

| Speed | Relative Frequency |
|---|---|
| 52 – under 54 | 0.057 |
| 54 – under 56 | 0.143 |
| 56 – under 58 | 0.171 |
| 58 – under 60 | 0.229 |
| 60 – under 62 | 0.200 |
| 62 – under 64 | 0.086 |
| 64 – under 66 | 0.057 |
| 66 – under 68 | 0.029 |
| 68 – under 70 | 0.000 |
| 70 – under 72 | 0.000 |
| 72 – under 74 | 0.000 |
| 74 – under 76 | 0.029 |
|  | 1.000 |

(c) The frequency histogram in Figure (a) is constructed using the frequency distribution obtained in part (a) of this exercise  Column 1 demonstrates that the data are grouped using classes with class widths of 2. Suitable candidates for vertical axis units in the frequency histogram are the integers within the range 0 through 8, since these are representative of the magnitude and spread of the frequencies presented in column 2. Also, the height of each bar in the frequency histogram matches the respective frequency in column 2.

(d) The relative-frequency histogram in Figure (b) is constructed using the relative-frequency distribution obtained in part (b) of this exercise. It has the same horizontal axis as the frequency histogram. We notice that the relative frequencies presented in column 3 range in size from 0.000 to 0.229. Thus, suitable candidates for vertical axis units in the relative-frequency histogram are increments of 0.05 (5%), from zero to 0.25 (25%). The height of each bar in the relative-frequency histogram matches the respective relative frequency in column 2.

Figure (a)                                    Figure (b)

 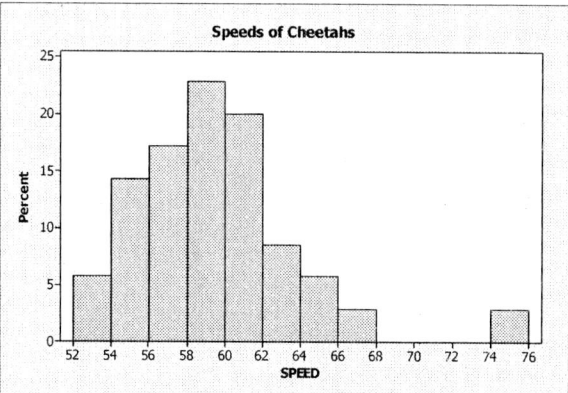

**2.91** (a) The first class to construct is 0 - under 1. Since all classes are to be of equal width 1, the second class is 1 - under 2. All of the classes are presented in column 1. The last class to construct is 7 - under 8, since the largest single data value is 7.6. Having established the classes, we tally the fuel tank capacities into their respective classes. These results are presented in column 2, which lists the frequencies.

| Oxygen Distribution | Frequency |
|---|---|
| 0 - under 1 | 1 |
| 1 - under 2 | 10 |
| 2 - under 3 | 5 |
| 3 - under 4 | 4 |
| 4 - under 5 | 0 |
| 5 - under 6 | 0 |
| 6 - under 7 | 1 |
| 7 - under 8 | 1 |
|  | 22 |

(b) Dividing each frequency by the total number of observations, which is 22, results in each class's relative frequency. The relative frequencies for all classes are presented in column 2

| Oxygen Distribution | Relative Frequency |
|---|---|
| 0 - under 1 | 0.045 |
| 1 - under 2 | 0.455 |
| 2 - under 3 | 0.227 |
| 3 - under 4 | 0.182 |
| 4 - under 5 | 0.000 |
| 5 - under 6 | 0.000 |
| 6 - under 7 | 0.045 |
| 7 - under 8 | 0.045 |
|  | 1.000 |

(c) The frequency histogram in Figure (a) is constructed using the frequency distribution obtained in part (a) of this exercise  Column 1 demonstrates that the data are grouped using classes with class widths of 2. Suitable candidates for vertical axis units in the frequency histogram are the integers within the range 0 through 10, since these are representative of the magnitude and spread of the frequencies presented in column 2. Also, the height of each bar in the frequency histogram matches the respective frequency in column 2.

(d) The relative-frequency histogram in Figure (b) is constructed using the relative-frequency distribution obtained in part (b) of this exercise. It has the same horizontal axis as the frequency histogram.

**40** Chapter 2

We notice that the relative frequencies presented in column 3 range in size from 0.000 to 0.455. Thus, suitable candidates for vertical axis units in the relative-frequency histogram are increments of 0.05 (5%), from zero to 0.50 (50%). The height of each bar in the relative-frequency histogram matches the respective relative frequency in column 2.

Figure (a)         Figure (b)

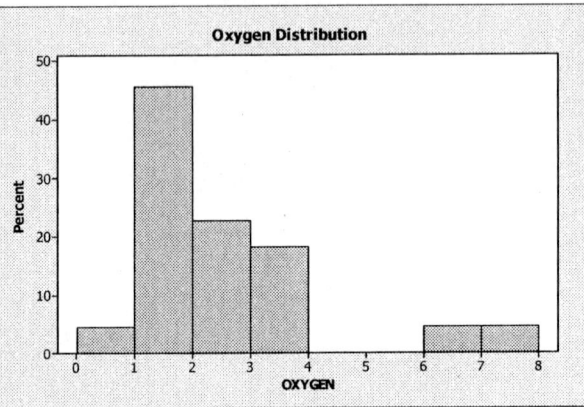

**2.93** The horizontal axis of this dotplot displays of range of possible ages of the passenger cars. To complete the dotplot, we go through the data set and record each age by placing a dot over the appropriate value on the horizontal axis.

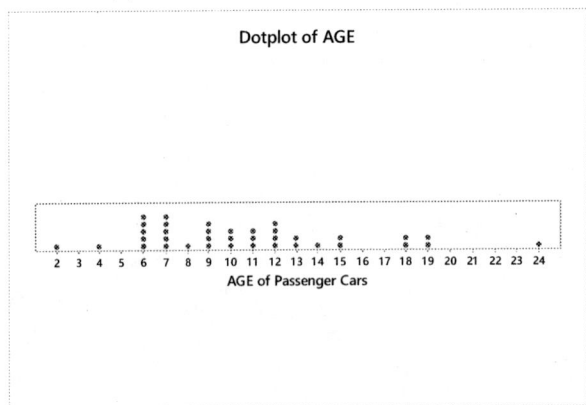

**2.95** (a) The data values range from 7 to 18, so the scale must accommodate those values. We stack dots above each value on two different lines using the same scale for each line. The result is

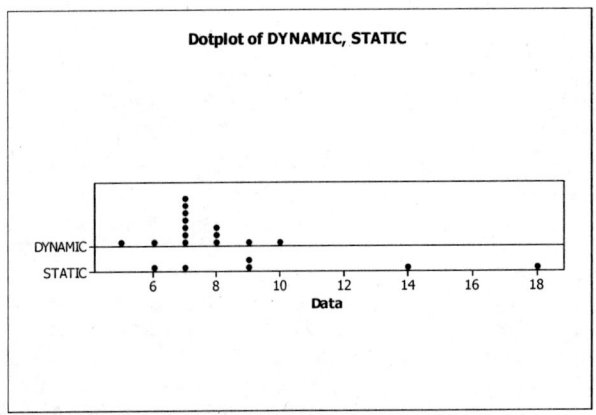

(b) The Dynamic system does seem to reduce acute postoperative days in the hospital on the average. The Dynamic data are centered at about 7 days, whereas the Static data are centered at about 11 days and are much more spread out than the Dynamic data.

**2.97** Since each data value consists of 2 digits, each beginning with 1, 2, 3, or 4, we will construct the stem-and-leaf diagram with these four values as the stems. The result is

```
1| 238
2| 1678899
3| 34459
4| 04
```

**2.99** (a) Since each data value lies between 2 and 93, we will construct the stem-and-leaf diagram with one line per stem. The result is

```
0| 2234799
1| 11145566689
2| 023479
3| 004555
4| 19
5| 5
6| 9
7| 9
8|
9| 3
```

(b) Using two lines per stem, the same data result in the following diagram:
```
0| 2234
0| 799
1| 1114
1| 5566689
2| 0234
2| 79
3| 004
3| 555
4| 1
4| 9
5|
5| 5
6|
6| 9
7|
7| 9
8|
8|
9| 3
```

(c) The stem with one line per stem is more useful. One gets the same impression regarding the shape of the distribution, but the two lines per stem version has numerous lines with no data, making it take up more space than necessary to interpret the data and giving it too many lines.

**2.101** (a) Since we have two digit numbers, the last digit becomes the leaf and the first digit becomes the stem. For this data, we have stems of 6, 7, and 8. Splitting the data into five lines per stem, we put the leaves 0-1 in the first stem, 2-3 in the second stem, 4-5 in the third stem, 6-7 in the fourth stem, and 8-9 in the fifth stem. The result is

**42** Chapter 2

```
6| 8
7| 11111111
7| 222222222233333
7| 4444555555
7| 66666777777
7| 8
8| 0
```

(b) Using one or two lines per stem would have given us too few lines.

**2.103** The graph indicates that:

(a) 20% of the patients have cholesterol levels between 205 and 209, inclusive.

(b) 20% are between 215 and 219; and 5% are between 220 and 224. Thus, 25% (i.e., 20% + 5%) have cholesterol levels of 215 or higher.

(c) 35% of the patients have cholesterol levels between 210 and 214, inclusive. With 20 patients in total, the number having cholesterol levels between 210 and 214 is 7 (i.e., 0.35 x 20).

**2.105** The graph indicates that:

(a) 1 of these 14 patients was under 45 years old at the onset of symptoms.

(b) 3 of these 14 patients were at least 65 years old at the onset of symptoms.

(c) 8 of these 14 patients were between 50 and 64 years old, inclusive, at the onset of symptoms.

**2.107** (a) Using Minitab, there is not a direct way to get a grouped frequency distribution. However, you can use an option in creating your histogram that will report the frequencies in each of the classes, essentially creating a grouped frequency distribution. Retrieve the data from the WeissStats Resource Site. Column 2 contains the length of the Beatles songs, in seconds. From the tool bar, select **Graph ▶ Histogram,** choose **Simple** and click **OK.** Double click on LENGTH to enter LENGTH in the **Graph variables** box. Click **Labels,** click the **Data Labels** tab, then check **Use y-value labels.** Click **OK** twice. The result is

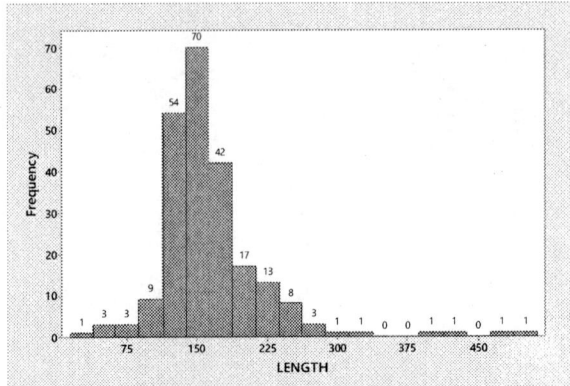

Above each bar is a label for each of the frequencies. Also, the labeling on the horizontal axis is the midpoint for class, where each class has a width of 25. To change the width of the bars, right Click on the bars and choose **Edit Bars,** click **Binning,** click **Midpoint/Cutpoint positions:,** type 0 50 100 150 200 250 300 350 400 450 500. The result is

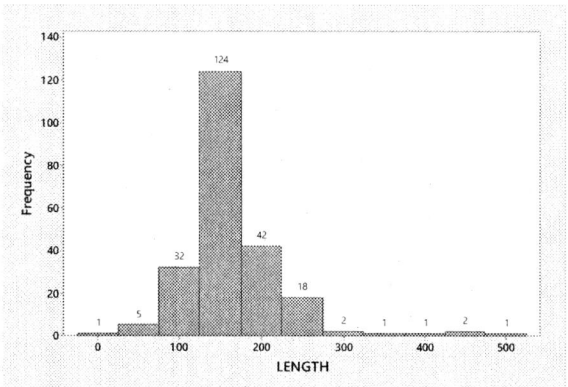

To get the relative-frequency distribution, follow the same steps as above, but also Click the **Scale** button, click the **Y-scale Type,** check **Percent,** then click **OK** twice. The result is

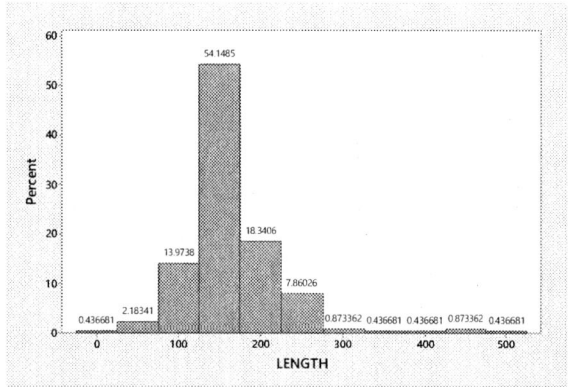

Above each bar is the percentage for that class, essentially creating a relative-frequency distribution. You could also transfer these results into a table.

(b) A frequency histogram and a relative frequency distribution were created in part (a).

(c) To obtain the dotplot, select **Graph ▶ Dotplot,** select **Simple** in the **One Y** row, and click **OK.** Double click on LENGTH to enter LEGNTH in the **Graph variables** box and click **OK.** The result is

(d) The graphs are similar, but not identical. This is because the dotplot preserves the raw data by plotting individual dots and the histogram

**44**  Chapter 2

looses the raw data because it groups observations into grouped classes. The overall impression, however, remains the same. They both are generally the same shape with outliers to the right.

**2.109** (a) After entering the data from the WeissStats Resource Site, in Minitab, select **Graph ▶ Stem-and-Leaf,** double click on PERCENT to enter PERCE T in the **Graph variables** box and enter a <u>10</u> in the **Increment** box, and click **OK**. The result is

```
Stem-and-leaf of PERCENT   N = 51
Leaf Unit = 1.0

   2    1  79
 (32)   2  01122333444455555566666777778999
  17    3  0001112223456668
   1    4  9
```

(b) Repeat part (a), but this time enter a <u>5</u> in the **Increment** box. The result is

```
Stem-and-leaf of PERCENT   N = 51
Leaf Unit = 1.0

   2    1  79
  14    2  011223334444
 (20)   2  55555566666777778999
  17    3  00011122234
   6    3  56668
   1    4
   1    4  9
```

(c) Repeat part (a) again, but this time enter a <u>2</u> in the **Increment** box. The result is

```
Stem-and-leaf of PERCENT   N = 51
Leaf Unit = 1.0

   1    1  7
   2    1  9
   5    2  011
  10    2  22333
  20    2  4444555555
 (10)   2  6666677777
  21    2  8999
  17    3  000111
  11    3  2223
   7    3  45
   5    3  666
   2    3  8
   1    4
   1    4
   1    4
   1    4
   1    4  9
```

(d) The second graph is the most useful. The third one has more classes than necessary to comprehend the shape of the distribution and has a number of empty stems. Typically, we like to have five to fifteen classes and the first and second diagrams satisfy that condition, but the second one provides a better idea of the shape of the distribution.

**2.111** (a) The classes are presented in column 1. With the classes established, we then tally the exam scores into their respective classes. These results are presented in column 2, which lists the frequencies. Dividing each frequency by the total number of exam scores, which is 20, results in each class's relative frequency. The relative frequencies for all classes are presented in column 3. The class mark of each class is the average of the lower and upper limits. The class marks for all classes are presented in column 4.

| Score | Frequency | Relative Frequency | Class Mark |
|-------|-----------|--------------------|-----------| 
| 30-39 | 2 | 0.10 | 34.5 |
| 40-49 | 0 | 0.00 | 44.5 |
| 50-59 | 0 | 0.00 | 54.5 |
| 60-69 | 3 | 0.15 | 64.5 |
| 70-79 | 3 | 0.15 | 74.5 |
| 80-89 | 8 | 0.40 | 84.5 |
| 90-100 | 4 | 0.20 | 95.0 |
|  | 20 | 1.00 |  |

(b) The first six classes have width 10; the seventh class had width 11.

(c) Answers will vary, but one choice is to keep the first six classes the same and make the next two classes 90-99 and 100-109. Another possibility is 31-40, 41-50, …, 91-100.

**2.113** Answers will vary, but by following the steps we first decide on the approximate number of classes. Since there are 37 observations, we should have 7-14 classes. This exercise states we should have approximately eight classes. Step 2 says that we calculate an approximate class width as $(278.8 - 129.2)/8 = 18.7$. A convenient class width close to 18.7 would be a class width of 20. Step 3 says that we choose a number for the lower cutpoint which is less than or equal to our minimum observation of 129.2. Let's choose 120. Beginning with a lower cutpoint of 120 and width of 20, we have a first class of 120 – under 140, a second class of 140 – under 160, a third class of 160 – under 180, a fourth class of 180 - under 200, a fifth class of 200 – under 220, a sixth class of 220 – under 240, a seventh class of 240 – under 260, and an eighth class of 260 – under 280. This would be our last class since the largest observation is 278.8.

**2.115** Consider columns 1 and 2 of the energy-consumption data given in Exercise 2.56 part (b). Compute the class mark for each class presented in column 1. Pair each class mark with its corresponding relative frequency found in column 2. Construct a horizontal axis, where the units are in terms of class marks and a vertical axis where the units are in terms of relative frequencies. For each class mark on the horizontal axis, plot a point whose height is equal to the relative frequency of the class. Then join the points with connecting lines. The result is a relative-frequency polygon.

**46** Chapter 2

**2.117** In single value grouping the horizontal axis would be labeled with the value of each class.

**2.119** (a) Consider parts (a) and (b) of the Cheetah speed data given in Exercise 2.61. The classes are now reworked to present just the lower cutpoint of each class. The frequencies are reworked to sum the frequencies of all classes representing values less than the specified lower cutpoint. These successive sums are the cumulative frequencies. The relative frequencies are reworked to sum the relative frequencies of all classes representing values less than the specified cutpoints. These successive sums are the cumulative relative frequencies. (Note: The cumulative relative frequencies can also be found by dividing the each cumulative frequency by the total number of data values.)

| Less than | Cumulative Frequency | Cumulative Relative Frequency |
|---|---|---|
| 52 | 0 | 0.000 |
| 54 | 2 | 0.057 |
| 56 | 7 | 0.200 |
| 58 | 13 | 0.371 |
| 60 | 21 | 0.600 |
| 62 | 28 | 0.800 |
| 64 | 31 | 0.886 |
| 66 | 33 | 0.943 |
| 68 | 34 | 0.971 |
| 70 | 34 | 0.971 |
| 72 | 34 | 0.971 |
| 74 | 34 | 0.971 |
| 76 | 35 | 1.000 |

(b) Pair each cutpoint with its corresponding cumulative relative frequency found in column 3. Construct a horizontal axis, where the units are in terms of the cutpoints and a vertical axis where the units are in terms of cumulative relative frequencies. For each cutpoint on the horizontal axis, plot a point whose height is equal to the cumulative relative frequency. Then join the points with connecting lines. The result, presented in the following figure, is an *ogive* using cumulative relative frequencies. (Note: A similar procedure could be followed using cumulative frequencies.)

**2.121** Minitab used truncation.  Note that there was a data point of 5.8 in the sample.  It would have been plotted with a stem of 0 and a leaf of 6 if it had been rounded.  Instead Minitab plotted the observation with a stem of 0 and a leaf of 5.

### Section 2.4
**2.123** Sample data are the values of a variable for a sample of the population.
**2.125** A sample distribution is the distribution of sample data.
**2.127** A distribution of a variable is the same as a population distribution.
**2.129** A large simple random sample from a bell-shaped distribution would be expected to have roughly a bell-shaped distribution since more sample values should be obtained, on average, from the middle of the distribution.
**2.131** Three distribution shapes that are symmetric are bell-shaped, triangular, and Uniform (or rectangular), shown in that order below.  It should be noted that there are others as well.

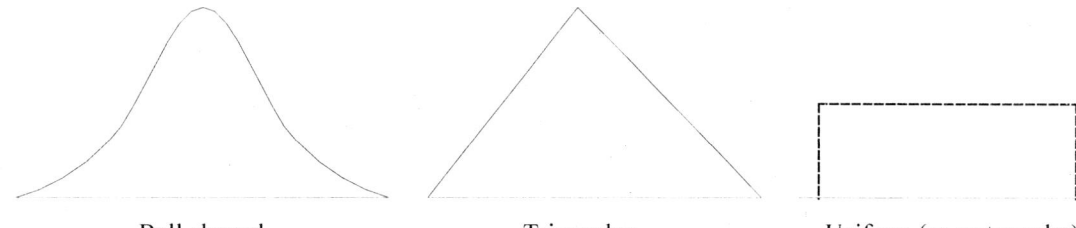

          Bell-shaped           Triangular           Uniform (or rectangular)

**2.133** (a) The shape of the distribution is unimodal.
     (b) The shape of the distribution is symmetric.
**2.135** (a) The shape of the distribution is unimodal.
     (b) The shape of the distribution is not symmetric.
     (c) The shape of the distribution is left-skewed.
**2.137** (a) The shape of the distribution is unimodal.
     (b) The shape of the distribution is not symmetric.
     (c) The shape of the distribution is left-skewed.
**2.139** (a) The shape of the distribution is multimodal.
     (b) The shape of the distribution is not symmetric.
**2.141** Except for the one data value between 74 and 76, this distribution is close to bell-shaped.  That one value makes the distribution slightly right skewed.
**2.143** The distribution of depths of the burrows is left skewed.
**2.145** The distribution of PCB concentration is symmetric.
**2.147** The distribution of cholesterol levels appears to be slightly left skewed.

**48** Chapter 2

**2.149** The distribution of length of stay is right skewed.

**2.151** (a) The distributions for Year 1 and Year 2 are both unimodal.
(b) Both distributions are not symmetric.
(c) Both distributions are right skewed.
(d) The distribution for Year 1 has a longer right tail indicating more variation than the distribution for Year 2. They are also not centered in the same place.

**2.153** (a) After entering the data from the WeissStats Resource Site, in Minitab, select **Graph ▶ Histogram,** select **Simple** and click **OK**. Double click on LENGTH to enter LENGTH in the **Graph variables** box and click **OK**. Our result is as follows. Results may vary depending on the type of technology used and graph obtained.

The distribution of LENGTH is unimodal and not symmetric.

(b) The distribution is right skewed.

**2.155** (a) In Exercise 2.109, we used Minitab to obtain a stem-and-leaf diagram using 2 lines per stem. That diagram is shown below.

```
Stem-and-leaf of PERCENT  N = 51
Leaf Unit = 1.0

    2   1  79
   14   2  011223334444
  (20)  2  55555566666777778999
   17   3  00011122234
    6   3  56668
    1   4
    1   4  9
```

The overall shape of this distribution is unimodal and roughly symmetric.

**2.157** (a) After entering the data from the WeissStats Resource Site, in Minitab, select **Graph ▶ Histogram,** select **Simple** and click **OK**. Double click on LENGTH to enter LENGTH in the **Graph variables** box and click **OK**. Our result is as follows. Results may vary depending on the type of technology used and graph obtained. The distribution of LENGTH is approximately unimodal and symmetric.

**2.159** The precise answers to this exercise will vary from class to class or individual to individual. Thus your results will likely differ from our results shown below.

(a) We obtained 50 random digits from a table of random numbers. The digits were

4 5 4 6 8 9 9 7 7 2 2 2 9 3 0 3 4 0 0 8 8 4 4 5 3

9 2 4 8 9 6 3 0 1 1 0 9 2 8 1 3 9 2 5 8 1 8 9 2 2

(b) Since each digit is equally likely in the random number table, we expect that the distribution would look roughly uniform.

(c) Using single value classes, the frequency distribution is given by the following table. The relative frequency histogram is shown below.

| Value | Frequency | Relative-Frequency |
|-------|-----------|--------------------|
| 0 | 5 | .10 |
| 1 | 4 | .08 |
| 2 | 8 | .16 |
| 3 | 5 | .10 |
| 4 | 6 | .12 |
| 5 | 3 | .06 |
| 6 | 2 | .04 |
| 7 | 2 | .04 |
| 8 | 7 | .14 |
| 9 | 8 | .16 |

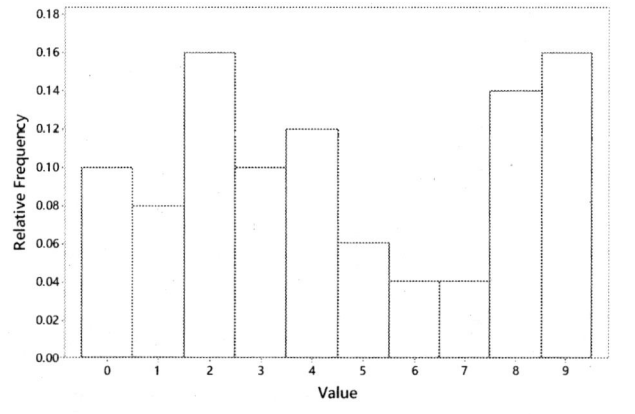

We did not expect to see this much variation.

**50** Chapter 2

- (d) We would have expected a histogram that was a little more 'even', more like a uniform distribution, but the relatively small sample size can result in considerable variation from what is expected.
- (e) We should be able to get a more uniformly distributed set of data if we choose a larger set of data.
- (f) Class project.

**2.161** (a) Your result will differ from, but be similar to, the one below which was obtained using Minitab.  Choose **Calc ▶ Random Data ▶ Normal...**, type 3000 in the **Generate rows of data** text box, click in the **Store in column(s)** text box and type C1, and click **OK**.

(b) Then choose **Graph ▶ Histogram**, choose the **Simple** version, click **OK**, enter C1 in the **Graph variables** text box, click on the **Scale** button and then on the **Y-Scale Type** tab.  Check the **Percent** box and click **OK** twice.

(c) The histogram in part (b) has the bell, unimodal and symmetric distribution.  The sample of 3000 is representative of the population from which the sample was taken.

### Section 2.5

**2.163** (a) A truncated graph is one for which the vertical axis starts at a value other than its natural starting point, usually zero.

(b) A legitimate motivation for truncating the axis of a graph is to place the emphasis on the ups and downs of the distribution rather than on the actual height of the graph.

(c) To truncate a graph and avoid the possibility of misinterpretation, one should start the axis at zero and put slashes in the axis to indicate that part of the axis is missing.

**2.165** (a) A large lower portion of the graph is eliminated.  When this is done, differences between district and national averages appear greater than in the original figure.

(b) Even more of the graph is eliminated.  Differences between district and national averages appear even greater than in part (a).

(c) The truncated graphs give the misleading impression that, in 2013, the district average is much greater relative to the national average than it actually is.

**2.167** (a) The problem with the bar chart is that it is truncated.  That is, the vertical axis, which should start at $0 (in trillions), starts with

$7.6 (in trillions) instead. The part of the graph from $0 (in trillions) to $7.6 (in trillions) has been cut off. This truncation causes the bars to be out of correct proportion and hence creates the misleading impression that the money supply is changing more than it actually is.

(b) A version of the bar chart with an untruncated and unmodified vertical axis is presented in Figure (a). Notice that the vertical axis starts at $0.00 (in trillions). Increments are in trillion dollars. In contrast to the original bar chart, this one illustrates that the changes in money supply from week to week are not that different. However, the "ups" and "downs" are not as easy to spot as in the original, truncated bar chart.

(c) A version of the bar chart in which the vertical axis is modified in an acceptable manner is presented in Figure (b). Notice that the special symbol "//" is used near the base of the vertical axis to signify that the vertical axis has been modified. Thus, with this version of the bar chart, not only are the "ups" and "downs" easy to spot but the reader is also aptly warned by the slashes that part of the vertical axis between $0.00 (in trillions) and $7.6 (in trillions) has been removed.

Figure (a)     Figure (b)

 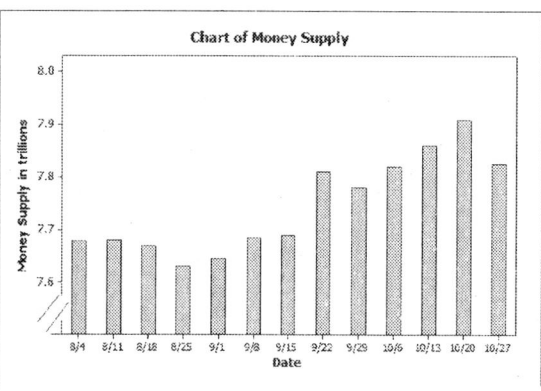

**2.169** (b) Without the vertical scale, it would appear that the happiness score dropped about 25% between the ages of 15 and 20.
(c) The actual drop was from a happiness score of about 5.5 to a happiness score of 5.25 between the ages of 15 and 20. This is a drop of 0.25/5.5 = 0.045 or about 4.5%.
(d) Without the vertical scale, it would look like the happiness scale has dropped about 50% from 5.5 to 5.0 where actually it really is closer to a 20% drop. The peak at 74 looks like a bigger significant increase without the vertical scale as well. Because the scale on the horizontal axis is not evenly spaced, the time between events is misleading.
(e) The graph could be made less potentially misleading by either making the vertical scale range from zero to 6.0 and by making the scale on the vertical axis the same width.

**2.171** (a) The brochure shows a "new" ball with twice the radius of the "old" ball. The intent is to give the impression that the "new" ball lasts roughly twice as long as the "old" ball. However, if the "new" ball has twice the radius of the "old" ball, the "new" ball will have eight times the volume of the "old" ball (since the volume of a sphere is proportional to the cube of its radius, or the radius $2^3 = 8$). Thus, the scaling is improper because it gives the impression that the "new" ball lasts eight times as long as the "old" ball rather than merely two times as long.

**52** Chapter 2

Old Ball             New Ball

(b) One possible way for the manufacturer to illustrate the fact that the "new" ball lasts twice as long as the "old" ball is to present pictures of two balls, side by side, each of the same magnitude as the picture of the "old" ball and to label this set of two balls "new ball". This will illustrate the point that a purchaser will be getting twice as much for the money.

Old Ball             New Ball

### Review Problems For Chapter 2

1. (a) A variable is a characteristic that varies from one person or thing to another.
   (b) Variables are quantitative or qualitative.
   (c) Quantitative variables can be discrete or continuous.
2. (d) Data are values of a variable.
   (e) The data type is determined by the type of variable being observed.
3. A frequency distribution of qualitative data is a table that lists the distinct values of data and their frequencies. It is useful to organize the data and make it easier to understand. A relative-frequency distribution of qualitative data is a table that lists the distinct values of data and a ratio of the class frequency to the total number of observations, which is called the relative frequency.
4. For both quantitative and qualitative data, the frequency and relative-frequency distributions list the values of the distinct classes and their frequencies and relative frequencies. For single value grouping of quantitative data, it is the same as the distinct classes for qualitative data. For class limit and cutpoint grouping in quantitative data, we create groups that form distinct classes similar to qualitative data.
5. The two main types of graphical displays for qualitative data are the bar chart and the pie chart.
6. The bars do not abut in a bar chart because there is not any continuity between the categories. Also, this differentiates them from histograms.
7. Answers will vary. One advantage of pie charts is that it shows the proportion of each class to the total. One advantage of bar charts is that it emphasizes each individual class in relation to each other. One disadvantage of pie charts is that if there are too many classes, the chart

8. Single value grouping is appropriate when the data are discrete with relatively few distinct observations.

9. (a) The second class would have lower and upper limits of 9 and 14. The class mark of this class would be the average of these limits and equal 11.5.

   (b) The third class would have lower and upper limits of 15 and 20.

   (c) The fourth class would have lower and upper limits of 21 and 26. This class would contain an observation of 23.

10. (a) If the width of the class is 5, then the class limits will be four whole numbers apart. Also, the average of the two class limits is the class mark of 8. Therefore, the upper and lower class limits must be two whole numbers above and below the number 8. The lower and upper class limits of the first class are 6 and 10.

    (b) The second class will have lower and upper class limits of 11 and 15. Therefore, the class mark will be the average of these two limits and equal 13.

    (c) The third class will have lower and upper class limits of 16 and 20.

    (d) The fourth class has limits of 21 and 25, the fifth class has limits of 26 and 30. Therefore, the fifth class would contain an observation of 28.

11. (a) The common class width is the distance between consecutive cutpoints, which is 15 - 5 = 10.

    (b) The midpoint of the second class is halfway between the cutpoints 15 and 25, and is therefore 20.

    (c) The sequence of the lower cutpoints is 5, 15, 25, 35, 45, ... Therefore, the lower and upper cutpoints of the third class are 25 and 35.

    (d) Since the third class has lower and upper cutpoints of 25 and 35, an observation of 32.4 would belong to this class.

12. (a) The midpoint is halfway between the cutpoints. Since the class width is 8, 10 is halfway between 6 and 14.

    (b) The class width is also the distance between consecutive midpoints. Therefore, the second midpoint is at 10 + 8 = 18.

    (c) The sequence of cutpoints is 6, 14, 22, 30, 38, ... Therefore the lower and upper cutpoints of the third class are 22 and 30.

    (d) An observation of 22 would go into the third class since that class contains data greater than or equal to 22 and strictly less than 30.

13. (a) If lower class limits are used to label the horizontal axis, the bars are placed between the lower class limit of one class and the lower class limit of the next class.

    (b) If lower cutpoints are used to label the horizontal axis, the bars are placed between the lower cutpoint of one class and the lower cutpoint of the next class.

    (c) If class marks are used to label the horizontal axis, the bars are placed directly above and centered over the class mark for that class.

    (d) If midpoints are used to label the horizontal axis, the bars are placed directly above and centered over the midpoint for that class.

**14.** (a) Bell-shaped  (b) Triangular  (c) Reverse J shape  (d) Uniform

**15.** Answers will vary but here is one possibility.

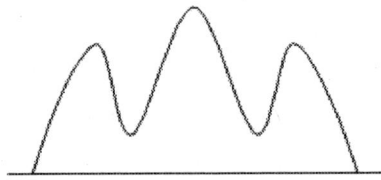

**16.** (a) The distribution of the large simple random sample will reflect the distribution of the population, so it would be left-skewed as well.

(b) No. The randomness in the samples will almost certainly produce different sets of observations resulting in shapes that are not identical.

(c) Yes. We would expect both of the simple random samples to reflect the shape of the population and be left-skewed.

**17.** (a) The first column ranks the hydroelectric plants. Thus, it consists of *quantitative, discrete* data.

(b) The fourth column provides measurements of capacity. Thus, it consists of *quantitative, continuous* data.

(c) The third column provides nonnumerical information. Thus, it consists of *qualitative* data.

**18.** (a) A single value frequency histogram for the prices of DVD players would be identical to the dotplot in example 2.16 because the classes would be 197 through 224 and the height for each bar in the frequency histogram would reflect the number of dots above each observation in the dotplot.

(b) No. The frequency histogram with cutpoint or class limit grouping would combine some of the single values together which would change the frequencies and the heights of the bars corresponding to those classes.

**19.** (a) The first class to construct is 40-44. Since all classes are to be of equal width, and the second class begins with 45, we know that the width of all classes is 45 - 40 = 5. All of the classes are presented in column 1 of the grouped-data table in the figure below. The last class to construct does not go beyond 65-69, since the largest single data value is 69. The sum of the relative frequency column is 0.999 due to rounding.

| Age at Inauguration | Frequency | Relative Frequency | Class Mark |
|---|---|---|---|
| 40-44 | 2 | 0.045 | 42 |
| 45-49 | 7 | 0.159 | 47 |
| 50-54 | 13 | 0.295 | 52 |
| 55-59 | 12 | 0.273 | 57 |
| 60-64 | 7 | 0.159 | 62 |
| 65-69 | 3 | 0.068 | 67 |
|  | 44 | 0.999 |  |

(b) By averaging the lower and upper limits for each class, we arrive at the class mark for each class. The class marks for all classes are presented in column 4.

(c) Having established the classes, we tally the ages into their respective classes. These results are presented in column 2, which lists the frequencies. Dividing each frequency by the total number of observations, which is 44, results in each class's relative frequency. The relative frequencies for all classes are presented in column 3.

(d) The frequency histogram presented below is constructed using the frequency distribution presented above; i.e., columns 1 and 2. Notice that the lower cutpoints of column 1 are used to label the horizontal axis of the frequency histogram. Suitable candidates for vertical-axis units in the frequency histogram are the even integers within the range 2 through 14, since these are representative of the magnitude and spread of the frequencies presented in column 2. The height of each bar in the frequency histogram matches the respective frequency in column 2.

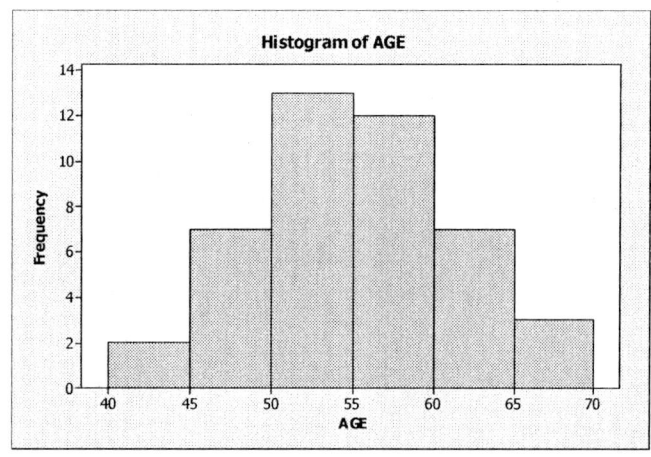

(e) The shape of the inauguration ages is unimodal and symmetric.

20. The horizontal axis of this dotplot displays a range of possible ages for the 44 Presidents of the United States. To complete the dotplot, we go through the data set and record each age by placing a dot over the appropriate value on the horizontal axis.

# 56  Chapter 2

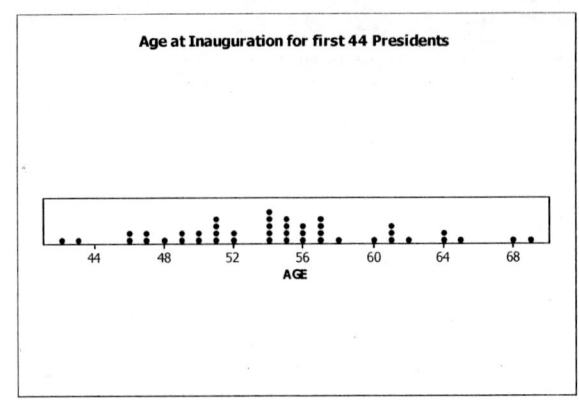

21. (a) Using *one* line per stem in constructing the ordered stem-and-leaf diagram means vertically listing the numbers comprising the stems *once*. The leaves are then placed with their respective stems in order. The ordered stem-and-leaf diagram using one line per stem is presented in the following figure.

    ```
    4| 236677899
    5| 00111122444445555566677778
    6| 0111244589
    ```

    (b) Using *two* lines per stem in constructing the ordered stem-and-leaf diagram means vertically listing the numbers comprising the stems *twice*. In turn, if the leaf is one of the digits 0 through 4, it is ordered and placed with the first of the two stem lines. If the leaf is one of the digits 5 through 9, it is ordered and placed with the second of the two stem lines. The ordered stem-and-leaf diagram using two lines per stem is presented in the following figure.

    ```
    4| 23
    4| 6677899
    5| 0011112244444
    5| 555566677778
    6| 0111244
    6| 589
    ```

    (c) The stem-and-leaf diagram in part (b) corresponds to the frequency distribution of Problem 19.

22. (a) Using *one* line per stem in constructing the ordered stem-and-leaf diagram means vertically listing the numbers comprising the stems *once*. The leaves are then placed with their respective stems in order. The ordered stem-and-leaf diagram using one line per stem is presented in the following figure.

    ```
    3| 4778
    4| 0467
    5| 446677778
    6| 0
    ```

    (b) Using *two* lines per stem in constructing the ordered stem-and-leaf diagram means vertically listing the numbers comprising the stems *twice*. In turn, if the leaf is one of the digits 0 through 4, it is ordered and placed with the first of the two stem lines. If the leaf is one of the digits 5 through 9, it is ordered and placed with the second of the two stem lines. The ordered stem-and-leaf diagram using two lines per stem is presented in the following figure.

```
3 | 4
3 | 778
4 | 04
4 | 67
5 | 44
5 | 6677778
6 | 0
```

(c) The second graph is the most useful. Typically, we like to have five to fifteen classes and the second one provides a better idea of the shape of the distribution.

23. (a) The frequency and relative-frequency distribution presented below is constructed using classes based on a single value. Since each data value is one of the integers 0 through 6, inclusive, the classes will be 0 through 6, inclusive. These are presented in column 1. Having established the classes, we tally the number of busy tellers into their respective classes. These results are presented in column 2, which lists the frequencies. Dividing each frequency by the total number of observations, which is 25, results in each class's relative frequency. The relative frequencies for all classes are presented in column 3.

| Number Busy | Frequency | Relative Frequency |
|---|---|---|
| 0 | 1 | 0.04 |
| 1 | 2 | 0.08 |
| 2 | 2 | 0.08 |
| 3 | 4 | 0.16 |
| 4 | 5 | 0.20 |
| 5 | 7 | 0.28 |
| 6 | 4 | 0.16 |
|   | 25 | 1.00 |

(b) The following relative-frequency histogram is constructed using the relative-frequency distribution presented in part (a); i.e., columns 1 and 3. Column 1 demonstrates that the data are grouped using classes based on a single value. These single values in column 1 are used to label the horizontal axis of the relative-frequency histogram. We notice that the relative frequencies presented in column 3 range in size from 0.04 to 0.28 (4% to 28%). Thus, suitable candidates for vertical axis units in the relative-frequency histogram are increments of 0.05, starting with zero and ending at 0.30. The middle of each histogram bar is placed directly over the single numerical value represented by the class. Also, the height of each bar in the relative-frequency histogram matches the respective relative frequency in column 3.

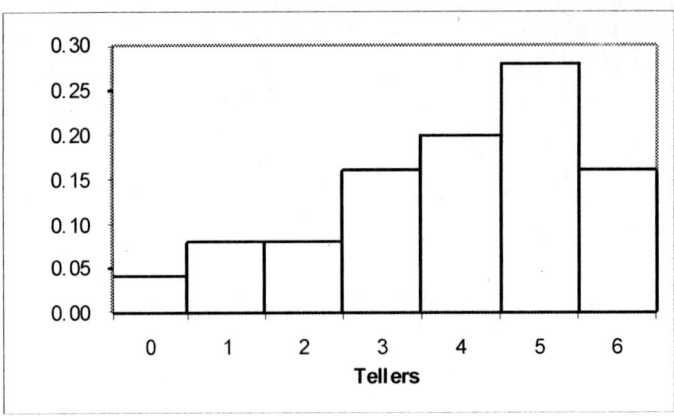

(c) The distribution is unimodal.

(d) The distribution is left skewed.

(e)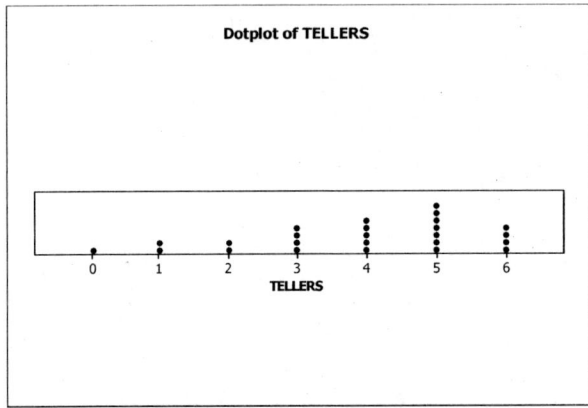

(f) Since both the histogram and the dotplot are based on single value grouping, they both convey exactly the same information.

24. (a) The classes will begin with the class 60 – under 65. The second class will be 65 – under 70. The classes will continue like this until the last class, which will be 90 – under 95, since the largest observation is 93.1. The classes can be found in column 1 of the frequency distribution in part (c).

(b) The midpoints of the classes are the averages of the lower and upper cutpoint for each class. For example, the first midpoint will be 62.5, the second midpoint will be 67.5. The midpoints will continue like this until the last midpoint, which will be 92.5. The midpoints can be found in column 4 of the frequency distribution in part (c).

(c) The first and second columns of the following table provide the frequency distribution. The first and third columns of the following table provide the relative-frequency distribution for percentages of on-time arrivals for the airlines.

| Percent On-Time | Frequency | Relative Frequency | Midpoint |
|---|---|---|---|
| 60 - under 65 | 1 | 0.0625 | 62.5 |
| 65 - under 70 | 5 | 0.3125 | 67.5 |
| 70 - under 75 | 5 | 0.3125 | 72.5 |
| 75 - under 80 | 3 | 0.1875 | 77.5 |
| 80 - under 85 | 0 | 0.0000 | 82.5 |
| 85 - under 90 | 1 | 0.0625 | 87.5 |
| 90 - under 95 | 1 | 0.0625 | 92.5 |
|  | 16 | 1.0000 |  |

(d) The frequency histogram is constructed using columns 1 and 2 from the frequency distribution from part (c). The cutpoints are used to label the horizontal axis. The vertical axis is labeled with the frequencies that range from 0 to 5.

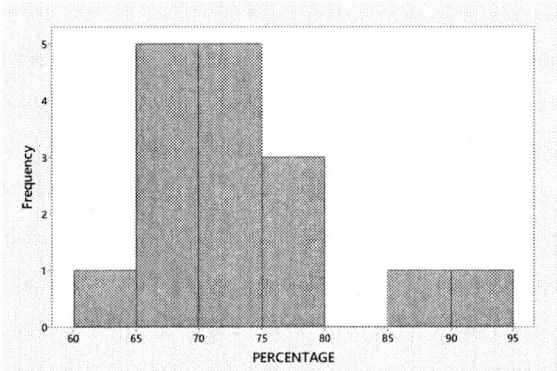

(e) After rounding each observation to the nearest whole number, the stem-and-leaf diagram with two lines per stem for the rounded percentages is

```
6| 2
6| 669
7| 0011334
7| 678
8|
8| 8
9| 3
```

(f) After obtaining the greatest integer in each observation, the stem-and-leaf diagram with two lines per stem is

```
6| 1
6| 56999
7| 01233
7| 677
8|
8| 7
9| 3
```

(g) The stem-and-leaf diagram in part (f) corresponds to the frequency distribution in part (d) because the observations weren't rounded in constructing the frequency distribution.

25. (a) The dotplot of the ages of the oldest player on each major league baseball team is

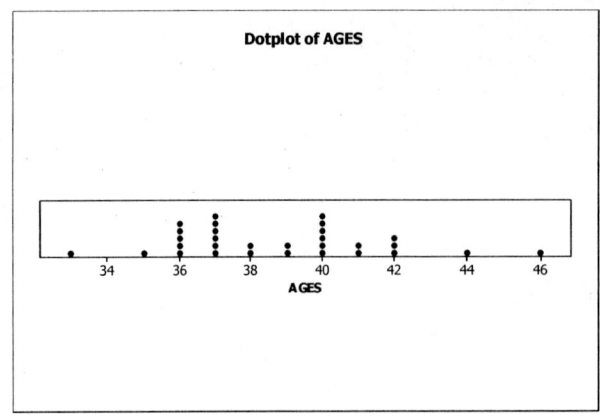

(b) The overall shape of the distribution of ages is bimodal and roughly symmetric.

26. (a) The classes are the types of evidence and are presented in column 1. The frequency distribution of the champions is presented in column 2. Dividing each frequency by the total number of submissions, which is 3436, results in each class's relative frequency. The relative frequency distribution is presented in column 3.

| Evidence Type | Frequency | Relative Frequency |
|---|---|---|
| Firearm | 289 | 0.084 |
| Magazine | 161 | 0.047 |
| Live Cartridge | 2727 | 0.794 |
| Spent Cartridge | 259 | 0.075 |
|  | 3436 | 1.000 |

(b) We multiply each of the relative frequencies by 360 degrees to obtain the portion of the pie represented by each evidence type. The result is

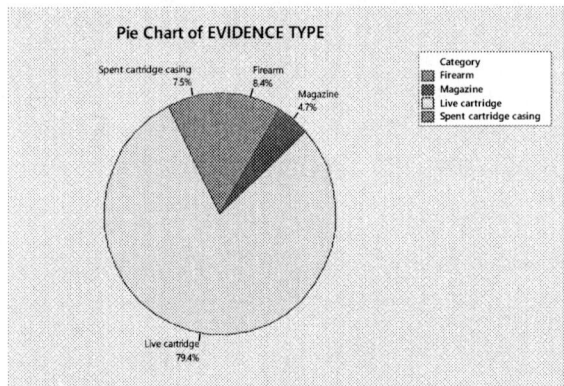

(c) We use the bar chart to show the relative frequency with which each EVIDENCE TYPE occurs. The result is

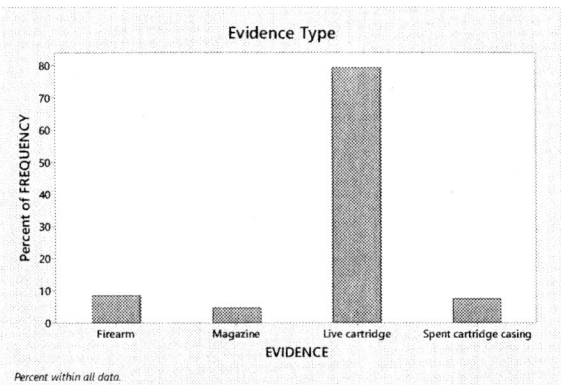

(d) The most common type of evidence type submitted to crime labs is a live cartridge which was the type submitted 79.4% of the time.

27. (a) The population consists of the states in the United States. The variable is the division.

(b) The frequency and relative frequency distribution for Region is shown below.

| Region | Frequency | Relative Frequency |
|---|---|---|
| East North Central | 5 | 0.10 |
| East South Central | 4 | 0.08 |
| Middle Atlantic | 3 | 0.06 |
| Mountain | 8 | 0.16 |
| New England | 6 | 0.12 |
| Pacific | 5 | 0.10 |
| South Atlantic | 8 | 0.16 |
| West North Central | 7 | 0.14 |
| West South Central | 4 | 0.08 |
|  | 50 | 1.000 |

(c) The pie chart for Region is shown below.

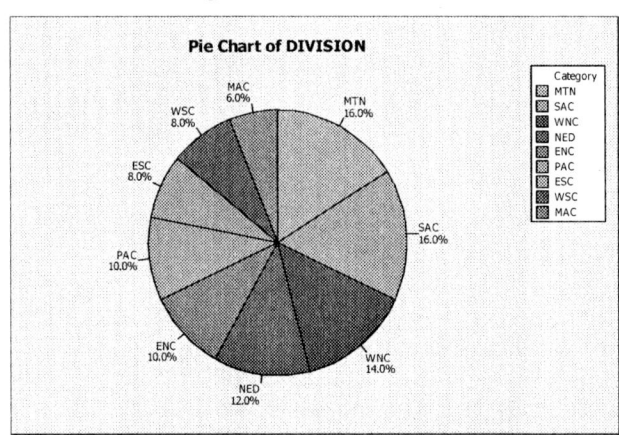

(d) The bar chart for Region is shown below.

(e) The distribution of Region seems to be almost uniformly distributed between the categories. The smallest region is the Middle Atlantic at a frequency of 3, the largest region is the Mountain and South Atlantic at a frequency of 8.

28. (a) The break in the third bar is to emphasize that the bar as shown is not as tall as it should be.

(b) The bar for the space available in coal mines is at a height of about 30 billion tonnes. To accurately represent the space available in saline aquifers (10,000 billion tonnes), the third bar would have to be over 300 times as high as the first bar and more than 10 times as high as the second bar. If the first two bars were kept at the sizes shown, there wouldn't be enough room on the page for the third bar to be shown at its correct height. If a reasonable height were chosen for the third bar, the first bar wouldn't be visible. The only apparent solution is to present the third bar as a broken bar.

29. (a) Covering up the numbers on the vertical axis totally obscures the percentages.

(b) Having followed the directions in part (a), we might conclude that the percentage of women in the labor force for 2000 is about three and one-half times that for 1960.

(c) Not covering up the vertical axis, we find that the percentage of women in the labor force for 2000 is about 1.8 times that for 1960.

(d) The graph is potentially misleading because it is truncated. Notice that vertical axis units begin at 30 rather than at zero.

(e) To make the graph less potentially misleading, we can start it at zero instead of 30.

30. (a) Using Minitab, retrieve the data from the WeissStats Resource Site. Column 2 contains the eye color and column 3 contains the hair color for the students. From the tool bar, select **Stat ▶ Tables ▶ Tally Individual Variables**, double-click on EYES and HAIR in the first box so that both EYES and HAIR appear in the **Variables** box, put a check mark next to Counts and Percents under Display, and click **OK**. The results are

| EYES | Count | Percent | HAIR | Count | Percent |
|---|---|---|---|---|---|
| Blue | 215 | 36.32 | Black | 108 | 18.24 |
| Brown | 220 | 37.16 | Blonde | 127 | 21.45 |
| Green | 64 | 10.81 | Brown | 286 | 48.31 |
| Hazel | 93 | 15.71 | Red | 71 | 11.99 |
| N= | 592 | | N= | 592 | |

(b) Using Minitab, select **Graph ▶ Pie Chart**, check Chart counts of unique values, double-click on EYES and HAIR in the first box so that EYES and HAIR appear in the Categorical Variables box. Click Pie Options, check decreasing volume, click OK. Click Multiple Graphs, check On the Same Graphs, Click OK. Click Labels, click Slice Labels, check Category Name, Percent, and Draw a line from label to slice, Click OK twice. The results are

(c) Using Minitab, select **Graph ▶ Bar Chart**, select Counts of unique values, select Simple option, click OK. Double-click on EYES and HAIR in the first box so that EYES and HAIR appear in the Categorical Variables box. Select Chart Options, check decreasing Y, check show Y as a percent, click OK. Click OK twice. The results are

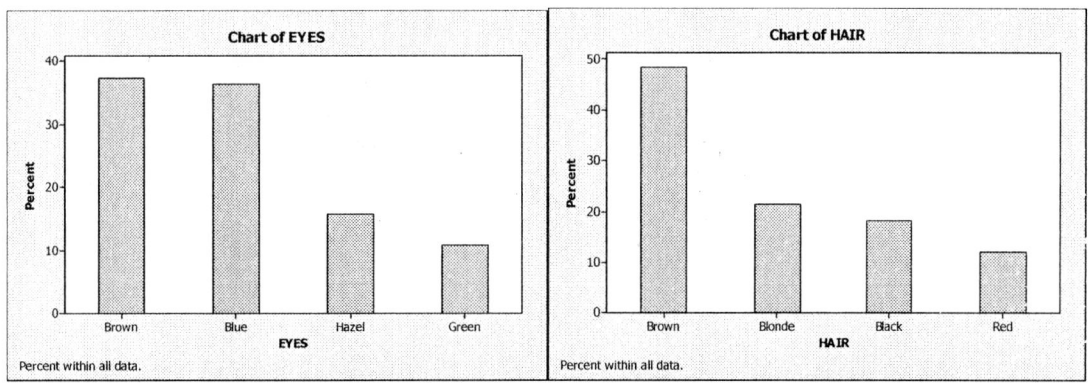

31. (a) The population consists of the states of the U.S. and the variable under consideration is the value of the exports of each state.

(b) Using Minitab, we enter the data from the WeissStats Resource Site, choose **Graph ▶ Histogram**, click on **Simple** and click **OK**. Then double click on VALUE to enter it in the **Graph variables** box and click **OK**. The result is

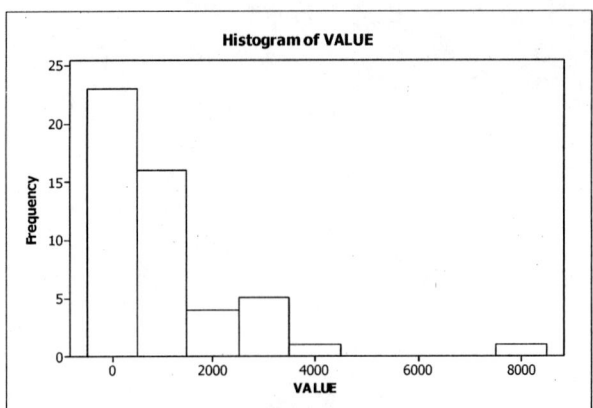

(c) For the dotplot, we choose **Graph ▶ Dotplot**, click on **Simple** from the **One Y** row and click **OK**. Then double click on VALUE to enter it in the **Graph variables** box and click **OK**. The result is

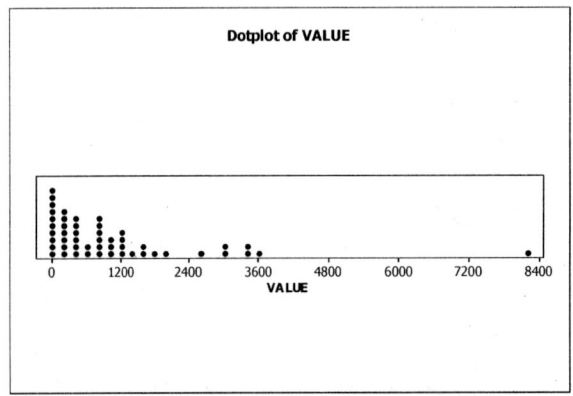

(d) For the stem-and-leaf plot, we choose **Graph ▶ Stem-and-Leaf**, double click on VALUE to enter it in the **Graph variables** box and click **OK**. The result is

```
Stem-and-leaf of VALUE   N = 50
Leaf Unit = 100

    23   0   00000000001112222344444
   (10)  0   5677888899
    17   1   012224
    11   1   5679
     7   2
     7   2   69
     5   3   034
     2   3   6
     1   4
     1   4
     1   5
     1   5
     1   6
     1   6
     1   7
     1   7
     1   8   2
```

(e) The overall shape of the distribution is unimodal and not symmetric.

(f) The distribution is right skewed.

32. (a) The population consists of countries of the world, and the variable under consideration is the expected life in years for people in those countries.

(b) Using Minitab, we enter the data from the WeissStats Resource Site, choose **Graph ▶ Histogram**, click on **Simple** and click **OK**. Then double click on YEARS to enter it in the **Graph variables** box and click **OK**. The result is

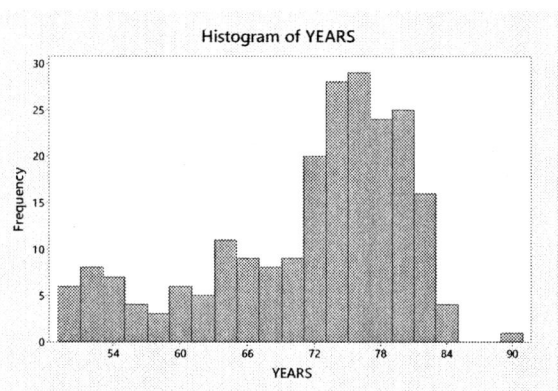

(c) For the dotplot, we choose **Graph ▶ Dotplot**, click on **Simple** from the **One Y** row and click **OK**. Then double click on YEARS to enter it in the **Graph variables** box and click **OK**. The result is

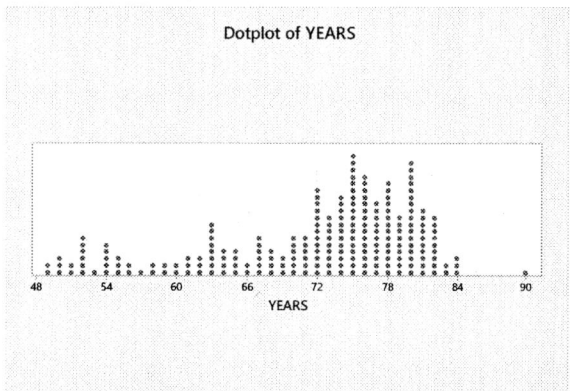

(d) For the stem-and-leaf plot, we choose **Graph ▶ Stem-and-Leaf**, double click on YEARS to enter it in the **Graph variables** box and click **OK**. The result is

```
Stem-and-leaf of YEARS   N  = 223
Leaf Unit = 1.0

     3    4  999
    21    5  000112222223344444
    30    5  556677899
    50    6  00001112233333333444
    73    6  5555666667778888899999
   (51)   7  000111111111122222222233333333333444444444444444444
    99    7  555555555555566666666666666777777777788888888888999999999999
    33    8  000000000000111111111111122223444
     1    8  9
```

**66** Chapter 2

(e) The overall shape of the distribution is unimodal and not symmetric.
(f) This distribution is classified as left skewed.

**33.** (a) The population consists of cities in the U.S., and the variables under consideration are their annual average maximum and minimum temperatures.

(b) Using Minitab, we enter the data from the WeissStats Resource Site, choose **Graph ▶ Histogram**, click on **Simple** and click **OK**. Double click on HIGH to enter it in the **Graph variables** box, and double click on LOW to enter it in the **Graph variables** box. Now click on the **Multiple graphs** button and click to **Show Graph Variables on separate graphs** and also check both boxes under **Same Scales for Graphs**, and click **OK** twice. The result is

 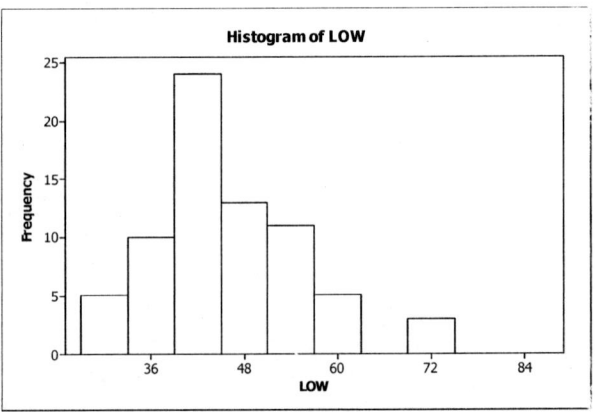

(c) For the dotplot, we choose **Graph ▶ Dotplot**, click on **Simple** from the **Multiple Y's** row and click **OK**. Then double click on HIGH and then LOW to enter them in the **Graph variables** box and click **OK**. The result is

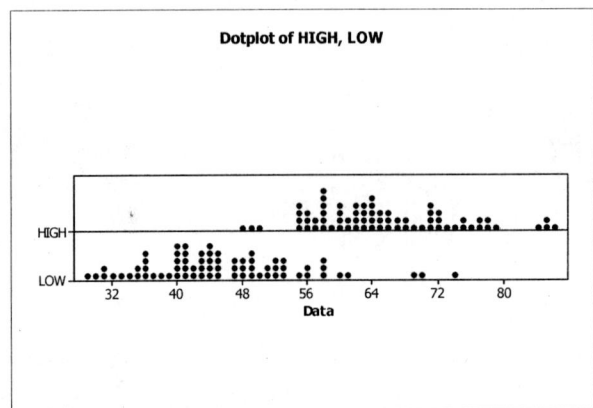

(d) For the stem-and-leaf diagram, we choose **Graph ▶ Stem-and-Leaf**, double click on HIGH and then on LOW to enter then in the **Graph variables** box and click **OK**. The result is

Stem-and-leaf of HIGH   N = 71
Leaf Unit = 1.0

```
    3   4  789
    6   5  444
   21   5  555677777888999
  (17)  6  00001122222333444
   33   6  55556677779
   22   7  00001122234
   11   7  5577889
    4   8  444
    1   8  5

Stem-and-leaf of LOW   N  = 71
Leaf Unit = 1.0

    1   2  9
    7   3  001234
   19   3  555556779999
  (20)  4  00011111223333344444
   32   4  5567777888899
   19   5  11122223
   11   5  5667889
    4   6  1
    3   6  9
    2   7  04
```

(e) Both variables have distributions that are unimodal. HIGH is close to symmetric and LOW is not symmetric.

(f) LOW is slightly right skewed.

# CHAPTER 3 SOLUTIONS

## Exercises 3.1

**3.1** The purpose of a measure of center is to indicate where the center or most typical value of a data set lies.

**3.3** Of the mean, median, and mode, only the mode is appropriate for use with qualitative data.

**3.5** (a) The mean is the sum of the values (45) divided by n (9). The result is 5. The median is the middle value in the ordered list and is also 5.

(b) The mean is the sum of the values (135) divided by n (9). The result is 15. The median is the middle value in the ordered list and is 5, as previously. The median is more typical of most of the data than is the mean and works better here.

(c) The mean does not have the property of being resistant to the influence of extreme observations.

**3.7** The median is more appropriate as a measure of central tendency than the mean because, unlike the mean, the median is not affected strongly by the relatively few homes that have extremely large or small areas.

**3.9** (a) The mean is calculated first by summing all 3 data values presented, which is 9, and then dividing this sum by 3. Thus the mean is 3.

(b) The median requires ordering the data from the smallest to the largest values: 0, 4, 5. Since the number of data values is odd, the median will be the middle data value in the ordered list. The middle value is the second ordered observation. Thus, the median is 4.

(c) The mode is the most frequently occurring data value or values. In these data, no value occurs more than once, so there is no mode.

**3.11** (a) The mean is calculated first by summing all 4 data values presented, which is 11, and then dividing this sum by 4. Thus the mean is 2.75.

(b) The median requires ordering the data from the smallest to the largest values: 1, 2, 4, 4. Since the number of data values is even, the median will be the mean of the two middle data values in the ordered list. The two middle values are the second and third ordered observation. Thus, the median is 3.

(c) The mode is the most frequently occurring data value or values. In these data, the mode is 4 because it occurs twice.

**3.13** (a) The mean is calculated first by summing all 5 data values presented, which is 25, and then dividing this sum by 5. Thus the mean is 5.

(b) The median requires ordering the data from the smallest to the largest values: 1, 3, 4, 8, 9. Since the number of data values is odd, the median will be the middle data value in the ordered list. The middle value is the third ordered observation. Thus, the median is 4.

(c) The mode is the most frequently occurring data value or values. In these data, no value occurs more than once, so there is no mode.

**3.15** (a) $\sum$ represents the word 'summation'. It indicates that one should sum the values of the quantity that follows it.

(b) n represents the number data values

(c) $\bar{x}$ represents the sample mean

**3.17** (a) n = number of data values = 5

(b) $\sum x_i = 1 + 7 + 4 + 5 + 10 = 27$

**70** Chapter 3

(c) $\bar{x} = \sum x_i/n = 27/5 = 5.4$

**3.19** The mean number of values is calculated first by summing all seven values presented, which results in 51 days, and then dividing this by 7. Carrying this result to one more decimal place than the original data, the mean is 7.3 days.

Calculating the median requires ordering the data from the smallest to the largest values: 5 5 6 **6** 7 11 11. Since the number of data values is odd, the median will be the observation exactly in the middle of this ordering, i.e., in the fourth ordered position. Thus, the median number of days is 6.

The mode is the most frequently occurring data value or values. For these data, there are three modes because 5, 6, and 11 each occur twice.

**3.21** The mean number of tornado touchdowns is calculated by first summing all 12 data values presented, which results in 941, and then dividing this sum by 12. Thus, the mean is 78.4. This figure is already rounded to one more decimal place than the original data.

Calculating the median requires ordering the data from the smallest to the largest values: 2, 3, 47, 57, 62, **68, 86**, 97, 98, 99, 118, 204. Since the number of data values is even, the median will be the mean of the two middle data values in the ordered list. The two middle values are the sixth and seventh ordered observations, or 68 and 86. Thus, the median age is found as (68 + 86)/2 = 77.0.

The mode is the most frequently occurring data value or values. In these data, no value occurs more than once, so there is no mode.

**3.23** The mean wealth is calculated first by summing all 10 data values presented, which results in 347.1, and then dividing this sum by 10. Thus, the mean is $34.71 billion. This figure is already rounded to one more decimal place than the original data.

Calculating the median requires ordering the data from the smallest to the largest values: 25.0, 26.1, 26.3, 26.8, 27.9, 31.0, 31.0, 41.0, 46.0, 66.0. Since the number of data values is even, the median will be the mean of the two middle data values in the ordered list. The two middle values are the fifth and sixth ordered observations. Thus, the median wealth is (27.9 + 31.0)/2 = $29.45 billion dollars.

The mode is the most frequently occurring data value or values. In these data, $31.0 billion dollars is the mode because it occurs twice.

**3.25** The mean is calculated first by summing all 12 data values presented, which results in 175, and then dividing this sum by 12. Thus, the mean is 14.6 songs.

Calculating the median requires ordering the data from the smallest to the largest values. Since the number of data values is even, the median will be the mean of the two middle data values in the ordered list. The two middle values are the sixth and seventh ordered observations. Thus, the median number of songs is (14 + 14)/2 = 14.0.

The mode is the most frequently occurring data value or values. In these data, the mode is 14 songs since 14 occurs 6 times.

**3.27** (a) The mean number of cremation burials is calculated first by summing all 17 data values presented, which results in 4977, and then dividing this sum by 17. Thus, the mean is 292.8. This figure is already rounded to one more decimal place than the original data.

Calculating the median requires ordering the data from the smallest to the largest values: 21, 34, 35, 46, 46, 48, 51, 64, **83**, 86, 119, 258,

265, 385, 429, 523, 2484. Since the number of data values is odd, the median will be the middle data value in the ordered list. The middle value is the ninth ordered observation. Thus, the median is 83 cremation burials.

The mode is the most frequently occurring data value or values. In these data, the mode is 46 since 46 occurs two times, more than any other value.

(b) The median works best. The mean is highly affected by the largest value and the mode is well to the left of the distribution's center.

**3.29** (a) The mean number of accidents is calculated first by summing all 7 data values presented, which results in 619, and then dividing this sum by 7. Thus, the mean is 88.4. This figure is already rounded to one more decimal place than the original data.

Calculating the median requires ordering the data from the smallest to the largest values: 69, 76, 85, **88**, 98, 100, 103. Since the number of data values is odd, the median will be the middle data value in the ordered list. The middle value is the fourth ordered observation. Thus the median is 88.

The mode is the most frequently occurring data value or values. In these data, there is no mode since no number occurs more than once.

(b) The mean number of accidents is calculated first by summing all 7 data values presented, which results in 492, and then dividing this sum by 7. Thus, the mean is 70.3. This figure is already rounded to one more decimal place than the original data.

Calculating the median requires ordering the data from the smallest to the largest values: 53, 56, 58, **59**, 70, 94, 102. Since the number of data values is even, the median will be the middle data value in the ordered list. The middle value is the fourth ordered observation. Thus the median is 59.

(c) For built-up roads, the modal day is Friday since Friday would appear 103 times in the list, more than any other day. For non-built-up roads the modal day is Sunday since Sunday would appear 102 times in the list, more than any other day.

(d) Perhaps on Sunday, motorcycles are being used more for pleasure and sightseeing, possibly on unfamiliar roads. This explanation seems reasonable since Saturday is the day with the second most number of accidents on non-built-up roads. On the weekdays, which include the modal day Friday, motorcycles are perhaps being used more for transportation to and from work. This traffic is more likely to be on built-up roads.

**3.31** (a) n = number of data values = 10

(b) $\sum x_i = 23.3$

(c) $\bar{x} = \sum x_i / n = 23.3/10 = 2.33$, which is given to two decimal places.

**3.33** (a) n = number of data values = 9

(b) $\sum x_i = 607$

(c) $\bar{x} = \sum x_i / n = 607/9 = 67.4444$, which, rounded to one decimal place, is 67.4 years.

**3.35** (a) The mode is defined as the data value or values that occur most frequently. In this exercise, the data values are the NCAA wrestling champion universities. During the sampled period, Iowa was champion twelve times. Since Iowa is the data value that occurs most frequently, it is the mode.

(b) Again, the data values are the champions. There is no way to compute a mean or median for such nonnumerical data. Thus, it would not be appropriate to use either the mean or median here. The mode is the only measure of center that can be used for qualitative data.

**3.37** (a) The mode is defined as the data value or values that occur most frequently. In this exercise, the data values are the law schools of the justices of the U.S Supreme Court. For this sample, Harvard was the law school for five of the justices. Since Harvard is the data value that occurs most frequently, it is the mode.

(b) Again, the data values are the law schools. There is no way to compute a mean or median for such nonnumerical data. Thus, it would not be appropriate to use either the mean or median here. The mode is the only measure of center that can be used for qualitative data.

**3.39** (a) The mode is defined as the data value or values that occur most frequently. In this exercise, the data values are the political view of the freshmen. During the year, Moderate was the most frequently occurring political view for the freshmen with a frequency of 237. Since moderate is the data value that occurs most frequently, it is the mode.

(b) Again, the data values are the three different political views. There is no way to compute a mean or median for such nonnumerical data. Thus, it would not be appropriate to use either the mean or median here. The mode is the only measure of center that can be used for qualitative data.

**3.41** (a) The mode is defined as the data value or values that occur most frequently. In this exercise, the data values are the colors that the Roulette ball landed on. During 200 trials, black was the most frequently occurring color with a frequency of 102. Since black is the data value that occurs most frequently, it is the mode.

(b) Again, the data values are the three different colors. There is no way to compute a mean or median for such nonnumerical data. Thus, it would not be appropriate to use either the mean or median here. The mode is the only measure of center that can be used for qualitative data.

**3.43** Using Minitab, retrieve the data from the WeissStats Resource Site. Since the data are qualitative, the only appropriate measure of center is the mode. Select **Stat ▶ Tables ▶ Tally Individual Variables**, then enter TYPE in the **Variables** box and click **OK**. The result is

| TYPE | Count |
|------|-------|
| NPC | 2919 |
| IOC | 889 |
| SLC | 1119 |
| FGH | 221 |
| NLT | 129 |
| NFP | 451 |
| HUI | 19 |
| N= | 5747 |

The mode is the most occurring value. We see that NPC (Nongovernment not-for-profit community hospitals) occurred 2919 times, more than any other value. Therefore, NPC is the mode.

**3.45** Using Minitab, retrieve the data from the WeissStats Resource Site. Since both of the variables are qualitative, the only appropriate measure of center for preference and highest degree obtained is the mode. Select **Stat** ▶ **Tables** ▶ **Tally Individual Variables**, then enter PREFERENCE and DEGREE in the **Variables** box and click **OK**. The result is

```
PREFERENCE  Count      DEGREE  Count
      Both    112          MA    167
     Email    239       Other     11
      Mail     86         PhD    388
       N/A    129          N=    566
        N=    566
```

The mode is the most occurring value. For PREFERENCE, we see that email occurred 239 times, more than any other value. Therefore, email is the mode for PREFERENCE. For DEGREE, we see that PhD occurred 388 times, more than any other value. Therefore, PhD is the mode for DEGREE.

**3.47** Using Minitab, retrieve the data from the WeissStats Resource Site. Since the data are quantitative, any of the three measures of center could be used. To obtain the mode, select **Stat** ▶ **Tables** ▶ **Tally Individual Variables**, then enter LENGTH in the **Variables** box and click **OK**. The mode is the most occurring value, which, in this case, is 158 (occurring 9 times).

To obtain the mean and median, select **Stat** ▶ **Basic Statistics** ▶ **Display Descriptive Statistics**, then enter LENGTH in the **Variables** box and click **OK**. The result is

```
Variable     N   N*    Mean   SE Mean   StDev   Minimum     Q1   Median      Q3   Maximum
LENGTH     229    0   163.75     3.80   57.57     23.00  132.50  153.00  181.50    493.00
```

From this display, we see that the mean is 163.75 seconds and the median is 153.00 seconds. In this case, the median is the most useful statistic since it tells us that half of the songs are longer than 153.00 seconds. The mean is heavily influenced by a few songs that are longer than the others. The median is more representative of the data as a whole.

**3.49** Using Minitab, retrieve the data from the WeissStats Resource Site. Since the data are quantitative, any of the three measures of center could be used. To obtain the mode, select **Stat** ▶ **Tables** ▶ **Tally Individual Variables**, then enter PERCENT in the **Variables** box and click **OK**. The result is

```
PERCENT  Count
     17      1         35      1
     19      1         36      3
     20      1         38      1
     21      2         49      1
     22      2         N=     51
     23      3
     24      4
     25      6
     26      5
     27      5
     28      1
     29      3
     30      3
     31      3
     32      3
     33      1
     34      1
```

**74** Chapter 3

The mode is the most occurring value, which is 25 (occurring 6 times).

To obtain the mean and median, select **Stat ▶ Basic Statistics ▶ Display Descriptive Statistics**, then enter PERCENT in the **Variables** box and click **OK**. The result is

| Variable | N | N* | Mean | SE Mean | StDev | Minimum | Q1 | Median | Q3 | Maximum |
|---|---|---|---|---|---|---|---|---|---|---|
| PERCENT | 51 | 0 | 27.706 | 0.788 | 5.626 | 17.000 | 24.000 | 27.000 | 31.000 | 49.000 |

From this display, we see that the mean is 27.706% and the median is 27.000%. In this case, the median is the most useful statistic since it tells us that half of the states had a bachelor's completion percentage of 27% or greater. The mean is not a true national crime rate since all states are counted equally in computing this mean, whereas states with greater populations should have more effect on the mean than states with smaller populations.

3.51 (a) In Minitab, open the worksheet from the WeissStats Resource Site. This will enter the data values in columns C1 and C2 under the names INTEGRATED and STANDARD. Select **Stat ▶ Basic Statistics ▶ Display Descriptive Statistics**, enter INTEGRATED and STANDARD in the **Variables** text box and click **OK**. The output appears in the Session Window as

| Variable | N | N* | Mean | SE Mean | StDev | Minimum | Q1 | Median | Q3 |
|---|---|---|---|---|---|---|---|---|---|
| INTEGRATED | 227 | 0 | 24.916 | 0.300 | 4.513 | 11.000 | 22.000 | 25.000 | 28.000 |
| STANDARD | 192 | 0 | 23.000 | 0.521 | 7.221 | 6.000 | 18.000 | 23.000 | 28.750 |

| Variable | Maximum |
|---|---|
| INTEGRATED | 35.000 |
| STANDARD | 40.000 |

From the output, we see that the mean score on client satisfaction for the integrated treatment is 24.916 and the median score is 25.000. The mean score on client satisfaction for the standard treatment is 23.000 and the median score is 23.000.

(b) It would appear from this data that the integrated treatment group had a slightly higher mean and median score than the standard treatment group. However, we can also see from the data that the standard treatment group had a higher maximum score than the integrated group.

3.53 (a) The mean of the ratings is 2.5. This is derived by summing the 14 individual ratings, which yields 35, and dividing by 14.

(b) The median of the data is the average of the two observations appearing in the middle positions after ordering the data from the smallest to the largest values: 1, 1, 2, 2, 2, 2, **2, 2**, 3, 3, 3, 4, 4, 4. The median is the average of the seventh and eighth ordered observations, or (2 + 2)/2 = 2.

(c) In this case, the median provides a better descriptive summary of the data than does the mean. Notice that six of the fourteen values are 2s and these are more in line with the median 2 than with the mean 2.5. Furthermore, the data are not actually measurements, but reflect the opinions of the participants, each of whom must decide what a '2' or a '3' really means.

3.55 The expression $(\sum x_i)^2$ represents the square of the sum of the data, whereas $\sum x_i^2$ represents the sum of the squares of the data. To show that these two quantities are generally unequal, consider three observations on x as presented in column 1 of the following table. Also, consider each observation's square, as presented in column 2.

| X | $x^2$ |
|---|---|
| 1 | 1 |
| 3 | 9 |
| 5 | 25 |
| 9 | 35 |

From the bottom row, we see that $(\sum x_i)^2 = 9^2 = 81$ and $\sum x_i^2 = 35$. Thus, $(\sum x_i)^2 \neq \sum x_i^2$.

### Exercises 3.2

**3.57** The purpose of a measure of variation is to show the amount of variation or spread in a data set. Any value of central tendency does not adequately characterize the elements of a data set. Measures of central tendency merely describe the center of the observations, but do not show how observations differ from each other. A measure of variation is intended to capture the degree to which observations differ among themselves.

**3.59** When we use the standard deviation as a measure of variation, the reference point is the mean since all of the differences used in computing the standard deviation are between the data values and the mean.

**3.61** (a) The mean of this data set is 5. Thus

$\sum (x_i - \bar{x})^2 = (-4)^2 + (-3)^2 + (-2)^2 + (-1)^2 + 0^2 + 1^2 + 2^2 + 3^2 + 4^2 = 60$

Dividing by n - 1 = 8, $s^2$ = 60/8 = 7.5. The standard deviation, s, is the square root of 7.5, or 2.739.

(b) The mean of this data set is 15. Thus

$\sum (x_i - \bar{x})^2 = (-14)^2 + (-13)^2 + (-12)^2 + (-11)^2 + 10^2 + (-9)^2 + (-8)^2 + (-7)^2 + 84^2 = 7980$

Dividing by n - 1 = 8, $s^2$ = 7980/8 = 997.5. The standard deviation, s, is the square root of 997.5, or 31.583.

(c) Changing the 9 to 99 greatly increases the standard deviation, illustrating that it lacks the property of resistance to the influence of extreme data values.

**3.63** (a) The range is 54 - 9 = 45.

(b) The defining formula for s is $s = \sqrt{\dfrac{\sum (x_i - \bar{x})^2}{n-1}}$.

The first three columns of the following table present the calculations that are needed to compute s by the defining formula:

| x | x-x̄ | (x-x̄)² | x² |
|---|---|---|---|
| 21 | -15 | 225 | 441 |
| 54 | 18 | 324 | 2916 |
| 9 | -27 | 729 | 81 |
| 45 | 9 | 81 | 2025 |
| 51 | 15 | 225 | 2601 |
| 180 | 0 | 1584 | 8064 |

With n = 5 and using the bottom figure of column 1, we find that $\bar{x}$ = $\sum x_i/n$ = 180/5 = 36. This value is needed to construct the differences presented in column 2. Column 3 squares these differences. The figure presented at the bottom of column 3 is the computation needed for the numerator of the defining formula for s. Thus,

$$s = \sqrt{\frac{1584}{5-1}} = 19.900$$

(c) The computing formula for s is

$$s = \sqrt{\frac{\sum x_i^2 - (\sum x_i)^2/n}{n-1}}$$

The computing formula for s with figures used from the bottom of Columns 1 and 4 of the previous table yields

$$s = \sqrt{\frac{8064 - 180^2/5}{5-1}} = 19.900$$

(d) The computing formula was a time-saver since only one column of intermediate numbers needed to be calculated, whereas the defining formula required three additional columns.

**3.65** (a) The range is 5 - 0 = 5.

(b) The defining formula for s is $s = \sqrt{\frac{\sum(x_i - \bar{x})^2}{n-1}}$.

The following table presents the calculations that are needed to compute s by the defining formula:

| X | x-x̄ | (x-x̄)² |
|---|---|---|
| 4 | 1.0 | 1.0 |
| 0 | -3.0 | 9.0 |
| 5 | 2.0 | 4.0 |
| 9 | 0.0 | 14.0 |

With n = 3 and using the bottom figure of column 1, we find that $\bar{x}$ = $\sum x_i/n$ = 9/3 = 3.0. This value is needed to construct the differences presented in column 2. Column 3 squares these differences. The figure presented at the bottom of column 3 is the computation needed for the numerator of the defining formula for s. Thus,

$$s = \sqrt{\frac{14.0}{3-1}} = 2.6$$

**3.67** (a) The range is 4 - 1 = 3.

(b) The defining formula for s is $s = \sqrt{\dfrac{\sum(x_i - \bar{x})^2}{n-1}}$.

The following table presents the calculations that are needed to compute s by the defining formula:

| X | x-x̄ | (x-x̄)² |
|---|------|--------|
| 1 | -1.75 | 3.06 |
| 2 | -0.75 | 0.56 |
| 4 | 1.25 | 1.56 |
| 4 | 1.25 | 1.56 |
| 11 | 0.00 | 6.74 |

With n = 4 and using the bottom figure of column 1, we find that $\bar{x}$ = $\sum x_i/n$ = 11/4 = 2.75. This value is needed to construct the differences presented in column 2. Column 3 squares these differences. The figure presented at the bottom of column 3 is the computation needed for the numerator of the defining formula for s. Thus,

$$s = \sqrt{\dfrac{6.74}{4-1}} = 1.5$$

**3.69** (a) The range is 9 - 1 = 8.

(b) The defining formula for s is $s = \sqrt{\dfrac{\sum(x_i - \bar{x})^2}{n-1}}$.

The following table presents the calculations that are needed to compute s by the defining formula:

| X | x-x̄ | (x-x̄)² |
|---|------|--------|
| 1 | -4.0 | 16.0 |
| 9 | 4.0 | 16.0 |
| 8 | 3.0 | 9.0 |
| 4 | -1.0 | 1.0 |
| 3 | -2.0 | 4.0 |
| 25 | 0.00 | 46.0 |

With n = 5 and using the bottom figure of column 1, we find that $\bar{x}$ = $\sum x_i/n$ = 25/5 = 5.0. This value is needed to construct the differences presented in column 2. Column 3 squares these differences. The figure presented at the bottom of column 3 is the computation needed for the numerator of the defining formula for s. Thus,

$$s = \sqrt{\dfrac{46.00}{5-1}} = 3.4$$

**3.71** The range is 11 - 5 = 6.

The defining formula for s is $s = \sqrt{\dfrac{\sum(x_i - \bar{x})^2}{n-1}}$.

The following table presents the calculations that are needed to compute s by the defining formula:

| x | x-x̄ | (x-x̄)² |
|---|---|---|
| 6 | -1.3 | 1.69 |
| 7 | -0.3 | 0.09 |
| 11 | 3.7 | 13.69 |
| 6 | -1.3 | 1.69 |
| 5 | -2.3 | 5.29 |
| 5 | -2.3 | 5.29 |
| 11 | 3.7 | 13.69 |
| 51 | -0.1 | 41.43 |

With n = 7 and using the bottom figure of column 1, we find that $\bar{x} = \sum x_i/n$ = 51/7 = 7.3. This value is needed to construct the differences presented in column 2. Column 3 squares these differences. The figure presented at the bottom of column 3 is the computation needed for the numerator of the defining formula for s. Thus,

$$s = \sqrt{\frac{41.43}{7-1}} = 2.63$$

**3.73** The range of a data set is defined to be the difference between the largest and smallest data values in the data set. The largest number of tornado touchdowns is 204. The smallest number is 2. The range is 204 - 2 = 202 tornadoes. The defining formula for s is $s = \sqrt{\frac{\sum(x_i - \bar{x})^2}{n-1}}$.

The first three columns of the following table present the calculations that are needed to compute s by the defining formula:

| x | x-x̄ | (x-x̄)² | x² |
|---|---|---|---|
| 3 | -75.4 | 5685.16 | 9 |
| 2 | -76.4 | 5836.96 | 4 |
| 47 | -31.4 | 985.96 | 2204 |
| 118 | 39.6 | 1568.16 | 13924 |
| 204 | 125.6 | 15775.36 | 41616 |
| 97 | 18.6 | 345.96 | 9409 |
| 68 | -10.4 | 108.16 | 4624 |
| 86 | 7.6 | 57.76 | 7396 |
| 62 | -16.4 | 268.96 | 3844 |
| 57 | -21.4 | 457.96 | 3249 |
| 98 | 19.6 | 384.16 | 9604 |
| 99 | 20.6 | 424.36 | 9801 |
| 941 | 0.2 | 31898.92 | 105689 |

$$s = \sqrt{\frac{31898.92}{12-1}} = 53.85$$

The computing formula for s is

$$s = \sqrt{\frac{\sum x_i^2 - (\sum x_i)^2/n}{n-1}}$$

Columns 1 and 4 of the previous table present the calculations needed to compute s by the computing formula. Using the figures presented at the bottom of each of these columns and substituting into the formula itself, we get:

$$s = \sqrt{\frac{105689 - 941^2/12}{12-1}} = 53.85$$

This is, of course, the same result that we obtained by the defining formula.

**3.75** The range of a data set is defined to be the difference between the largest and smallest data values in the data set. The highest value is 66 billion and the lowest value is 25 billion. The range is 66 - 25 = 41 billion.

The defining formula for s is $s = \sqrt{\frac{\sum(x_i - \bar{x})^2}{n-1}}$.

The following table presents the calculations that are needed to compute s by the defining formula:

| x | x-x̄ | (x-x̄)² |
|---|---|---|
| 66.0 | 31.29 | 979.06 |
| 46.0 | 11.29 | 127.46 |
| 41.0 | 6.29 | 39.56 |
| 31.0 | -3.71 | 13.76 |
| 31.0 | -3.71 | 13.76 |
| 27.9 | -6.81 | 46.38 |
| 26.8 | -7.91 | 62.57 |
| 26.3 | -8.41 | 70.73 |
| 26.1 | -8.61 | 74.13 |
| 25.0 | -9.71 | 94.28 |
| 347.1 | 0.00 | 1521.71 |

With n = 10 and using the bottom figure of column 1, we find that x̄ = $\sum x_i/n$ = 347.1/10 = 34.71. This value is needed to construct the differences in column 2. Column 3 squares the differences presented in column 2. The figure presented at the bottom of column 3 is the computation needed for the numerator of the defining formula s. Thus,

$$s = \sqrt{\frac{1521.71}{10-1}} = 13.00$$

**3.77** The range of a data set is defined to be the difference between the largest and smallest data values in the data set. The highest overall value is 30 and the lowest overall value is 6. The range is 30 - 6 = 24.

**80** Chapter 3

The defining formula for s is $s = \sqrt{\dfrac{\sum(x_i - \bar{x})^2}{n-1}}$.

The following table presents the calculations that are needed to compute s by the defining formula:

| x | x−x̄ | (x−x̄)² |
|---|------|---------|
| 14 | −0.58 | 0.3364 |
| 14 | −0.58 | 0.3364 |
| 13 | −1.58 | 2.4964 |
| 14 | −0.58 | 0.3364 |
| 14 | −0.58 | 0.3364 |
| 14 | −0.58 | 0.3364 |
| 14 | −0.58 | 0.3364 |
| 13 | −1.58 | 2.4964 |
| 30 | 15.42 | 237.7764 |
| 6 | −8.58 | 73.6164 |
| 17 | 2.42 | 5.8564 |
| 12 | −2.58 | 6.6564 |
| 175 | 0.00 | 330.9168 |

With n = 12 and using the bottom figure of column 1, we find that $\bar{x} = \sum x_i / n$ = 175/12 = 14.58. This value is needed to construct the differences in column 2. Column 3 squares the differences presented in column 2. The figure presented at the bottom of column 3 is the computation needed for the numerator of the defining formula s. Thus,

$$s = \sqrt{\dfrac{330.9168}{12-1}} = 5.5$$

**3.79** (a) We will use the computing formula to find s. The following table presents the calculations that are needed to compute s by the computing formula:

| x | x² |
|---|-----|
| 83 | 6889 |
| 46 | 2116 |
| 64 | 4096 |
| 385 | 148225 |
| 46 | 2116 |
| 21 | 441 |
| 48 | 2304 |
| 86 | 7396 |
| 523 | 273529 |
| 429 | 184041 |

| | |
|---|---|
| 35 | 1225 |
| 51 | 2601 |
| 34 | 1156 |
| 258 | 66564 |
| 265 | 70225 |
| 119 | 14161 |
| 2484 | 6170256 |
| 4977 | 6957341 |

The computing formula for s is

$$s = \sqrt{\frac{\sum x_i^2 - (\sum x_i)^2/n}{n-1}}$$

The previous table presents the calculations needed to compute s by the computing formula. Using the figures presented at the bottom of the columns and substituting into the formula itself, we get:

$$s = \sqrt{\frac{6957341 - 4977^2/17}{17-1}} = 586.3$$

(b) No. The standard deviation is based on deviations from the mean. However, these data are right skewed and contain a very large outlier (2484) that greatly increases the value of the mean. The deviation of 2484 from the mean accounts for about 87% of the total of the squared deviations from the mean, so the influence of that data value on both the mean and the standard deviation is very great.

**3.81** (a) We would guess that the non-built up roads would have the greater variation due to more variation in the recreational use of those roads at various times of the week.

(b) The range for built-up roads is 103 - 69 = 34, and the range for non-built-up roads is 102 - 53 = 49.

For built-up roads, n = 17, $\sum x_i = 619$, $\sum x_i^2 = 55719$, and

$$s = \sqrt{\frac{55719 - 619^2/7}{7-1}} = 12.79$$

For non-built-up-roads, n = 7, $\sum x_i = 492$, $\sum x_i^2 = 36930$, and

$$s = \sqrt{\frac{36930 - 492^2/7}{7-1}} = 19.79$$

Thus both measures of variation confirm that our guess is correct. The range and standard deviation are greater for non-built-up roads than for built-up roads.

**3.83** The data in this set are qualitative (the hospital type for U.S. registered hospitals). Being nonnumerical, it is not possible to compute a range or standard deviation.

**3.85** The data in this set are qualitative (ballot preference and highest degree obtained). Being nonnumerical, it is not possible to compute a range or standard deviation.

**3.87** Using Minitab, retrieve the data from the WeissStats Resource Site. Select **Stat ▶ Basic Statistics ▶ Display Descriptive Statistics**, enter LENGTH in the **Variables** text box. Now click on the **Statistics** button and check the range to add it to the list of statistics to be computed. Click **OK** twice. The output appears in the Session Window as

```
Variable    N    N*    Mean    SE Mean   StDev   Minimum      Q1   Median      Q3
LENGTH    229    0   163.75      3.80    57.57     23.00   132.50   153.00   181.50

Variable   Maximum   Range
LENGTH      493.00   470.00
```

The range (the difference between the maximum and the minimum data values) is 470.00 and the standard deviation is 57.57. Roughly speaking, on the average, the lengths differ from the overall mean length by 57.57 minutes. For these data, which are right skewed and contain a few very large values, the standard deviation may not be a very good measure of variation due to the large influence of those large values. Typically, the range will be about 4 to 6 times the standard deviation. In this case, it is about 8 times the standard deviation, indicating that there may be outliers or highly influential data values.

**3.89** Using Minitab, retrieve the data from the WeissStats Resource Site. Select **Stat ▶ Basic Statistics ▶ Display Descriptive Statistics**, enter PERCENT in the **Variables** text box. Now click on the **Statistics** button and check the range to add it to the list of statistics to be computed. Click **OK** twice. The output appears in the Session Window as

```
Variable   N   N*    Mean   SE Mean   StDev   Minimum     Q1   Median      Q3
PERCENT   51   0   27.706    0.788    5.626    17.000  24.000  27.000   31.000
Maximum   Range
 49.000   32.000
```

The range (the difference between the maximum and the minimum data values) is 32% and the standard deviation is 5.626%. Roughly speaking, on the average, the completion percetages differ from the overall mean completion percentage by 5.626%. One should keep in mind, however, that the overall mean computed by Minitab assumes that the completion percentages for each state count equally. In fact, the true overall mean completion percentage would take into account that there are different numbers of students in each state.

**3.91** (a) Using Minitab, retrieve the data from the WeissStats Resource Site. Select **Stat ▶ Basic Statistics ▶ Display Descriptive Statistics**, enter INTEGRATED and STANDARD in the **Variables** text box. Now click on the **Statistics** button and check the range to add it to the list of statistics to be computed. Click **OK** twice. The output appears in the Session Window as

```
Variable       N    N*    Mean   SE Mean   StDev   Minimum     Q1   Median      Q3
INTEGRATED   227    0   24.916    0.300   4.513    11.000  22.000  25.000   28.000
STANDARD     192    0   23.000    0.521   7.221     6.000  18.000  23.000   28.750

Variable      Maximum   Range
INTEGRATED     35.000   24.000
STANDARD       40.000   34.000
```

The range of the questionnaire scores for the integrated treatment is 24. The range of the questionnaire scores for the standard treatment is 34. The standard deviation of the questionnaire scores for the integrated treatment is 4.513 and 7.221 for the standard treatment.

(b) Both the range and the standard deviation are greater for the standard treatment. This indicates that the standard treatment had more variation in its observations than the integrated treatments observations.

**3.93** (a) We will compute $s$ for each data set using the computing formula. The computing formula requires the following column manipulations.

Data Set I

| x | $x^2$ |
|---|---|
| 0 | 0 |
| 0 | 0 |
| 10 | 100 |
| 12 | 144 |
| 14 | 196 |
| 14 | 196 |
| 14 | 196 |
| 15 | 225 |
| 15 | 225 |
| 15 | 225 |
| 16 | 256 |
| 17 | 289 |
| 23 | 529 |
| 24 | 576 |
| 189 | 3,157 |

Data Set II

| x | $x^2$ |
|---|---|
| 10 | 100 |
| 12 | 144 |
| 14 | 196 |
| 14 | 196 |
| 14 | 196 |
| 15 | 225 |
| 15 | 225 |
| 15 | 225 |
| 16 | 256 |
| 17 | 289 |
| 142 | 2,052 |

For each data set, the calculations for $s$ are presented in column 2 of the following table.

| Data Set | s | Range |
|---|---|---|
| I | $s = \sqrt{\dfrac{3157 - 189^2/14}{14 - 1}} = 6.8$ | 24 |
| II | $s = \sqrt{\dfrac{2052 - 142^2/10}{10 - 1}} = 0.20.$ | 7 |

(b) The range for Data Set I is 24 - 0 = 24. The range for Data Set II is 17 - 10 = 7. These are recorded in column 3 of the previous table.

(c) Outliers increase the variation in a data set; in other words, removing the outliers from a data set results in a decrease in the variation.

**3.95** (a) We create a table with columns headed by x, f, xf, $(x-\bar{x})$, $(x-\bar{x})^2$, and $(x-\bar{x})^2 f$ to estimate the sample mean and standard deviation of the days-to-maturity data.

| Class Mark x | Frequency f | xf | x-x̄ | (x-x̄)² | f(x-x̄)² |
|---|---|---|---|---|---|
| 34.5 | 3 | 103.5 | -33.5 | 1122.25 | 3366.75 |
| 44.5 | 1 | 44.5 | -23.5 | 552.25 | 552.25 |
| 54.5 | 8 | 436.0 | -13.5 | 182.25 | 1458.00 |
| 64.5 | 10 | 645.0 | -3.5 | 12.25 | 122.50 |
| 74.5 | 7 | 521.5 | 6.5 | 42.25 | 295.75 |
| 84.5 | 7 | 591.5 | 16.5 | 272.25 | 1905.75 |
| 94.5 | 4 | 378.0 | 26.5 | 702.25 | 2809.00 |
| | 40 | 2720.0 | | | 10510.00 |

The mean $\bar{x} = 2720/40 = 68$. This is subtracted from each of the class marks in column 1 to get the entries in column 4. The grouped data formula yields the following estimate of the standard deviation:

$$s = \sqrt{\frac{\sum(x-\bar{x})^2 f}{n-1}} = \sqrt{\frac{10510}{40-1}} = 16.4$$

(b) There is a small discrepancy between the estimated mean and estimated standard deviation and the actual values. These discrepancies occur because in the grouped data formulas, every actual data value in a given class is replaced by the class mark even though most values in the class are not equal to the class mark. For example, every data value in the first class is represented by the class mark 34.5, but the three data values are 36, 38, and 39, all of them larger than the class mark. This introduces a small error in each difference between an observation and the mean. It is unlikely that all of these small errors will cancel each other out causing both the mean and the standard deviation to exhibit small errors.

### Exercises 3.3

**3.97** The empirical rule improves on the estimates given by Chebyshev's rule.

**3.99** (a) If $k = 4$, using Chebyshev's rule, at least $1 - 1/k^2 = 1 - 1/16 = 15/16$, or 93.75%, of the observations lie within 4 standard deviations to either side of the mean.

(b) If $k = 2.5$, using Chebyshev's rule, at least $1 - 1/k^2 = 1 - 1/6.25 = 0.84$, or 84%, of the observations lie within 2.5 standard deviations to either side of the mean.

**3.101** If $k = 5$, using Chebyshev's rule, at least $1 - 1/k^2 = 1 - 1/25 = 24/25$, or 96%, of the observations lie within 5 standard deviations to either side of the mean.

**3.103** (a) The mean of the data set is 83.0 and the standard deviation is 7.85. Locate the mean on the horizontal axis and measure intervals equal in length to the standard deviation. The dotplot of the data set is as follows.

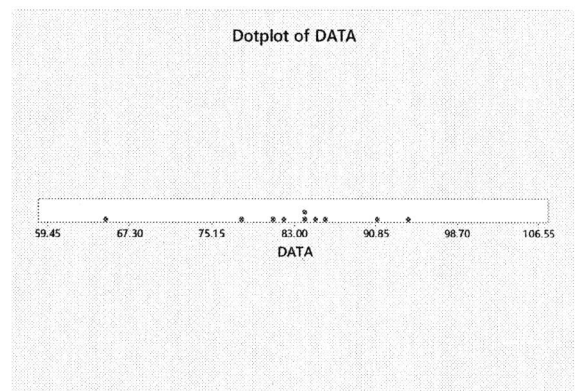

(b) Two standard deviations to either side of the mean would be 83.00 − 2(7.85) to 83.00 + 2(7.85) or 67.3 to 98.70. 9 of the 10 observations are within this range, which is 90% of the observations. If $k = 2$, using Chebyshev's rule, at least $1 - 1/k^2 = 1 - 1/4 = 3/4$, or 75%, of the observations are estimated to lie within 2 standard deviations to either side of the mean. The actual percentage is higher than the Chebyshev's estimated percentage.

(c) Three standard deviations to either side of the mean would be 83.00 − 3(7.85) to 83.00 + 3(7.85) or 59.45 to 106.55. All of the 10 observations are within three standard deviations to either side of the mean. This is 100% of the observations. If $k = 3$, using Chebyshev's rule, at least $1 - 1/k^2 = 1 - 1/9 = 8/9$, or 88.9%, of the observations is estimated to lie within 3 standard deviations to either side of the mean. The actual percentage is higher than the Chebyshev's estimated percentage.

**3.105** (a) The shape of the distribution is bell-shaped. Therefore application of the empirical rule is appropriate.

(b) The shape of the distribution is very right-skewed. Therefore application of the empirical rule is not appropriate.

(c) The shape of the distribution is left skewed. Therefore, application of the empirical rule is not appropriate.

**3.107** (a) Chebyshev's rule estimates that, for any distribution, at least 75% of the observations will be within two standard deviations to either side of the mean. The empirical rule states that in a roughly bell-shaped distribution, 95% of the observations will be within two standard deviations to either side of the mean. The empirical rule improved on the estimated percentages of Chebyshev's rule.

(b) Chebyshev's rule estimates that, for any distribution, at least 89% of the observations will be within three standard deviations to either side of the mean. The empirical rule states that in a roughly bell-shaped distribution, 99.7% of the observations will be within three standard deviations to either side of the mean. The empirical rule improved on the estimated percentages of Chebyshev's rule.

**3.109** (a) 89% corresponds to $k = 3$ for Chebyshev's rule. At least 89% of the observations lie between 25 − 3(5) = <u>10</u> and 25 + 3(5) = <u>40</u>.

(b) Since 15 and 35 are two standard deviations to either side of the mean, Chebyshev's rule estimates that at least <u>75%</u> of the observations lie within 15 and 35.

**86　Chapter 3**

**3.111** 10 and 50 are five standard deviations to either side of the mean. If $k = 5$, using Chebyshev's rule, at least $1 - 1/k^2 = 1 - 1/25 = 24/25$, or 96%, of the observations lie within 5 standard deviations to either side of the mean. Therefore, at least 96% of the observations lie within 10 and 50.

**3.113** If $k = 2$, using Chebyshev's rule, at least $1 - 1/k^2 = 1 - 1/4 = 3/4$, or 75%, of the observations lie within 2 standard deviations to either side of the mean. 75% of 80 is 60. Therefore, at least 60 observations lie within two standard deviations to either side of the mean.

**3.115** If $k = 3$, using Chebyshev's rule, at least $1 - 1/k^2 = 1 - 1/9 = 8/9$, or 89%, of the observations lie within 3 standard deviations to either side of the mean. 89% of 50 is 44.5. Therefore, at least 45 observations lie within three standard deviations to either side of the mean.

**3.117** 68 and 132 are two standard deviations to either side of the mean. If $k = 2$, using Chebyshev's rule, at least $1 - 1/k^2 = 1 - 1/4 = 3/4$, or 75%, of the observations lie within 2 standard deviations to either side of the mean. Furthermore, 75% of 60 is 45. Therefore, at least 45 of the observations lie within 68 and 132.

**3.119** 23 and 47 are three standard deviations to either side of the mean. If $k = 3$, using Chebyshev's rule, at least $1 - 1/k^2 = 1 - 1/9 = 8/9$, or 89%, of the observations lie within 3 standard deviations to either side of the mean. Furthermore, 89% of 150 is 133.5. Therefore, at least 134 of the observations lie within 23 and 47.

**3.121** (a) The empirical rule states that 68% of the observations will be within one standard deviation to either side of the mean. Therefore, approximately 68% of the observations lie between 25 − 5 = 20 and 25 + 5 = 30.

(b) The empirical rule states that 95% of the observations will be within two standard deviations to either side of the mean. Therefore, approximately 95% of the observations lie between 25 − 2(5) = 15 and 25 + 2(5) = 35.

(c) The empirical rule states that 99.7% of the observations will be within three standard deviations to either side of the mean. Therefore, approximately 99.7% of the observations lie between 25 − 3(5) = 10 and 25 + 3(5) = 40.

**3.123** 22 and 38 are two standard deviations to either side of the mean. Using the empirical rule, 95% of the observations lie within 2 standard deviations to either side of the mean.

**3.125** Using the empirical rule, 95% of the observations lie within 2 standard deviations to either side of the mean. 95% of 80 is 76. Therefore, approximately 76 of the observations are estimated to be within 2 standard deviations to either side of the mean.

**3.127** Using the empirical rule, 68% of the observations lie within 1 standard deviation to either side of the mean. 68% of 50 is 34. Therefore, approximately 34 of the observations are estimated to be within 1 standard deviation to either side of the mean.

**3.129** 52 and 148 are three standard deviations to either side of the mean. Using the empirical rule, 99.7% of the observations lie within 3 standard deviations to either side of the mean. 99.7% of 250 is 249.25. Therefore, approximately 249 of the observations are estimated to be between 52 and 148.

**3.131** 31 and 39 are one standard deviation to either side of the mean. Using the empirical rule, 68% of the observations lie within 1 standard deviation to either side of the mean. 68% of 150 is 102. Therefore, approximately 102 of the observations are estimated to be between 31 and 39.

**3.133** (a) Chebyshev's rule with $k = 2$ says that at least 75% of the observations lie within two standard deviations to either side of the mean. The sample mean and sample standard deviation are 11.9 and 6.5, respectively. Two standard deviations to either side of the mean would be $11.9 - 2(6.5) = -1.1$ and $11.9 + 2(6.5) = 24.9$. Thirty-one out of the thirty-three observations are within this range. Therefore, actually 93.9% of the observations lie within two standard deviations to either side of the mean. This is greater than the Chebyshev's estimate.

(b) Chebyshev's rule with $k = 3$ says that at least 89% of the observations lie within three standard deviations to either side of the mean The sample mean and sample standard deviation are 11.9 and 6.5, respectively. Three standard deviations to either side of the mean would be $11.9 - 3(6.5) = -7.6$ and $11.9 + 3(6.5) = 31.4$. All of the thirty observations are within this range. Therefore, actually 100% of the observations lie within three standard deviations to either side of the mean. This is greater than the Chebyshev's estimate.

(c) The actual percentage of observations within 2 and within 3 standard deviations to either side of the mean were greater that each of the Chebyshev's estimates. Chebychev's estimates are not necessarily precise.

**3.135** (a) Based on the fact that the sample mean is 14.7 mg and the sample standard deviation is 3.1 mg, we constructed the following figure. Note, for instance, that two standard deviations to the left is $14.7 - 2(3.1) = 8.5$.

| $\bar{x}-3s$ | $\bar{x}-2s$ | $\bar{x}-s$ | $\bar{x}$ | $\bar{x}+s$ | $\bar{x}+2s$ | $\bar{x}+3s$ |
|---|---|---|---|---|---|---|
| 5.4 | 8.5 | 11.6 | 14.7 | 17.8 | 20.9 | 24.0 |

(b) Chebyshev's rule with $k = 2$ says that at least 75% of the observations lie within two standard deviations to either side of the mean. 75% of 45 is 33.75. Therefore, at least 34 people in this sample of females under the age of 51 have intakes between 8.5 and 20.9 mg of iron during a 24-hour period.

(c) Chebyshev's rule with $k = 3$ says that at least 89% of the observations lie within three standard deviations to either side of the mean. 89% of 45 is 40.05. Therefore, at least 41 people in this sample of females under the age of 51 have intakes between 5.4 and 24.0 mg of iron during a 24-hour period.

**3.137** (a) Based on the fact that the sample mean is 98.1°F and the sample standard deviation is 0.65°F, we constructed the following figure. Note, for instance, that two standard deviations to the left is $98.1 - 2(0.65) = 96.8$.

| $\bar{x}-3s$ | $\bar{x}-2s$ | $\bar{x}-s$ | $\bar{x}$ | $\bar{x}+s$ | $\bar{x}+2s$ | $\bar{x}+3s$ |
|---|---|---|---|---|---|---|
| 96.15 | 96.8 | 97.45 | 98.1 | 98.75 | 99.4 | 100.05 |

(b) Chebyshev's rule with $k = 2$ says that at least 75% of the observations lie within two standard deviations to either side of the mean. 75% of 93 is 69.75. Therefore, at least 70 people in this sample of healthy humans have body temperatures between 96.8°F and 99.4°F.

**88** Chapter 3

(c) Chebyshev's rule with $k = 3$ says that at least 89% of the observations lie within three standard deviations to either side of the mean. 89% of 93 is 82.77. Therefore, at least 83 people in this sample of healthy humans have body temperatures between 96.15°F and 100.05°F.

**3.139** (a) The data set is approximately bell-shaped. It is reasonable to use the empirical rule.

(b) The empirical rule estimates that 68% of the observations will lie within one standard deviation to either side of the mean, 95% will lie within two standard deviations to either side of the mean, and 99.7% will lie within three standard deviations to either side of the mean.

(c) Based on the fact that the sample mean is 45.3 kg and the sample standard deviation is 4.16 kg, we constructed the following figure.

One standard deviation to either side of the mean would be the range of 45.3 - 4.16 = 41.14 to 45.3 + 4.16 = 49.46 kg. Forty-four of the observations lie within this range. Therefore, the exact percentage of observations within one standard deviation to either side of the mean is 44/60 * 100 = 73.3%. Two standard deviations to either side of the mean would be the range of 45.3 - 2(4.16) = 36.98 to 45.3 + 2(4.16) = 53.62 kg. Fifty-seven of the observations lie within this range. Therefore, the exact percentage of observations within two standard deviations to either side of the mean is 57/60 * 100 = 95%. Three standard deviations to either side of the mean would be the range of 45.3 - 3(4.16) = 32.82 to 45.3 + 3(4.16) = 57.78. All of the observations lie within this range. Therefore, the exact percentage of observations that lie within three standard deviations to either side of the mean is 100%.

(d) The actual percentages calculated in part (c) are very close to the estimated empirical percentages in part (b). Due to the fact that the shape was close to bell-shaped, it is expected that the empirical rule will give good estimates.

**3.141** (a) Based on the fact that the sample mean is 15.83 and the sample standard deviation is 1.74, we constructed the following figure.

(b) According to Property 1 of the empirical rule, approximately 68% of the fifth grade classes in the sample have student-to-faculty ratios within one standard deviation to either side of the mean. According to the figure in part (a), one standard deviation to either side of the mean would be from 14.09 to 17.57. Also, 68% of 81 is 55.08. Therefore, approximately 55 of the fifth grade classes in the sample have student-to-faculty ratios between 14.09 and 17.57.

(c) According to Property 2 of the empirical rule, approximately 95% of the fifth grade classes in the sample have student-to-faculty ratios within two standard deviations to either side of the mean. According to the figure in part (a), two standard deviations to either side of the mean would be from 12.35 to 19.31. Also, 95% of 81 is 76.95. Therefore, approximately 77 of the fifth grade classes in the sample have student-to-faculty ratios between 12.35 and 19.31.

(d) According to Property 3 of the empirical rule, approximately 99.7% of the fifth grade classes in the sample have student-to-faculty ratios within three standard deviations to either side of the mean. According to the figure in part (a), three standard deviations to either side of the mean would be from 10.61 to 21.05. Also, 99.7% of 81 is 80.757. Therefore, approximately all of fifth grade classes in the sample have student-to-faculty ratios between 10.61 and 21.05.

**3.143** (a) Based on the fact that the sample mean is 1.40 kg and the sample standard deviation is 0.11 kg, we constructed the following figure.

| $\bar{x}-3s$ | $\bar{x}-2s$ | $\bar{x}-s$ | $\bar{x}$ | $\bar{x}+s$ | $\bar{x}+2s$ | $\bar{x}+3s$ |
|---|---|---|---|---|---|---|
| 1.07 | 1.18 | 1.29 | 1.40 | 1.51 | 1.62 | 1.73 |

(b) According to Property 1 of the empirical rule, approximately 68% of the Swedish men in the sample have brain weights within one standard deviation to either side of the mean. According to the figure in part (a), one standard deviation to either side of the mean would be from 1.29 to 1.51 kg. Also, 68% of 225 is 153. Therefore, approximately 153 of the Swedish men in the sample have brain weights between 1.29 and 1.51 kg.

(c) According to Property 2 of the empirical rule, approximately 95% of the Swedish men in the sample have brain weights within two standard deviations to either side of the mean. According to the figure in part (a), two standard deviations to either side of the mean would be from 1.18 to 1.62 kg. Also, 95% of 225 is 213.75. Therefore, approximately 214 of the Swedish men in the sample have brain weights between 1.18 and 1.62 kg.

(d) According to Property 3 of the empirical rule, approximately 99.7% of the Swedish men in the sample have brain weights within three standard deviations to either side of the mean. According to the figure in part (a), three standard deviations to either side of the mean would be from 1.07 to 1.73 kg. Also, 99.7% of 225 is 224.325. Therefore, approximately 224 of the Swedish men in the sample have brain weights between 1.07 and 1.73 kg.

**3.145** We are trying to solve for $k$ in Chebyshev's rule such that it estimates at least 99% of the observations will lie within $k$ standard deviations to either side of the mean. Setting Chebyshev's rule equal to 0.99 and solving for $k$, we have

$$1 - 1/k^2 = 0.99$$
$$0.01 = 1/k^2$$
$$k^2 = 1/0.01 = 100$$
$$k = 10$$

Therefore, Chebyshev's rule estimates that at least 99% of the observations for any data set will lie within 10 standard deviations to either side of the mean.

**3.147** (a) If you added $2m^2-1$ zeros, one $-m$ and one $m$, you would get a sum of zero. Therefore, the sample mean would be 0. Using the defining formula to calculate the sample standard deviation, we would get the following.

| x | x-x̄ | (x-x̄)² |
|---|------|--------|
| 0 | 0 | 0 |
| 0 | 0 | 0 |
| 0 | 0 | 0 |
| etc. | etc. | etc. |
| -m | -m | m² |
| m | m | m² |
| 0 | 0.00 | 2m² |

The defining formula for s is $s = \sqrt{\dfrac{\sum(x-\bar{x})^2}{n-1}}$. There are $2m^2 - 1 + 2 = 2m^2 + 1$ observations in the sample. Plugging this in for $n$ in the defining formula, we have $s = \sqrt{\dfrac{\sum(x-\bar{x})^2}{n-1}} = \sqrt{\dfrac{2m^2}{(2m^2+1)-1}} = \sqrt{\dfrac{2m^2}{2m^2}} = 1$. The sample mean is 0 and the sample standard deviation is 1.

(b) Since the sample standard deviation is 1 and the sample mean is 0, the number 5 would be 5 standard deviations from the mean. Therefore, the observation $m$ would be $m$ standard deviations from the mean.

(c) If $m \geq 4$, then $m$ and $-m$ would be at least four standard deviations from the mean. The remaining observations of the data set are all equal to zero. These remaining observations are all equal to the sample mean of 0 and therefore all within three standard deviations from the mean. All but two observations are within three standard deviations of the mean. Since there are $2m^2 + 1$ observations and $2m^2 - 1$ zeros, this would be $\left(\dfrac{2m^2-1}{2m^2+1}\right)*100$ percent of observations that lie within three standard deviations to either side of the mean.

### Exercises 3.4

**3.149** The median and the interquartile range have the advantage over the mean and standard deviation that they are not sensitive to extreme values; they are said to be resistant.

**3.151** No. An extreme observation may be an outlier, but it may also be an indication of skewness.

**3.153** (a) The interquartile range is a descriptive measure of variation.

(b) It measures the spread of the middle two quarters (50%) of the data.

**3.155** The adjacent values are just the minimum and maximum observations when there are no potential outliers.

**3.157** Roughly, when arranged in increasing order, the middle 50% of a data set are found between $Q_1$ and $Q_3$.

**3.159** (a) Following Procedure 3.1, Step 1: First arrange the 4 data values in increasing order and note the two middle values:

1 **2 3** 4

Step 2: The median (second quartile) is the average of the two middle values, 2 and 3, and thus equals (2 + 3)/2 = 2.5.

Step 3 and 4: The first quartile $Q_1$ is the median of the lower half of the ordered data set 1 2 and thus is the average of the first and second values, 1 and 2. $Q_1$ = (1 + 2)/2 = 1.5

Step 5: The third quartile $Q_3$ is the median of the upper half of the ordered data set 3 4 and thus is the average of the first and second values, 3 and 4. $Q_3$ = (3 + 4)/2 = 3.5

Step 6: The quartiles are $Q_1$ = 1.5, $Q_2$ = 2.5, and $Q_3$ = 3.5.

(b) The interquartile range is $Q_3 - Q_1$ = 3.5 - 1.5 = 2.0.

(c) The five-number summary consists of the Minimum, $Q_1$, Median, $Q_3$, and the Maximum. Respectively, these are 1, 1.5, 2.5, 3.5, and 4.

**3.161** (a) Following Procedure 3.1, Step 1: First arrange the 5 data values in increasing order:

1 2 3 4 5

Step 2: Since this data set has an odd number of values (5), the median is the middle value (3rd) in the ordered list. The median, or second quartile, is 3.

Step 3 and 4: The first quartile $Q_1$ is the median of the lower half of the ordered data set (including the median) 1 **2** 3 and thus is the second value. $Q_1$ = 2

Step 5: The third quartile $Q_3$ is the median of the upper half of the ordered data set (including the median) 3 **4** 5 and thus is the second value. $Q_3$ = 4

Step 6: The quartiles are $Q_1$ = 2, $Q_2$ = 3, and $Q_3$ = 4.

(b) The interquartile range is $Q_3 - Q_1$ = 4 - 2 = 2.

(c) The five-number summary consists of the Minimum, $Q_1$, Median, $Q_3$, and the Maximum. Respectively, these are 1, 2, 3, 4, and 5.

**3.163** (a) Following Procedure 3.1, Step 1: First arrange the 6 data values in increasing order and note the two middle values:

1 2 **3 4** 5 6

Step 2: The median (second quartile) is the average of the two middle values, 3 and 4, and thus equals (3 + 4)/2 = 3.5.

Step 3 and 4: The first quartile $Q_1$ is the median of the lower half of the ordered data set 1 **2** 3 and thus is the second value. $Q_1$ = 2

Step 5: The third quartile $Q_3$ is the median of the upper half of the ordered data set 4 **5** 6 and thus is second values. $Q_3$ = 5

Step 6: The quartiles are $Q_1$ = 2, $Q_2$ = 3.5, and $Q_3$ = 5.

(b) The interquartile range is $Q_3 - Q_1$ = 5 - 2 = 3.

(c) The five-number summary consists of the Minimum, $Q_1$, Median, $Q_3$, and the Maximum. Respectively, these are 1, 2, 3.5, 5, and 6.

**92** Chapter 3

**3.165** (a) Following Procedure 3.1, Step 1: First arrange the 7 data values in increasing order:

1 2 3 4 5 6 7

Step 2: Since this data set has an odd number of values (7), the median is the middle value (4th) in the ordered list. The median, or second quartile, is 4.

Step 3 and 4: The first quartile $Q_1$ is the median of the lower half of the ordered data set (including the median) 1 **2 3** 4 and thus is the mean of the second and third values. $Q_1 = (2 + 3)/2 = 2.5$

Stpe 5: The third quartile $Q_3$ is the median of the upper half of the ordered data set (including the median) 4 **5 6** 7 and thus is the mean of the second and third values. $Q_3 = (5 + 6)/2 = 5.5$

Step 6: The quartiles are $Q_1 = 2.5$, $Q_2 = 4$, and $Q_3 = 5.5$.

(b) The interquartile range is $Q_3 - Q_1 = 5.5 - 2.5 = 3$.

(c) The five-number summary consists of the Minimum, $Q_1$, Median, $Q_3$, and the Maximum. Respectively, these are 1, 2.5, 4, 5.5, and 7.

**3.167** (a) Following Procedure 3.1, Step 1: First arrange the 20 data values in increasing order and note the two middle values:

45 48 64 70 73 74 74 78 78 **79 79** 80 80 80 80 80 80 81 82 82

Step 2: The median (second quartile) is the average of the two middle values, 79 and 79, and thus equals $(79 + 79)/2 = 79$.

Step 3 and 4: The first quartile $Q_1$ is the median of the lower half of the ordered data set 45 48 64 70 **73 74** 74 78 78 79 and thus is the average of the fifth and sixth values, 73 and 74. $Q_1 = (73 + 74)/2 = 73.5$

Step 5: The third quartile $Q_3$ is the median of the upper half of the ordered data set 79 80 80 80 **80 80** 80 81 82 82 and thus is the average of the fifth and sixth values, 80 and 80. $Q_3 = (80 + 80)/2 = 80.0$

Step 6: The quartiles are $Q_1 = 73.5$, $Q_2 = 79$, and $Q_3 = 80$.

Thus 25% of the data values are less than or equal to 73.5, the next 25% are between 73.5 and the median 79, the next 25% are between 79 and 80, and the upper 25% are greater than or equal to 80.

(b) The interquartile range is $Q_3 - Q_1 = 80.0 - 73.5 = 6.5$.

(c) The five-number summary consists of the Minimum, $Q_1$, Median, $Q_3$, and the Maximum. Respectively, these are 45, 73.5, 79.0, 80.0, and 82. In addition to the explanation of the quartiles in part (a), all of the values are greater than or equal to the Minimum, and all of the values are less than or equal to the Maximum.

(d) To identify any potential outliers, we first find the lower and upper limits. These are found by subtracting 1.5 x IQR from $Q_1$ and adding 1.5 x IQR to $Q_3$. The lower limit is $73.5 - 1.5(6.5) = 63.75$ and the upper limit is $80.0 + 1.5(6.5) = 89.75$. Potential outliers consist of any values that are less than 63.75 or greater than 89.75. We see that 45 and 48 are both less than 63.75, so these two values are potential outliers.

(e) We will construct a modified boxplot so as to show the potential outliers. The boxplot consists of a rectangle with its ends at the first and third quartiles, a line through it at the median, and two "whiskers" that extend from the quartiles outward to the adjacent values. The adjacent values are the most extreme data values that do not lie outside the lower and upper limits. For these data, the adjacent values are 64 and 82. The potential outliers, 45 and 48, are plotted individually. The modified boxplot, prepared using Minitab, is

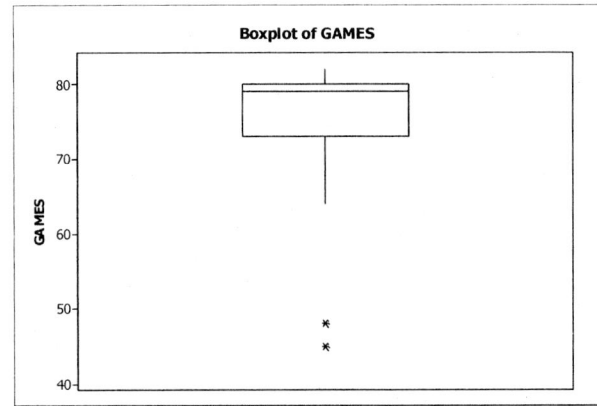

We see that the data is very left skewed, even without considering the two potential outliers. This is because, in most seasons, Gretzky played almost every game, but in a few seasons, he played considerably fewer than the maximum number of games, probably due to injuries.

**3.169** (a) Following Procedure 3.1, Step 1: First arrange the data in increasing order:

1  1  3  3  4  4  5  6  6  7  **7**  9  9  10  12  12  13  15  18  23  55

Step 2: Since this data set has an odd number of values (21), the median is the middle value (11$^{th}$) in the ordered list. The median, or second quartile, is 7.

Step 3 and 4: The first quartile $Q_1$ is the median of the lower half of the ordered data set (including the median) 1  1  3  3  4  **4**  5  6  6  7  7. Since this data set has an odd number of values (11), $Q_1$ is the sixth value, so $Q_1 = 4$.

Step 5: The third quartile $Q_3$ is the median of the upper half of the ordered data set (including the median) 7  9  9  10  12  **12**  13  15  18  23  55. Since this data set has an odd number of values (11), $Q_3$ is the sixth value, so $Q_3 = 12$.

Step 6: The quartiles are $Q_1 = 4$, $Q_2 = 7$, and $Q_3 = 12$.

Thus 25% of the data values are less than or equal to 4, the next 25% are between 4 and the median 7, the next 25% are between 7 and 12, and the upper 25% are greater than or equal to 12.

(b) The interquartile range is $Q_3 - Q_1 = 12 - 4 = 8$.

(c) The five-number summary consists of the Minimum, $Q_1$, Median, $Q_3$, and the Maximum. Respectively, these are 1, 4, 7, 12, and 55. In addition to the explanation of the quartiles in part (a), all of the values are greater than or equal to the Minimum, and all of the values are less than or equal to the Maximum.

(d) To identify any potential outliers, we first find the lower and upper limits. These are found by subtracting 1.5 x IQR from $Q_1$ and adding 1.5 x IQR to $Q_3$. Thus the lower limit is $4 - 1.5(8) = -8$ and the upper limit is $12 + 1.5(8) = 24$. Potential outliers consist of any values that are less than -8 or greater than 24. We see that 55 is the only potential outlier.

(e) We will construct a modified boxplot, which consists of a rectangle with its ends at the first and third quartiles, a line through it at the median, and two "whiskers" that extend from the quartiles outward to the adjacent values. The adjacent values are the most extreme data values that do not lie outside the lower and upper limits. For these data, the adjacent values are 1 and 23. The potential outlier, 55, is plotted individually. Thus the modified boxplot, prepared using Minitab, is

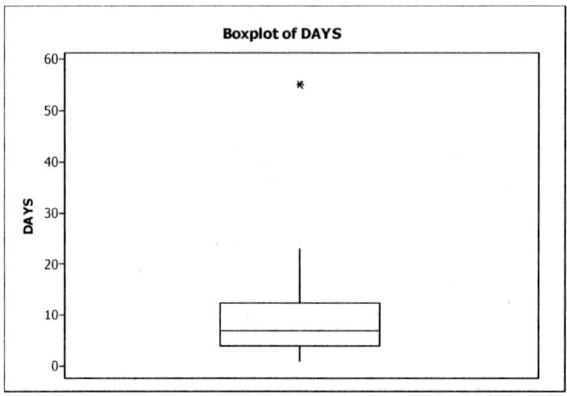

We see that the data are slightly right skewed if the potential outlier is not considered. The potential outlier is far from the other values, making it important to investigate the observation further.

**3.171** (a) Following Procedure 3.1, Step 1: First arrange the data in increasing order:

57   66   88   96   **116**   **147**   147   154   154   175

Step 2: Since this data set has an even number of values (10), the median is the mean of the middle two values ($5^{th}$ and $6^{th}$) in the ordered list. Thus the median, or second quartile, is $(116 + 147)/2 = 131.5$.

Step 3 and 4: The first quartile $Q_1$ is the median of the lower half of the ordered data set 57   66   **88**   96   116. Since this data set has an odd number of values (5), $Q_1$ is the middle value ($3^{rd}$) in the ordered list = 88.

Step 5: The third quartile $Q_3$ is the median of the upper half of the ordered data set 147   147   **154**   154   175. Since this data set has an odd number of values (5), $Q_3$ is the middle value in the ordered list = 154.

Step 6: The quartiles are $Q_1 = 88$, $Q_2 = 131.5$, and $Q_3 = 154$.

Thus 25% of the data values are less than or equal to 88, the next 25% are between 88 and the median 131.5, the next 25% are between 131.5 and 154, and the upper 25% are greater than or equal to 154.

(b) The interquartile range is $Q_3 - Q_1 = 154 - 88 = 66$.

(c) The five-number summary consists of the Minimum, $Q_1$, Median, $Q_3$, and the Maximum. Respectively, these are 57, 88, 131.5, 154, and 175. In addition to the explanation of the quartiles in part (a), all of the values are greater than or equal to the Minimum, and all of the values are less than or equal to the Maximum.

(d) To identify any potential outliers, we first find the lower and upper limits. These are found by subtracting 1.5 x IQR from $Q_1$ and adding 1.5 x IQR to $Q_3$. Thus the lower limit is 88 − 1.5(66) = −11 and the upper limit is 154 + 1.5(66) = 253. Potential outliers consist of any values that are less than −11 or greater than 253. We see that there are no potential outliers.

(e) We will construct a boxplot, which consists of a rectangle with its ends at the first and third quartiles, a line through it at the median, and two "whiskers" that extend from the quartiles outward to the minimum and maximum values. Following is the boxplot,

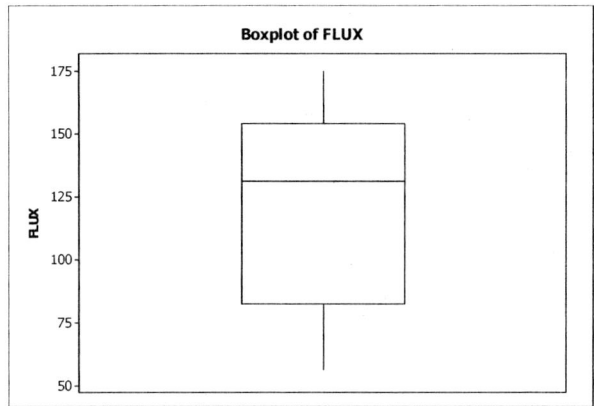

We see that the data are roughly symmetrical with no potential outliers.

**3.173** (a) Following Procedure 3.1, Step 1: First arrange the data in increasing order:

21   70   125   195   389   649   656   664   682   1006   1300   1403   1433   **1800**
1982   2205   2515   3027   3634   4200   5299   5947   7886   8543   9310   11189   17341

Step 2: Since this data set has an odd number of values (27), the median is the middle value (14th) in the ordered list. The median, or second quartile, is 1800.

Step 3 and 4: The first quartile $Q_1$ is the median of the lower half of the ordered data set (including the median) 21   70   125   195   389   649   **656   664**   682   1006   1300   1403   1433   1800. Since this data set has an even number of values (14), $Q_1$ is the average of the seventh and eighth values, so $Q_1$ = (656 + 664)/2 = 660.0.

Step 5: The third quartile $Q_3$ is the median of the upper half of the ordered data set (including the median) 1800   1982   2205   2515   3027   3634   **4200   5299**   5947   7886   8543   9310   11189   17341. Since this data set has an even number of values (14), $Q_3$ is the average of the seventh and eighth values, so $Q_3$ = (4200 + 5299)/2 = 4749.5.

Step 6: The quartiles are $Q_1$ = 660.0, $Q_2$ = 1800.0, and $Q_3$ = 4749.5.

Thus 25% of the data values are less than or equal to 660.0, the next 25% are between 660.0 and the median 1800, the next 25% are between 1800 and 1719.5, and the upper 25% are greater than or equal to 4749.5.

(b) The interquartile range is $Q_3 − Q_1$ = 4749.5 − 660.0 = 4089.5

(c) The five-number summary consists of the Minimum, $Q_1$, Median, $Q_3$, and the Maximum. Respectively, these are 21, 660, 1800, 4749.5, and 17341. In addition to the explanation of the quartiles in part (a), all of the values are greater than or equal to the Minimum, and all of the values are less than or equal to the Maximum.

(d) To identify any potential outliers, we first find the lower and upper limits. These are found by subtracting 1.5 x IQR from $Q_1$ and adding 1.5 x IQR to $Q_3$. Thus the lower limit is 660.0 − 1.5(4089.5) = −5474.25 and the upper limit is 4749.5 + 1.5(4089.5) = 10883.75. Potential outliers consist of any values that are less than −5474.25 or greater than 10883.75. We see that 11189 and 17341 are potential outliers.

(e) We will construct a modified boxplot that consists of a rectangle with its ends at the first and third quartiles, a line through it at the median, and two "whiskers" that extend from the quartiles outward to the adjacent values. The adjacent values are the most extreme data values that do not lie outside the lower and upper limits. For these data, the adjacent values are 21 and 9310. The potential outliers, 11189 and 17341, are plotted individually. Thus the boxplot, prepared using Minitab, is

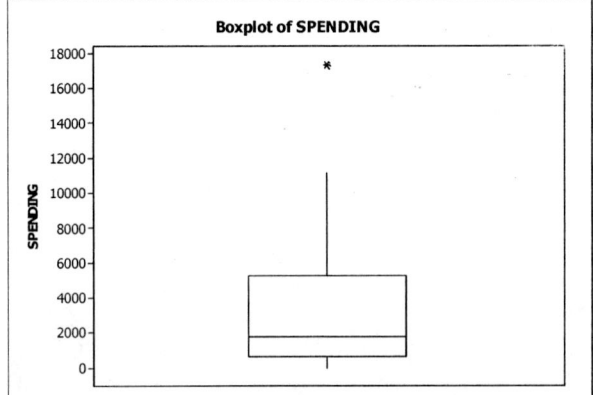

We have shown you the modified boxplot from Minitab to illustrate that software may produce a slightly different boxplot than we expected from our own computations. Minitab uses a different definition of the quartiles, which leads to a larger value of the IQR (4643). This, in turn, results in upper and lower limits that are farther from the median than ours are. Consequently, 11189 shows up as an adjacent value, not an outlier, in Minitab's modified boxplot. The overall picture of the data being right skewed with outlier(s) remains much the same.

**3.175** (a) Following Procedure 3.1, Step 1: First arrange the data in increasing order:

26.09  26.16  26.26  26.38  26.50  26.54  26.55  26.61  26.64  26.67
26.70  26.70  26.71  26.79  26.83  26.87  26.97  27.04  27.04  27.05
**27.12**  27.12  27.14  27.27  27.45  27.52  28.32  28.44  28.78  28.79
28.84  28.90  28.97  29.14  29.17  29.19  29.23  29.32  29.33  29.33
29.40

Step 2: Since this data set has an odd number of values (41), the median is the middle value (21st) in the ordered list. The median, or second quartile, is 27.12.

Step 3 and 4: The first quartile $Q_1$ is the median of the lower half of the ordered data set (including the median) 26.09  26.16  26.26  26.38  26.50  26.54  26.55  26.61  26.64  26.67  **26.70**  26.70  26.71  26.79  26.83  26.87  26.97  27.04  27.04  27.05  27.12  . Since this data set has an odd number of values (21), $Q_1$ is the middle value, so $Q_1$ = 26.70.

Step 5: The third quartile $Q_3$ is the median of the upper half of the ordered data set (including the median) 27.12  27.12  27.14  27.27  27.45  27.52  28.32  28.44  28.78  28.79  **28.84**  28.90  28.97  29.14  29.17  29.19  29.23  29.32  29.33  29.33  29.40.  Since this data set has an odd number of values (21), $Q_3$ is the middle value, so $Q_3$ = 28.84.

Step 6: The quartiles are $Q_1$ = 26.70, $Q_2$ = 27.12, and $Q_3$ = 28.84.

Thus 25% of the data values are less than or equal to 26.70, the next 25% are between 26.70 and the median 27.12, the next 25% are between 27.12 and 28.84, and the upper 25% are greater than or equal to 28.84.

(b) The interquartile range is $Q_3 - Q_1$ = 28.84 - 26.70 = 2.14.

(c) The five-number summary consists of the Minimum, $Q_1$, Median, $Q_3$, and the Maximum. Respectively, these are 26.09, 26.70, 27.12, 28.84, and 29.40. In addition to the explanation of the quartiles in part (a), all of the values are greater than or equal to the Minimum, and all of the values are less than or equal to the Maximum.

(d) To identify any potential outliers, we first find the lower and upper limits. These are found by subtracting 1.5 x IQR from $Q_1$ and adding 1.5 x IQR to $Q_3$. Thus the lower limit is 26.70 - 1.5(2.14) = 23.49, and the upper limit is 28.84 + 1.5(2.14) = 29.94. Potential outliers consist of any values that are less than 23.49 or greater than 29.94. We see that there are no potential outliers.

(e) We will construct a boxplot, which consists of a rectangle with its ends at the first and third quartiles, a line through it at the median, and two "whiskers" that extend from the quartiles outward to the minimum and maximum values. Thus the boxplot, prepared using Minitab, is

We see that the data are roughly symmetric with a slight right skew. Also, there are no potential outliers.

**3.177** (a) Following Procedure 3.1, Step 1: First, arrange the data in increasing order

6  7  8  8  8  8  9  **9**  9  10  10  10  10  10  10

**98** Chapter 3

Step 2: Since there is an odd number of data values (15), the median is the middle (8th) value. Thus the median is 9.

Step 3 and 4: The first quartile $Q_1$ is the median of the lower half of the ordered data set (including the median) 6 7 8 **8 8** 8 9 9. Since this data set has an even number of values (8), $Q_1$ is the average of the fourth and fifth values, so $Q_1 = (8 + 8)/2 = 8$.

Step 5: The third quartile $Q_3$ is the median of the upper half of the ordered data set (including the median) 9 9 10 **10 10** 10 10 10. Since this data set has an even number of values (8), $Q_3$ is the average of the fourth and fifth values, so $Q_3 = (10 + 10)/2 = 10$.

Step 6: The quartiles are $Q_1 = 8$, $Q_2 = 9$, and $Q_3 = 10$.

(b) The quartiles are not particularly useful for this set of data due to the small range of the data and the large numbers of identical values at 8, 9, and 10. For example, $Q_3$ and the maximum are equal as a result of the highest six values all being 10.

**3.179** The center of the distribution of weight loss, as measured by the median, is very similar for the two groups. However, group 1 has less variation in weight loss, both in the middle 50% and overall. Group 2 has a larger range than group 1. The distribution of weight loss for both groups is approximately symmetric.

**3.181** The median hemoglobin level is much lower in the HB SS patients than in the other two types of patients (which were roughly the same). The least variation in hemoglobin levels occurred in the HB SC patients, while the greatest variation occurred in the HB ST patients  All three distributions were roughly symmetric about their medians.

**3.183** (a) Using Minitab, retrieve the data from the WeissStats Resource Site. Then choose **Stat ▶ Basic Statistics ▶ Display Descriptive Statistics**, enter PUPS in the **Variables** text box, and click **OK**. The output appears in the Session Window as

```
Variable   N   N*   Mean   SE Mean   StDev   Minimum    Q1   Median    Q3   Maximum
PUPS      80   0   7.125    0.211    1.885    3.000   6.000  7.000   8.000   12.000
```

We see from the output that $Q_1 = 6.0$, the median = 7.0, and $Q_3 = 8.0$. Thus 25% of the values are less than or equal to 6.0, another 25% are between 6.0 and 7.0, the next 25% are between 7.0 and 8.0, and the top 25% are greater than 8.0.

(b) The interquartile range is $Q_3 - Q_1 = 8.0 - 6.0 = 2.0$. The middle 50% of the data values span an interval of 2.0.

(c) The five-number summary consists of the Minimum, $Q_1$, median, $Q_3$, and the Maximum. These are, respectively, 3, 6, 7, 8, and 12. In addition to the interpretation in part (a), all of the values are greater than or equal to the minimum, 3, and all are less than or equal to the maximum, 12.

(d) To identify any potential outliers, we first find the lower and upper limits. These are found by subtracting 1.5 x IQR from $Q_1$ and adding 1.5 x IQR to $Q_3$. Thus the lower limit is $6 - 1.5(2) = 3$, and the upper limit is $8 + 1.5(2) = 11$. Potential outliers consist of any values that are less than 3 or greater than 11. We see that 12 is a potential outlier. The maximum, 128, is greater than the upper limit, and it occurs once, so there is one potential outlier.

(e) To obtain a boxplot, we choose **Graph ▶ Boxplot**, select **Simple** in the **One Y** row and click **OK**. Enter PUPS in the **Graph Variables** text box and click **OK**. The result is

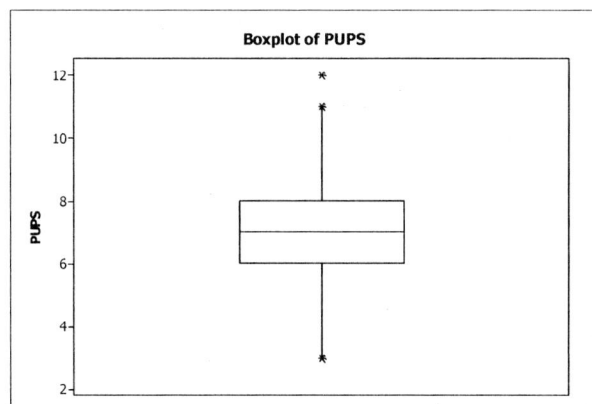

With the exception of the one potential outlier, the distribution of the data is very symmetric, approximately bell-shaped.

**3.185** (a) Using Minitab, retrieve the data from the WeissStats Resource Site. Then choose **Stat ▶ Basic Statistics ▶ Display Descriptive Statistics**, enter PERCENT in the **Variables** text box, and click **OK**. The output appears in the Session Window as

```
Variable    N   N*    Mean  SE Mean  StDev  Minimum     Q1  Median      Q3
PERCENT    51   0   86.902    0.474  3.384   80.000 84.000  87.000  90.000

Variable  Maximum
PERCENT    92.000
```

We see from the output that $Q_1$ = 84.0, the median = 87.0, and $Q_3$ = 90.0. Thus 25% of the values are less than or equal to 84.0, another 25% are between 84.0 and 87.0, the next 25% are between 87.0 and 90.0, and the top 25% are greater than 90.0.

(b) The interquartile range is $Q_3 - Q_1$ = 90.0 - 84.0 = 6.0. The middle 50% of the data values span an interval of 6.0.

(c) The five-number summary consists of the Minimum, $Q_1$, median, $Q_3$, and the Maximum. These are, respectively, 80.0, 84.0, 87.0, 90.0, and 92.0. In addition to the interpretation in part (a), all of the values are greater than or equal to the minimum, 80.0, and all are less than or equal to the maximum, 92.0.

(d) To identify any potential outliers, we first find the lower and upper limits. These are found by subtracting 1.5 x IQR from $Q_1$ and adding 1.5 x IQR to $Q_3$. Thus the lower limit is 84.0 - 1.5(6.0) = 75.0, and the upper limit is 90.0 + 1.5(6.0) = 99. Potential outliers consist of any values that are less than 75 or greater than 99. We see that there are no potential outliers.

(e) To obtain a boxplot, we choose **Graph ▶ Boxplot**, select **Simple** in the **One Y** row and click **OK**. Enter PERCENT in the **Graph Variables** text box and click **OK**. The result is

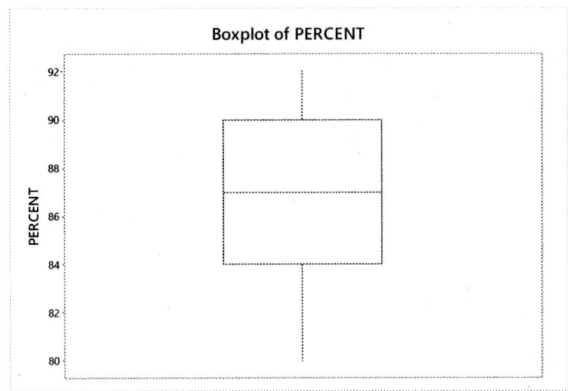

The distribution of the data is roughly symmetrical. The whiskers on the graph extend out to the minimum and maximum values.

**3.187** (a) Using Minitab, retrieve the data from the WeissStats Resource Site. Then choose **Stat ▶ Basic Statistics ▶ Display Descriptive Statistics**, enter BODYTEMP in the **Variables** text box, and click **OK**. The output appears in the Session Window as

```
Variable    N   N*    Mean   SE Mean   StDev   Minimum      Q1   Median      Q3
BODYTEMP   93   0   98.124    0.0671   0.647    96.700   97.650   98.200   98.600

Variable   Maximum
BODYTEMP    99.400
```

We see from the output that $Q_1$ = 97.65, the median = 98.20, and $Q_3$ = 98.60. Thus 25% of the values are less than or equal to 97.65, another 25% are between 97.65 and 98.20, the next 25% are between 98.20 and 98.60, and the top 25% are greater than 98.60.

(b) The interquartile range is $Q_3 - Q_1$ = 98.60 - 97.65 = 0.95. The middle 50% of the data values span an interval of 0.95.

(c) The five-number summary consists of the Minimum, $Q_1$, median, $Q_3$, and the Maximum. These are, respectively, 96.70, 97.65, 98.20, 98.60, and 99.40. In addition to the interpretation in part (a), all of the values are greater than or equal to the minimum, 96.70, and all are less than or equal to the maximum, 99.40.

(d) To identify any potential outliers, we first find the lower and upper limits. These are found by subtracting 1.5 x IQR from $Q_1$ and adding 1.5 x IQR to $Q_3$. Thus the lower limit is 97.65 - 1.5(.095) = 96.225, and the upper limit is 98.60 + 1.5(0.95) = 100.025. Potential outliers consist of any values that are less than 96.225 or greater than 100.025. We see that there are no potential outliers.

(e) To obtain a boxplot, we choose **Graph ▶ Boxplot,** select **Simple** in the **One Y** row and click **OK**. Enter RATE in the **Graph Variables** text box and click **OK**. The result is

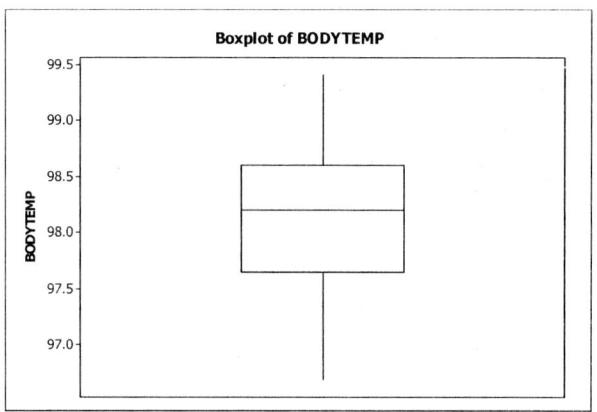

We see that the distribution of the data is roughly symmetrical with no outliers.

**3.189** (a) Using Minitab, retrieve the data from the WeissStats Resource Site. To obtain a boxplot choose **Graph ▶ Boxplot,** select **Simple** in the **Multiple Y's** row and click **OK**. Enter ITALIANS and ETRUSCANS in the **Graph Variables** text box and click **OK**. The result is

(b) The overall shape of the distributions for the Italian and Etruscan head breadths are very similar. They are both symmetric with a single lower potential outlier and similarly spread out quartiles. However, the distribution for the Etruscans is greater than the Italians. The median, quartiles, maximum, etc. are all larger for the Etruscans than the Italians.

**3.191** (a) Using Minitab, retrieve the data from the WeissStats Resource Site. Then choose **Graph ▶ Boxplot,** select **Simple** in the **Multiple Y's** row and click **OK**. Enter STOMACH, BRONCHUS, COLON, OVARY, and BREAST in the **Graph Variables** text box and click **OK**. The result is

**102** Chapter 3

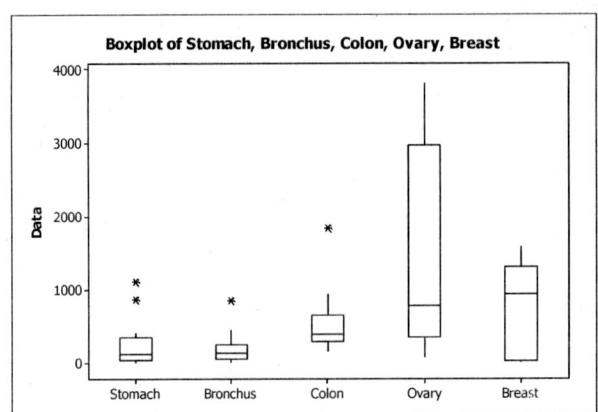

(b) The median survival times were greatest for breast cancer patients, followed by those with cancer of the ovary, colon, bronchus, and stomach. The variation in survival times for colon, bronchus, and stomach cancer patients, as measured by the interquartile range, were considerably smaller than the variation experienced by the ovary and breast cancer patients. The first three also contained outliers in the data, whereas the last two did not. All of the data distributions except for breast cancer are right skewed, whereas the breast cancer survival times are slightly left skewed.

## Exercises 3.5

**3.193** The ultimate objective of dealing with sample data in inferential studies is to describe the entire population.

**3.195** (a) A standardized variable always has mean 0 and standard deviation 1.

(b) The z-score corresponding to an observed value of a variable tells you how many standard deviations the observation is from the mean and, by its sign, what direction it is from the mean.

(c) A positive z-score indicates that the observation is greater than the mean, whereas a negative z-score indicates that the observation is less than the mean.

**3.197** The number 413.1 is a parameter since it is the mean of the entire population of lengths of holes at Augusta National Golf Club.

**3.199** If we consider the heights as a population:

(a) $\mu = \dfrac{\sum x_i}{N} = \dfrac{375}{5} = 75$ inches

(b) $\sigma = \sqrt{\dfrac{\sum (x_i - \bar{x})^2}{N}} = \sqrt{\dfrac{156}{5}} = 5.6$ inches

**3.201** (a) The population mean is $\mu = \dfrac{\sum x_i}{N} = \dfrac{9}{3} = 3$.

(b) We will use the computing formula to find the standard deviation. For that, we need $\sum x_i^2 = 4^2 + 0^2 + 5^2 = 41$. The standard deviation is then

$\sigma = \sqrt{\dfrac{\sum x_i^2}{N} - \mu^2} = \sqrt{\dfrac{41}{3} - 3^2} = 2.2$.

**3.203** (a) The population mean is $\mu = \frac{\sum x_i}{N} = \frac{11}{4} = 2.75$.

(b) We will use the computing formula to find the standard deviation. For that, we need $\sum x_i^2 = 1^2 + 2^2 + 4^2 + 4^2 = 37$. The standard deviation is then $\sigma = \sqrt{\frac{\sum x_i^2}{N} - \mu^2} = \sqrt{\frac{37}{4} - 2.75^2} = 1.3$.

**3.205** (a) The population mean is $\mu = \frac{\sum x_i}{N} = \frac{25}{5} = 5$.

(b) We will use the computing formula to find the standard deviation. For that, we need $\sum x_i^2 = 1^2 + 9^2 + 8^2 + 4^2 + 3^2 = 171$. The standard deviation is then $\sigma = \sqrt{\frac{\sum x_i^2}{N} - \mu^2} = \sqrt{\frac{171}{5} - 5^2} = 3.0$.

**3.207** (a) The population consists of all people living in the United States. The variable of interest is the age of each person.

(b) The median is the mean of the two middle ages when they are listed in increasing order as 7, 9, **29, 45**, 51, 54. Thus the median is (29 + 45)/2 = 37.0. This is a statistic since it is derived from a sample. We write our results as M = 37.0.

(c) This median is a parameter since it is derived from the entire population. Thus we write η = 37.2.

**3.209** (a) The population mean is $\mu = \frac{\sum x_i}{N} = \frac{970}{10} = 97.0$.

(b) We will use the computing formula to find the standard deviation. For that we need $\sum x_i^2 = 85^2 + 100^2 + ... + 115^2 = 95800$. The standard deviation is then $\sigma = \sqrt{\frac{\sum x_i^2}{N} - \mu^2} = \sqrt{\frac{95800}{10} - 97.0^2} = 13.1$.

(c) The median is the mean of the two middle observations in the following ordered list of 10 observations: 80  80  85  90  **90  100**  105  110  115  115. The mean of the two middle observations is (90 + 100)/2 = 95.0, so η = 95.0.

(d) The mode is the value(s) that occurs more often than any other value, so the modes are 80, 90, and 115.

(e) The first quartile is the median of the first five values in the ordered list of part (c), so it is the third value. Thus $Q_1$ = 85. The third quartile is the median of the last five values in the ordered list of part (c), so $Q_3$ = 110. The IQR is the difference between $Q_1$ and $Q_3$. Thus IQR = $Q_3 - Q_1$ = 110 - 85 = 25.

**3.211** (a) For Orlando, the population mean is $\mu = \frac{\sum x_i}{N} = \frac{2327}{5} = 465.4$. For Cincinnati, the population mean is $\mu = \frac{\sum x_i}{N} = \frac{1329}{5} = 265.8$.

**104** Chapter 3

(b) The range of values for Cincinnati is about 400, while it is only a little under 200 for Orlando, so Orlando is likely to have the smaller standard deviation.

(c) We will use the computing formula to find the standard deviation. For Orlando, we need $\sum x_i^2 = 583^2 + 460^2 + ... + 485^2 = 1,106,059$. The standard deviation is then $\sigma = \sqrt{\dfrac{\sum x_i^2}{N} - \mu^2} = \sqrt{\dfrac{1106059}{5} - 465.4^2} = 67.93$. For Cincinnati, we need $\sum x_i^2 = 77^2 + 105^2 ... + 436^2 = 492,835$. The standard deviation is then $\sigma = \sqrt{\dfrac{\sum x_i^2}{N} - \mu^2} = \sqrt{\dfrac{492835}{5} - 265.8^2} = 167.08$.

(d) Yes. We expected the standard deviation to be smaller for Orlando than for Cincinnati, and that turned out to be the case.

**3.213** (a) The standardized version of x is z = (x - 32.9)/17.9.

(b) The mean and standard deviation of z are 0 and 1 respectively.

(c) For x = 81.3, the z score is (81.3 - 32.9)/17.9 = 2.70

For x = 20.8, the z score is (20.8 - 32.9)/17.9 = -0.68

(d) The value 81.3 is 2.70 standard deviations above the mean 32.9.

The value 20.8 is 0.68 standard deviations below the mean 32.9.

(e)

```
    x̄-3s   x̄-2s   x̄-s    x̄    x̄+s   x̄+2s  x̄+3s

    -20.8  -2.9   15.0  32.9  50.8  68.7   86.6    x
     -3     -2     -1     0     1     2      3     z
```

**3.215** (a) The standardized version of x is z = (x - 6.71)/0.67.

(b) For x = 5.2, the z score is (5.2 - 6.71)/0.67 = -2.25

For x = 8.1, the z score is (8.1 - 6.71)/0.67 = 2.07

The value 5.2 is 2.25 standard deviations below the mean 6.71.

The value 8.1 is 2.07 standard deviations above the mean 6.71.

**3.217** (a) z = (21.4 - 25)/1.15 = -3.13.

(b) Yes. If the gas mileages have a distribution that is roughly bell-shaped, we would expect more than 99% of cars to get mileage within three standard deviations from the mean of 25 mpg. Your car's mpg is more than three standard deviations below the mean. Even if the distribution is not bell-shaped, Chebychev's Rule says that at least 89% of the mileages must be within three standard deviations.

**3.219** The formulas used here to compute s and σ are

$$s = \sqrt{\dfrac{\sum(x_i - \bar{x})^2}{n-1}} \quad \text{and} \quad \sigma = \sqrt{\dfrac{\sum(x_i - \mu)^2}{N}}$$

For each data set, the relevant calculations are:

Section 3.5  **105**

| Data Set | Number of Observations | $\Sigma(x_i-\text{mean})^2$ | s | σ |
|---|---|---|---|---|
| 1 | 4 | 14.00 | 2.16 | 1.87 |
| 2 | 7 | 24.86 | 2.04 | 1.88 |
| 3 | 10 | 36.40 | 2.01 | 1.91 |

(a) The sample standard deviations are found in the fourth column of the previous table.

(b) The population standard deviations are found in the fifth column of the previous table.

(c) Comparing s and σ for a given data set, the two measures will tend to be closer together if the data set is large.

**3.221** (a) Using Minitab, retrieve the data from the WeissStats Resource Site. Then choose **Stat ▶ Basic Statistics ▶ Display Descriptive Statistics**, enter WEIGHT in the **Variables** text box, and click **OK**. The output appears in the Session Window as

```
Variable   N  N*    Mean  SE Mean  StDev  Minimum      Q1  Median      Q3
WEIGHT    21   0  214.05     5.86  26.88   176.00  190.00  211.00  237.50

Maximum
270.00
```

The sample mean is calculated the same as the population mean, so the population mean is 214.05 pounds.

(b) Unfortunately, this procedure produces a sample standard deviation instead of a population standard deviation, so you than have to convert s to σ using the formula derived in Exercise 3.220. The sample standard deviation is computed using Minitab in part (a) to be 26.88. In Exercise 3.220, the conversion formula is $\sigma = s\sqrt{\dfrac{m-1}{m}}$. Thus the population standard deviation would be $\sigma = 26.88\sqrt{\dfrac{21-1}{21}} = 26.23$ pounds.

**3.223** z = (205500 - 220258)/5237 = -2.82. Applying Chebychev's rule to that z-score with k = 82.82, we conclude that at least $100(1 - 1/2.82^2)\%$ = 87.4% of the sale prices lie within 2.82 standard deviations to either side of the mean of $220,258. Therefore, the $205,500 price of the home you are contemplating buying is lower than at least 87.4% of the prices of comparable homes. It appears that this home is bargain.

**3.225** (a) In the population of lowland copperhead snakes, for x = 850, the z score is (850 - 812.07)/330.24 = 0.115. In the population of tiger snakes, for x = 850, the z score is (850 - 743.65)/336.36 = 0.316.

(b) If the distributions of the two types of snakes are approximately the same shape, it would be reasonable to use z-scores to compare the relative standings of the weights of the two snakes.

(c) The value 850 is 0.115 standard deviations above the mean for lowland copperhead snakes. The value 850 is 0.316 standard deviations above the mean for tiger snakes. The tiger snake of 850g is larger relative to other tiger snakes than the copperhead of 850g is to other copperheads.

## REVIEW PROBLEMS FOR CHAPTER 3

1. (a) Descriptive measures are numbers used to describe data sets.

   (b) Measures of center indicate where the center or most typical value of a data set lies.

   (c) Measures of variation indicate how much variation or spread the data set has.

2. The two most commonly used measures of center for quantitative data are the mean and the median. The mean uses all of the data, but can be influenced by the presence of a few outliers. The median is computed from the one or two center values in an ordered list of the data. It does not make use of all of the data, but it has the advantage that it is not influenced by the presence of a few outliers.

3. The only measure of center, among those we discussed, that is appropriate for qualitative data is the mode.

4. (a) standard deviation

   (b) interquartile range

5. (a) $\bar{x}$  (b) s  (c) $\mu$  (d) $\sigma$

6. (a) This is not necessarily true.

   (b) This is necessarily true.

7. Almost all of the observations in any data set lie within 3 standard deviations to either side of the mean.

8. (a) If $k = 6$, using Chebyshev's rule, at least $1 - 1/k^2 = 1 - 1/36 = 35/36$, or 97.2%, of the observations lie within 6 standard deviations to either side of the mean.

   (b) If $k = 1.5$, using Chebyshev's rule, at least $1 - 1/k^2 = 1 - 1/2.25 = 0.556$, or 55.6%, of the observations lie within 1.5 standard deviations to either side of the mean.

9. 60 and 100 are two standard deviations to either side of the mean. Chebyshev's rule states that at least 75% of the observations for any data set lie within two standard deviations to either side of the mean. 75% of 87 is 65.25. Therefore, at least 66 observations in the data set will be between 60 and 100.

10. 33 and 57 are one standard deviation to either side of the mean. The empirical rule states that approximately 68% of the observations in a data set that is bell shaped will lie within one standard deviation to either side of the mean. Therefore, approximately 68% of the observations in this data set will lie between 33 and 57.

11. 17 and 33 are two standard deviations to either side of the mean. The empirical rule states that approximately 95% of the observations in a data set that is bell shaped will lie within two standard deviations to either side of the mean. 95% of 152 is 144.4. Therefore, approximately 144 of the observations in this data set will lie between 17 and 33.

12. (a) The components of the five-number summary are the minimum, $Q_1$, median, $Q_3$, and the maximum.

    (b) The median is a measure of the center. The interquartile range, which is found as $Q_3 - Q_1$, and the range, which is the difference between the maximum and the minimum, are measures of variation.

    (c) The boxplot is based on the five-number summary.

13. (a) An outlier is an observation that falls well outside the overall pattern of the data.

    (b) First compute the interquartile range IQR = $Q_3 - Q_1$. Then compute two limits as $Q_1 - 1.5\,IQR$ and $Q_3 + 1.5\,IQR$. Observations that are lower than the lower limit or higher than the upper limit are potential outliers and require further study.

14. (a) A z-score for an observation x is obtained by subtracting the mean of the data set from x and then dividing the result by the standard deviation; i.e., $z = (x - \mu)/\sigma$ for population data or $z = (x - \bar{x})/s$ for sample data.

    (b) A z-score indicates how many standard deviations an observation is below or above the mean of the data set.

    (c) An observation with a z-score of 2.9 is likely to be greater than all or almost all of the other data values.

15. (a) The mean is calculated as $\bar{x} = \Sigma x_i/n$. For the sample of 20 guests, $\bar{x}$ = 47/20 = 2.35 alcoholic drinks.

    The median is found by ordering the observations from lowest to highest and finding the average of the two observations in the middle. The list is

    0 0 1 1 1 1 2 2 **2 2** 2 3 3 4 4 4 4 5 5

    The median is the average of the 10th and 11th values, and is therefore (2 + 2)/2 = 2.

    The mode is the value that occurs the most times. In this data set, 1 and 2 both occur 5 times, and so the modes are 1 and 2.

    (b) If the purpose is to help in estimating the cost of the party, the mean is the best since it takes into account the drinking practices of all of the guests. If the purpose is to describe the average guest, the median is more typical of the list of values. The mode is less useful for this data since there are two modes.

16. Many more marriages are characterized as being of short duration than other durations. Since the median is ordinarily preferred for data sets that have exceptional (very large or small) values, the median is more appropriate than the mean as a measure of central tendency for data on the duration of marriages.

17. Causes of death are qualitative. There is no way to compute a mean or median for such data. The mode is the only measure of central tendency that can be used for qualitative data.

18. The mean is $\bar{x} = \dfrac{\Sigma x_i}{n} = \dfrac{305.3}{10} = 30.53$; since n is even (10), the median is the mean of the two middle observations in the ordered list of the data values:
    17.4  21.0  17.4  31.5  **32.0  33.0**  33.0  34.5  37.5  38.0. Thus the median is (32 + 33)/2 = 32.5. The mode is the most frequently occurring value. Since 33.0 occurs twice and no other value occurs more than once, 33.0 is the mode.

19. (a) $\bar{x} = \dfrac{\Sigma x_i}{n} = \dfrac{457}{10} = 45.7$ kilograms

    (b) Range = 54 − 37 = 17 kilograms

**108** Chapter 3

(c) $s = \sqrt{\dfrac{\sum x^2 - (\sum x)^2/n}{n-1}} = \sqrt{\dfrac{21109 - 457^2/10}{9}} = 5.0$ kg

**20.** (a)

(b) Almost all of the ages of the 36 millionaires are between <u>18.3</u> and <u>98.7</u> years old.

**21.** (a) 15.1 and 36.3 are two standard deviations to either side of the mean. Chebyshev's rule states that at least 75% of the observations in any data set will lie within two standard deviations to either side of the mean. 75% of 10,215 is 7,661.25. Therefore, at least 7,662 women in the sample will have a BMI between 15.1 and 36.3.

(b) Chebyshev's rule states that at least 89% of the observations in any data set will lie within three standard deviations to either side of the mean. Therefore, at least 89% of the women in the sample have BMIs between 25.7 − 3(5.3) = <u>9.8</u> and 25.7 + 3(5.3) = <u>41.6</u>.

**22.** (a) The Empirical rule states that for a roughly bell shaped distribution, approximately 95% of the observations will lie within two standard deviations to either side of the mean. Therefore, at least 95% of the mobile homes in the sample have prices between 63.3 − 2(7.9) = $<u>47.5</u> thousand and 63.3 + 2(7.9) = $<u>79.1</u> thousand.

(b) 39.6 and 87.0 are three standard deviations to either side of the mean. The empirical rule states that approximately 99.7% of the observations in a data set that is bell-shaped will lie within three standard deviations to either side of the mean. Therefore, approximately <u>99.7%</u> of the mobile homes in the sample have prices between $39.6 thousand and $87.0 thousand.

(c) 55.4 and 71.2 are one standard deviation to either side of the mean. The empirical rule states that approximately 68% of the observations in a data set that is bell-shaped will lie within one standard deviation to either side of the mean. 68% of 250 is 170. Therefore, approximately 170 mobile homes in the sample will have prices between $55.4 thousand and $71.2 thousand.

**23.** (a) Following Procedure 3.1, Step 1: The data set is already ordered.

Step 2: The second quartile is the median of the entire data set. The number of pieces of data is 36, and so the position of the median is (36 + 1)/2 = 18.5, halfway between the eighteenth and nineteenth data values. Thus the median of the entire data set is (59 + 60)/2 = 59.5. That is, $Q_2$ = 59.5.

Step 3 and 4: The first quartile is the median of the lower half of the ordered list. Since there is a total of 36 ages, the first 18 are in the lower half and the median of these is the average of the middle two, the 9th and 10th values. Thus $Q_1$ = (48 + 48)/2 = 48.

Step 5: The third quartile is the median of the upper half of the ordered list. Since there is a total of 36 ages, the last 18 are in the upper half and the median of these is the average of the middle two, the 9th and 10th values. Thus $Q_3 = (68 + 69)/2 = 68.5$.

Step 6: $Q_1 = 48$, $Q_2 = 59.5$, and $Q_3 = 68.5$.

Interpreting our results, we conclude that 25% of the ages are less than 48 years; 25% of the ages are between 48 and 59.5 years; 25% of the ages are between 59.5 and 68.5 years; and 25% of the ages are greater than 68.5 years.

(b) The IQR = $Q_3 - Q_1$ = 68.5 - 48 = 20.5. Thus, the middle 50% of the ages has a range of 20.5 years.

(c) Min = 31, $Q_1$ = 48, $Q_2$ = 59.5, $Q_3$ = 68.5, Max = 79

(d) The limits are given by:

Lower limit = $Q_1$ - 1.5(IQR) = 48.0 - 1.5(20.5) = 17.25

Upper limit = $Q_3$ + 1.5(IQR) = 68.5 + 1.5(20.5) = 99.25

(e) There are no values below 17.25 or above 99.25, so there are no potential outliers.

(f) A boxplot is constructed easily using the information in part (a).

   (i)   low data value = 31 years
   (ii)  high data value = 79 years
   (iii) $Q_1$ = 48.0 years
   (iv)  $Q_3$ = 68.5 years
   (v)   median = $Q_2$ = 59.5 years.

These values are used to construct the boxplot as follows:

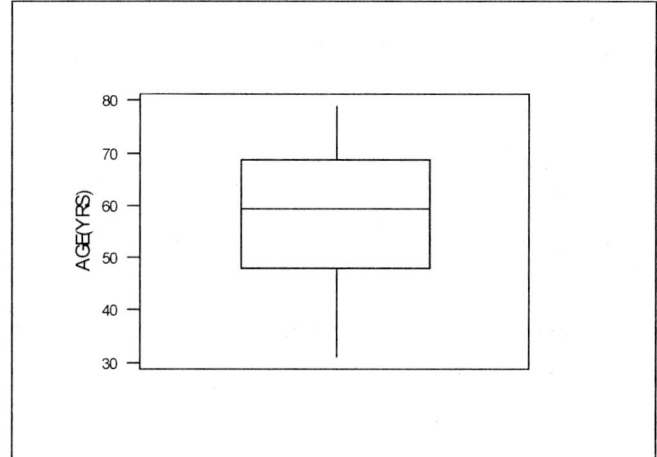

24. (a) Following Procedure 3.1, Step 1: The order list of the data is 0.7 1.0 1.1 1.1 1.2 1.5 1.8 1.8 1.8 1.8 **1.9 2.0** 2.0 2.0 2.3 2.7 3.3 3.4 3.6 3.8 6.7 7.6.

Step 2: Since n is even (22), the median is mean of the two middle values, the 11th and 12th. Thus the median is (1.9 + 2.0)/2 = 1.95.

Step 3 and 4: The first quartile is the median of the first 11 values in the list, so $Q_1$ = the 6th value = 1.5.

Step 5: The third quartile is the median of the last 11 values in the list, so $Q_3$ = the 6th value from the upper end = 3.3.

Step 6: Thus, the five-number summary is Minimum = 0.7, $Q_1$ = 1.5, Median = 1.95, $Q_3$ = 3.3, and Maximum = 7.6.

(b) The IQR = 3.3 - 1.95 = 1.35. The lower limit = Q1 - 1.5 x IQR = 1.5 - 1.5(1.35) = -0.525, while the upper limit = Q3 + 1.5 x IQR = 3.3 + 1.5(1.35) = 5.325. There are no values below the lower limit, but 6.7 and 7.6 lie above the upper limit and are therefore potential outliers.

(c) The boxplot consists of a rectangle with its ends at 1.5 and 3.3 and a line across the middle at 1.95. Whiskers extend from the ends of the box out to the adjacent values (the most extreme values on each end of the box that are not beyond the limits). Thus the lower whisker extends to 0.7, and the upper whisker extends to 3.8. The two potential outliers are plotted individually. The result is

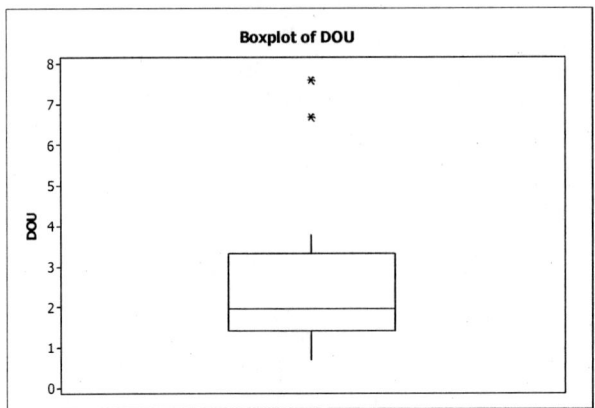

**25.** The median distance traveled for golf balls hit off the wooden tee is about 227 yards while the median distance traveled when hit off the Stinger tee is about 240 yards. In fact, the maximum distance with the regular wooden tee is shorter than the minimum distance with the Stinger tee. Both distributions have roughly a similar amount of variation (Stinger perhaps slightly more) and both are roughly bell-shaped (Stinger perhaps slightly right-skewed).

**26.** (a) The population mean is $\mu = \dfrac{\sum x_i}{N} = \dfrac{659}{12} = 54.9$ minutes.

(b) We will use the computing formula to find the standard deviation. For that we need $\sum x_i^2 = 53^2 + 58^2 + \ldots + 16^2 = 39311$. The standard deviation is then $\sigma = \sqrt{\dfrac{\sum x_i^2}{N} - \mu^2} = \sqrt{\dfrac{39311}{12} - 54.9^2} = 16.1$ minutes.

**27.** (a) $\mu = \dfrac{\sum x_i}{n} = \dfrac{183.1}{9} = 20.34$ thousand students.

(b) Squaring each data value and then summing, we find that $\sum x_i^2 = 4094.39$. Then the population standard deviation is $\sigma = \sqrt{\dfrac{\sum x_i^2}{N} - \mu^2} = \sqrt{\dfrac{4094.39}{9} - 20.34^2} = 6.41$ thousand students.

(c) z = (x - μ)/σ = (x - 20.34)/6.41.

(d) The mean of z is 0, and the standard deviation of z is 1. All standardized variables have mean 0 and standard deviation 1.

(e) Converting each x value (dotplot on the left) to its corresponding z-score results in the dotplot on the right.

The relative positions of the points in the two plots are the same, but the scales are different.

(f)

| x | z = (x − μ)/σ |
|---|---|
| 27.7 | (27.7 − 20.34)/6.41 = 1.15 |
| 18.6 | (18.6 − 20.34)/6.41 = −0.27 |

The enrollment at Los Angeles is 1.15 standard deviations above the mean of 27.7 thousand students and the enrollment at Riverside is 0.27 standard deviations below the mean.

28. (a) The mean price given here is a sample mean. This is because it is the mean price per gallon for the sample of 800 gasoline service stations.

(b) The symbol used to designate the mean of $3.547 is $\bar{x}$.

(c) The mean price given here is a statistic. It is a descriptive measure.

29. Using Minitab, retrieve the data from the WeissStats Resource Site. Then choose **Stat ▶ Tables ▶ Tally Individual Variables**. Enter REGION and DIVISION in the **Variables** text box and click **OK**. The result is the table below.

```
      REGION  Count              DIVISION  Count
     Midwest     12     East North Central     5
   Northeast      9     East South Central     4
       South     16         Middle Atlantic    3
        West     13                Mountain    8
          N=     50             New England    6
                                   Pacific    5
                             South Atlantic    8
                         West North Central    7
                         West South Central    4
                                        N=   50
```

(a) We see from the first column of the above table that the South region occurs 16 times, more than any other region, so South is the mode of the regions.

(b) We see from the second column of the above table that the Mountain and the South Atlantic divisions each occur 8 times, more than any other region, so Mountain and South Atlantic are both modal divisions.

**112** Chapter 3

30. (a) Using Minitab, retrieve the data from the WeissStats Resource Site. Then choose **Stat ▶ Basic Statistics ▶ Display Descriptive Statistics**. Click **Statistics** and check **Mode** and click **OK**. Enter VALUE in the **Variables** text box and click **OK**. The result is

    ```
    Variable   N  N*  Mean  SE Mean  StDev  Minimum  Q1   Median  Q3    Maximum
    VALUE     50   0  1563  281      1983   4        234  995     2118  11302
    
                       N for
    Variable  Mode    Mode
    VALUE     *       0
    ```

    We see that the mean is 1563 and the Median is 995. Examining the data, we find that no value occurs more than once, so there is no mode. If we wanted to estimate the total agricultural exports for the nation, the mean would be the best measure of center to use since we could just multiply the mean by 50 to get the total. If we wanted to know what value half of the states are below, then the median is the best measure of center to use. Another reason for using the median is that there are several outliers that lie above the median, including one that is very large (11301.7 for California). Since the mean is quite sensitive to outliers such as this one, the median is the better choice for describing these data.

    (b) From the table above, Range = 11302 − 4 = 11298, and the sample standard deviation s = 1983.

    (c) The five-number summary consists of the minimum, $Q_1$, the median, $Q_3$, and the maximum. These are, respectively, 4, 234, 995, 2118, and 11302. The interquartile range IQR = $Q_3$ − $Q_1$ = 2118 − 234 = 1884.

    (d) We first find the lower limit as $Q_1$ − 1.5 x IQR = 234 − 1.5(1884) = −2592. The upper limit is $Q_3$ + 1.5 x IQR = 2118 + 1.5(1884) = 4944. Any data values below the lower limit or above the upper limit are potential outliers. Thus 5198.6, 5246.8, and 11301.7 are potential outliers.

    (e) To obtain a boxplot of the data, choose **Graph ▶ Boxplot,** select **Simple** from the **One Y** row and click **OK**. Enter VALUE in the **Variables** text box and click **OK**. The result is

    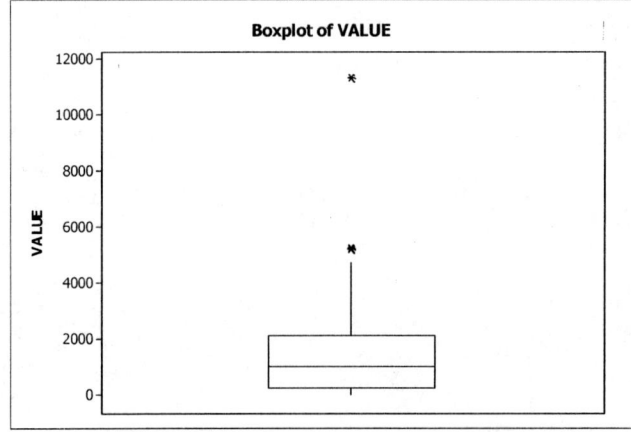

    The data distribution is quite right skewed with a median of 995. The plot shows potential outliers on the high side of the data.

31. (a) Using Minitab, retrieve the data from the WeissStats Resource Site. Then choose **Stat ▶ Basic Statistics ▶ Display Descriptive Statistics**.

Enter 'LIFE EXP' in the **Variables** text box and click **OK**. The result is

```
Variable           N   N*   Mean   SE Mean   StDev   Minimum    Q1   Median
LIFE EXPECTANCY  193    0  69.968   0.708    9.839    47.800  64.400  73.300

   Q3   Maximum
76.700   83.400
```

We see that the mean is 69.968 and the median is 73.300. Finding the mode is easiest if we choose **Stat ▶ Tables ▶ Tally Individual Variables** and enter 'LIFE EXP' in the **Variables** text box. Examining the resulting table (not shown here), we find that seventeen values each occur three times, more than any other values, so there are seventeen modes. The mean is not appropriate for these data since each value is a mean life expectancy based on population sizes that are different for different countries. The median is the best measure of center to use since it tells us that half of the listed countries have life expectancies below this number. Another reason for using the median is that the distribution is left skewed with several potential outliers at the lower end, as will be seen in parts (d) and (e).

(b) From the table above, Range = 83.4 - 47.8 = 35.6, and the sample standard deviation s = 9.839. [Note: The sample standard deviation is based on a calculation that uses the sample mean. Since the sample mean being used does not account for the fact that the different countries have different populations, the true sample standard deviation may differ somewhat from this value.]

(c) The five-number summary consists of the minimum, $Q_1$, the median, $Q_3$, and the maximum. These are, respectively, 47.8, 64.4, 73.3, 76.7, 83.4. The interquartile range IQR = $Q_3$ - $Q_1$ = 76.7 - 64.4 = 12.3.

(d) We first find the lower limit as $Q_1$ - 1.5 x IQR = 64.4 - 1.5(12.3) = 45.95. The upper limit is $Q_3$ + 1.5 x IQR = 76.7 + 1.5(12.3) = 95.15. Any data values below the lower limit or above the upper limit are potential outliers. Thus there are no potential outliers.

(e) To obtain a boxplot of the data, choose **Graph ▶ Boxplot,** select **Simple** from the **One Y** row and click **OK**. Enter 'LIFE EXP' in the **Variables** text box and click **OK**. The result is

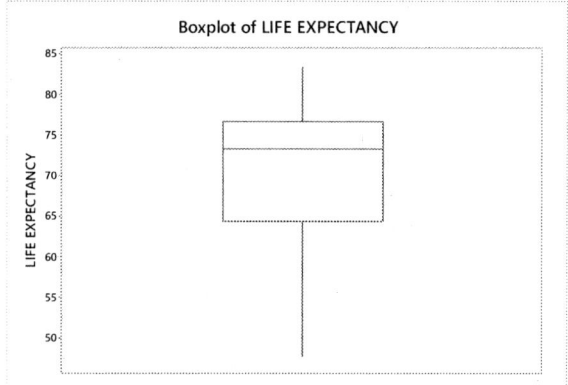

We see that the data are slightly left-skewed with the median at about 73.

**114  Chapter 3**

**32.** (a) Using Minitab, retrieve the data from the WeissStats Resource Site. Then choose **Stat ▶ Basic Statistics ▶ Display Descriptive Statistics**. Enter HIGH and LOW in the **Variables** text box and click **OK**. The result is

```
Variable   N   N*   Mean  SE Mean  StDev  Minimum    Q1  Median     Q3  Maximum
HIGH      71   0   65.33    1.04    8.72    47.60  58.40  64.00  71.10    85.50
LOW       71   0   45.52    1.11    9.34    29.30  39.80  44.20  51.40    74.20
```

We see that the mean HIGH temperature is 65.33 and the median HIGH is 64.00. The mean LOW temperature is 45.52 and the median LOW is 44.20.

Finding the mode is easiest if we choose **Stat ▶ Tables ▶ Tally Individual Variables** and enter HIGH and LOW in the **Variables** text box. Examining the resulting table (not shown here), we find for HIGH that 54.5, 55.9, 57.6, 59.8, 62.6, 63.6, 65.1, 67.4, 67.8, and 70.6 all occur twice, more than any other values, so there are ten modes. For LOW, the values 35.8, 39.8, 41.2, 43.2, 44.8, 47.4, 48.6, and 52.5 all occur twice, more than any other values, so there are eight modes. The median is the best measure of center to use since it tells us that half of the listed cities have minimums (or maximums) below this number. Another reason for using the median is that the distribution for LOW has several potential outliers at the upper end, as will be seen in parts (d) and (e).

(b) From the table above, for HIGH, Range = 85.5 − 47.6 = 37.9, and the sample standard deviation s = 8.72. For LOW, Range = 74.20 − 29.30 = 44.90 and the sample standard deviation s = 9.34.

(c) The five-number summary consists of the minimum, $Q_1$, the median, $Q_3$, and the maximum. For HIGH, these are, respectively, 47.60, 58.40, 64.00, 71.10, and 85.50. The interquartile range for HIGH is IQR = $Q_3 - Q_1$ = 71.10 − 58.40 = 12.70. For LOW, the five-number summary is 29.30, 39.80, 44.20, 51.40, and 74.20. IQR = $Q_3 - Q_1$ = 51.40 − 39.80 = 11.60.

(d) For HIGH, we first find the lower limit as $Q_1$ − 1.5 x IQR = 58.40 − 1.5(12.70) = 39.35. The upper limit is $Q_3$ + 1.5 x IQR = 71.10 + 1.5(12.70) = 90.15. Any data values below the lower limit or above the upper limit are potential outliers. Examining the data, we find that there are no potential outliers for HIGH.

For LOW, we first find the lower limit as $Q_1$ − 1.5 x IQR = 39.80 − 1.5(11.60) = 22.40. The upper limit is $Q_3$ + 1.5 x IQR = 51.40 + 1.5(11.60) = 68.80. Any data values below the lower limit or above the upper limit are potential outliers. Examining the data, we see that 69.1, 70.2, and 74.2 are potential outliers. If you are looking for someplace that is warm year around, these values are for Miami, Honolulu, and San Juan.

(e) To obtain a boxplot of the data, choose **Graph ▶ Boxplot,** select **Simple** from the **Multiple Y's** row and click **OK**. Enter HIGH and LOW in the **Variables** text box and click **OK**. The result is

The distribution of HIGH temperatures is quite symmetrical, while the distribution of LOW temperatures is also quite symmetrical except for three outliers. The three cities with high LOW temperatures are all affected by warm southern ocean waters that keep the temperature quite moderate year around.

33. (a) To obtain a boxplot of the height data, choose **Graph ▶ Boxplot,** select **Simple** from the **Multiple Y's** row and click **OK**. Enter 'VEGETARIAN' and 'OMNIVORE' in the **Variables** text box and click **OK**. The result is

(b) The range from minimum to maximum seems to be similar for the vegetarian and omnivore daily protein intake. Also, the IQR seem to be similar in value for the two groups. The biggest difference is that the quartiles (including the median) for the omnivore daily protein intake is greater than that for the vegetarians. There were not any outliers in either group.

## CHAPTER 4 SOLUTIONS

### Exercises 4.1

**4.1** (a) $y = b_0 + b_1 x$      (b) Constants are $b_0$, $b_1$; variables are x,y.
(c) The independent variable is x and the dependent variable is y.

**4.3** (a) $b_0$ is the y-intercept; it is the value of y where the line crosses the y-axis.
(b) $b_1$ is the slope; it indicates the change in the value of y for every 1 unit increase in the value of x.

**4.5** (a) $b_0 = 3$, $b_1 = 4$
(b) slopes upward
(c)

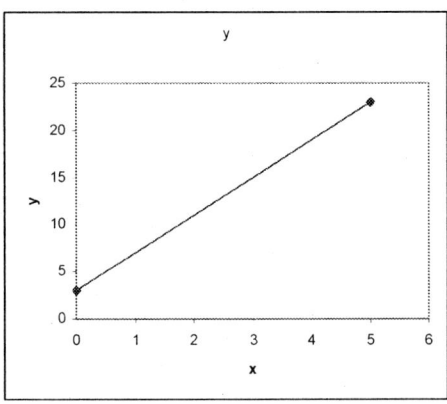

**4.7** (a) $b_0 = 6$, $b_1 = -7$
(b) slopes downward
(c)

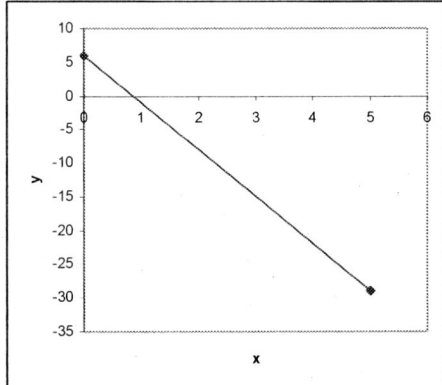

**4.9**  (a)  $b_0 = -2$, $b_1 = 0.5$
(b)  slopes upward
(c)

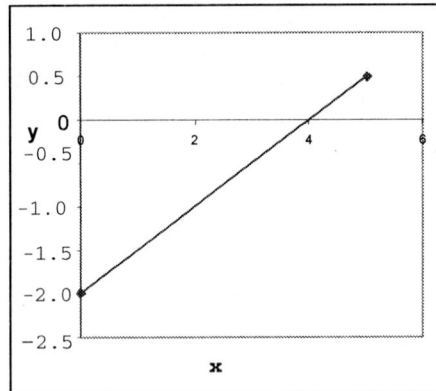

**4.11** (a)  $b_0 = 2$, $b_1 = 0$
(b)  horizontal
(c)

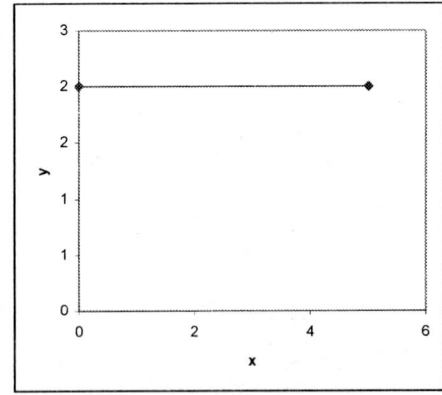

**4.13** (a)  $b_0 = 0$, $b_1 = 1.5$
(b)  slopes upward
(c)

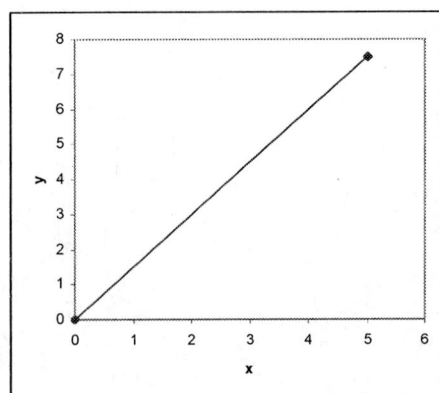

**4.15** (a) slopes upward
(b) y = 5 + 2x
(c)
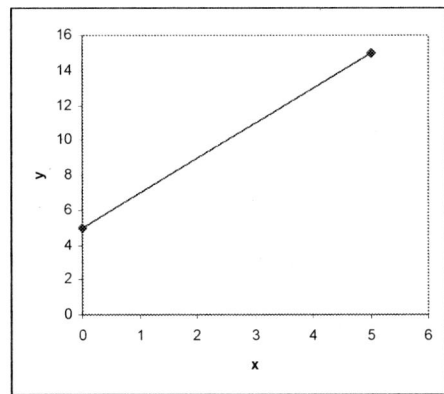

**4.17** (a) slopes downward
(b) y = -2 - 3x
(c)
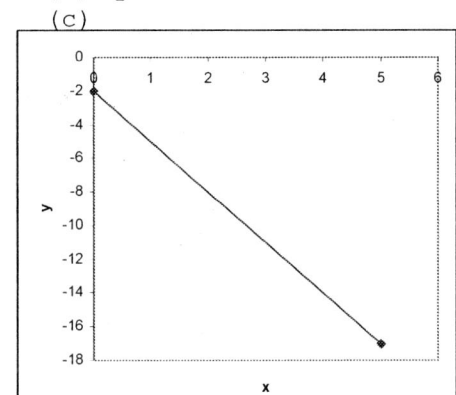

**4.19** (a) slopes downward
(b) y = -0.5x
(c)
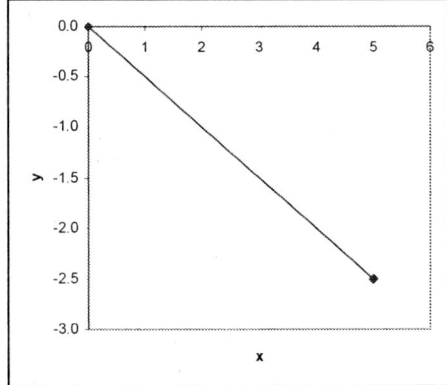

**120** Chapter 4

**4.21** (a) horizontal
(b) y = 3
(c)

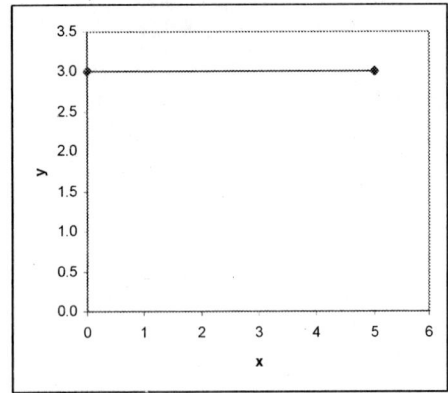

**4.23** (a) y = 68.22 + 0.25x  (b) $b_0$ = 68.22, $b_1$ = 0.25
(c)           (d)

| Miles<br>x | Cost ($)<br>Y |
|---|---|
| 50 | 68.22 + 0.25(50) = 80.72 |
| 100 | 68.22 + 0.25(100) = 93.22 |
| 250 | 68.22 + 0.25(250) = 130.72 |

(e) The visual cost estimate of driving the car 150 miles is a little more than $100. The exact cost is
y = 68.22 + 0.25(150) = 105.72

**4.25** (a) $b_0$ = 32, $b_1$ = 1.8
(b)           (c)

| X(°C) | y(°F) |
|---|---|
| -40 | 32 + 1.8(-40) = -40 |
| 0 | 32 + 1.8(0) = 32 |
| 20 | 32 + 1.8(20) = 68 |
| 100 | 32 + 1.8(100) = 212 |

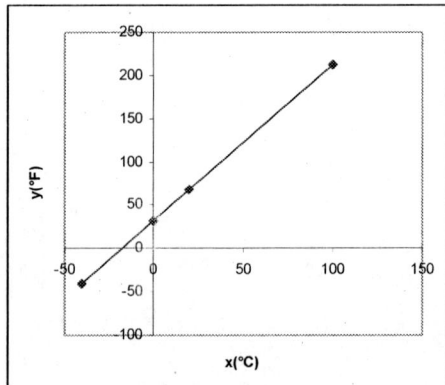

(d) The visual Fahrenheit temperature estimate corresponding to a Celsius temperature of 28° is about 80°F. The exact temperature is y = 32 + 1.8(28) = 82.4°F.

**4.27** (a) $b_0$ = 68.22, $b_1$ = 0.25
(b) The y-intercept $b_0$ = 68.22 gives the y-value at which the straight line y = 68.22 + 0.25x intersects the y-axis. The slope $b_1$ = 0.25 indicates

that the y-value increases by 0.25 units for every increase in x of one unit.
- (c) The y-intercept $b_0$ = 68.22 is the cost (in dollars) for driving the car zero miles. The slope $b_1$ = 0.25 represents the fact that the cost per mile is $0.25; it is the amount the total cost increases for each additional mile driven.

**4.29**
- (a) $b_0$ = 32, $b_1$ = 1.8
- (b) The y-intercept $b_0$ = 32 gives the y-value where the line y = 32 + 1.8x intersects the y-axis. The slope $b_1$ = 1.8 indicates that the y-value increases by 1.8 units for every increase in x of one unit.
- (c) The y-intercept $b_0$ = 32 is the Fahrenheit temperature corresponding to 0°C. The slope $b_1$ = 1.8 represents the fact that Fahrenheit temperature increases by 1.8° for every increase of the Celsius temperature of 1°.

**4.31** The line defined by the equation must pass through the two points (x,F) - (2,32 and (3,16). We need to solve for k and $x_0$. We have

$$\begin{cases} 32 = -k(2-x_0) \\ 16 = -k(3-x_0) \end{cases} \text{ or } \begin{cases} 32 = -2k + kx_0 \\ 16 = -3k + kx_0 \end{cases}$$

Subtracting the second equation from the first, we obtain 16 = k. Substituting 16 for k in the first equation, we find
$32 = -2(16) + 16x_0$ or $64 = 16 x_0$., from which we find $x_0$ = 4.
- (a) Thus, the linear equation that relates the force exerted to the length compressed is F = -16(x - 4).
- (b) The spring constant k = 16.
- (c) The natural length of the spring is $x_0$ = 4 feet.

**4.33**
- (a) If we can express a line in the form y = $b_0$ + $b_1$x, then there will be one and only one y-value corresponding to each x-value. However, that is not the case for a vertical line; one value of x results in an infinite number of y-values.
- (b) The form of the equation of a vertical line is x = $x_0$, where $x_0$ is the x-coordinate of the vertical line.
- (c) For a linear equation, the slope indicates how much the y-value on the line increases (or decreases) when the x-value increases by one unit. We cannot apply this concept to a vertical line, since the x-value is *not* permitted to change. Thus, a vertical line has no slope.

**Exercises 4.2**

**4.35**
- (a) The criterion used to decide on the line that best fits a set of data points is called the least squares criterion.
- (b) The criterion is that the line that best fits a set of data points is the one that has the smallest possible sum of the squares of the errors (errors are the differences between and actual and predicted y values).

**4.37**
- (a) The dependent variable is called the response variable.
- (b) The independent variable is called the predictor variable or the explanatory variable.

**4.39**
- (a) In the context of regression, an outlier is a data point that lies far from the regression line, relative to the other data points.
- (b) In regression analysis, an influential observation is a data point whose removal causes the regression equation to change considerably.

**4.41** The idea behind finding a regression line is based on the assumption that the data points are actually scattered about a line. Only the second data set appears to be scattered about a line. Thus, it is reasonable to determine a regression line only for the second set of data.

**4.43** (a)     Line A: $y = -1 + 3x$            Line B: $y = 1 + 2x$

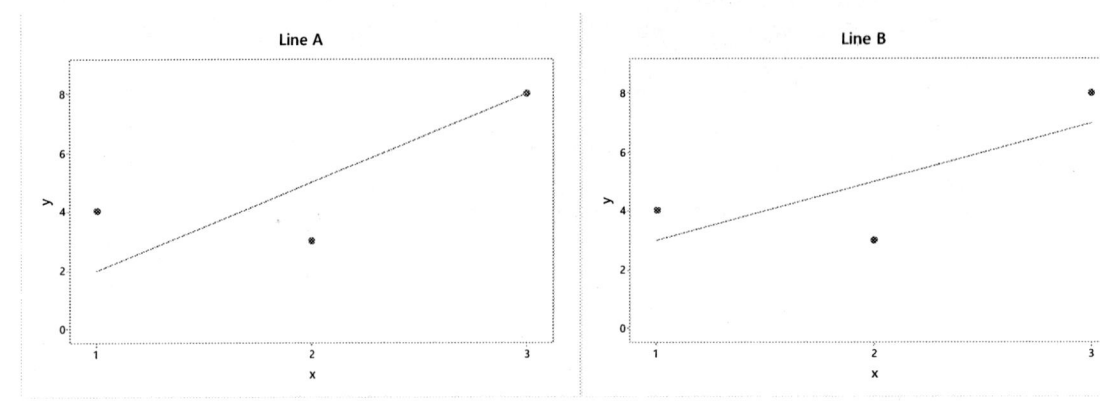

(b)

Line A: $y = -1 + 3x$

| x | y | $\hat{y}$ | e | $e^2$ |
|---|---|---|---|---|
| 1 | 4 | 2 | 2 | 4 |
| 2 | 3 | 5 | -2 | 4 |
| 3 | 8 | 8 | 0 | 0 |
|   |   |   |   | 8 |

Line B: $y = 1 + 2x$

| x | y | $\hat{y}$ | e | $e^2$ |
|---|---|---|---|---|
| 1 | 4 | 3 | 1 | 1 |
| 2 | 3 | 5 | -2 | 4 |
| 3 | 8 | 7 | 1 | 1 |
|   |   |   |   | 6 |

(c) According to the least-squares criterion, Line B fits the set of data points better than Line A. This is because the sum of squared errors, i.e., $\sum e^2$, is smaller for Line B than for Line A.

**4.45** (a)     Line A: $y = 3 - 0.6x$          Line B: $y = 4 - x$

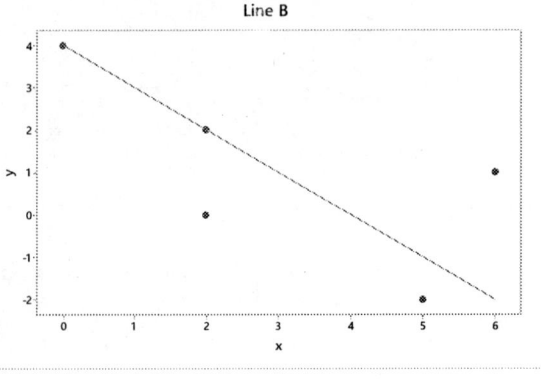

(b)

Line A: $y = 3 - 0.6x$

| x | y | $\hat{y}$ | e | $e^2$ |
|---|---|---|---|---|
| 0 | 4 | 3.0 | 1.0 | 1.00 |
| 2 | 2 | 1.8 | 0.2 | 0.04 |
| 2 | 0 | 1.8 | -1.8 | 3.24 |
| 5 | -2 | 0.0 | -2.0 | 4.00 |
| 6 | 1 | -0.6 | 1.6 | 2.56 |
|   |   |   |   | 10.84 |

Line B: $y = 4 - x$

| x | y | $\hat{y}$ | e | $e^2$ |
|---|---|---|---|---|
| 0 | 4 | 4 | 0 | 0 |
| 2 | 2 | 2 | 0 | 0 |
| 2 | 0 | 2 | -2 | 4 |
| 5 | -2 | -1 | -1 | 1 |
| 6 | 1 | -2 | 3 | 9 |
|   |   |   |   | 14 |

(c) According to the least-squares criterion, Line A fits the set of data points better than Line B. This is because the sum of squared errors, i.e., $\sum e^2$, is smaller for Line A than for Line B.

**4.47** (a)

| x | y | xy | $x^2$ |
|---|---|----|----|
| 2 | 1 | 2  | 4  |
| 4 | 3 | 12 | 16 |
| 6 | 4 | 14 | 20 |

Using Formula 4.1,

$$S_{xx} = 20 - 6^2/2 = 2,\ S_{xy} = 14 - (6)(4)/2 = 2,\ b_1 = S_{xy}/S_{xx} = 2/2 = 1$$
$$b_0 = \bar{y} - b_1 \bar{x} = 2 - 1(3) = -1$$

Thus, the regression line is $\hat{y} = -1 + x$.

Without using Formula 4.1, both points must be on the line $y = b_0 + b_1 x$, so $\begin{cases} 1 = b_0 + b_1 2 \\ 3 = b_0 + b_1 4 \end{cases}$. Subtracting the first equation from the second, we get $2 = b_1 2$, or $b_1 = 1$. Substituting 1 back in the first equation for $b_1$, $1 = b_0 + 1(2)$, so $b_0 = -1$, and the equation of the line is y = -1 + x.

(b)

| x | y  | xy  | $x^2$ |
|---|----|-----|----|
| 1 | 3  | 3   | 1  |
| 5 | -3 | -15 | 25 |
| 6 | 0  | -12 | 26 |

Using Formula 4.1,

$$S_{xx} = 20 - 6^2/2 = 2,\ S_{xy} = -12 - (6)(0)/2 = -12,\ b_1 = S_{xy}/S_{xx} = -12/8 = -1.5$$
$$b_0 = \bar{y} - b_1 \bar{x} = 0 - (-1.5)(3) = 4.5$$

Thus, the regression line is $\hat{y} = 4.5 - 1.5x$.

Without using Formula 4.1, both points must be on the line $y = b_0 + b_1 x$, so $\begin{cases} 3 = b_0 + b_1 1 \\ -3 = b_0 + b_1 5 \end{cases}$. Subtracting the first equation from the second, we get $-6 = 4b_1$, or $b_1 = -1.5$. Substituting -1.5 back in the first equation for $b_1$, $3 = b_0 + (-1.5)(1)$, so $b_0 = 4.5$, and the equation of the line is y = 4.5 - 1.5x.

**4.49**

| x | y  | xy  | $x^2$ |
|---|----|-----|----|
| 3 | -4 | -12 | 9  |
| 1 | 0  | 0   | 1  |
| 2 | -5 | -10 | 4  |
| 6 | -9 | -22 | 14 |

## 124 Chapter 4

(a)
$$S_{xx} = \sum x^2 - \left(\sum x\right)^2 / n = 14 - 6^2/3 = 2;$$
$$S_{xy} = \sum xy - \left(\sum x\right)\left(\sum y\right)/n = -22 - (6)(-9)/3 = -4$$
$$b_1 = S_{xy}/S_{xx} = -4/2 = -2; \quad b_0 = \bar{y} - b_1\bar{x} = -3 - (-2)(2) = 1$$
$$\hat{y} = 1 - 2x$$

(b) Applying the regression equation for x = 1 and x = 3,

| x | $\hat{y}$ |
|---|---|
| 1 | −1 |
| 3 | −5 |

### 4.51

| x | y | xy | x² |
|---|---|---|---|
| 3 | 4 | 12 | 9 |
| 4 | 5 | 20 | 16 |
| 1 | 0 | 0 | 1 |
| 2 | −1 | −2 | 4 |
| 10 | 8 | 30 | 30 |

(a)
$$S_{xx} = \sum x^2 - \left(\sum x\right)^2 / n = 30 - 10^2/4 = 5;$$
$$S_{xy} = \sum xy - \left(\sum x\right)\left(\sum y\right)/n = 30 - (10)(8)/4 = 10$$
$$b_1 = S_{xy}/S_{xx} = 10/5 = 2; \quad b_0 = \bar{y} - b_1\bar{x} = 2 - (2)(2.5) = -3$$
$$\hat{y} = -3 + 2x$$

(b) Applying the regression equation for x = 1 and x = 4

| x | $\hat{y}$ |
|---|---|
| 1 | −1 |
| 4 | 5 |

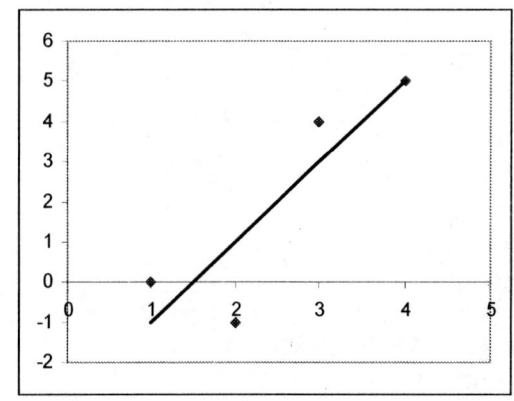

**4.53**

| x | y | xy | x² |
|---|---|----|-----|
| 2 | 3 | 6 | 4 |
| 2 | 4 | 8 | 4 |
| 3 | 0 | 0 | 9 |
| 4 | 2 | 8 | 16 |
| 4 | 1 | 4 | 16 |
| 15 | 10 | 26 | 49 |

(a)
$S_{xx} = \sum x^2 - (\sum x)^2 / n = 49 - 15^2/5 = 4;$
$S_{xy} = \sum xy - (\sum x)(\sum y)/n = 26 - (15)(10)/5 = -4$
$b_1 = S_{xy}/S_{xx} = -4/4 = -1;\quad b_0 = \bar{y} - b_1 \bar{x} = 2 - (-1)(3) = 5$
$\hat{y} = 5 - x$

(b) Applying the regression equation for x = 2 and x = 4

| x | $\hat{y}$ |
|---|---|
| 2 | 3 |
| 4 | 1 |

**4.55**

| x | y | xy | x² |
|---|---|----|-----|
| 1 | 4 | 4 | 1 |
| 2 | 3 | 6 | 4 |
| 3 | 8 | 24 | 9 |
| 6 | 15 | 34 | 14 |

(a)
$S_{xx} = \sum x^2 - (\sum x)^2 / n = 14 - 6^2/3 = 2;$
$S_{xy} = \sum xy - (\sum x)(\sum y)/n = 34 - (6)(15)/3 = 4$
$b_1 = S_{xy}/S_{xx} = 4/2 = 2;\quad b_0 = \bar{y} - b_1 \bar{x} = 5 - (2)(2) = 1$
$\hat{y} = 1 + 2x$

(b) Applying the regression equation for x = 1 and x = 3

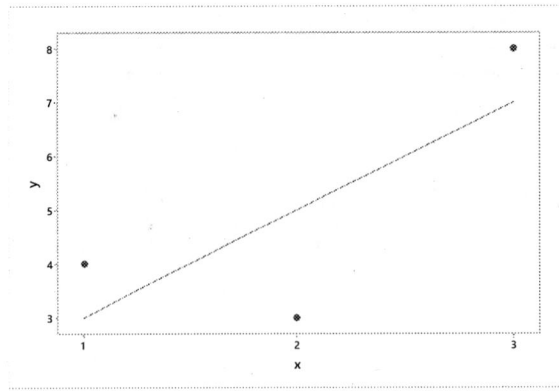

| x | $\hat{y}$ |
|---|---|
| 1 | 3 |
| 3 | 7 |

**4.57**

| x | y | xy | x2 |
|---|---|---|---|
| 0 | 4 | 0 | 0 |
| 2 | 2 | 4 | 4 |
| 2 | 0 | 0 | 4 |
| 5 | -2 | -10 | 25 |
| 6 | 1 | 6 | 36 |
| 15 | 5 | 0 | 69 |

(a)

$S_{xx} = \sum x^2 - (\sum x)^2/n = 69 - 15^2/5 = 24;$

$S_{xy} = \sum xy - (\sum x)(\sum y)/n = 0 - (15)(5)/5 = -15$

$b_1 = S_{xy}/S_{xx} = -15/24 = -0.625; \quad b_0 = \bar{y} - b_1 \bar{x} = 1 - (-0.625)(3) = 2.875$

$\hat{y} = 2.875 - 0.625x$

(b) Applying the regression equation for x = 0 and x = 6

| x | $\hat{y}$ |
|---|---|
| 0 | 2.875 |
| 6 | –0.875 |

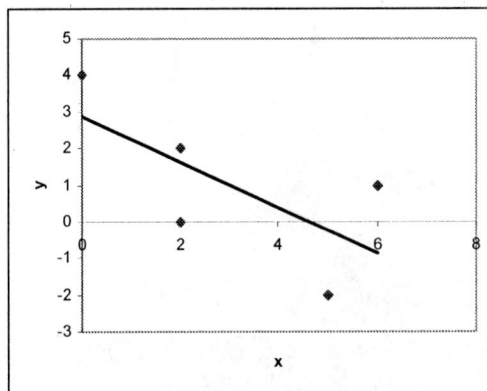

**4.59** (a) Formulas for the slope ($b_1$) and intercept ($b_0$) of the regression equation are, respectively:

$b_1 = \dfrac{\sum xy - \sum x \sum y/n}{\sum x^2 - (\sum x)^2/n} \qquad b_0 = \dfrac{1}{n}(\sum y - b_1 \sum x)$

To compute $b_0$ and $b_1$, construct a table of values for x, y, xy, $x^2$, and their sums:

| x | y | xy | $x^2$ |
|---|---|---|---|
| 6 | 290 | 1740 | 36 |
| 6 | 280 | 1680 | 36 |
| 6 | 295 | 1770 | 36 |
| 2 | 425 | 850 | 4 |
| 2 | 384 | 768 | 4 |
| 5 | 315 | 1575 | 25 |
| 4 | 355 | 1420 | 16 |
| 5 | 328 | 1640 | 25 |
| 1 | 425 | 425 | 1 |
| 4 | 325 | 1300 | 16 |
| 41 | 3422 | 13168 | 199 |

Thus: $b_1 = \dfrac{13168 - 41(3422)/10}{199 - 41^2/10} = -27.9029 \quad b_0 = \dfrac{1}{10}(3422 - (-27.9029)41) = 456.6019$

and the regression equation is $\hat{y} = 456.6019 - 27.9029x$

(b) Begin by selecting two x-values within the range of the x-data. For the x-values 1 and 6, the calculated values for y (in $hundreds) are, respectively:

$\hat{y} = 456.6019 - 27.9029(1) = 428.6990$

$\hat{y} = 456.6019 - 27.9029(6) = 289.1845$

The regression equation can be graphed by plotting the pairs (1, 428.699) and (6, 289.185) and connecting these points with a line. This equation, the original set of 10 data points, and all of the predicted values are shown in the following graph:

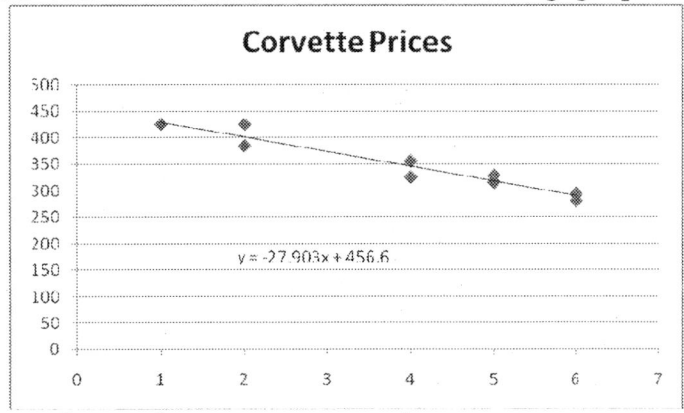

(c) Price tends to decrease as age increases.
(d) Corvettes depreciate an estimated $2,790.29 per year.
(e) The predictor variable is age. The response variable is price.
(f) There are no outliers or potential influential observations.
(g) Price of two-year old Corvette: $\hat{y} = 456.6019 - 27.9029(2) = 400.7961$

or $40,080.

Price of three-year old Corvette: $\hat{y} = 456.6019 - 27.9029(3) = 372.8932$

or $37,289.

**4.61** (a) Formulas for the slope ($b_1$) and intercept ($b_0$) of the regression equation are, respectively:

$$b_1 = \frac{\sum xy - \sum x \sum y / n}{\sum x^2 - (\sum x)^2 / n} \qquad b_0 = \frac{1}{n}\left(\sum y - b_1 \sum x\right)$$

To compute $b_0$ and $b_1$, construct a table of values for x, y, xy, $x^2$, and their sums:

| x | y | xy | $x^2$ |
|---|---|---|---|
| 57 | 8.0 | 456.0 | 3249 |
| 85 | 22.0 | 1870.0 | 7225 |
| 57 | 10.5 | 598.5 | 3249 |
| 65 | 22.5 | 1462.5 | 4225 |
| 52 | 12.0 | 624.0 | 2704 |
| 67 | 11.5 | 770.5 | 4489 |
| 62 | 7.5 | 465.0 | 3844 |
| 80 | 13.0 | 1040.0 | 6400 |
| 77 | 16.5 | 1270.5 | 5929 |
| 53 | 21.0 | 1113.0 | 2809 |
| 68 | 12.0 | 816.0 | 4624 |
| 723 | 156.5 | 10486.0 | 48747 |

$$b_1 = \frac{10486 - 723(1565)/11}{48747 - 723^2/11} = 0.16285 \qquad b_0 = \frac{1}{11}(1565 - 0.16285(723)) = 3.52369$$

and the regression equation is $\hat{y} = 3.52369 + 0.16285x$.

(b) Begin by selecting two x-values within the range of the x-data. For the x-values 52 and 85, the calculated values for y are, respectively:

$$\hat{y} = 3.52369 + 0.16285(52) = 11.992$$
$$\hat{y} = 3.52369 + 0.16285(85) = 17.366$$

The regression equation can be graphed by plotting the pairs (52, 12.000) and (85, 17.379) and connecting these points with a line. This equation and the original set of 11 data points are presented as follows:

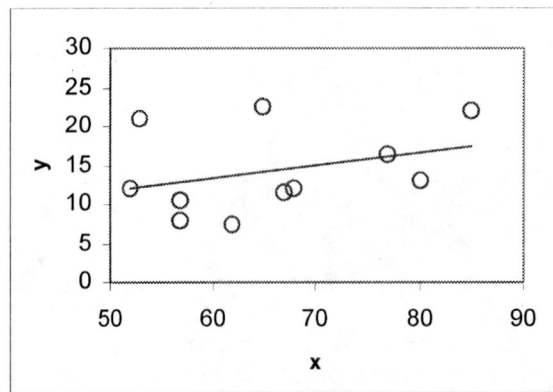

(c) The quantity of volatile compounds emitted tends to decrease as plant weight increases.

(d) The quantity of volatile compounds emitted increases by an estimated 0.163 hundred nanograms for each increase in plant weight of one gram.

(e) The predictor variable is plant weight. The response variable is hundreds of nanograms of volatile compounds emitted by the plant.
(f) There are no outliers or potential influential observations.
(g) The quantity of volatile compounds emitted by a 75 gram plant:
$\hat{y} = 3.52369 + 0.16285(75) = 15.737$ hundred nanograms or 1574 nanograms.

**4.63** (a) Formulas for the slope ($b_1$) and intercept ($b_0$) of the regression equation are, respectively:

$$b_1 = \frac{\sum xy - \sum x \sum y / n}{\sum x^2 - (\sum x)^2 / n} \qquad b_0 = \frac{1}{n}(\sum y - b_1 \sum x)$$

To compute $b_0$ and $b_1$, construct a table of values for x, y, xy, $x^2$, and their sums:

| x | y | xy | $x^2$ |
|---|---|------|------|
| 10 | 92 | 920 | 100 |
| 15 | 81 | 1215 | 225 |
| 12 | 84 | 1008 | 144 |
| 20 | 74 | 1480 | 400 |
| 8 | 85 | 680 | 64 |
| 16 | 80 | 1280 | 256 |
| 14 | 84 | 1176 | 196 |
| 22 | 80 | 1760 | 484 |
| 117 | 660 | 9519 | 1869 |

$$b_1 = \frac{9519 - 117(660)/8}{1869 - 117^2/8} = -0.84561 \qquad b_0 = \frac{1}{8}(660 - (-0.84561)(117)) = 94.86698$$

and the regression equation is $\hat{y} = 94.86698 - 0.84561x$.

(b) Begin by selecting two x-values within the range of the x-data. For the x-values 10 and 28, the calculated values for y are, respectively:
$\hat{y} = 94.86698 - 0.84561(10) = 86.41$
$\hat{y} = 94.86698 - 0.84561(22) = 76.26$

The regression equation can be graphed by plotting the pairs (10,68.40) and (28, 278.16) and connecting these points with a line. This equation and the original set of 10 data points are presented as follows:

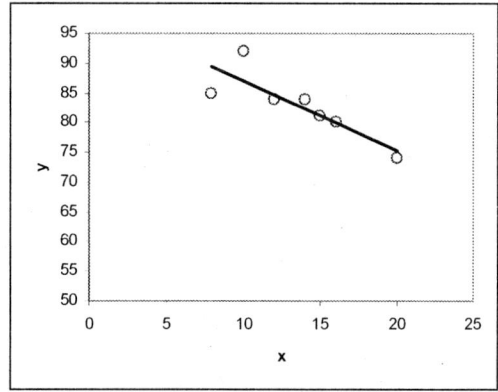

(c) The exam scores tend to decrease as the number of study hours increases.
(d) The exam score decreases, on average, about 0.85 points for each hour of study time.

(e) The predictor variable is study time, in hours. The response variable is the exam score.
(f) There may be two outliers, (8,85) and (10,92). There are no potential influential observations.
(g) For a student who studies 15 hours the predicted exam score is
$\hat{y} = 94.86698 - 0.84561(15) = 82.18$

**4.65** (a) It is acceptable to use the regression equation to predict the price of a four-year-old Corvette since that age lies within the range of the ages in the sample data. It is not acceptable (and would be extrapolation) to use the regression equation to predict the price of a 10-year-old Corvette since that age falls outside the range of the ages in the sample data.
(b) It is reasonable to use the regression equation to predict price for ages between one and six years old, inclusive.

**4.67** One possibility is that there are students who understand the material well after only a short period of study. These students do not need to study as long as students who have a more difficult time comprehending the material and possibly never reach the same level of understanding. There are other possibilities. Perhaps some students are more organized and do their studying when they are wide awake while others put it off until they are tired and less efficient.

**4.69** (a)

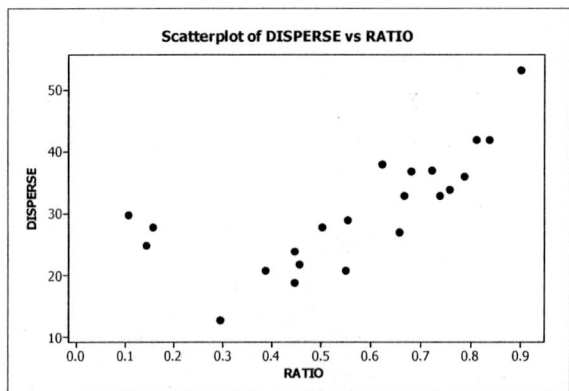

(b) No. The data obviously do not fall in a line.

**4.71** (a) Using Minitab, with the data in columns named INAUGURATION and DEATH, choose **Graph ▶ Scatterplot...**, select the **Simple** version and click **OK**. Enter <u>DEATH</u> in Row 1 of the **Y variables** column and <u>INAUGURATION</u> in the **X variables** column, and click **OK**. The result is

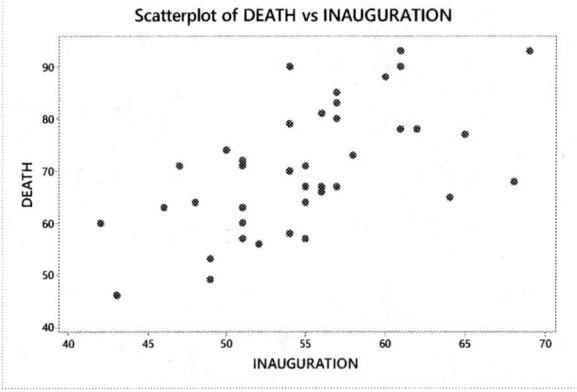

(b) A line appears to be reasonable, although there may be an outlier or two and an influential observation. There does not appear to be any curvature in the data.

(c) Choose **Stat ▶ Regression ▶ Regression,** enter DEATH in the **Response** text box, and INAUGURATION in the **Predictors** text box. Click **OK**. The results are

```
Analysis of Variance

Source             DF   Adj SS   Adj MS   F-Value   P-Value
Regression          1    2021   2020.57    21.45     0.000
  INAUGURATION      1    2021   2020.57    21.45     0.000
Error              37    3485     94.19
  Lack-of-Fit      19    2179    114.68     1.58     0.168
  Pure Error       18    1306     72.56
Total              38    5506

Model Summary

      S     R-sq   R-sq(adj)   R-sq(pred)
9.70514   36.70%     34.99%       28.71%


Coefficients

Term             Coef   SE Coef   T-Value   P-Value   VIF
Constant          5.7      14.1      0.40     0.690
INAUGURATION    1.179     0.255      4.63     0.000   1.00

Regression Equation
```

**DEATH = 5.7 + 1.179 INAUGURATION**

```
Fits and Diagnostics for Unusual Observations
                          Std
Obs   DEATH    Fit   Resid   Resid
 31   90.00  69.35   20.65    2.16   R
 39   93.00  87.04    5.96    0.67          X

R  Large residual
X  Unusual X
```

The regression equation (bold-faced in the output) indicates that the age at death is 5.7 years plus an increase of 1.179 years for every year later that inauguration occurs.

(d) There is one outlier, the point (54,90) and a potential influential observation (69,93)

(e) After removing the outlier, the regression equation becomes
   **DEATH = 4.4 + 1.193 INAUGURATION**
The effect of removing the outlier is to reduce the constant and increase the slope, both slightly.

(f) After replacing the outlier and removing the potential influential observation (69,93), the regression equation becomes
   **DEATH = 9.3 + 1.110 INAUGURATION**
Removing the influential observation increases the constant and reduces the slope, both significantly. In addition, part of the output is

```
      Obs   DEATH    Fit   Resid  Std Resid
        9   68.00  84.77  -16.77     -1.88      X
       31   90.00  69.23   20.77      2.15   R
```

This indicates that the point (54,90) is still an outlier, and there is now a new potential influential observation at (68,68).

**4.73** (a) Using Minitab, with the data in columns named LOT SIZE and VALUE EXPECTANCY, choose **Graph ▶ Scatterplot...**, select the **Simple** version and click **OK**. Enter VALUE in Row 1 of the **Y variables** column and 'LOT SIZE' in the **X variables** column, and click **OK**. The result is

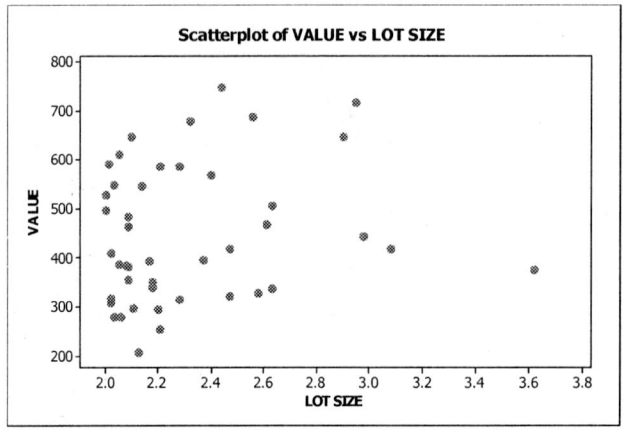

(b) The data points do not have a linear pattern, there is at least one influential observation, and there are likely some outliers. Therefore, find a regression line for these data is not reasonable.

**4.75** (a) Using Minitab, with the data in columns named HIGH and LOW, choose **Graph ▶ Scatterplot...**, select the **Simple** version and click **OK**. Enter LOW in Row 1 of the **Y variables** column and HIGH in the **X variables** column, and click **OK**. The result is

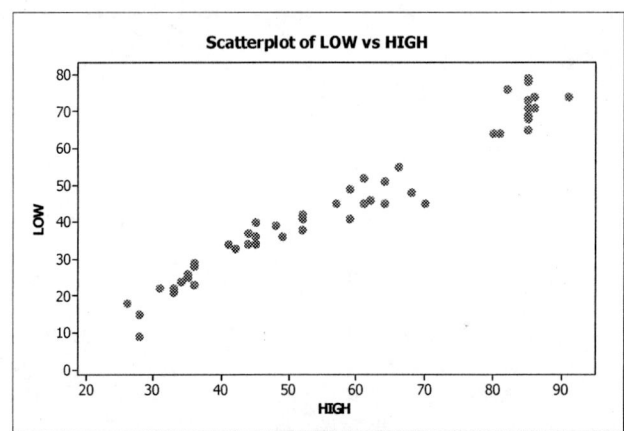

(b) It is reasonable to find a regression line for these data. There is a linear pattern without influential observations.

(c) Choose **Stat ▶ Regression ▶ Regression,** enter LOW in the **Response** text box, and HIGH in the **Predictors** text box. Click **OK**. The results are

```
The regression equation is
LOW = - 7.57 + 0.917 HIGH
```

```
Predictor       Coef    SE Coef       T       P
Constant      -7.572      1.786   -4.24   0.000
HIGH         0.91685    0.02967   30.91   0.000

S = 4.15111    R-Sq = 95.2%    R-Sq(adj) = 95.1%

Analysis of Variance

Source          DF       SS      MS       F       P
Regression       1    16459   16459  955.17   0.000
Residual Error  48      827      17
Total           49    17286

Unusual Observations

Obs  HIGH    LOW      Fit   SE Fit  Residual  St Resid
 15  70.0  45.000   56.607   0.705   -11.607    -2.84R
 17  82.0  76.000   67.610   0.949     8.390     2.08R
 30  85.0  79.000   70.360   1.021     8.640     2.15R
 45  28.0   9.000   18.100   1.038    -9.100    -2.26R

R denotes an observation with a large standardized residual.
```

As one would expect, there is positive linear relationship between the high and low average temperatures.

(d) Observations 15, 17, 30 and 45 are all potential outliers. There are no influential observations.

(e) Delete the four data points mentioned in part (d), starting with number 45 and ending with number 15. Repeat the regression procedure. The result is

```
The regression equation is
LOW = - 5.66 + 0.884 HIGH

Predictor       Coef    SE Coef       T       P
Constant      -5.655      1.442   -3.92   0.000
HIGH         0.88407    0.02433   36.34   0.000

S = 3.18323    R-Sq = 96.8%    R-Sq(adj) = 96.7%

Analysis of Variance

Source          DF       SS      MS        F       P
Regression       1    13381   13381  1320.50   0.000
Residual Error  44      446      10
Total           45    13826

Unusual Observations

Obs  HIGH    LOW      Fit   SE Fit  Residual  St Resid
 32  68.0  48.000   54.462   0.552    -6.462    -2.06R
 37  85.0  78.000   69.491   0.847     8.509     2.77R

R denotes an observation with a large standardized residual.
```

Deleting the outliers resulted in an increase in the intercept and a slight decrease in the slope of the regression line. It also resulted in two new potential outliers.

(f) Not applicable

**4.77** (a) Using Minitab, with the data in columns named INCOME and BEER, choose **Graph ▶ Scatterplot...**, select the **Simple** version and click **OK**.

**134** Chapter 4

Enter BEER in Row 1 of the **Y variables** column and INCOME in the **X variables** column, and click **OK**. The result is

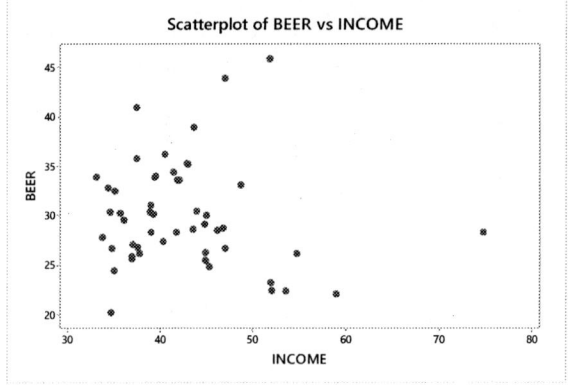

(b) There is no linear pattern in these data, so it is inappropriate to carry out a linear regression. Parts (c)-(f) are omitted.

**4.79** (a) Using Minitab, with the data in columns named MANAGERS and ASSETS, choose **Graph ▶ Scatterplot...**, select the **Simple** version and click **OK**. Enter MANAGERS in Row 1 of the **Y variables** column and ASSETS in the **X variables** column, and click **OK**. When you see the scatterplot, you might wonder just where the regression line will go. To see, put the mouse cursor on the graph and right click. Select Add ▶ Regression fit…, select the button for Linear and check the box for Fit intercept. Click OK twice. The result is (You could reverse the roles of ASSETS and MANAGERS, but the basic conclusion will remain the same.)

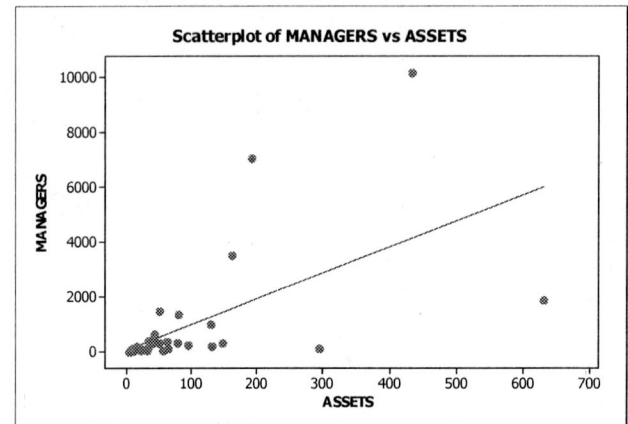

(b) Clearly, there will be outliers and influential data points if we fit a line to these data. A linear regression fit will be of little use, so we omit parts (c-f). [You can do them if you wish just to confirm this opinion.]

**4.81** (a) From Exercise 4.45 and 4.57,

$$n = 5; \ \sum x_i = 15; \ \sum y_i = 5; \ \sum x_i y_i = 0; \ \sum x_i^2 = 69$$

Therefore, $s_{xy} = \dfrac{\sum x_i y_i - (\sum x_i)(\sum y_i)/n}{n-1} = \dfrac{0 - (15)(5)/5}{4} = -15/4 =$ Covariance

(b) $s_x^2 = \dfrac{\sum x_i^2 - (\sum x_i)^2/n}{n-1} = \dfrac{69 - (15)^2/5}{4} = 6$

$b_1 = s_{xy}/s_x^2 = (-15/4)/(6) = -0.625;\ b_0 = \bar{y} - b_1\bar{x} = 1 - (-0.625)(3) = 2.875$

$\hat{y} = 2.875 - 0.625x$

This is the same result as was obtained in Exercise 4.57.

### Exercises 4.3

**4.83** (a) The coefficient of determination is a descriptive measure of the utility of the regression equation for making predictions. The symbol for the coefficient of determination is $r^2$.

(b) Two interpretations of the coefficient of determination are
(i) It represents the percentage reduction obtained in the total squared error by using the regression equation to predict the observed values, instead of simply using the mean of the y-values.
(ii) It represents the percentage of variation in the observed y-values that is explained by the regression.

**4.85** A measure of the amount of variation in the observed values of the response variable that is explained by the regression is the <u>regression sum of squares</u>. The mathematical abbreviation for it is <u>SSR</u>.

**4.87** (a) The coefficient of determination is $r^2$ = SSR/SST = 7626.6/8291.0 = 0.9199. 91.99% of the variation in the observed values of the response variable is explained by the regression. The fact that $r^2$ is near 1 indicates that the regression equation is extremely useful for making predictions.

(b) SSE = SST - SSR = 8291.0 - 7626.6 = 664.4

**4.89** To use the defining formulas, we begin with the following table.

| x | y | $(y-\bar{y})$ | $(y-\bar{y})^2$ | $\hat{y}$ | $(\hat{y}-\bar{y})$ | $(\hat{y}-\bar{y})^2$ | $(y-\hat{y})$ | $(y-\hat{y})^2$ |
|---|---|---|---|---|---|---|---|---|
| 3 | -4 | -1.0 | 1 | -5 | -2.0 | 4.00 | 1 | 1 |
| 1 | 0 | 3.0 | 9 | -1 | 2.0 | 4.00 | 1 | 1 |
| 2 | -5 | -2.0 | 4 | -3 | 0.0 | 0.00 | -2 | 4 |
| 6 | -9 | 0.0 | 14 |   | 0.0 | 8.00 | 0 | 6 |

Each $\hat{y}$ value is obtained by substituting the respective value of x into the regression equation $\hat{y} = 1 - 2x$.

(a) $SST = \sum(y-\bar{y})^2 = 14 \qquad SSR = \sum(\hat{y}-\bar{y})^2 = 8 \qquad SSE = \sum(y-\hat{y})^2 = 6$

(b) 14 = 8 + 6

(c) $r^2 = 1 - \dfrac{SSE}{SST} = 1 - \dfrac{6}{14} = \dfrac{4}{7} = 0.571 \quad \text{or} \quad r^2 = \dfrac{SSR}{SST} = \dfrac{8}{14} = 0.571$

(d) The percentage of variation in the observed y-values that is explained by the regression is 57.1%.

(e) Based on the answer in part (d), the regression equation is somewhat useful for making predictions.

**4.91** To use the defining formulas, we begin with the following table.

| x | y | $(y-\bar{y})$ | $(y-\bar{y})^2$ | $\hat{y}$ | $(\hat{y}-\bar{y})$ | $(\hat{y}-\bar{y})^2$ | $(y-\hat{y})$ | $(y-\hat{y})^2$ |
|---|---|---|---|---|---|---|---|---|
| 3 | 4 | 2.0 | 4 | 3 | 1.0 | 1.00 | 1 | 1 |
| 4 | 5 | 3.0 | 9 | 5 | 3.0 | 9.00 | 0 | 0 |
| 1 | 0 | -2.0 | 4 | -1 | -3.0 | 9.00 | 1 | 1 |
| 2 | -1 | -3.0 | 9 | 1 | -1.0 | 1.00 | -2 | 4 |
| 10 | 8 | 0.0 | 26 | | 0.0 | 20.00 | 0 | 6 |

Each $\hat{y}$ value is obtained by substituting the respective value of x into the regression equation $\hat{y} = -3 + 2x$.

(a) $SST = \sum(y-\bar{y})^2 = 26$  $SSR = \sum(\hat{y}-\bar{y})^2 = 20$  $SSE = \sum(y-\hat{y})^2 = 6$

(b) 46 = 40 + 6

(c) $r^2 = 1 - \dfrac{SSE}{SST} = 1 - \dfrac{6}{26} = \dfrac{10}{13} = 0.769$  or  $r^2 = \dfrac{SSR}{SST} = \dfrac{20}{26} = 0.769$

(d) The percentage of variation in the observed y-values that is explained by the regression is 76.9%.

(e) Based on the answer in part (d), the regression equation is useful for making predictions.

**4.93** To use the defining formulas, we begin with the following table.

| x | y | $(y-\bar{y})$ | $(y-\bar{y})^2$ | $\hat{y}$ | $(\hat{y}-\bar{y})$ | $(\hat{y}-\bar{y})^2$ | $(y-\hat{y})$ | $(y-\hat{y})^2$ |
|---|---|---|---|---|---|---|---|---|
| 2 | 3 | 1 | 1 | 3 | 1 | 1 | 0 | 0 |
| 2 | 4 | 2 | 4 | 3 | 1 | 1 | 1 | 1 |
| 3 | 0 | -2 | 4 | 2 | 0 | 0 | -2 | 4 |
| 4 | 2 | 0 | 0 | 1 | -1 | 1 | 1 | 1 |
| 4 | 1 | -1 | 1 | 1 | -1 | 1 | 0 | 0 |
| 15 | 10 | 0 | 10 | | 0 | 4 | 0 | 6 |

Each $\hat{y}$ value is obtained by substituting the respective value of x into the regression equation $\hat{y} = 5 - x$.

(a) $SST = \sum(y-\bar{y})^2 = 10$  $SSR = \sum(\hat{y}-\bar{y})^2 = 4$  $SSE = \sum(y-\hat{y})^2 = 6$

(b) 10 = 4 + 6

(c) $r^2 = 1 - \dfrac{SSE}{SST} = 1 - \dfrac{6}{10} = 0.4$  or  $r^2 = \dfrac{SSR}{SST} = \dfrac{4}{10} = 0.4$

(d) The percentage of variation in the observed y-values that is explained by the regression is 40%.

(e) Based on the answer in part (c), the regression equation is moderately useful for making predictions.

**4.95** To use the defining formulas, we begin with the following table.

| x | y | $(y-\bar{y})$ | $(y-\bar{y})^2$ | $\hat{y}$ | $(\hat{y}-\bar{y})$ | $(\hat{y}-\bar{y})^2$ | $(y-\hat{y})$ | $(y-\hat{y})^2$ |
|---|---|---|---|---|---|---|---|---|
| 1 | 4 | -1 | 1 | 3 | -2 | 4 | 1 | 1 |
| 2 | 3 | -2 | 4 | 5 | 0 | 0 | -2 | 4 |
| 3 | 8 | 3 | 9 | 7 | 2 | 4 | 1 | 1 |
| 6 | 15 | 0 | 14 |   | 0 | 8 | 0 | 6 |

Each $\hat{y}$ value is obtained by substituting the respective value of x into the regression equation $\hat{y}=1+2x$.

(a) $SST = \sum(y-\bar{y})^2 = 14$   $SSR = \sum(\hat{y}-\bar{y})^2 = 8$   $SSE = \sum(y-\hat{y})^2 = 6$

(b) 14 = 8 + 6

(c) $r^2 = 1 - \dfrac{SSE}{SST} = 1 - \dfrac{6}{14} = 0.571$   or   $r^2 = \dfrac{SSR}{SST} = \dfrac{8}{14} = 0.571$

(d) The percentage of variation in the observed y-values that is explained by the regression is 57.1%.

(e) Based on the answer in part (c), the regression equation is moderately useful for making predictions.

**4.97** To use the defining formulas, we begin with the following table.

| x | Y | $(y-\bar{y})$ | $(y-\bar{y})^2$ | $\hat{y}$ | $(\hat{y}-\bar{y})$ | $(\hat{y}-\bar{y})^2$ | $(y-\hat{y})$ | $(y-\hat{y})^2$ |
|---|---|---|---|---|---|---|---|---|
| 0 | 4 | 3 | 9 | 2.875 | 1.875 | 3.51563 | 1.125 | 1.26563 |
| 2 | 2 | 1 | 1 | 1.625 | 0.625 | 0.39063 | 0.375 | 0.14063 |
| 2 | 0 | -1 | 1 | 1.625 | 0.625 | 0.39063 | -1.625 | 2.64062 |
| 5 | -2 | -3 | 9 | -0.250 | -1.250 | 1.56250 | -1.750 | 3.06250 |
| 6 | 1 | 0 | 0 | -0.875 | -1.875 | 3.51563 | 1.875 | 3.51563 |
| 15 | 5 | 0 | 20 |   | 0 | 9.375 | 0 | 10.625 |

Each y value is obtained by substituting the respective value of x into the regression equation $\hat{y} = 2.875 - 0.625x$.

(a) $SST = \sum(y-\bar{y})^2 = 20$   $SSR = \sum(\hat{y}-\bar{y})^2 = 9.375$   $SSE = \sum(y-\hat{y})^2 = 10.625$

(b) 20 = 9.365 + 10.625

(c) $r^2 = 1 - \dfrac{SSE}{SST} = 1 - \dfrac{10.625}{20} = 0.46875$   or   $r^2 = \dfrac{SSR}{SST} = \dfrac{6.375}{20} = 0.46875$

(d) The percentage of variation in the observed y-values that is explained by the regression is 46.875%.

(e) Based on the answer in part (c), the regression equation appears to be moderately useful for making predictions.

**4.99** To use the computing formulas, we begin with the following table. Notice that columns 1-4 are presented in the solution to part (a) of Exercise 4.59, so that only the column for y² needs to be developed.

| x | y | xy | x² | y² |
|---|---|---|---|---|
| 6 | 290 | 1740 | 36 | 84100 |
| 6 | 280 | 1680 | 36 | 78400 |
| 6 | 295 | 1770 | 36 | 87025 |
| 2 | 425 | 850 | 4 | 180625 |
| 2 | 384 | 768 | 4 | 147456 |
| 5 | 315 | 1575 | 25 | 99225 |
| 4 | 355 | 1420 | 16 | 126025 |
| 5 | 328 | 1640 | 25 | 107584 |
| 1 | 425 | 425 | 1 | 180625 |
| 4 | 325 | 1300 | 16 | 105625 |
| 41 | 3422 | 13168 | 199 | 1196690 |

(a) Using the last row of the table and Formula 4.2 of the text, we obtain the three sums of squares as follows.

$$SST = S_{yy} = \sum y^2 - (\sum y)^2/n = 1196690 - 3422^2/10 = 25681.6$$

$$SSR = \frac{S_{xy}^2}{S_{xx}} = \frac{[\sum xy - (\sum x)(\sum y)/n]^2}{\sum x^2 - (\sum x)^2/n} = \frac{[13168 - (41)(3422)/10]^2}{199 - 41^2/10} = 24057.9$$

$$SSE = S_{yy} - \frac{S_{xy}^2}{S_{xx}} = 25681.6 - 24057.9 = 1623.7$$

(b) $r^2 = \dfrac{SSR}{SST} = \dfrac{24057.9}{25681.6} = 0.9368$

(c) The percentage of variation in the observed y-values that is explained by the regression is 93.68%. In words, 93.68% of the variation in the tax efficiency data is explained by the percentage of the mutual funds investments that are in energy securities.

(d) Based on the answers to parts (b) and (c), the regression equation appears to be very useful for making predictions.

**4.101** To use the computing formulas, we begin with the following table. Notice that columns 1-4 are presented in the solution to part (a) of Exercise 4.61, so that only the column for y² needs to be developed.

| x | y | xy | x² | y² |
|---|---|---|---|---|
| 57 | 8.0 | 456.0 | 3249 | 64.00 |
| 85 | 22.0 | 1870.0 | 7225 | 484.00 |
| 57 | 10.5 | 598.5 | 3249 | 110.25 |
| 65 | 22.5 | 1462.5 | 4225 | 506.25 |
| 52 | 12.0 | 624.0 | 2704 | 144.00 |
| 67 | 11.5 | 770.5 | 4489 | 132.25 |
| 62 | 7.5 | 465.0 | 3844 | 56.25 |
| 80 | 13.0 | 1040.0 | 6400 | 169.00 |
| 77 | 16.5 | 1270.5 | 5929 | 272.25 |
| 53 | 21.0 | 1113.0 | 2809 | 441.00 |
| 68 | 12.0 | 816.0 | 4624 | 144.00 |
| 723 | 156.5 | 10486.0 | 48747 | 2523.25 |

(a) Using the last row of the table and Formula 4.2 of the text, we obtain the three sums of squares as follows.

$$SST = S_{yy} = \sum y^2 - (\sum y)^2/n = 2523.25 - 156.5^2/11 = 296.68$$

$$SSR = \frac{S_{xy}^2}{S_{xx}} = \frac{[\sum xy - (\sum x)(\sum y)/n]^2}{\sum x^2 - (\sum x)^2/n} = \frac{[10486 - (723)(1565)/11]^2}{48747 - 723^2/11} = 32.52$$

$$SSE = S_{yy} - \frac{S_{xy}^2}{S_{xx}} = 296.68 - 32.52 = 264.16$$

(b) $r^2 = \dfrac{SSR}{SST} = \dfrac{32.52}{7296.68} = 0.1096$

(c) The percentage of variation in the observed y-values that is explained by the regression is 10.96%. In words, only 10.96% of the variation in the price data is explained by size.

(d) Based on the answers to parts (b) and (c), the regression equation appears to be useless for making predictions.

**4.103** To use the computing formulas, we begin with the following table. Notice that columns 1-4 are presented in the solution to part (a) of Exercise 4.63, so that only the column for y² needs to be developed.

| x | y | xy | x² | y² |
|---|---|----|----|----|
| 10 | 92 | 920 | 100 | 8464 |
| 15 | 81 | 1215 | 225 | 6561 |
| 12 | 84 | 1008 | 144 | 7056 |
| 20 | 74 | 1480 | 400 | 5476 |
| 8 | 85 | 680 | 64 | 7225 |
| 16 | 80 | 1280 | 256 | 6400 |
| 14 | 84 | 1176 | 196 | 7056 |
| 22 | 80 | 1760 | 484 | 6400 |
| 117 | 660 | 9519 | 1869 | 54638 |

(a) Using the last row of the table and Formula 4.2 of the text, we obtain the three sums of squares as follows.

$$SST = S_{yy} = \sum y^2 - (\sum y)^2/n = 54638 - 660^2/8 = 188$$

$$SSR = \frac{S_{xy}^2}{S_{xx}} = \frac{[\sum xy - (\sum x)(\sum y)/n]^2}{\sum x^2 - (\sum x)^2/n} = \frac{[9519 - (117)(660)/8]^2}{1869 - 117^2/8} = 112.89$$

$$SSE = S_{yy} - \frac{S_{xy}^2}{S_{xx}} = 188 - 112.89 = 75.11$$

(b) $r^2 = \dfrac{SSR}{SST} = \dfrac{112.89}{188} = 0.6005$

(c) The percentage of variation in the observed y-values that is explained by the regression is 60.05%. In words, 60.05% of the variation in the score data is explained by study-time.

(d) Based on the answers to parts (b) and (c), the regression equation appears to be somewhat useful for making predictions.

**4.105** (a) In Exercise 4.71, we saw from the scatterplot that finding a regression line for these data was reasonable.

(b) From the regression output in that exercise, the coefficient of determination (shown as R-Sq) was found to be 0.367.

(c) The percentage of variation in the observed scores explained by the regression line is 36.7%; i.e., 36.7% of the variation in the death ages was explained by inauguration ages.

(d) Although the pattern is linear, the variation around the fitted line is so great that the regression equation is only marginally useful for making predictions.

**4.107** (a) In Exercise 4.73, we saw from the scatterplot that finding a regression line for these data was not reasonable. Parts (b-d) are omitted.

**140** Chapter 4

**4.109** (a) In Exercise 4.75, we saw from the scatterplot that finding a regression line for these data was very reasonable.
(b) From the regression output in that exercise, the coefficient of determination (shown as R-Sq) was found to be 0.952.
(c) The percentage of variation in the observed values explained by the regression line is 95.2%; i.e., 95.2% of the variation in the average low temperatures was explained by the average high temperatures.
(d) The regression equation is very useful for making predictions.

**4.111** (a) Using Minitab, with the data in columns named INCOME and BEER, we choose **Graph ▶ Scatterplot...**, select the **Simple** version and click **OK**. Enter BEER in the first row of the **Y variables** column and INCOME in the first row of the **X variables** column. Click **OK**. The result is

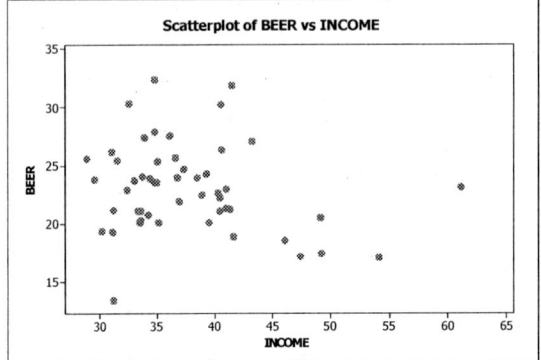

There does not appear to be a linear relationship between beer consumption and income. It is not reasonable to find a regression line. Parts (b-d) are omitted.

**4.113** (a) In Exercise 4.80, we saw from the scatterplot that finding a regression line for these data was not reasonable because the data exhibited an increasing concave upward curved patternn. Parts (b-d) are omitted.

**4.115** (a) Using Minitab, with the data in columns named ESTRIOL and WEIGHT, we choose **Graph ▶ Scatterplot...**, select the **Simple** version and click **OK**. Enter WEIGHT in the first row of the **Y variables** column and ESTRIOL in the first row of the **X variables** column. Click **OK**. The result is

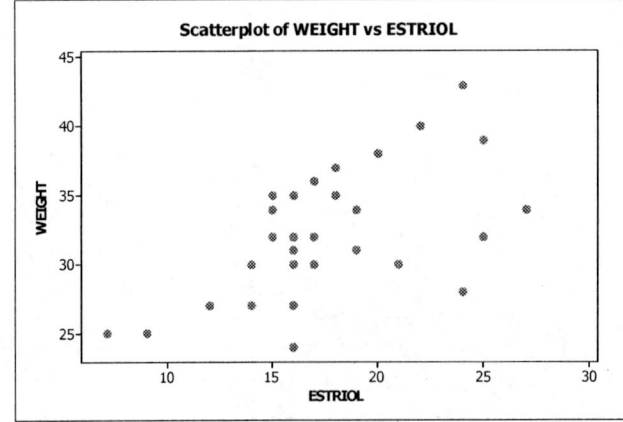

It appears that there is a linear relationship between Estriol and Weight. It is reasonable to find a linear regression equation.

(b) Using Minitab, with the data in columns named ESTRIOL and WEIGHT, we choose **Stat ▶ Regression ▶ Regression...**, select WEIGHT in the **Response** text box, select ESTRIOL in the **Predictors** text box, and click **OK**. The results are

Copyright © 2016 Pearson Education, Inc.

```
The regression equation is
WEIGHT = 21.5 + 0.608 ESTRIOL

Predictor        Coef         StDev            T            P
Constant       21.523         2.620         8.21        0.000
ESTRIOL         0.6082        0.1468        4.14        0.000

S = 3.821      R-Sq = 37.2%    R-Sq(adj) = 35.0%

Analysis of Variance

Source             DF          SS           MS            F          P
Regression          1       250.57       250.57        17.16      0.000
Residual Error     29       423.43        14.60
Total              30       674.00
```
Thus the coefficient of determination is $r^2 = 0.372$.

(c) Only 37.2% of the variation in the observed values of birth weight is explained by a linear relationship with the predictor variable, estriol level.

(d) Based solely on the coefficient of determination, the regression equation is not very useful for making predictions.

**4.117** (a) $r^2 = 1 - SSE/SST = (SST - SSE)/SST$. If the mean were used to predict the observed values of the response variable, the total squared error would be SST. By using the regression line to predict the observed values of the response variable, the sum of squares of the differences between the predicted values and the mean is $SSR = SST - SSE$. This is a reduction in the error sum of squares. Dividing this quantity by SST (and converting to a percentage) gives us the percentage reduction in the squared error when we use the regression equation instead of the mean to predict the observed values of the response variable.

(b) From Exercise 4.99, $r^2 = 0.9368$, so the percentage reduction obtained in the total squared error by using the regression equation instead of the mean of observed prices to predict the observed prices is 93.68%.

### Exercises 4.4

**4.119** Pearson product moment correlation coefficient

**4.121** strong

**4.123** $\pm 1$

**4.125** positively

**4.127** uncorrelated

**4.129** negative

**4.131** False. It is possible that both variables are effects from a third variable. Correlation does not imply causation.

**4.133** The sign of the slope and of r are always the same. This can be seen from $r = \dfrac{S_{xy}}{\sqrt{S_{xx}S_{yy}}} = \dfrac{S_{xy}}{S_{xx}}\dfrac{\sqrt{S_{xx}}}{\sqrt{S_{yy}}} = b_1 \dfrac{\sqrt{S_{xx}}}{\sqrt{S_{yy}}}$.

Since the ratio of square roots in the last term is always positive, r will have the same sign as $b_1$. Since the slope of the regression line is $-3.58$, r will also be negative. Therefore, $r = -\sqrt{r^2} = -\sqrt{0.709} = -0.842$

**4.135** (a) To compute the linear correlation coefficient using the defining formula, begin with the following table.

| $x$ | $y$ | $x - \bar{x}$ | $y - \bar{y}$ | $(x - \bar{x})^2$ | $(y - \bar{y})^2$ | $(x - \bar{x})(y - \bar{y})$ |
|---|---|---|---|---|---|---|
| 3 | -4 | 1 | -1 | 1 | 1 | -1 |
| 1 | 0 | -1 | 3 | 1 | 9 | -3 |
| 2 | -5 | 0 | -2 | 0 | 4 | 0 |
| 6 | -9 | | | 2 | 14 | -4 |

**142** Chapter 4

The columns after the first two have used $\bar{x} = 2$ and $\bar{y}=-3$.
We will need $s_x$ and $s_y$. These are given by

$$s_x = \sqrt{\frac{\sum(x_i - \bar{x})^2}{n-1}} = \sqrt{\frac{2}{2}} = 1 \text{ and } s_y = \sqrt{\frac{\sum(y_i - \bar{y})^2}{n-1}} = \sqrt{\frac{14}{2}} = 2.646$$

$$r = \frac{\frac{1}{n-1}\sum(x_i - \bar{x})(y_i - \bar{y})}{s_x s_y} = \frac{\frac{1}{2}(-4)}{(1)(2)} = \frac{-2}{2.646} = -0.76$$

(b) To compute the linear correlation coefficient using the computing formula, begin with the following table.

| $x$ | $y$ | $xy$ | $x^2$ | $y^2$ |
|---|---|---|---|---|
| 3 | -4 | -12 | 9 | 16 |
| 1 | 0 | 0 | 1 | 0 |
| 2 | -5 | -10 | 4 | 25 |
| 6 | -9 | -22 | 14 | 41 |

$$r = \frac{\sum x_i y_i - (\sum x_i)(\sum y_i)/n}{\sqrt{\left[\sum x_i^2 - (\sum x_i)^2/n\right]\left[\sum y_i^2 - (\sum y_i)^2/n\right]}}$$

$$= \frac{-22 - (6)(-9)/3}{\sqrt{[14 - 6^2/3][41 - (-9)^2/3]}} = \frac{-4}{\sqrt{28}} = -0.76$$

The answers to both the defining and computing formula are equivalent.

**4.137** (a) To compute the linear correlation coefficient using the defining formula, begin with the following table.

| $x$ | $y$ | $x-\bar{x}$ | $y-\bar{y}$ | $(x-\bar{x})^2$ | $(y-\bar{y})^2$ | $(x-\bar{x})(y-\bar{y})$ |
|---|---|---|---|---|---|---|
| 3 | 4 | 0.5 | 2 | 0.25 | 4 | 1.0 |
| 4 | 5 | 1.5 | 3 | 2.25 | 9 | 4.5 |
| 1 | 0 | -1.5 | -2 | 2.25 | 4 | 3.0 |
| 2 | -1 | -0.5 | -3 | 0.25 | 9 | 1.5 |
| 10 | 8 | | | 5.00 | 26 | 10.0 |

The columns after the first two have used $\bar{x} = 2.5$ and $\bar{y}=2$.
We will need $s_x$ and $s_y$. These are given by

$$s_x = \sqrt{\frac{\sum(x_i - \bar{x})^2}{n-1}} = \sqrt{\frac{5}{3}} = 1.291 \text{ and } s_y = \sqrt{\frac{\sum(y_i - \bar{y})^2}{n-1}} = \sqrt{\frac{26}{3}} = 2.944$$

$$r = \frac{\frac{1}{n-1}\sum(x_i - \bar{x})(y_i - \bar{y})}{s_x s_y} = \frac{\frac{1}{3}(10)}{(1.291)(2.944)} = \frac{10/3}{3.801} = 0.877$$

(b) To compute the linear correlation coefficient using the computing formula, begin with the following table.

| $x$ | $y$ | $xy$ | $x^2$ | $y^2$ |
|---|---|---|---|---|
| 4 | 5 | 20 | 16 | 25 |
| 3 | 4 | 12 | 9 | 16 |
| 1 | 0 | 0 | 1 | 0 |
| 2 | -1 | -2 | 4 | 1 |
| 10 | 8 | 30 | 30 | 42 |

$$r = \frac{\sum x_i y_i - (\sum x_i)(\sum y_i)/n}{\sqrt{[\sum x_i^2 - (\sum x_i)^2/n][\sum y_i^2 - (\sum y_i)^2/n]}}$$

$$= \frac{30 - (10)(8)/4}{\sqrt{[30 - 10^2/4][42 - (8)^2/4]}} = \frac{10}{\sqrt{130}} = 0.877$$

The answers to both the defining and computing formula are equivalent.

**4.139** (a) To compute the linear correlation coefficient using the defining formula, begin with the following table.

| $x$ | $y$ | $x - \bar{x}$ | $y - \bar{y}$ | $(x - \bar{x})^2$ | $(y - \bar{y})^2$ | $(x - \bar{x})(y - \bar{y})$ |
|---|---|---|---|---|---|---|
| 2 | 3 | -1 | 1 | 1 | 1 | -1 |
| 2 | 4 | -1 | 2 | 1 | 4 | -2 |
| 3 | 0 | 0 | -2 | 0 | 4 | 0 |
| 4 | 2 | 1 | 0 | 1 | 0 | 0 |
| 4 | 1 | 1 | -1 | 1 | 1 | -1 |
| 15 | 10 |  |  | 4 | 10 | -4 |

The columns after the first two have used $\bar{x} = 3$ and $\bar{y} = 2$. We will need $s_x$ and $s_y$. These are given by

$$s_x = \sqrt{\frac{\sum(x_i - \bar{x})^2}{n-1}} = \sqrt{\frac{4}{4}} = 1.0 \text{ and } s_y = \sqrt{\frac{\sum(y_i - \bar{y})^2}{n-1}} = \sqrt{\frac{10}{4}} = 1.581$$

$$r = \frac{\frac{1}{n-1}\sum(x_i - \bar{x})(y_i - \bar{y})}{s_x s_y} = \frac{\frac{1}{4}(-4)}{(1.0)(1.581)} = -0.633$$

(b) To compute the linear correlation coefficient using the computing formula, begin with the following table.

| $x$ | $y$ | $xy$ | $x^2$ | $y^2$ |
|---|---|---|---|---|
| 2 | 3 | 6 | 4 | 9 |
| 2 | 4 | 8 | 4 | 16 |
| 3 | 0 | 0 | 9 | 0 |
| 4 | 2 | 8 | 16 | 4 |
| 4 | 1 | 4 | 16 | 1 |
| 15 | 10 | 26 | 49 | 30 |

$$r = \frac{\sum x_i y_i - (\sum x_i)(\sum y_i)/n}{\sqrt{[\sum x_i^2 - (\sum x_i)^2/n][\sum y_i^2 - (\sum y_i)^2/n]}}$$

$$= \frac{26 - (15)(10)/5}{\sqrt{[49 - 15^2/5][30 - (10)^2/5]}} = -0.632$$

The answers to the defining and computing formula are only slightly different due to rounding.

**4.141** (a) To compute the linear correlation coefficient using the defining formula, begin with the following table.

| $x$ | $y$ | $x - \bar{x}$ | $y - \bar{y}$ | $(x - \bar{x})^2$ | $(y - \bar{y})^2$ | $(x - \bar{x})(y - \bar{y})$ |
|---|---|---|---|---|---|---|
| 1 | 4 | -1 | -1 | 1 | 1 | 1 |
| 2 | 3 | 0 | -2 | 0 | 4 | 0 |
| 3 | 8 | 1 | 3 | 1 | 9 | 3 |
| 6 | 15 |  |  | 2 | 14 | 4 |

The columns after the first two have used $\bar{x}=2$ and $\bar{y}=5$.
We will need $s_x$ and $s_y$. These are given by

$$s_x = \sqrt{\frac{\sum(x_i-\bar{x})^2}{n-1}} = \sqrt{\frac{2}{2}} = 1.0 \text{ and } s_y = \sqrt{\frac{\sum(y_i-\bar{y})^2}{n-1}} = \sqrt{\frac{14}{2}} = 2.646$$

$$r = \frac{\frac{1}{n-1}\sum(x_i-\bar{x})(y_i-\bar{y})}{s_x s_y} = \frac{\frac{1}{2}(4)}{(1.0)(2.646)} = 0.756$$

(b) To compute the linear correlation coefficient using the computing formula, begin with the following table.

| $x$ | $y$ | $xy$ | $x^2$ | $y^2$ |
|---|---|---|---|---|
| 1 | 4 | 4 | 1 | 16 |
| 2 | 3 | 6 | 4 | 9 |
| 3 | 8 | 24 | 9 | 64 |
| 6 | 15 | 34 | 14 | 89 |

$$r = \frac{\sum x_i y_i - (\sum x_i)(\sum y_i)/n}{\sqrt{\left[\sum x_i^2 - (\sum x_i)^2/n\right]\left[\sum y_i^2 - (\sum y_i)^2/n\right]}}$$

$$= \frac{34-(6)(15)/3}{\sqrt{[14-6^2/3][89-(15)^2/3]}} = 0.756$$

The answers to the defining and computing formula are equivalent.

**4.143** (a) To compute the linear correlation coefficient using the defining formula, begin with the following table.

| $x$ | $y$ | $x-\bar{x}$ | $y-\bar{y}$ | $(x-\bar{x})^2$ | $(y-\bar{y})^2$ | $(x-\bar{x})(y-\bar{y})$ |
|---|---|---|---|---|---|---|
| 0 | 4 | -3 | 3 | 9 | 9 | -9 |
| 2 | 2 | -1 | 1 | 1 | 1 | -1 |
| 2 | 0 | -1 | -1 | 1 | 1 | 1 |
| 5 | -2 | 2 | -3 | 4 | 9 | -6 |
| 6 | 1 | 3 | 0 | 9 | 0 | 0 |
| 15 | 5 | | | 24 | 20 | -15 |

The columns after the first two have used $\bar{x}=3$ and $\bar{y}=1$.
We will need $s_x$ and $s_y$. These are given by

$$s_x = \sqrt{\frac{\sum(x_i-\bar{x})^2}{n-1}} = \sqrt{\frac{24}{4}} = 2.449 \text{ and } s_y = \sqrt{\frac{\sum(y_i-\bar{y})^2}{n-1}} = \sqrt{\frac{20}{4}} = 2.236$$

$$r = \frac{\frac{1}{n-1}\sum(x_i-\bar{x})(y_i-\bar{y})}{s_x s_y} = \frac{\frac{1}{4}(-15)}{(2.449)(2.236)} = \frac{-3.75}{5.476} = -0.685$$

(b) To compute the linear correlation coefficient using the computing formula, begin with the following table.

| $x$ | $y$ | $xy$ | $x^2$ | $y^2$ |
|---|---|---|---|---|
| 0 | 4 | 0 | 0 | 16 |
| 2 | 2 | 4 | 4 | 4 |
| 2 | 0 | 0 | 4 | 0 |
| 5 | -2 | -10 | 25 | 4 |
| 6 | 1 | 6 | 36 | 1 |
| 15 | 5 | 0 | 69 | 25 |

$$r = \frac{\sum x_i y_i - (\sum x_i)(\sum y_i)/n}{\sqrt{\left[\sum x_i^2 - (\sum x_i)^2/n\right]\left[\sum y_i^2 - (\sum y_i)^2/n\right]}}$$

$$= \frac{0 - (15)(5)/5}{\sqrt{[69 - 15^2/5][25 - (5)^2/5]}} = \frac{-15}{\sqrt{480}} = -0.685$$

The answers to both the defining and computing formula are equivalent.

**4.145** To compute the linear correlation coefficient, return to the table presented at the beginning of the solution to Exercise 4.99. Use the last row of this table to perform the calculations in part (a).

(a) The linear correlation coefficient r is computed using the formula in Formula 4.3 of the text:

$$r = \frac{S_{xy}}{\sqrt{S_{xx} S_{yy}}} = \frac{\sum xy - (\sum x)(\sum y)/n}{\sqrt{[\sum x^2 - (\sum x)^2/n][\sum y^2 - (\sum y)^2/n]}}$$

$$= \frac{13168 - (41)(3422)/10}{\sqrt{[199 - 41^2/10][1196690 - 3422^2/10]}} = -0.967872$$

(b) The value of r in part (a) suggests a strong negative linear relationship between age and price of Corvettes.

(c) Data points are clustered closely about the regression line.

(d) r² = (-0.967872)² = 0.9368. This matches the coefficient of determination that was calculated in part (b) of Exercise 4.99.

**4.147** To compute the linear correlation coefficient, return to the table presented at the beginning of the solution to Exercise 4.101. Use the last row of this table to perform the calculations in part (a).

(a) The linear correlation coefficient r is computed using the formula in Formula 4.3 of the text:

$$r = \frac{S_{xy}}{\sqrt{S_{xx} S_{yy}}} = \frac{\sum xy - (\sum x)(\sum y)/n}{\sqrt{[\sum x^2 - (\sum x)^2/n][\sum y^2 - (\sum y)^2/n]}}$$

$$= \frac{10486 - (723)(156.5)/11}{\sqrt{[48747 - 723^2/11][2523.252 - 156.5^2/11]}} = 0.331067$$

(b) The value of r in part (a) suggests a weak positive linear relationship between potato plant weight and quantity of volatile emissions.

(c) Data points are clustered very loosely about the regression line.

(d) r² = (0.331037)² = 0.1096. This matches the coefficient of determination that was calculated in part (b) of Exercise 4.101.

**4.149** To compute the linear correlation coefficient, return to the table presented at the beginning of the solution to Exercise 4.103. Use the last row of this table to perform the calculations in part (a).

(a) The linear correlation coefficient r is computed using the formula in Formula 4.3 of the text:

$$r = \frac{S_{xy}}{\sqrt{S_{xx}S_{yy}}} = \frac{\sum xy - (\sum x)(\sum y)/n}{\sqrt{[\sum x^2 - (\sum x)^2/n][\sum y^2 - (\sum y)^2/n]}}$$

$$= \frac{9519 - (117)(660)/8}{\sqrt{[1869 - 117^2/8][54638 - 660^2/8]}} = -0.7749$$

(b) The value of r in part (a) suggests a fairly strong negative linear relationship between study-time and score for calculus students.
(c) Data points are clustered fairly closely about the regression line.
(d) $r^2 = (-0.7749)^2 = 0.6005$. This matches the coefficient of determination that was calculated in part (b) of Exercise 4.103.

**4.151** To compute the linear correlation coefficient, begin with the following table.

| x | y | xy | x² | y² |
|---|---|----|----|----|
| -3 | 9 | -27 | 9 | 81 |
| -2 | 4 | -8 | 4 | 16 |
| -1 | 1 | -1 | 1 | 1 |
| 0 | 0 | 0 | 0 | 0 |
| 1 | 1 | 1 | 1 | 1 |
| 2 | 4 | 8 | 4 | 16 |
| 3 | 9 | 27 | 9 | 81 |
| 0 | 28 | 0 | 28 | 196 |

(a) The linear correlation coefficient r is computed using the formula in Formula 4.3 of the text:

$$r = \frac{S_{xy}}{\sqrt{S_{xx}S_{yy}}} = \frac{\sum xy - (\sum x)(\sum y)/n}{\sqrt{[\sum x^2 - (\sum x)^2/n][\sum y^2 - (\sum y)^2/n]}}$$

$$= \frac{0 - (0)(282)/7}{\sqrt{[28 - 0^2/7][196 - 28^2/7]}} = 0.0$$

(b) We cannot conclude from the result in part (a) that x and y are unrelated. We can conclude only that there is no *linear* relationship between x and y.
(c)                                              Graph for part (e)

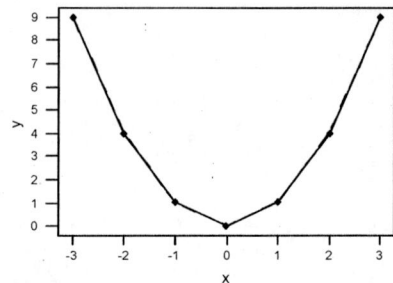

(d) It is not appropriate to use the linear correlation coefficient as a descriptive measure for the data because the data points are not scattered about a line.

(e) For each data point (x, y), we have y = x². (See columns 1 and 2 of the previous table.) See the graph above at the right.

**4.153** (a) From the data, we obtain the following sums:
$$\sum x_i = 21; \quad \sum y_i = 91; \quad \sum x_i^2 = 91; \quad \sum y_i^2 = 2275; \quad \sum x_i y_i = 441$$
Then
$$r = \frac{S_{xy}}{\sqrt{S_{xx} S_{yy}}} = \frac{\sum xy - (\sum x)(\sum y)/n}{\sqrt{[\sum x^2 - (\sum x)^2/n][\sum y^2 - (\sum y)^2/n]}}$$
$$= \frac{441 - (21)(91)/7}{\sqrt{[91 - (21)^2/7][2275 - (91)^2/7]}} = 0.961$$

(b) No. A graphical analysis is also needed before you can conclude that they are linearly related.

(c)
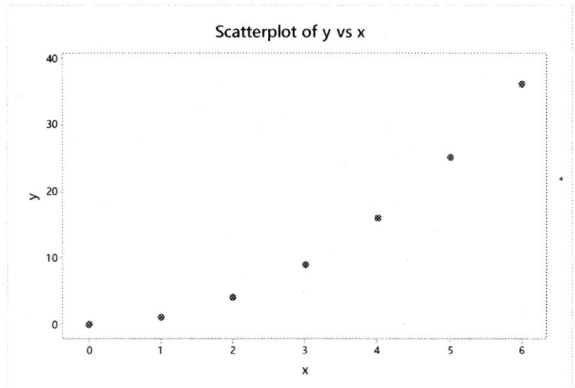

(d) No. The scatterplot reveals that the graph is not linear.

(e) Each of the y-values is the solution of the respective x-value plugged into the equation. For example, plugging in x = 3 in the equation yields (3)^2 = 9.

(f) The following is the graph of the equation.

**4.155** (a) In Exercise 4.71, we saw that a linear relationship between the age at inauguration and the age at death was reasonable. The scatterplot showed that the line relating the two variables had a positive slope.

(b) From the output in Exercise 4.71, the coefficient of determination was 0.367. Since this is $r^2$ and r is positive because of the positive slope of the line, $r = \sqrt{0.367} = 0.606$. You can also obtain the correlation coefficient in Minitab by choosing **Stat ▶ Basic Statistics ▶ Correlation...**, select INAUGURATION and DEATH in the **Variables** text box, and click **OK**. The result is

Pearson correlation of INAUGURATION and DEATH = 0.606

(c) The correlation coefficient is positive, indicating that the linear relationship has a positive slope. The coefficient is not very close to 1, indicating that the linear relationship is not very strong.

**4.157** (a) In Exercise 4.73, we saw that a linear relationship between LOT SIZE and VALUE was not reasonable. Parts (b-c) are omitted.

**4.159** (a) In Exercise 4.75, we saw that a linear relationship between HIGH and LOW was reasonable.

(b) In Minitab, choose **Stat ▶ Basic Statistics ▶ Correlation...**, select HIGH and LOW in the **Variables** text box, and click **OK**. The result is

Pearson correlation of HIGH and LOW = 0.976

(c) The correlation coefficient is positive, indicating that the linear relationship has a positive slope. The coefficient is very close to 1, indicating that the linear relationship is very strong.

**4.161** (a) In Exercise 4.77, we saw that a linear relationship between INCOME and BEER was not reasonable as there was no linear pattern in the data. Parts (b-c) are omitted.

**4.163** (a) In Exercise 4.80, we saw that there was an increasing relationship between diameter and volume, but the relationship was concave upward, not linear. Parts (b-c) are omitted.

**4.165** (a) In Exercise 4.115, we saw that there was an increasing relationship between estriol levels of pregnant women and birth weights of their children. Use of the correlation coefficient is appropriate as a descriptive measure of the strength of the linearity of that relationship.

(b) Using Minitab, with the data in columns named ESTRIOL and WEIGHT, we choose **Stat ▶ Basic Statistics ▶ Correlation...**, select ESTRIOL and WEIGHT in the **Variables** text box, and click **OK**. The result is

Pearson correlation of ESTRIOL and WEIGHT = 0.610, P-Value = 0.000

(c) There is a moderately positive linear relationship between estriol concentration and birth weight. The data points will be moderately clustered about the regression line.

**4.167** (a) No. We only know that the linear correlation coefficient is the square root of the coefficient of determination or it is the negative of that square root.

(b) No. The slope is positive if r is positive and negative if r is negative, but we can not determine the sign of r.

(c) Yes. $r = -\sqrt{0.716} = -0.846$

(d) Yes. $r = \sqrt{0.716} = 0.846$

**4.169** (a) Using Minitab, choose **Graph ▶ Scatterplot**, select the **Simple** version, enter SCORE in the first row of the **Y variables** column and TIME in the first row of the **X variables** column, and click **OK**. The result is

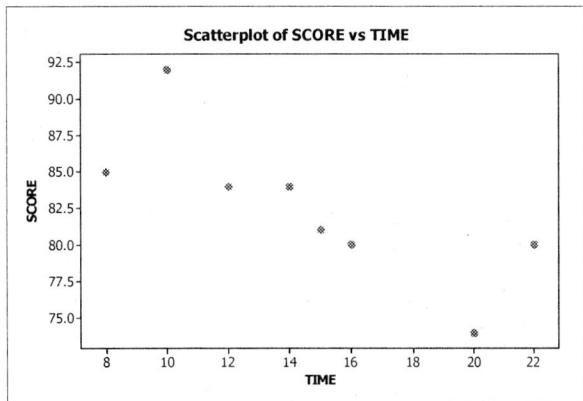

(b) Score decreases as time increases, so the using the rank correlation coefficient is reasonable.
(c) Score decreases linearly as time increases, so the using the linear correlation coefficient is reasonable.
(d) To find the rank correlation coefficient using Minitab, we must first find the ranks for each of the data values in the SCORE and TIME columns. This is done for TIME by choosing **Data ▶ Rank**, entering TIME in the **Rank data in** text box and RANKTIME in the **Store ranks in** text box and clicking **OK**. Then choose **Data ▶ Rank**, enter SCORE in the **Rank data in** text box and RANKSCORE in the **Store ranks in** text box and click **OK**. Now choose **Stat ▶ Basic Statistics ▶ Correlation...**, select RANKTIME and RANKSCORE in the **Variables** text box, and click **OK**. The result is
    Pearson correlation of RANKTIME and RANKSCORE = -0.928
    P-Value = 0.001
The rank correlation coefficient is -0.928, indicating that the ranks are strongly negatively correlated and that SCORE decreases as TIME increases.

### Review Problems for Chapter 4

1. (a) x   (b) y   (c) $b_1$   (d) $b_0$
2. (a) It intersects the y-axis when x = 0. Therefore y = 4.
   (b) It intersects the x-axis when y = 0. Therefore x = 4/3.
   (c) slope = -3
   (d) The y-value decreases by 3 units when x increases by 1 unit.
   (e) The y-value increases by 6 units when x decreases by 2 units.
3. True. The y-intercept is determined by $b_0$ and that value is independent of $b_1$, the slope.
4. False. A horizontal line has a slope of zero.
5. True. If the slope is positive, the x-values and y-values increase and decrease together.
6. Scatterplot or scatter diagram
7. A regression equation can be used to make predictions of the response variable for specific values of the predictor variable within the range of the observed values of the predictor variable.
8. (a) predictor variable or explanatory variable
   (b) response variable
9. Based on the least-squares criterion, the line that best fits a set of data points is the one have the smallest possible sum of squared errors.
10. The line that best fits a set of data points according to the least-squares criterion is called the regression line.

11. Using a regression equation to make predictions for values of the predictor variable outside the range of the observed values of the predictor variable is called extrapolation.
12. (a) An outlier is a data point that lies far from the regression line relative to the other data points.
    (b) An influential observation is a data point whose removal causes the regression equation to change considerably. Often this is a data point for which the x-value is considerably to the left or right of the rest of the data points.
13. The coefficient of determination is the percentage of the total variation in the y-values that is explained by the regression equation.
14. (a) SST is the total sum of squares and measures the variation in the observed values of the response variable.
    (b) SSR is the regression sum of squares and measures the variation in the observed values of the response variable that is explained by the regression. It can also be thought of as the variation in the predicted values of the response variable corresponding to the x-values in the data points.
    (c) SSE is the error sum of squares and measures the variation in the observed values of the response variable that is not explained by the regression.
15. One use of the linear correlation coefficient is as a descriptive measure of the strength of the linear relationship between two variables.
16. A positive linear relationship between two variables means that one variable tends to increase linearly as the other increases.
17. A value of r close to -1 suggests a strong negative linear relationship between the variables.
18. A value of r close to zero suggests at most a weak linear relationship between the variables.
19. True. It is quite possible that both variables are strongly affected by one (or more) other variables (called lurking variables).
20. (a) y = 72 - 12x   (b)  $b_0 = 72$, $b_1 = -12$
    (c) The line slopes downward since $b_1 < 0$.
    (d) After two years: y = 72 - 12(2) = $48 hundred = $4800
        After five years: y = 72 - 12(5) = $12 hundred = $1200.
    (e)

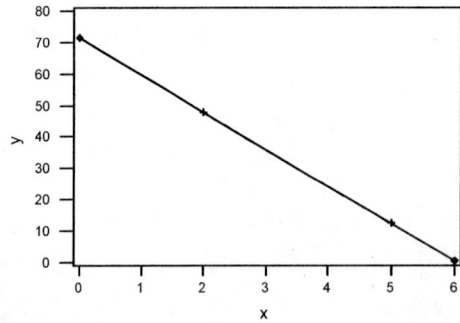

    (f) From the graph, we estimate the value to be about $2500 after 4 years.
        The actual value is y = 7200 - 1200(4) = $2400.

**21.** (a)

(b) It is moderately reasonable to find a regression line for the data because the data points appear to be scattered about a line. This perception is, however, heavily influenced by the single data point at x = 10.

(c) The regression equation can be determined by calculating its slope and intercept. Formulas for the slope ($b_1$) and intercept ($b_0$) of the regression equation are, respectively:

$$b_1 = \frac{\sum xy - \sum x \sum y / n}{\sum x^2 - (\sum x)^2 / n} \qquad b_0 = \frac{1}{n}(\sum y - b_1 \sum x)$$

To compute $b_0$ and $b_1$, construct a table of values for x, y, xy, $x^2$, and their sums. (Note: A column for $y^2$ is also presented. This is used in Problems 22 and 23.)

| x | y | xy | $x^2$ | $y^2$ |
|---|---|---|---|---|
| 16 | 45 | 720 | 256 | 2,025 |
| 20 | 55 | 1,100 | 400 | 3,025 |
| 17 | 70 | 1,190 | 289 | 4,900 |
| 19 | 50 | 950 | 361 | 2,500 |
| 22 | 47 | 1,034 | 484 | 2,209 |
| 17 | 46 | 782 | 289 | 2,116 |
| 17 | 50 | 850 | 289 | 2,500 |
| 17 | 66 | 1,122 | 289 | 4,356 |
| 10 | 26 | 260 | 100 | 676 |
| 18 | 60 | 1,080 | 324 | 3,600 |
| 173 | 515 | 9,088 | 3,081 | 27,907 |

Thus,

$$b_1 = \frac{9088 - 173(515)/10}{3081 - 173^2/10} = 2.02611 \qquad b_0 = \frac{1}{11}(515 - 2.02611(173)) = 16.448$$

and the regression equation is: $\hat{y} = 16.448 + 2.02611x$.

To graph the regression equation, begin by selecting two x-values within the range of the x-data. For the x-values 10 and 22, the calculated values for y are, respectively:

$$\hat{y} = 16.448 + 2.02611(10) = 36.71$$
$$\hat{y} = 16.448 + 2.02611(22) = 61.02$$

The regression equation can be graphed by plotting the pairs (10, 36.71) and (22, 61.02) and connecting these points with a line. This equation and the original set of 10 data points are presented as follows:

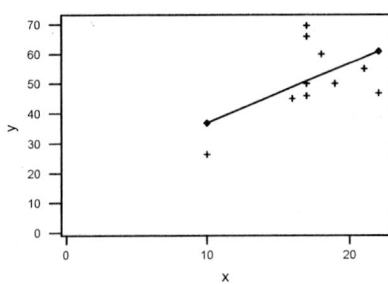

(d) Graduation rate tends to increase as the student-to-faculty ratio increases.
(e) Graduation rate increases an estimated 2.026% for each additional 1 unit increase in the student-to-faculty ratio.
(f) The predicted graduation rate for a university with a student-to-faculty ratio of 17 is 16.448 + 2.0261(17) = 50.89%.
(g) There is one potential influential observation, (10, 26) and no outliers. Looking at the scatterplot in part (c), we suspect that there may be little relationship between the student-to-faculty ratio and the graduation rate if the influential observation is removed from the data.

22. (a) To compute SST, SSR, and SSE using the computing formulas, begin with the table presented in Problem 21(c). Using the last row of this table and the formula in Definition 4.7 of the text, we obtain the three sums of squares and the coefficient of determination as follows.

$$SST = S_{yy} = \sum y^2 - (\sum y)^2/n = 27907 - 515/10 = 1384.5$$

$$SSR = \frac{S_{xy}^2}{S_{xx}} = \frac{[\sum xy - (\sum x)(\sum y)/n]^2}{\sum x^2 - (\sum x)^2/n} = \frac{[9088 - (173)(515)/10]^2}{3081 - 173^2/10} = 361.66$$

$$SSE = S_{yy} - \frac{S_{xy}^2}{S_{xx}} = 1384.50 - 361.66 = 1022.846$$

$$r^2 = \frac{SSR}{SST} = \frac{361.66}{1384.50} = 0.261$$

(b) The percentage reduction obtained in the total squared error by using the regression equation, instead of the sample mean y, to predict the observed costs is 26.1%.
(c) The percentage of the variation in the graduation rate that is explained by the student-to-faculty ratio is 26.1%.
(d) The regression equation appears to be not very useful for making predictions.

23. (a) To compute the linear correlation coefficient, begin with the table presented in Problem 21(c). Using the last row of this table and the formula in Formula 4.3 of the text, we get

$$r = \frac{S_{xy}}{\sqrt{S_{xx}S_{xy}}} = \frac{\sum xy - (\sum x)(\sum y)/n}{\sqrt{[\sum x^2 - (\sum x)^2/n][\sum y^2 - (\sum y)^2/n]}}$$

$$= \frac{9088 - (173)(515)/10}{\sqrt{[3081 - 173^2/10][27907 - 515^2/10]}} = 0.511$$

   (b) The value of r suggests a weak to moderate positive linear correlation.
   (c) Data points are clustered about the regression line, but not very closely.
   (d) $r^2 = (0.511)^2 = 0.261$

24. Using Minitab, with the data in three columns named POPULATION, AREA, and PLANTS, choose **Stat ▶ Basic statistics ▶ Correlation,** enter POPULATION, AREA, and PLANTS in the **Variables** text **box.** Make sure that the **Display P-values** box is checked. **Click OK.** The output is

```
              POPULATION      AREA
   AREA         0.109
                0.453

   PLANTS       0.721        -0.309
                0.000         0.029
   Cell Contents: Pearson correlation
                  P-Value
```

   (a) The correlation coefficient between population and area is 0.109.
   (b) The correlation coefficient between population and number of exotic plants is 0.721.
   (c) The correlation coefficient between area and number of exotic plants is -0.309.
   (d) Any linear association between population and area is very weak since the correlation coefficient is near zero. There is a moderately strong positive linear relationship between population and number of exotic plants. There is a fairly weak negative linear association between area and number of exotic plants. One possible explanation of the stronger positive relationship between population and number of exotic plants might be that the more people there are, the greater the possibility that exotic plants are introduced into the state by those people, intentionally or unintentionally.

25. (a) Using Minitab, with the data in columns named IMR and LE, choose **Graph ▶ Scatterplot...,** select the **Simple** version, enter 'LE' in row 1 of the **Y variables** column and IMR in the **X variables** column and click **OK.**

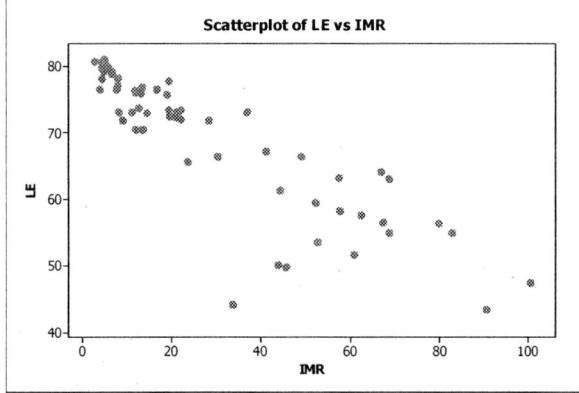

Life expectancy appears to decrease linearly as IMR increases.

(b) It appears to be reasonable to find a regression line for the data, although we expect that there will be both outliers and influential observations.

(c) Choose **Stat ▶ Regression ▶ Regression**, enter 'LE' in the **Response box** and IMR in the **Predictor box.** Click on the Options button and enter 30 in the Prediction intervals for new observations text box. Click **OK** twice. The results are

The regression equation is
**LE = 79.4 - 0.354 IMR**

```
Predictor        Coef    SE Coef       T        P
Constant       79.396      1.009   78.72    0.000
IMR           -0.35374    0.02601  -13.60    0.000

S = 5.17598    R-Sq = 76.1%    R-Sq(adj) = 75.7%
```

Analysis of Variance

```
Source           DF      SS       MS       F        P
Regression        1   4956.2   4956.2   185.00    0.000
Residual Error   58   1553.9     26.8
Total            59   6510.1
```

Unusual Observations

```
Obs  IMR     LE      Fit    SE Fit  Residual  St Resid
 11   46  49.890   63.251   0.795   -13.361    -2.61R
 25  100  47.430   43.894   1.971     3.536     0.74 X
 34   34  44.280   67.418   0.680   -23.138    -4.51R
 48   44  50.160   63.828   0.773   -13.668    -2.67R
 52   91  43.450   47.365   1.734    -3.915    -0.80 X
```

R denotes an observation with a large standardized residual.
X denotes an observation whose X value gives it large leverage.

Predicted Values for New Observations

```
New Obs    Fit    SE Fit       95% CI             95% PI
   1     68.784   0.669   (67.445, 70.122)   (58.337, 79.231)
```

Values of Predictors for New Observations

```
New Obs    IMR
   1      30.0
```

The regression equation, **LE = 79.4 - 0.354 IMR,** indicates that life expectancy decreases as the infant mortality rate increases.

(d) For a country with an IMR = 30, the predicted life expectancy is 79.4 - 0.354(30) = 68.78 years. The output above also has the predicted value of 68.784 years

(e) Choose **Stat ▶ Basic statistics ▶ Correlation,** enter 'LE' and IMR in the **Variables** text **box.** Click **OK.** The results are

Pearson correlation of IMR and LE = -0.873
P-Value = 0.000

Life expectancy and IMR are highly negatively correlated. Thus, there is strong negative linear association between the two variables.

(f) The output shown in part (c) indicates that observations 11, 34, and 48 are potential outliers and observations 25 and 52 are influential observations.

**26.** (a) Using Minitab, with the data in columns named HIGH and PRECIPITAION, choose **Graph ▶ Scatterplot…**, select the **Simple** version, enter 'PRECIPITATIONN' in row 1 of the **Y variables** column and HIGH in the **X variables** column and click **OK**.

The graph at the right does not show any relationship between average July precipitation and average July high temperature.

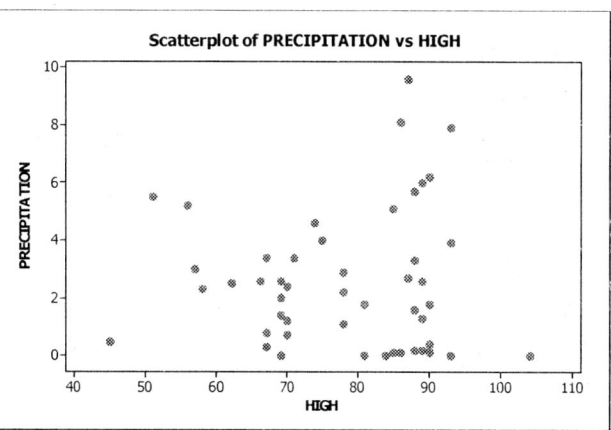

(b) The graph in part (a) does not indicate that there is any linear relationship between average precipitation and average high temperature in July for 48 cities. Parts (c-f) are omitted.

**27.** (a) Using Minitab, with the data in columns named FIRST and SECOND, choose **Graph ▶ Scatterplot…**, select the **Simple** version, enter SECOND in row 1 of the **Y variables** column and FIRST in the **X variables** column and click **OK**.

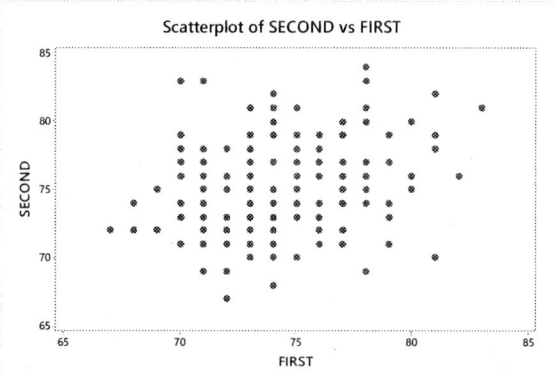

(b) Obtaining a linear regression does not seem reasonable. There is no noticeable linear relationship in the data. Therefore, we will skip steps (c) - (f).

# CHAPTER 5 SOLUTIONS

## Exercises 5.1

**5.1**  An experiment is an action the result of which cannot be predicted with certainty. An event is a specified result that may or may not occur when the experiment is performed.

**5.3**  There is no difference.

**5.5**  An experiment has 20 possible outcomes, all equally likely. An event can occur in five ways. The probability that the event occurs is 5/20 = <u>0.25</u>.

**5.7**  The frequentist interpretation of probability is that the probability of an event is the proportion of times the event occurs in a large number of repetitions of the experiment.

**5.9**  (a) In a large number of Texas hold'em poker hands, a pocket pair will be dealt approximately 5.9% of the time.

(b) If three balanced dimes are tossed a large number of times, it will come up all three heads approximately 12.5% of the time.

**5.11**  The following could not possibly be a probability:

(b) 3.5: A probability cannot exceed 1.

**5.13**  (a)  G,L,S,A   G,L,S,T   G,L,A,T   G,S,A,T   L,S,A,T

(b) Two samples include the governor, attorney general, and treasurer. Therefore, the probability is f/N = 2/5.

(c) Three samples include the governor and treasurer. Therefore, the probability is f/N = 3/5.

(d) Four samples include the governor. Thus, the probability is f/N = 4/5.

**5.15**  (a) There are 3 chips that are red. Therefore, the probability is f/N = 3/12 = 1/4.

(b) There are 3 + 4 = 7 chips that are red or white. Therefore, the probability is f/N = 7/12.

(c) There are 3 + 5 = 8 chips that are not white. Therefore, the probability is f/N = 8/12 = 2/3.

**5.17**  (a) There was a total of 70,686,784 votes cast. Therefore, the probability that a randomly selected voter voted for Vladimir Putin is f/n = 45,513,001/70,686,784 = 0.644.

(b) There were 4,448,959 + 2,755,642 = 7,204,601 votes cast for Zhirinovsky or Mironov. Therefore, the probability that a randomly selected voter voted for Zhirinovsky or Mironov is f/n = 7,204,601/70,686,784 = 0.102.

(c) The number of votes not for Putin is 70,686,784 − 45,513,001 = 25,173,783. Thus, the probability that a randomly selected voter did not vote for Putin is 25,173,783/70,686,784 = 0.356.

**5.19**  The total number of units (in thousands) is N = 132,418.

(a) The probability that the unit has 4 rooms is f/N = 23636/132418 = 0.178 (to three decimal places).

(b) The probability that the unit has more than 4 rooms is f/N = (30440 + 27779 + 17868 + 19257)/132418 = 95344/132418 = 0.720.

(c) The probability that the unit has 1 or 2 rooms is f/N = (601 + 1404)/132418 = 2005/132418 = 0.015.

(d) The probability that the unit has fewer than one room is f/N = 0/132418 = 0.000.

**158** Chapter 5

- (e) The probability that the unit has one or more rooms is f/N = 132418/132418 = 1.000.

**5.21** The total number of graduate students is 49,088.

- (a) The probability that his occupation is service is f/N = 9274/49088 = 0.189.
- (b) The probability that his occupation is administrative is f/N = (2197 + 6450)/49088 = 8647/49088 = 0.176.
- (b) The probability that his occupation is manufacturing is f/N = (2197 + 2166 + 1640 + 5721)/49088 = 11724/49088 = 0.239
- (c) The number whose occupation is not manufacturing is 49088 - 11724 = 37364. The probability that his occupation is not manufacturing is f/N = 37364/49088 = 0.761.

**5.23**
- (a) The total number of graduate science students in doctorate-granting institutions is 349.3 thousand. Of these, 45.6 thousand are in psychology. Thus, the probability that a randomly selected graduated student in science is in the field of psychology is f/N = 45.6/349.3 = 0.131.
- (b) There are 38.7 + 97.2 = 135.9 thousand graduate students in physical or social science. Thus, the probability that a randomly selected graduated student in science is in physical or social science is f/N = 135.9/349.3 = 0.389.
- (c) There are 46.1 thousand graduate students in computer science and 349.3 - 46.1 = 303.2 thousand graduate students not in computer science. Thus, the probability that a randomly selected graduated student in science is not in computer science is f/N = 303.2/349.3 = 0.868.

**5.25**
- (a) There are five ways [(1,5), (2,4), (3,3), (4,2), (5,1)] among the 36 possibilities that sum to 6. Thus, f/N = 5/36 = 0.139.
- (b) There are 18 ways among the 36 possibilities that provide a sum that is even. Thus, f/N = 18/36 = 0.500.
- (c) There are six ways among the 36 possibilities that sum to 7 and two ways among the 36 possibilities that sum to 11. Thus, f/N = (6 + 2)/36 = 8/36 = 0.222.
- (d) There is one way among the 36 possibilities that sums to 2, two ways in which the sum is 3, and one way in which the sum is 12. Thus, f/N = (1 + 2 + 1)/36 = 4/36 = 0.111.

**5.27**
- (a) The event in part (e), that the housing unit has one or more rooms, is a certainty. The event in part (d), that the housing unit has less than one room, is impossible.
- (b) The event that is a certainty has a probability of 1. The event that is impossible has a probability of 0.

**5.29** Answers will vary

**5.31** In a large sample of horse races, approximately 66.7% of the favorites will finish in the money. Therefore, 66.7% of 500 is 333.5. Approximately 334 of the favorites will finish in the money.

**5.33** When rolling two balanced dice and observing the sum, there are 36 possible equally likely outcomes, not 11. The probability that the sum is 12 equals 1/36, not 1/11.

**5.35** (a) Answers will vary. For my experiment, here are the outcomes from tossing a coin five times and keeping track of the running totals and running proportions.

| Toss | Outcome | Number of Heads | Proportion of heads |
|------|---------|-----------------|---------------------|
| 1 | H | 1 | 1.00 |
| 2 | H | 2 | 1.00 |
| 3 | T | 2 | 0.67 |
| 4 | T | 2 | 0.50 |
| 5 | T | 2 | 0.40 |

(b) Based on my five tosses, I would estimate that the probability of a head when this coin is tossed is 0.40 because that was my running total for the proportion of heads after tossing the coin five times.

(c)

| Toss | Outcome | Number of Heads | Proportion of heads |
|------|---------|-----------------|---------------------|
| 1 | H | 1 | 1.00 |
| 2 | H | 2 | 1.00 |
| 3 | T | 2 | 0.67 |
| 4 | T | 2 | 0.50 |
| 5 | T | 2 | 0.40 |
| 6 | H | 3 | 0.50 |
| 7 | T | 3 | 0.43 |
| 8 | H | 4 | 0.50 |
| 9 | T | 4 | 0.44 |
| 10 | H | 5 | 0.50 |

Based on my ten tosses, I would estimate that the probability of a head when this coin is tossed is 0.50 because that was my running total for the proportion of heads after tossing the coin ten times.

(d)

| Toss | Outcome | Number of Heads | Proportion of heads |
|------|---------|-----------------|---------------------|
| 1 | H | 1 | 1.00 |
| 2 | H | 2 | 1.00 |
| 3 | T | 2 | 0.67 |
| 4 | T | 2 | 0.50 |
| 5 | T | 2 | 0.40 |
| 6 | H | 3 | 0.50 |
| 7 | T | 3 | 0.43 |
| 8 | H | 4 | 0.50 |
| 9 | T | 4 | 0.44 |
| 10 | H | 5 | 0.50 |
| 11 | T | 5 | 0.45 |
| 12 | T | 5 | 0.42 |
| 13 | T | 5 | 0.38 |

| 14 | T | 5 | 0.36 |
| 15 | H | 6 | 0.40 |
| 16 | T | 6 | 0.38 |
| 17 | H | 7 | 0.41 |
| 18 | T | 7 | 0.39 |
| 19 | H | 8 | 0.42 |
| 20 | H | 9 | 0.45 |

Based on my twenty tosses, I would estimate that the probability of a head when this coin is tossed is 0.45 because that was my running total for the proportion of heads after tossing the coin twenty times.

(d) For my answers in part (b)-(d), the frequency interpretation cannot be used as the definition of probability because the number of trials was small. The frequency interpretation requires a large number of trials before it can be used as the definition of probability.

**5.37** If p is the probability that a randomly selected adult woman believes that a "cyber affair" is cheating, then p = 0.75. The odds against selecting such a woman are in the ratio of 1-p to p, or 0.25 to 0.75. This is normally expressed in integers as 1 to 3.

**5.39** If p is the probability that a randomly selected person age 18-34 years has cursed at their computer, then p = 0.46. The odds against selecting such a person are in the ratio of 1-p to p, or 0.54 to 0.46. This is normally expressed in integers and reduced as 27 to 23.

### Exercises 5.2

**5.41** Venn diagrams are useful for portraying events and relationships between events.

**5.43** The shaded areas correspond to the desired events in the diagrams below.

(a) A&B

(b) A&B&C

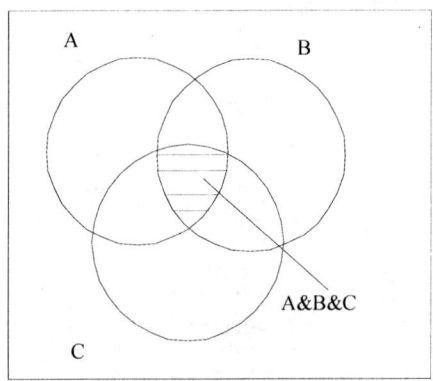

**5.45** The shaded areas correspond to the desired events in the diagrams below.

(a) (A & (not B))   (b) ((A or B) & (not (A & B)))

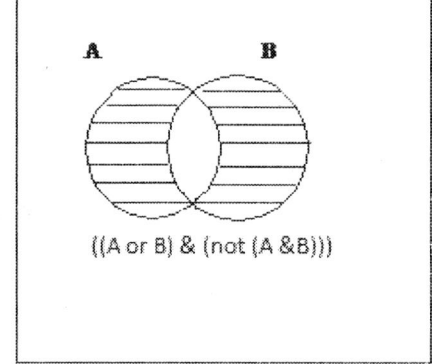

**5.47** (a) At least six would be 6, 7, 8, 9, and 10.

(b) At most 3 would be 1, 2, and 3.

(c) Between 2 and 5, inclusive, would be 2, 3, 4, and 5.

**5.49** Three events are mutually exclusive if no two of them can occur at the same time, i.e., no pair of the events has any outcomes in common.

**5.51** False. C could have outcomes in common with A or with B even though A and B have no outcomes in common. For example, if a die is tossed and A = even number, B = odd number, C = divisible by 3, then A and B are mutually exclusive, but A and C have the outcome 6 in common while B and C have the number 3 in common.

**5.53** A = {2, 4, 6}; B = {4, 5, 6}; C = {1, 2}; D = {3}

**5.55** A contains the outcomes JM, JS, JH, JB, WM, WS, WH, WJ

B contains the outcomes HM, HS, HJ, HW

C contains the outcomes MW, SW, HW, JW

D contains the outcomes MS, MH, SM, SH, HM, HS

**5.57** (a) (not A) = {1, 3, 5} = the event the die comes up odd.

(b) (A & B) = {4, 6} = the event the die comes up four or six.

(c) (B or C) = {1, 2, 4, 5, 6} = the event the die does *not* come up three.

**5.59** (a) (not A) = (MS, MH, MJ, MW, SM, SH, SJ, SW, HM, HS, HJ, HW) = the event that a female is appointed chairperson

(b) (B & D) = (HM, HS) = the event that Holly is appointed chairperson and a woman is appointed secretary

(c) (B or C) = (HM, HS, HJ, HW, MW, SW, JW) = the event that Holly is appointed chairperson or Will is appointed secretary.

**5.61** (a) (not C) = the event that a state has a diabetes prevalence percentage less than 6% or is 13% or more. There are 1 + 1 = 2 states in (not C).

(b) (A&B) = the event that a state has a diabetes prevalence percentage at least 8% and is less than 7%. There are no states in this event.

(c) (C or D) = the event that a state has a diabetes prevalence percentage that is less than 13%. There are 49 states in this event.

(d) (C&B) = the event that a state has a diabetes prevalence percentage of at least 6% and less than 7%. There are 5 states in this event.

**162** Chapter 5

**5.63** (a) (not A) = the event that the World Series was decided in at least 5 games. There are 24 + 24 + 36 = 84 World Series in this event.

(b) (A&B) = the event that the World Series was decided in 4 games. There are 21 World Series in this event.

(c) (A or C) = the event that the World Series was decided in 4 or 7 games. There are 21 + 36 = 57 World Series in this event.

(d) (A&C) = the event that the World Series was decided in both 4 and 7 games. There are no World Series in this event. This is an impossible event.

**5.65** (a) (not A) is the event that the unit has more than 4 rooms. There are 30,440 + 27,779 + 17,868 + 19,257 = 95,344 (thousand) such units.

(b) (A&B) is the event that the unit has at most 4 rooms and at least 2 rooms, i.e., has 2 or 3 or 4 rooms. There are 1,404 + 11,433 + 23,636 = 36,473 (thousand) such units.

(c) The event (C or D) is the event that the unit has between 5 and 7 rooms inclusive or more than seven rooms, i.e., has 5 or more rooms. This is the same as the event (not A) and therefore there are 95,344 (thousand) such units.

**5.67** (a) Events A and B are not mutually exclusive. They have a four and a six in common.

(b) Events B and C are mutually exclusive. They have no outcomes in common.

(c) Events A, C, and D are not mutually exclusive. The outcome two is common to A and C.

(d) Among A, B, C, and D, there are three mutually exclusive events. These are B, C, and D. Among B, C, and D, there are no outcomes in common. There are not, however, four mutually exclusive events. The outcome two is common to A and C, four is common to A and B, and six is common to A and B.

**5.69** The groups that are mutually exclusive are A and C; A and D; C and D; and A, C, and D.

**5.71** (a) The first number in each sample space outcome is the number of dots that are face up after rolling the die. The possible outcomes for the first number are one through six. The second number in each sample space outcome is the total number of heads that are observed from a coin toss out of the number of trials that the die first showed. Therefore, the second number can be the numbers zero up to and including the number that the die first showed. The sample space is (1,0), (1,1), (2,0), (2,1), (2,2), (3,0), (3,1), (3,2), (3,3), (4,0), (4,1), (4,2), (4,3), (4,4), (5,0), (5,1), (5,2), (5,3), (5,4), (5,5), (6,0), (6,1), (6,2), (6,3), (6,4), (6,5), (6,6).

(b) The event that the total number of heads is even would consist of outcomes from the sample space that have an even second number (count zero as even). The event is (1,0), (2,0), (2,2), (3,0), (3,2), (4,0), (4,2), (4,4), (5,0), (5,2), (5,4), (6,0), (6,2), (6,4), (6,6).

**5.73** (a) If events *A* and (not *B*) are mutually exclusive, this means that *A* and (not *B*) never overlap. Thus, if *A* occurs (not *B*) does not occur. This means that *B* does occur.

(b) If *B* occurs whenever *A* occurs, this means that (not *B*) never occurs when *A* occurs. This implies that *A* and (not *B*) are mutually exclusive.

**5.75** To say that A, B, and C do not all occur simultaneously does not necessarily imply that A, B, and C are mutually exclusive. In the following diagram, C does not touch A or B. That is, we have a situation where A, B, and C do not occur simultaneously. However, these three events clearly are not mutually exclusive.

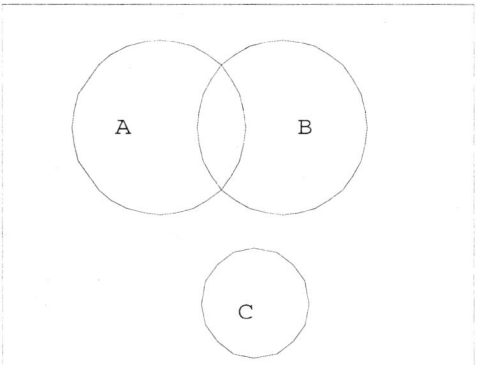

### Exercises 5.3

**5.77** There are 26 cards that are red. Therefore, the probability is f/N = 26/52 = 1/2. P(R) = 1/2.

**5.79** Since A and B are mutually exclusive, the Special Addition Rule applies. P(A or B) = P(A) + P(B) = 0.25 + 0.40 = 0.65.

**5.81** Using the Complementation rule, P(E) = 1 - P(not E), we have 0.35 = 1 - P(not E). Solving this equation results in P(not E) = 0.65.

**5.83** Since C and D are not mutually exclusive, the General Addition Rule applies. P(C or D) = P(C) + P(D) - P(C&D) = 0.35 + 0.40 - 0.30 = 0.45.

**5.85** (a) A and B are not mutually exclusive since P(A&B) = 1/10.

(b) Apply the General Addition Rule, which states P(A or B) = P(A) + P(B) - P(A&B). Plugging in the known values, we have 1/2 = 1/3 + P(B) - 1/10. Solve for P(B). P(B) = 1/2 + 1/10 - 1/3 = 4/15.

**5.87** (a) P(S) = f/N = (11 + 30 + 37)/100 = 0.78.

(b) S = (A or B or C)

(c) P(A) = 11/100 = 0.11;  P(B) = 30/100 = 0.30;

P(C) = 37/100 = 0.37.

(d) P(S) = P(A or B or C) = P(A) + P(B) + P(C) = 0.11 + 0.30 + 0.37 = 0.78. This result is identical to the one found in part (a).

**5.89** (a) The percentage who smoked within the last 30 days is 9.8 + 7.8 + 5.3 + 2.8 + 0.7 + 0.3 = 26.7. Therefore, the probability that a randomly selected twelfth grader smoked is 0.267.

(b) The percentage who smoked at least one cigarette per day within the last 30 days is 7.8 + 5.3 + 2.8 + 0.7 + 0.3 = 16.9. Therefore, the probability that a randomly selected twelfth grader smoked at least one cigarette per day is 0.169.

(d) The percentage who smoked between 6 and 34 cigarettes per day, inclusive, within the last 30 days is 5.3 + 2.8 + 0.7 = 8.8. Therefore, the probability that a randomly selected twelfth grader smoked between 6 and 34 cigarettes per day is 0.088.

**164** Chapter 5

**5.91** (a) P(Income under $50,000) = P(A or B) = P(A) + P(B) = 0.171 + 0.240 = 0.411.

(b) P(Income of $25,000 or above) = P (B or C or D or E) = P(B) + P(C) + P(D) +P(E) = 0.240 + 0.350 + 0.141 + 0.098 = 0.829.

(c) P(Income between $25,000 and $149,999, inclusive) = P(B or C or D) = P(B) + P(C) + P(D) = 0.240 + 0.350 + 0.141 = 0.731.

(d) Of households with Internet access, 41.1% have incomes under $50,000, 82.9% have incomes of $25,000 or above, and 73.1% have incomes between $25,000 and $149,999, inclusive.

**5.93** (a) P(senator is at least 50) = $\frac{30}{100}+\frac{37}{100}+\frac{20}{100}+\frac{2}{100}=0.89$.

This is accomplished more easily using the complementation rule:

P(senator is at least 50) = 1 - P(senator is under 50) = $1-\frac{11}{100}=0.89$.

(b) P(senator is under 70) = $\frac{11}{100}+\frac{30}{100}+\frac{37}{100}=\frac{78}{100}=0.78$.

This is accomplished more easily using the complementation rule:

P(senator is under 70) = 1 - P(senator is at least 70) =

$1-\left(\frac{20}{100}+\frac{2}{100}\right)=1-\frac{22}{100}=\frac{78}{100}=0.78$.

**5.95** (a) The probability that the individual obtained a loan balance between $10,001 and $100,000, inclusive, is 0.292 + 0.165 + 0.058 + 0.023 = 0.538. We are asked the probability of this OR the event of at most $75,000. The only loan balance of this second event that is not already included in the event between $10,001 and $100,000, inclusive, is $1 - $10,000. Adding this to our previous sum, we have 0.538 + 0.431 = 0.969. The event that the loan balance is between $10,001 and $100,000, inclusive, or at most $75,000 is 0.969.

(b) Let A = the event that the loan balance is between $10,001 and $100,000, inclusive. Let B = the event that the loan balance is at most $75,000. P(A) = 0.538 from part (a). P(B) = 0.431 + 0.292 + 0.165 + 0.058 = 0.946. P(A&B) = the event that the loan balance is between $10,001 and $75,000, inclusive. P(A&B) = 0.292 + 0.165 + 0.058 = 0.515. Using the General Addition Rule, P(A or B) = P(A) + P(B) - P(A&B) = 0.538 + 0.946 - 0.515 = 0.969

(c) Answers will vary.

**5.97** (a) $P(A)=\frac{6}{36}=0.167$  $P(B)=\frac{2}{36}=0.056$  $P(C)=\frac{1}{36}=0.028$  $P(D)=\frac{2}{36}=0.056$

$P(E)=\frac{1}{36}=0.028$  $P(F)=\frac{5}{36}=0.139$  $P(G)=\frac{6}{36}=0.167$

(b) P(7 or 11)= P(A) + P(B) = 0.167 + 0.056 = 0.223.

(c) P(2 or 3 or 12) = P(C) or P(D) or P(E).
= 0.028 + 0.056 + 0.028 = 0.112

(d) P(8 or doubles):  Using Figure 4.1: $\dfrac{10}{36} = 0.278$

(e) P(8) + P(doubles) - P(8 & doubles) = 0.139 + 0.167 - 0.028 = 0.278.

**5.99** P(Public school or college) = P(Public school) + P(College) - P(Public school & college) = 0.853 + 0.279 - 0.201 = 0.931. 93.1% of students attend either public school or college.

**5.101** Let H = the event that the household gets the *Herald*, E = the event that the household gets the *Examiner,* and T = the event that the household gets the *Times*. The event that a household gets at least one of the three major newspapers can be written (H or E or T).

The general addition rule for three events is

P(H or E or T) = P(H) + P(E) + P(T) - P(H&E) - P(H&T) - P(E&T) + P(H&E&T).

Substituting in the values given in the problem we have

P(H or E or T) = 0.334 + 0.346 + 0.470 - 0.104 - 0.119 - 0.151 + 0.048 = 0.824.

Therefore, the probability that a household gets at least one of the three major newspapers is 0.824.

### Exercises 5.4

**5.103** (a) probability

(b) probability

**5.105** The notation {X=3} denotes an event that occurs when X=3, whereas P(X=3) denotes the probability that the event {X=3} will occur. Another way of thinking about the difference is that the first is an event and the second is a number between zero and one.

**5.107** This table will resemble the probability distribution of the random variable.

**5.109** (a)

| x | Probability |
|---|---|
| 1 | 0.30 |
| 2 | 0.40 |
| 3 | 0.30 |

(b) (X = 2) is the event that X takes on the value 2, (X ≤ 2) is the event that X takes on a value of at most 2, and (X > 2) is the event that X takes on a value greater than 2.

(c) P(X = 2) = 0.40, P(X ≤ 2) = 0.70, P(X > 2) = 0.30
The probability that X takes on a value of 2 is 0.40. The probability that X takes on a value of at most 2 is 0.70. The probability that X takes on a value more than 2 is 0.30.

(d)

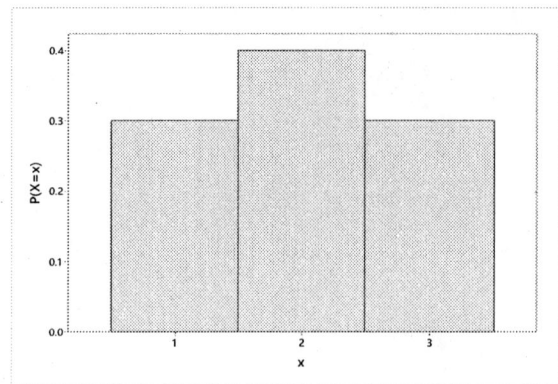

**5.111** (a)

| y | Probability |
|---|---|
| 0 | 0.36 |
| 1 | 0.28 |
| 4 | 0.16 |
| 6 | 0.20 |

(b) (Y = 3) is the event that Y takes on the value 3, (Y < 3) is the event that Y takes on a value less than 3, and (Y ≥ 3) is the event that Y takes on a value of at least 3.

(c) P(Y = 3) = 0, P(Y < 3) = 0.64, P(Y ≥ 3) = 0.36

The probability that Y takes on the value of 3 is 0. The probability that Y takes on a value less than 3 is 0.64. The probability that Y takes on a value of at least 3 is 0.36

(d)

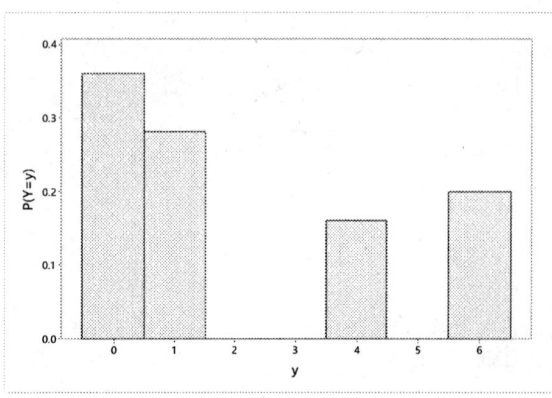

**5.113** (a) X = 2, 4, 5, 6, 7, 8

(b) {X = 7}

(c) The total number of shuttle missions is 135. P(X = 4) = 3/135 = 0.022, so 2.2% of the shuttle crews consist of exactly four people.

(d)

| Size of Crew | Probability |
|---|---|
| 2 | 0.030 |
| 3 | 0.000 |
| 4 | 0.022 |
| 5 | 0.267 |
| 6 | 0.207 |
| 7 | 0.467 |
| 8 | 0.007 |

(e)

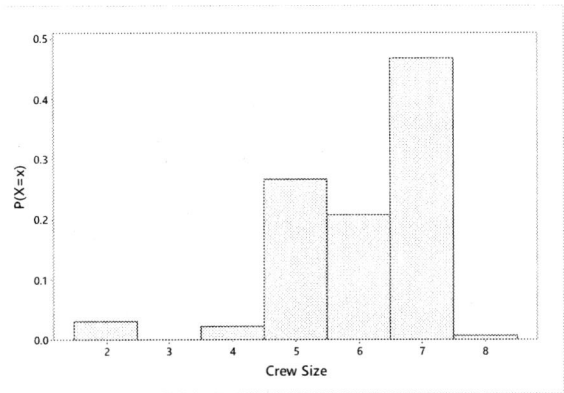

**5.115** (a) $\{Y \geq 1\}$     (b) $\{Y = 3\}$     (c) $\{2 \leq Y \leq 4\}$

(d) $P(Y \geq 1) = P(Y = 1) + P(Y = 2) + P(Y = 3) + P(Y = 4) + P(Y = 5) + P(Y = 6) + P(Y = 7) + P(Y = 8)$
$= 0.296 + 0.266 + 0.093 + 0.049 + 0.056 + 0.037 + 0.012 + 0.006 = 0.815$

(e) $P(Y = 3) = 0.093$

(f) $P(2 \leq y \leq 4) = P(Y = 2) + P(Y = 3) + P(Y = 4) = 0.266 + 0.093 + 0.049 = 0.408$

**5.117** (a) $Y = 2,3,4,5,6,7,8,9,10,11,12$

(b) $\{Y = 7\}$

(c) $P(Y = 7) = 6/36 = 1/6 = 0.167$

(d)

| Y | P(Y) |
|---|---|
| 2 | 1/36 |
| 3 | 2/36 |
| 4 | 3/36 |
| 5 | 4/36 |
| 6 | 5/36 |
| 7 | 6/36 |
| 8 | 5/36 |
| 9 | 4/36 |
| 10 | 3/36 |
| 11 | 2/36 |
| 12 | 1/36 |

(e)

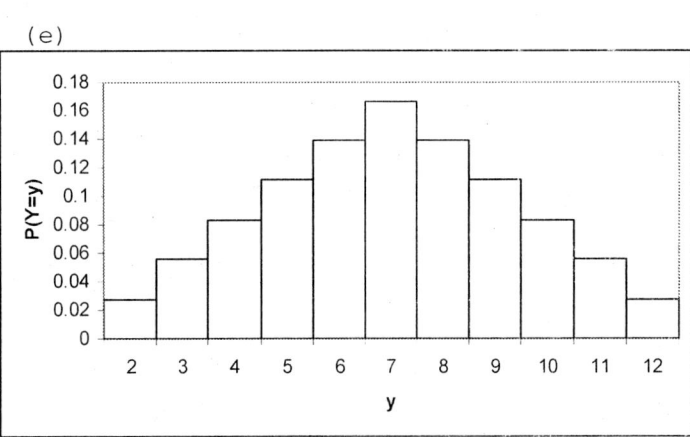

**168** Chapter 5

**5.119** (a) The area of the square is $6^2$ = 36 square feet. The area of the bull's eye is $\pi(1)^2$ = 3.1416 square feet. The area of the ring is the area of a circle with radius 2 less the area of the bull's eye, or $\pi(2)^2 - \pi(1)^2$ = 12.5664 − 3.1416 = 9.4248. The remaining area of the square outside the ring has area 36 − 12.5664 = 23.4336. Thus, P(S = 10) = 3.1416/36 = 0.087; P(S = 5) = 9.4248/36 = 0.262; and P(S = 0) = 23.3226/36 = 0.651. These are shown in the table below.

| S | P(S = s) |
|---|---|
| 0 | 0.651 |
| 5 | 0.262 |
| 10 | 0.087 |

Assuming that all points in the six foot square are equally likely, the archer has almost a 2/3 chance of scoring 0 points and about a 1/12 chance of scoring 10 points.

(b) P(S = 5) = 0.262

P(S > 0) = P(S = 5 or 10) = 0.262 + 0.087 = 0.349

P(S ≤ 7) = P(S = 0 or 5) = 0.651 + 0.262 = 0.913

P(5 < S ≤ 15) = P(S = 10) = 0.087

P(S < 15) = P(S = 0 or 5 or 10) = 1; This event always occurs.

P(S < 0) = 0.000; This event cannot occur.

**5.121** (a) The probabilities are only an estimate of the probability distribution, because the information was gathered from a sample.

(b) Let X = the number of young per litter for female Florida black bears. The probability that X is 2 or 3 is P(X = 2) + P(X = 3) = 0.606 + 0.152 = 0.758.

**5.123** P(Z ≤ 1.96) + P(Z > 1.96) = 1.

Since P(Z > 1.96) = 0.025,

P(Z ≤ 1.96) + 0.025 = 1, or

P(Z ≤ 1.96) = 1 − 0.025 = 0.975.

**5.125** (a) P(X ≤ c) + P(X > c) = 1.

Since P(X > c) = $\alpha$,

P(X ≤ c) + $\alpha$ = 1, or

P(X ≤ c) = 1 − $\alpha$.

(b) P(Y < −c) + P(−c ≤ Y ≤ c) + P(Y > c) = 1.

Since P(Y < −c) = P(Y > c) = $\alpha/2$,

$\alpha/2$ + P(−c ≤ Y ≤ c) + $\alpha/2$ = 1, or

P(−c ≤ Y ≤ c) = 1 − $\alpha/2$ − $\alpha/2$ = 1 − $\alpha$.

(c) P(T < −c) + P(−c ≤ T ≤ c) + P(T > c) = 1.

Since P(−c ≤ T ≤ c) = 1 − $\alpha$,

P(T < −c) + (1 − $\alpha$) + P(T > c) = 1, or

P(T < −c) + P(T > c) = 1 − 1 + $\alpha$ = $\alpha$.

Since P(T < −c) = P(T > c),

2·[P(T > c)] = $\alpha$    or    P(T > c) = $\alpha/2$.

## Exercises 5.5

**5.127** The mean of a discrete random variable generalizes the concept of a population mean.

**5.129** The required calculations are

| x | P(X=x) | xP(X=x) | x² | x² P(X=x) |
|---|---|---|---|---|
| 1 | 0.3 | 0.3 | 1.0 | 0.3 |
| 2 | 0.4 | 0.8 | 4.0 | 1.6 |
| 3 | 0.3 | 0.9 | 9.0 | 2.7 |
|   |   | 2.0 |   | 4.6 |

(a) $\mu_x = \sum xP(X=x) = 2.0$

(b) $\sigma = \sqrt{\sum x^2 P(X=x) - \mu_x^2} = \sqrt{4.6 - 2.0^2} = 0.8$

**5.131** The required calculations are

| x | P(X=x) | xP(X=x) | x² | x² P(X=x) |
|---|---|---|---|---|
| 0 | 0.36 | 0.00 | 0.0 | 0.00 |
| 1 | 0.28 | 0.28 | 1.0 | 0.28 |
| 4 | 0.16 | 0.64 | 16.0 | 2.56 |
| 6 | 0.20 | 1.20 | 36.0 | 7.20 |
|   |   | 2.12 |   | 10.04 |

(a) $\mu_x = \sum xP(X=x) = 2.12$

(b) $\sigma = \sqrt{\sum x^2 P(X=x) - \mu_x^2} = \sqrt{10.04 - 2.12^2} = 2.35$

**5.133** The required calculations are

| x | P(X=x) | xP(X=x) | x² | x² P(X=x) |
|---|---|---|---|---|
| 2 | 0.030 | 0.060 | 4.000 | 0.120 |
| 4 | 0.022 | 0.088 | 16.000 | 0.352 |
| 5 | 0.267 | 1.335 | 25.000 | 6.675 |
| 6 | 0.207 | 1.242 | 36.000 | 7.452 |
| 7 | 0.467 | 3.269 | 49.000 | 22.883 |
| 8 | 0.007 | 0.056 | 64.000 | 0.448 |
|   |   | 6.050 |   | 37.930 |

(a) $\mu_x = \sum xP(X=x) = 6.050$. The average number of persons in a shuttle crew is about 6.05.

(b) $\sigma = \sqrt{\sum x^2 P(X=x) - \mu_x^2} = \sqrt{37.930 - 6.05^2} = 1.15$

(c)

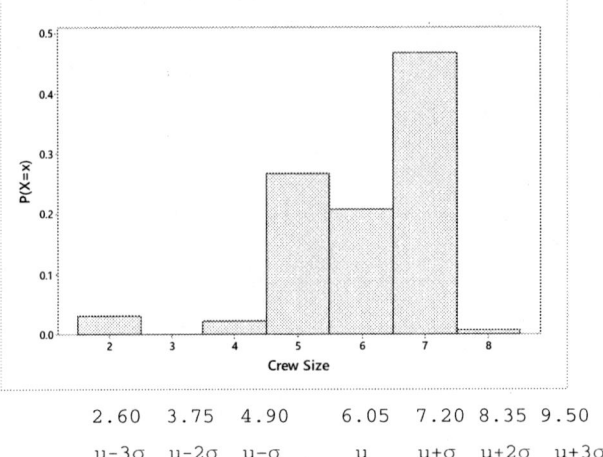

```
       2.60  3.75  4.90    6.05  7.20  8.35  9.50
       μ-3σ  μ-2σ  u-σ      μ    μ+σ   μ+2σ  μ+3σ
```

**5.135** The required calculations are

| y | P(Y=y) | yP(Y=y) | y² | y²P(Y=y) |
|---|--------|---------|----|----------|
| 0 | 0.185  | 0.000   | 0  | 0.000    |
| 1 | 0.296  | 0.296   | 1  | 0.296    |
| 2 | 0.266  | 0.532   | 4  | 1.064    |
| 3 | 0.093  | 0.279   | 9  | 0.837    |
| 4 | 0.049  | 0.196   | 16 | 0.784    |
| 5 | 0.056  | 0.280   | 25 | 1.400    |
| 6 | 0.037  | 0.222   | 36 | 1.332    |
| 7 | 0.012  | 0.084   | 49 | 0.588    |
| 8 | 0.006  | 0.048   | 64 | 0.384    |
|   | 1      | 1.937   |    | 6.685    |

(a) $\mu_y = \sum yP(Y=y) = 1.937$. The mean number of major hurricanes per year is 1.934.

(b) $\sigma = \sqrt{\sum y^2 P(Y=y) - \mu_y^2} = \sqrt{6.685 - 1.937^2} = 1.713$. Roughly speaking, the number of major hurricane in a given year is about 1.7 away from the mean of 1.937.

(c)

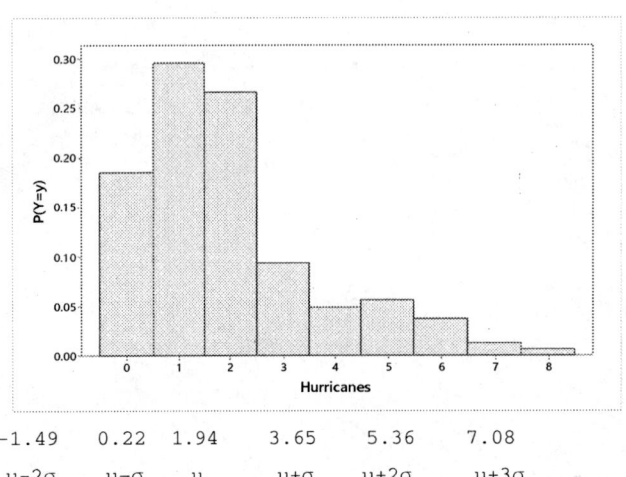

```
  -3.20  -1.49  0.22  1.94  3.65  5.36  7.08
   μ-3σ   μ-2σ  u-σ    μ    μ+σ   μ+2σ  μ+3σ
```

Section 5.5  171

**5.137** The required calculations are

| $y$ | $P(Y=y)$ | $yP(Y=y)$ | $y-\mu_y$ | $(y-\mu_y)^2$ | $(y-\mu_y)^2 P(Y=y)$ | $y^2$ | $y^2 P(Y=y)$ |
|---|---|---|---|---|---|---|---|
| 2 | 1/36 | 2/36 | -5 | 25 | 25/36 | 4 | 4/36 |
| 3 | 2/36 | 6/36 | -4 | 16 | 32/36 | 9 | 18/36 |
| 4 | 3/36 | 12/36 | -3 | 9 | 27/36 | 16 | 48/36 |
| 5 | 4/36 | 20/36 | -2 | 4 | 16/36 | 25 | 100/36 |
| 6 | 5/36 | 30/36 | -1 | 1 | 5/36 | 36 | 180/36 |
| 7 | 6/36 | 42/36 | 0 | 0 | 0/36 | 49 | 294/36 |
| 8 | 5/36 | 40/36 | 1 | 1 | 5/36 | 64 | 320/36 |
| 9 | 4/36 | 36/36 | 2 | 4 | 16/36 | 81 | 324/36 |
| 10 | 3/36 | 30/36 | 3 | 9 | 27/36 | 100 | 300/36 |
| 11 | 2/36 | 22/36 | 4 | 16 | 32/36 | 121 | 242/36 |
| 12 | 1/36 | 12/36 | 5 | 25 | 25/36 | 144 | 144/36 |
|  |  | 252/36=7 |  |  | 210/36 |  | 1,974/36 |

(a) $\mu_y = \sum yP(Y=y) = 252/36 = 7$. When rolling two balanced dice, the mean of their sum is 7.

(b) $\sigma = \sqrt{\sum(y-\mu_y)^2 P(Y=y)} = \sqrt{210/36} = 2.415$

$\sigma = \sqrt{\sum y^2 P(Y=y) - \mu_y^2} = \sqrt{\dfrac{1974}{36} - 7^2} = 2.415$

(c)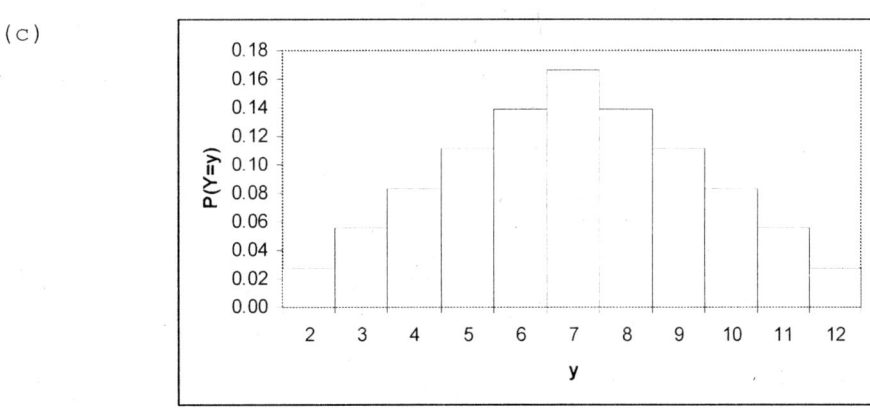

| -0.2 | 2.2 | 4.6 | 7 | 9.4 | 11.8 | 14.2 |
| μ-3σ | μ-2σ | μ-σ | μ | μ+σ | μ+2σ | μ+3σ |

**5.139** (a) The necessary calculations are

| $s$ | $P(S=s)$ | $sP(S=s)$ | $s^2$ | $s^2P(S=s)$ |
|---|---|---|---|---|
| 0 | 0.651 | 0.000 | 0 | 0.000 |
| 5 | 0.262 | 1.310 | 25 | 6.550 |
| 10 | 0.087 | 0.870 | 100 | 8.700 |
|  |  | 2.180 |  | 15.250 |

On average, the archer will score 2.180 points per arrow shot.

(b) $\sigma = \sqrt{\sum s^2 P(S=s) - \mu_s^2} = \sqrt{15.250 - 2.180^2} = 3.24$. Roughly speaking, on

172  Chapter 5

average, the number of points on a shot will differ from the mean by 3.24 points.

**5.141** (a)

$$P(X=1) = \frac{18}{38} = 0.474$$

$$P(X=-1) = \frac{20}{38} = 0.526; \quad 0.474 + 0.526 = 1.000$$

(b) $\mu = \sum x \cdot P(X=x) = (1)(0.474) + (-1)(0.526) = -0.052$
(c) On the average, you will lose 5.2¢ per play.
(d) If you bet $1 on red 100 times, you can expect to lose 100 · 0.052 = $5.20. If you bet $1 on red 1000 times, you can expect to lose $52.00.
(e) Roulette is not a profitable game for a person to play. Parts (c) and (d) demonstrate that, no matter how much you play, you can expect to lose. Also, part (a) shows that a higher probability is associated with losing rather than winning.

**5.143** (a) $\mu = \sum x \cdot P(X=x) = (0)(0.95) + (10)(0.045) + (50)(0.004) + 100(0.0009) + (200)(0.0001) = 0.76$. The expected annual claim amount per homeowner is 0.76 thousand dollars, or $760.
(b) They want an average net profit of $50 per policy, so they need to charge $50 more than the expected claim amount of $760. Therefore, they should charge $810.

**5.145**

| w | P(W=w) | w·P(w) | w² | w²·P(W=w) |
|---|--------|--------|-----|-----------|
| 0 | 0.80   | 0.00   | 0   | 0.00      |
| 1 | 0.15   | 0.15   | 1   | 0.15      |
| 2 | 0.05   | 0.10   | 4   | 0.20      |
|   |        | 0.25   |     | 0.35      |

(a)

$$\mu_w = \sum wP(W=w) = 0.25$$

$$\sigma_w = \sqrt{\sum w^2 P(W=w) - \mu_w^2} = \sqrt{0.350 - 0.25^2} = \sqrt{0.2875} = 0.536$$

(b) On the average, there are 0.25 breakdowns per day.
(c) Assuming 250 workdays per year, the number of breakdowns expected per year is (0.25)(250) = 62.5.

**5.147** (a) The necessary calculations are

| y | P(Y=y) | yP(Y=y) | y² | y²P(Y=y) |
|---|--------|---------|-----|----------|
| 0 | 0.424  | 0.000   | 0   | 0.000    |
| 1 | 0.161  | 0.161   | 1   | 0.161    |
| 2 | 0.134  | 0.268   | 4   | 0.536    |
| 3 | 0.111  | 0.333   | 9   | 0.999    |
| 4 | 0.093  | 0.372   | 16  | 1.488    |
| 5 | 0.077  | 0.385   | 25  | 1.925    |
|   |        | 1.519   |     | 5.109    |

Copyright © 2016 Pearson Education, Inc.

The mean of Y, the mean number of customers waiting is 1.519. On the
average, there are about 1.5 customers waiting at any given time.

(b) In a large number of independent observations, 1.519 customers will be
waiting, on average.

(c) We will use Minitab to simulate these observations. Enter the values
of Y in column C1 and their probabilities in column C2. Then select

**Calc ▶ Random data ▶ Discrete,** enter 500 in the **Generate Rows of data**
text box, enter Customers in the **Store in Column(s)** text box, then
enter C1 in the **Values in** text box and C2 in the **Probabilities in** text
box and click **OK**. This will produce 500 random values in a column
named Customeers. Now select **Stat ▶ Basic Statistics ▶ Display
Descriptive Statistics,** enter Customers in the **Variables** text box and
click **OK**. The result appears in the Sessions Window. Our result is

```
Variable      N    N*    Mean   SE Mean   StDev     Minimum        Q1       Median
Customers   500    0   1.5800   0.0767   1.7151   0.000000000   0.000000000   1.0000

Variable     Q3    Maximum
Customers  3.0000   5.0000
```

(d) The mean for our observations was 1.58, a little bit higher than the
exact mean of 1.519. Your data and mean will likely be different.

(e) Part (d) illustrates that simulations yield sample means that are close
to the population mean.

## Exercises 5.6

**5.149** Each repetition of an experiment is called a trial.

**5.151** The binomial coefficient equals the number of outcomes that contain exactly
x successes in n Bernoulli trials.

**5.153** The binomial distribution is the probability distribution for the number of
successes in a sequence of Bernoulli trials.

**5.155** (1) Randomly selected pieces of identical rope are tested by subjecting each
to a 1000 pound force. Each rope either breaks or it doesn't. (2) People
are selected at random by ticket numbers at a large convention and their
gender is noted. Each is either male or female.

**5.157** $3! = 3 \cdot 2 \cdot 1 = 6;\quad 7! = 7 \cdot 6 \cdot 5 \cdot 4 \cdot 3 \cdot 2 \cdot 1 = 5,040;$

$8! = 8 \cdot 7 \cdot 6 \cdot 5 \cdot 4 \cdot 3 \cdot 2 \cdot 1 = 40,320;\quad 9! = 9 \cdot 8 \cdot 7 \cdot 6 \cdot 5 \cdot 4 \cdot 3 \cdot 2 \cdot 1 = 362,880$

**5.159** (a) $\binom{5}{2} = \frac{5!}{2!(5-2)!} = \frac{5 \cdot 4 \cdot 3 \cdot 2 \cdot 1}{2 \cdot 1 \cdot 3 \cdot 2 \cdot 1} = 10$

(b) $\binom{7}{4} = \frac{7!}{4!(7-4)!} = \frac{7 \cdot 6 \cdot 5 \cdot 4 \cdot 3 \cdot 2 \cdot 1}{4 \cdot 3 \cdot 2 \cdot 1 \cdot 3 \cdot 2 \cdot 1} = 35$

(c) $\binom{10}{3} = \frac{10!}{3!(10-3)!} = \frac{10 \cdot 9 \cdot 8 \cdot 7 \cdot 6 \cdot 5 \cdot 4 \cdot 3 \cdot 2 \cdot 1}{3 \cdot 2 \cdot 1 \cdot 7 \cdot 6 \cdot 5 \cdot 4 \cdot 3 \cdot 2 \cdot 1} = 120$

(d) $\binom{12}{5} = \frac{12!}{5!(12-5)!} = \frac{12 \cdot 11 \cdot 10 \cdot 9 \cdot 8 \cdot 7 \cdot 6 \cdot 5 \cdot 4 \cdot 3 \cdot 2 \cdot 1}{5 \cdot 4 \cdot 3 \cdot 2 \cdot 1 \cdot 7 \cdot 6 \cdot 5 \cdot 4 \cdot 3 \cdot 2 \cdot 1} = 792$

**5.161** (a) $\binom{4}{1} = \dfrac{4!}{1!(4-1)!} = \dfrac{4\cdot 3\cdot 2\cdot 1}{1\cdot 3\cdot 2\cdot 1} = 4$

(b) $\binom{6}{2} = \dfrac{6!}{2!(6-2)!} = \dfrac{6\cdot 5\cdot 4\cdot 3\cdot 2\cdot 1}{2\cdot 1\cdot 4\cdot 3\cdot 2\cdot 1} = 15$

(c) $\binom{8}{3} = \dfrac{8!}{3!(8-3)!} = \dfrac{8\cdot 7\cdot 6\cdot 5\cdot 4\cdot 3\cdot 2\cdot 1}{3\cdot 2\cdot 1\cdot 5\cdot 4\cdot 3\cdot 2\cdot 1} = 56$

(d) $\binom{9}{6} = \dfrac{9!}{6!(9-6)!} = \dfrac{9\cdot 8\cdot 7\cdot 6\cdot 5\cdot 4\cdot 3\cdot 2\cdot 1}{6\cdot 5\cdot 4\cdot 3\cdot 2\cdot 1\cdot 3\cdot 2\cdot 1} = 84$

**5.163** (a) Since the graph is symmetric, p = 0.5.

(b) Since the graph is right skewed, p is less than 0.5.

**5.165** (a) Each trial consists of observing whether the child is cured of the pinworm infestation by pyrantel pamoate. Each is either cured or not cured. The results are independent assuming that the children are not in the same families. Since a success s is that the child is cured, the success probability p is 0.90.

(b)

| Outcome | Probability |
|---|---|
| sss | (0.90)(0.90)(0.90)=0.729 |
| ssf | (0.90)(0.90)(0.10)=0.081 |
| sfs | (0.90)(0.10)(0.90)=0.081 |
| sff | (0.90)(0.10)(0.10)=0.009 |
| fss | (0.10)(0.90)(0.90)=0.081 |
| fsf | (0.10)(0.90)(0.10)=0.009 |
| ffs | (0.10)(0.10)(0.90)=0.009 |
| fff | (0.10)(0.10)(0.10)=0.001 |

(c)

```
              First Child    Second Child    Third child    Outcome    Probability

                                                  S
                                                  ● sss        0.729
                                          0.9
                                 S
                             0.9 ●
                                          0.1   f
                                                ●  ssf        0.081
                     S
                  0.9●
                                                   S
                                                   ●  sfs        0.081
                                          0.9
                                 0.1
                                   f
                                   ●
                                          0.1   f
                                                ●  sff        0.009

                                                   S
                                                   ● fss        0.081
                                          0.9
                                 S
                             0.9 ●
                  0.1             0.1   f
                                        ●  fsf        0.009
                     f
                     ●
                                                   S
                                                   ●  ffs        0.009
                                          0.9
                                 0.1
                                   f
                                   ●
                                          0.1   f
                                                ●  fff        0.001
```

(d) The outcomes in which exactly two of the three children are cured are ssf, sfs, and fss.

(e) Each outcome in part (d) has the probability 0.081. This probability is the same for each outcome because each probability is obtained by multiplying two success probabilities of 0.90 and one failure probability of 0.10.

(f) P(exactly two children are cured) = P(ssf) + P(sfs) + P(fss)
= 0.081 + 0.081 + 0.081 = 0.243

(g) P(exactly one child is cured) = P(sff) + P(fsf) + P(ffs)
= 0.009 + 0.009 + 0.009 = 0.027

P(exactly three children are cured) = P(sss) = 0.729, and

P(exactly zero children are cured) = P(fff) = 0.001.

Thus the complete probability distribution of X, the number of children out of three that are cured is

| X | 0 | 1 | 2 | 3 |
|---|---|---|---|---|
| P(X=x) | 0.001 | 0.027 | 0.243 | 0.729 |

5.167 (a) Plugging n = 4, p = 0.3, and x = 2 into formula 5.4, we have

$$P(X=2) = \binom{4}{2}(0.3)^2(1-0.3)^{4-2} = 0.265.$$

(b) Using Table XI1 in Appendix A, look for n = 4, x = 2, and then navigate over to the column with p = 0.3. The answer is 0.265. This is the same answer as that in part (a).

**5.169** (a) Plugging n = 6, p = 0.5, and x = 4 into formula 5.4, we have

$$P(X=4) = \binom{6}{4}(0.5)^4(1-0.5)^{6-4} = 0.234.$$

(b) Using Table XI1 in Appendix A, look for n = 6, x = 4, and then navigate over to the column with p = 0.5. The answer is 0.234. This is the same answer as that in part (a).

**5.171** (a) Plugging n = 5, p = 3/4 = 0.75, and x = 4 into formula 5.4, we have

$$P(X=4) = \binom{5}{4}(0.75)^4(1-0.75)^{5-4} = 0.396.$$

(b) Using Table XI1 in Appendix A, look for n = 5, x = 4, and then navigate over to the column with p = 0.75. The answer is 0.396. This is the same answer as that in part (a).

**5.173** Step 1: A success is that a treated child is cured.

Step 2: The success probability is p = 0.90.

Step 3: The number of trials is n = 3.

Step 4: The formula for x successes is

$$P(X=x) = \binom{3}{x}(.90)^x(.10)^{3-x}.$$ For x = 0, 1, 2, and 3, the probabilities are

$$P(X=0) = \binom{3}{0}(.90)^0(.10)^3 = \frac{3!}{0!\,3!}(.90)^0(.10)^3 = 0.001$$

$$P(X=1) = \binom{3}{1}(.90)^1(.10)^2 = \frac{3!}{1!\,2!}(.90)^1(.10)^2 = 0.027$$

$$P(X=2) = \binom{3}{2}(.90)^2(.10)^1 = \frac{3!}{2!\,1!}(.90)^2(.10)^1 = 0.243$$

$$P(X=3) = \binom{3}{3}(.90)^3(.10)^0 = \frac{3!}{0!\,3!}(.90)^3(.10)^0 = 0.729$$

**5.175** Step 1: A success is that the coin will come up heads.

Step 2: The success probability is p = 0.5.

Step 3: The number of trials is n = 10.

Step 4: The formula for y successes is

$$P(Y=y) = \binom{10}{y}(.50)^y(.50)^{10-y}.$$

We want the probability of exactly half of 10, or 5 successes

$$P(Y=5) = \binom{10}{5}(.50)^5(.50)^{10-5} = 0.246$$

**5.177** The calculations required to answer all parts of this exercise are

$$P(0) = \binom{5}{0} \cdot (.67)^0(.33)^5 = 0.004 \quad P(3) = \binom{5}{3} \cdot (.67)^3(.33)^2 = 0.328$$

$$P(1) = \binom{5}{1} \cdot (.67)^1(.33)^4 = 0.040 \quad P(4) = \binom{5}{4} \cdot (.67)^4(.33)^1 = 0.332$$

$$P(2) = \binom{5}{2} \cdot (.67)^2(.33)^3 = 0.161 \quad P(5) = \binom{5}{5} \cdot (.67)^5(.33)^0 = 0.135$$

(a)   P(2) = 0.161        (b)   P(4) = 0.332

(c)   P(X ≥ 4) = P(4) + P(5) = 0.332 + 0.135 = 0.467 (actually 0.468 if rounding is done after the addition).

(d)   P(2 ≤ X ≤ 4) = P(2) + P(3) + P(4) = 0.161 + 0.328 + 0.332 = 0.821

(e)                                      (f)   Left skewed

| x | P(X=x) |
|---|--------|
| 0 | 0.004  |
| 1 | 0.040  |
| 2 | 0.161  |
| 3 | 0.328  |
| 4 | 0.332  |
| 5 | 0.135  |

(g)
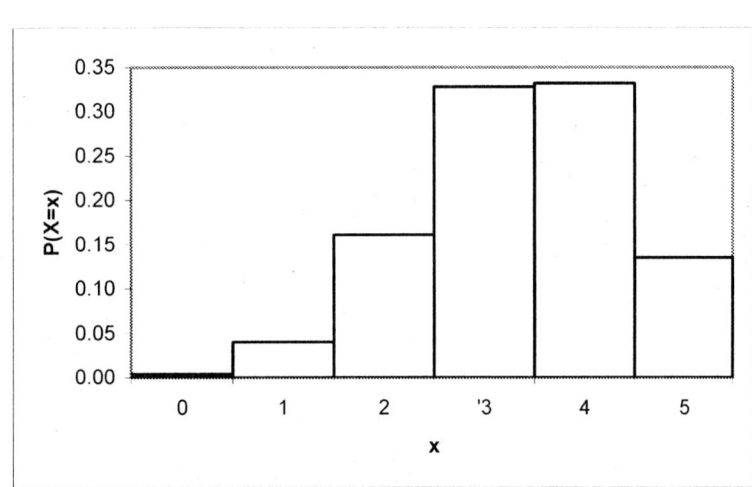

(h)

| X | P(X=x) | xP(X=x) | x² | x²P(X=x) |
|---|--------|---------|-----|----------|
| 0 | 0.004 | 0.000 | 0 | 0.000 |
| 1 | 0.040 | 0.040 | 1 | 0.040 |
| 2 | 0.161 | 0.322 | 4 | 0.644 |
| 3 | 0.328 | 0.984 | 9 | 2.952 |
| 4 | 0.332 | 1.328 | 16 | 5.312 |
| 5 | 0.135 | 0.675 | 25 | 3.375 |
|   |   | 3.349 |   | 12.323 |

$$\mu = 3.349$$
$$\sigma^2 = 12.323 - 3.349^2 = 1.107; \quad \sigma = \sqrt{1.107} = 1.052$$

(i) $\mu = np = 5(0.67) = 3.35$

$\sigma^2 = np(1-p) = 5(0.67)(0.33) = 1.1055$

$\sigma = \sqrt{1.1055} = 1.051$

(j) Out of any five races, the average number of favorites that finish in the money is 3.35.

**5.179** (a) n = 8; p = 0.40

$$P(\text{Exactly } 3) = \binom{8}{3}(0.4)^3(0.6)^5 = 0.0241$$

$$P(3 \text{ or more}) = 1 - P(0 \text{ or } 1 \text{ or } 2) = 1 - \binom{8}{0}(0.4)^0(0.6)^8 - \binom{8}{1}(0.4)^1(0.6)^7 - \binom{8}{2}(0.4)^2(0.6)^6$$

$$= 1 - 0.0168 - 0.0896 - 0.2090 = 0.6846$$

$$P(\text{At most } 3) = P(0) + P(1) + P(2) + P(3)$$

$$= \binom{8}{0}(0.4)^0(0.6)^8 + \binom{8}{1}(0.4)^1(0.6)^7 + \binom{8}{2}(0.4)^2(0.6)^6 + \binom{8}{3}(0.4)^3(0.6)^5$$

$$= 0.0168 + 0.0896 + 0.2090 + 0.2787 = 0.5941$$

(b)
$$P(2 \leq Y \leq 4) = P(2) + P(3) + P(4)$$
$$= \binom{8}{2}(0.4)^2(0.6)^6 + \binom{8}{3}(0.4)^3(0.6)^5 + \binom{8}{4}(0.4)^4(0.6)^4 = 0.2092 + 0.2787 + 0.2322 = 0.7201$$

(If rounding is done after the addition, you will get 0.7200.)

(c) The mean of Y is µ = np = 8(0.4) = 3.2. On average, of eight traffic fatalities, 3.2 will involve an intoxicated or alcohol-impaired driver or non-occupant.

(d) σ² = np(1 − p) = 8(0.4)(0.6) = 19.2, so σ = 4.3418.

**5.181** (a) $n = 9$, $p = 0.35$, so

$$P(\text{Exactly } 5) = \binom{9}{5}(0.35)^5(0.65)^4 = 0.1181$$

$$P(\text{At least } 5) = \binom{9}{5}(0.35)^5(0.65)^4 + \binom{9}{6}(0.35)^6(0.65)^3 + \binom{9}{7}(0.35)^7(0.65)^2$$
$$+ \binom{9}{8}(0.35)^8(0.65)^1 + \binom{9}{9}(0.35)^9(0.65)^1$$
$$= 0.1181 + 0.0424 + 0.0098 + 0.0013 + 0.0001 = 0.1717$$

$$P(\text{At most } 5) = \binom{9}{0}(0.35)^0(0.65)^9 + \binom{9}{1}(0.35)^1(0.65)^8 + \binom{9}{2}(0.35)^2(0.65)^7$$
$$+ \binom{9}{3}(0.35)^3(0.65)^6 + \binom{9}{4}(0.35)^4(0.65)^5 + \binom{9}{5}(0.35)^5(0.65)^4$$
$$= 0.0207 + 0.1004 + 0.2162 + 0.2716 + 0.2194 + 0.1181 = 0.9464$$

(b)

$$P(\text{At least 1 correct}) = 1 - P(0 \text{ Correct}) = 1 - \binom{9}{0}(0.35)^0(0.65)^9 = 1 - 0.0207 = 0.9793$$

$$P(\text{At most 1 correct}) = P(0) + P(1) = \binom{9}{0}(0.35)^0(0.65)^9 + \binom{9}{1}(0.35)^1(0.65)^8$$
$$= 0.0207 + 0.1004 = 0.1211$$

(c) $$P(6 \text{ or } 7 \text{ or } 8) = \binom{9}{6}(0.35)^6(0.65)^3 + \binom{9}{7}(0.35)^7(0.65)^2 + \binom{9}{8}(0.35)^8(0.65)^1$$
$$= 0.0424 + 0.0098 + 0.0013 = 0.0535$$

(d) The formula is $P(X = x) = \binom{9}{x}(0.35)^x(0.65)^{9-x}$ for $x = 0, 1, 2, ..., 9$

Applying this formula for each value of x results in the table below.

| x | P(X=x) |
|---|--------|
| 0 | 0.0207 |
| 1 | 0.1004 |
| 2 | 0.2162 |
| 3 | 0.2716 |
| 4 | 0.2194 |
| 5 | 0.1181 |
| 6 | 0.0424 |
| 7 | 0.0098 |
| 8 | 0.0013 |
| 9 | 0.0001 |

(e) The sampling was actually done without replacement, so the trials are not independent and the success probability changes very slightly from trial to trial. The exact probability distribution is called a hypergeometric distribution.

**5.183** (a) n = 7 and p = 0.09, so the formula for the probability distribution is

$$P(X=x) = \binom{7}{x}(0.09)^x(0.91)^{7-x} \text{ for } x = 0, 1, 2, \ldots 6, 7$$

Applying this formula for each value of x results in the table below

| x | P(X=x) |
|---|--------|
| 0 | 0.517 |
| 1 | 0.358 |
| 2 | 0.106 |
| 3 | 0.017 |
| 4 | 0.002 |
| 5 | 0.000 |
| 6 | 0.000 |
| 7 | 0.000 |

(b) The mean μ = np = 7(0.09) = 0.63. Thus, on the average, we would expect 0.63 youths out of seven children selected will be a PVGU.

(c) Yes. If, in fact, 9% of children in grades 3-8 are PVGUs, then finding that 3 or more out of 7 were PVGUs would mean that an event with probability 0.017 + 0.002 + 0 = 0.019 had just occurred. This is a fairly rare event if p is really 0.09. Rather than believe that we had just observed a rare event, we would be inclined to conclude that our assumption (p = 0.09) was incorrect and that p has increased from the 9% rate of 2011.

(d) No. If, in fact, 9% of children in grades 3-8 are PVGUs, then finding that two or more out of seven were PVGUs would mean that an event with probability 0.106 + 0.019 = 0.125 had just occurred. This is a fairly common event if p is really 0.09, so we would not be inclined to take 2 out of 7 as evidence that the PVGU rate has increased from the 9% rate in 2011.

**5.185** (a) n = 4, p = 18/38 = 0.4737

$$P(2) = \binom{4}{2}(0.4737)^2(0.5263)^2 = 0.3729$$

(b) $P(\text{At least } 1) = 1 - P(0) = 1 - \binom{4}{0}(0.4737)^0(0.5263)^4 = 1 - 0.0767 = 0.9233$

**5.187** (a) P(both are male) = (17/40)(17/40) = 289/1600 = 0.180625 if sampling is done with replacement.

(b) P(both are male) = (17/40)(16/39) = 272/1560 = 0.174359 if sampling is done without replacement.

(c) The probability without replacement is slightly smaller than the probability with replacement. There is about a 3% difference in the probabilities.

(d) P(both are male) = (170/400)(170/400) = 28900/160000 = 0.180625 if sampling is done with replacement. P(both are male) = (170/400)(169/399) = 28730/159600 = 0.180013 if sampling is done without replacement. The probability without replacement is just barely smaller than the probability with replacement. There is only about a 0.3% difference in the probabilities.

(e) In the second case, there is less difference between sampling without and with replacement. This is because removing the first male from the population makes less difference in the probability of the second person being a male when there are 400 people in the class than when there are only 40 people in the class. In other words, 169/399 is closer to 170/400 than 16/39 is to 17/40.

**5.189** (a) $p = 0.029$; $P(X = x) = 0.029(0.971)^{x-1}$

This formula gives the probability that a person will first win a prize in week x.

(b) $P(3) = 0.029(0.971)^2 = 0.0273$

$P(X \leq 3) = P(1) + P(2) + P(3) = 0.0290 + 0.0282 + 0.0273 = 0.0845$

$P(X \geq 3) = 1 - P(X \leq 2) = 1 - [0.0290 + 0.0282] = 1 - 0.0572$
$= 0.9428$

(c) On the average, it will take $\mu = 1/p = 1/0.029 = 34.5$ weeks until you win a prize.

## Review Problems for Chapter 5

1. Probability theory enables us to control and evaluate the likelihood that a statistical inference is correct, and it provides the mathematical basis for statistical inference.

2. (a) The equal-likelihood model is used for computing probabilities when an experiment has N possible outcomes, all equally likely.

   (b) If an experiment has N equally likely outcomes and an event can occur in f ways, then the probability of the event is f/N.

3. In the frequentist interpretation of probability, the probability of an event is the relative frequency of the event in a large number of experiments.

4. The numbers (b) -0.047 and (c) 3.5 cannot be probabilities because probabilities must always be between 0 and 1.

5. The Venn diagram is a common graphical technique for portraying events and relationships among events.

6. Two or more events are mutually exclusive if no two of the events have any outcomes in common, i.e., no two of the events can occur at the same time.

7. (a) P(E)

   (b) P(E) = 0.436

8. (a) False. For any two events, the probability that one or the other occurs equals the sum of the two individual probabilities minus the probability that both events occur.

   (b) True. Either the event occurs or it does not. Therefore, the probability that the event occurs plus the probability that it does not occur equals 1. Thus the probability that it occurs is 1 minus the probability that it does not occur.

**182** Chapter 5

9. Frequently, it is quicker to compute the probability of the complement of an event than it is to compute the probability of the event itself. The complement rule allows one to use the probability of the complement to obtain the probability of the event itself.

10. Since A, B, and C are mutually exclusive, the Special Addition Rule applies. P(A or B or C) = P(A) + P(B) + P(C) = 0.2 + 0.6 + 0.1 = 0.9.

11. Using the Complementation rule, P(E) = 1 - P(not E), we have P(E) = 1 - 0.4 = 0.6.

12. Since A and B are not mutually exclusive, the General Addition Rule applies. P(A or B) = P(A) + P(B) - P(A&B) = 0.2 + 0.6 - 0.1 = 0.7.

13. (a) random variable

    (b) finite (or countably infinite)

14. A probability distribution of a discrete distribution of a discrete random variable gives us a listing of the possible values of the random variable and their probabilities; or a formula for the probabilities.

15. Probability histogram

16. 1

17. (a) P(X = 2) = 0.386

    (b) 38.6%

    (c) 50(0.386) = 19.3 or about 19; 500(0.386) = 193

18. 3.6

19. X is more likely to take a value close to its mean because it has a smaller standard deviation and therefore less variation.

20. 0! = 1    3! = 3·2·1 = 6    4! = 4·3·2·1 = 24

    7! = 7·6·5·4·3·2·1 = 5040

21.

(a) $\binom{8}{3} = \frac{8!}{3!5!} = 56$    (b) $\binom{8}{5} = \frac{8!}{5!3!} = 56$    (c) $\binom{6}{6} = \frac{6!}{6!6!} = 1$

(d) $\binom{10}{2} = \frac{10!}{2!8!} = 45$    (e) $\binom{40}{4} = \frac{40!}{4!36!} = 91390$    (f) $\binom{100}{0} = \frac{100!}{0!100!} = 1$

22. Each trial must have the same two possible outcomes (success and failure), the trials must be independent, and they must have a probability of success p that remains constant for all trials.

23. The binomial distribution is a probability distribution for the number of successes in a sequence of n Bernoulli trials.

24. $\binom{10}{3} = \frac{10!}{3!7!} = \frac{10(9)(8)}{3(2)(1)} = 120$

25. (a) The success probability is p = 0.493.

    (b) The outcomes and their probabilities are shown in the table.

| Outcome | Probability |
|---------|-------------|
| WWW | 0.120 |
| WWL | 0.123 |
| WLW | 0.123 |
| WLL | 0.127 |
| LWW | 0.123 |
| LWL | 0.127 |
| LLW | 0.127 |
| LLL | 0.130 |

See the following tree diagram for the details of each probability.

(c)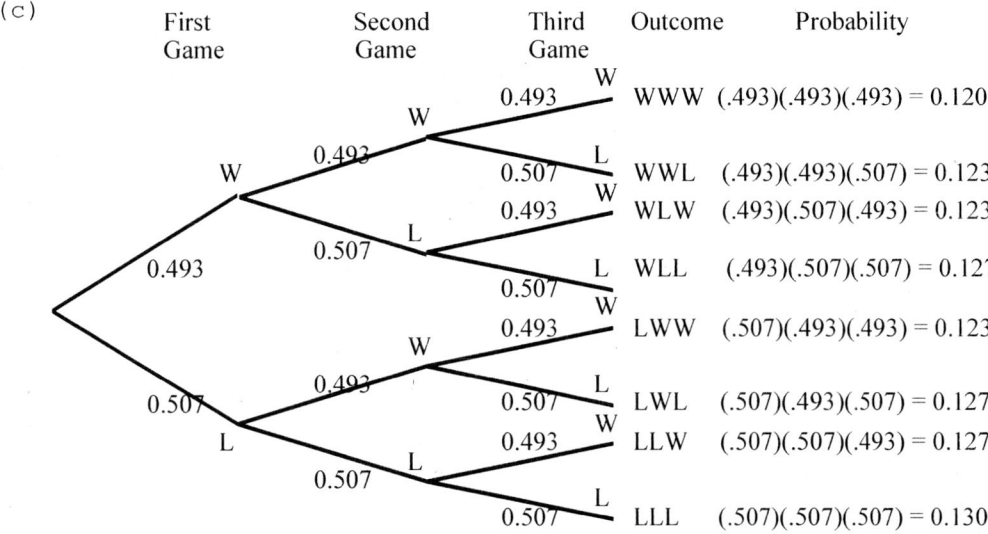

(d) The outcomes in which the player wins two out of the three times are WWL, WLW, and LWW.

(e) Each outcome in part (d) has probability 0.123. The probabilities are equal for each outcome because each probability is obtained by multiplying two WIN probabilities of 0.493 and one LOSS probability of 0.507.

(f) P(2 wins) = P(WWL) + P(WLW) + P(LWW) = 0.123 + 0.123 + 0.123
= 0.369

(g) P(0 wins) = P(LLL) = 0.130

P(1 win) = P(WLL) + P(LWL) + P(LLW) = 0.127 + 0.127 + 0.127 = 0.381

P(3 wins) = P(WWW) = 0.120

## 184 Chapter 5

| Y | P(Y=y) |
|---|---|
| 0 | 0.120 |
| 1 | 0.381 |
| 2 | 0.369 |
| 3 | 0.130 |

    (h) Binomial distribution with parameters n = 3 and p = 0.493.

26. (a) p < 0.5 since the distribution is left-skewed.

    (b) p = 0.5 since the distribution is symmetric.

27. (a) Binomial distribution

    (b) Hypergeometric distribution

    (c) The hypergeometric distribution may be approximated by the binomial distribution if the population size N is much greater than the sample size n. When n is no more than 5% of N, the probability of a success does not change much from trial to trial.

28. Other common discrete probability distributions are the Poisson, hypergeometric, and geometric distributions.

29. (a) The probability that a randomly selected house television is located in a bedroom is 0.24 + 0.19 = 0.43.

    (b) Applying the complementation rule, the probability that the television is not located in a bedroom is 1 − 0.43 = 0.57.

    (c) The probability that a randomly selected house television is located in the kitchen, dining room, basement, or garage is 0.05 + 0.05 = 0.10.

30. (a) $P(A) = \dfrac{26268}{142450} = 0.184$

    (b) $P(D \text{ or } E \text{ or } F) = P(D) + P(E) + P(F) = \dfrac{14554 + 11087 + 30926}{142450} = \dfrac{56567}{142450} = 0.397$

    (c)

| Event | Probability |
|---|---|
| A | 0.184 |
| B | 0.160 |
| C | 0.131 |
| D | 0.102 |
| E | 0.078 |
| F | 0.217 |
| G | 0.128 |

31. (a) (not J) is the event that the return selected shows an adjusted gross income of at least $100,000. There are 18,227 thousand such returns.

    (b) (H & I) is the event that the return selected shows an adjusted gross income of between $20,000 and $50,000. There are 44,251 thousand such returns.

    (c) (H or K) is the event that the return selected shows an adjusted gross income of at least $20,000. There are 93,404 thousand such returns.

(d) (H & K) is the event that the return selected shows an adjusted gross income of between $50,000 and $100,000. There are 30,926 thousand such returns.

**32.** (a) Events H and I are not mutually exclusive. Both have events C, D, and E in common.

(b) Events I and K are mutually exclusive. They have no events in common.

(c) Events H and (not J) are mutually exclusive. H = (C or D or E or F) while (not J) = G. H and (not J) have no events in common.

(d) Events H, (not J), and K are not mutually exclusive. While H and (not J) have no events in common, a part of K (i.e., event F) is common to event H, and the other part of K (i.e., event G) is common to event (not J).

**33.** (a)
$$P(H) = \frac{18610+14554+11087+30926}{142450} = \frac{75177}{142450} = 0.528$$
$$P(I) = 1 - \frac{30926+18227}{142450} = 1 - \frac{49153}{142450} = 0.655$$
$$P(J) = 1 - \frac{18227}{142450} = 0.872$$
$$P(K) = \frac{30926+18227}{142450} = \frac{49153}{142450} = 0.345$$

(b) H = (C or D or E or F)

I = (A or B or C or D or E)

J = (A or B or C or D or E or F)

K = (F or G)

(c) P(H) = P(C) + P(D) + P(E) + P(F) = 0.131 + 0.102 + 0.078 + 0.217
= 0.528

P(I) = P(A) + P(B) + P(C) + P(D) + P(E)
= 0.184 + 0.160 + 0.131 + 0.102 + 0.078 = 0.655

P(J) = P(A) + P(B) + P(C) + P(D) + P(E) + P(F)
= 0.184 + 0.160 + 0.131 + 0.102 + 0.078 + 0.217 = 0.872

P(K) = P(F) + P(G) = 0.217 + 0.128 = 0.345

**34.** (a)
$$P(not\ J) = \frac{18227}{142450} = 0.128$$
$$P(H\ \&\ I) = \frac{44251}{142450} = 0.311$$
$$P(H\ or\ K) = \frac{93404}{142450} = 0.656$$
$$P(H\ \&\ K) = \frac{30926}{142450} = 0.217$$

(b) P(J) = 1 - P(not J) = 1 - 0.128 = 0.872

(c) P(H or K) = P(H) + P(K) - P(H & K) = 0.528 + 0.345 - 0.217
= 0.656

**186** Chapter 5

(d) The answer in (c) agrees with that in (a).

**35.** (a) X = 1, 2, 3, 4      (b) {X = 3}

(c) P(X = 3) = 17,302/59,183 = 0.2923.

29.2% of the undergraduates at this university are juniors.

(d)

| Class level x | Probability P(X=x) |
|---|---|
| 1 | 0.1631 |
| 2 | 0.1878 |
| 3 | 0.2923 |
| 4 | 0.3568 |

(e)

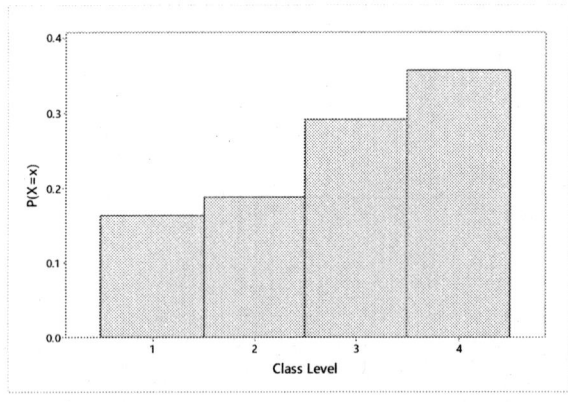

**36.** (a) {Y = 4}

(b) {Y ≥ 4}

(c) {2 ≤ Y ≤ 4}

(d) {Y ≥ 1}

(e) P(Y = 4) = 0.174

(f) P(Y ≥ 4) = P(4) + P(5) + P(6) = 0.174 + 0.105 + 0.043 = 0.322

(g) P(2 ≤ Y ≤ 4) = P(2) + P(3) + P(4) = 0.232 + 0.240 + 0.174 = 0.646

(h) P(Y ≥ 1) = 1 - P(Y ≤ 0) = 1 - 0.052 = 0.948

**37.** The required calculations are

| y | P(Y=y) | yP(Y=y) | $y^2$ | $y^2$P(Y=y) |
|---|---|---|---|---|
| 0 | 0.052 | 0.000 | 0 | 0.000 |
| 1 | 0.154 | 0.154 | 1 | 0.154 |
| 2 | 0.232 | 0.464 | 4 | 0.928 |
| 3 | 0.240 | 0.720 | 9 | 2.160 |
| 4 | 0.174 | 0.696 | 16 | 2.784 |
| 5 | 0.105 | 0.525 | 25 | 2.625 |
| 6 | 0.043 | 0.258 | 36 | 1.548 |
|   |   | 2.817 |   | 10.199 |

(a)  (a)  $\mu_Y = \sum yP(Y=y) = 2.817$.

(b)  On the average, the number of busy lines is about 2.817.

(c)
$$\sigma_y = \sqrt{\sum (Y-\mu_y)^2 P(Y=y)} = \sqrt{2.2635} = 1.504, \text{ or}$$
$$\sigma_y = \sqrt{\sum Y^2 P(Y=y) - \mu_y^2} = \sqrt{10.199 - 2.817^2} = 1.504$$

(d)

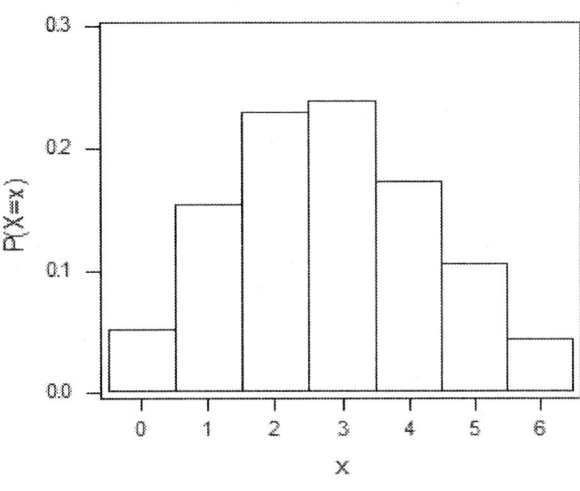

|  -1.7 | -0.2 | 1.3 | 2.8 | 4.3 | 5.8 | 7.3 |
| $\mu-3\sigma$ | $\mu-2\sigma$ | $\mu-\sigma$ | $\mu$ | $\mu+\sigma$ | $\mu+2\sigma$ | $\mu-3\sigma$ |

**38.** Using the Binomial formula with $n = 3$ and $p = 0.493$, the probability distribution for Y, the number of times out of three that the player wins is given by the following table.

| x | P(X=x) |
|---|---|
| 0 | $\binom{3}{0} 0.493^0 0.507^3 = 0.130$ |
| 1 | $\binom{3}{1} 0.493^1 0.507^2 = 0.380$ |
| 2 | $\binom{3}{2} 0.493^2 0.507^1 = 0.370$ |
| 3 | $\binom{3}{3} 0.493^3 0.507^0 = 0.120$ |

**39.** (a) Let X be a random variable equal to the number of penalty kicks by professional soccer players in a sample of 15 that are successful. X is a Binomial random variable with $n = 15$ and $p = 0.85$. The probability that all of the kicks in the sample are successful is

$$P(X=15) = \binom{15}{15} 0.85^{15} 0.15^0 = 0.087.$$

(b) The Probability that at least 13 are successful is
$$P(X \geq 13) = P(X=13) + P(X=14) + P(X=15)$$
$$= \binom{15}{13} 0.85^{13} 0.15^2 + \binom{15}{14} 0.85^{14} 0.15^1 + \binom{15}{15} 0.85^{15} 0.15^0 = 0.286 + 0.231 + 0.087 = 0.604$$

(c) The mean number of successful kicks is $\mu = np = 12.75$. The standard deviation of successful kicks is $\sigma = \sqrt{np(1-p)} = \sqrt{15(0.85)(0.15)} = 1.38$. In a sample of 15 penalty kicks by professional soccer players, the expected average number of successful kicks is 12.75. And, roughly speaking, on average, the number out of 15 penalty kicks that will be successful will differ from the mean number of 12.75 by 1.38.

**40.** The calculations required to answer all parts of this exercise are

$$P(0) = \binom{4}{0} \cdot (0.6)^0 (0.4)^4 = 0.0256 \qquad P(1) = \binom{4}{1} \cdot (0.6)^1 (0.4)^3 = 0.1536$$

$$P(2) = \binom{4}{2} \cdot (0.6)^2 (0.4)^2 = 0.3456 \qquad P(3) = \binom{4}{3} \cdot (0.6)^3 (0.4)^1 = 0.3456$$

$$P(4) = \binom{4}{4} \cdot (0.6)^4 (0.4)^0 = 0.1296$$

(a) P(3) = 0.3456, P(X ≥ 3) = P(3) + P(4) = 0.3456 + 0.1296 = 0.4752, P(X ≤ 3) = P(0) + P(1) + P(2) + P(3) = 0.0256 + 0.1536 + 0.3456 + 0.3456 = 0.8704

(b)

| Number with Pets x | Probability P(X=x) |
|---|---|
| 0 | 0.0256 |
| 1 | 0.1536 |
| 2 | 0.3456 |
| 3 | 0.3456 |
| 4 | 0.1296 |

(c) Left skewed, since p > 0.5.

**41.** (a)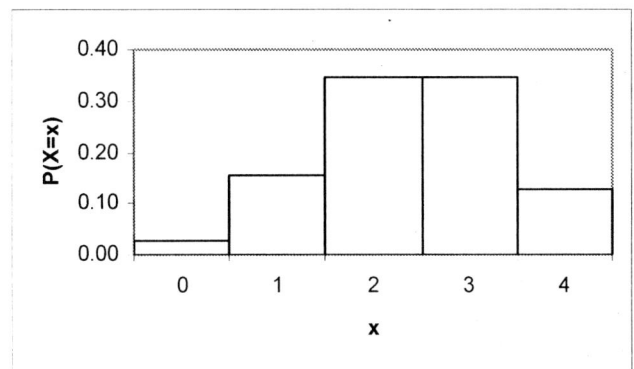

(b) The distribution is only approximate for several reasons: the actual distribution is hypergeometric, based on sampling without replacement; and the success probability p = 0.60 is probably based on a sample.

(c) Yes, because the sample size most likely does not exceed 5% of the population side.

# CHAPTER 6 SOLUTIONS

## Exercises 6.1

**6.1** A density curve is a smooth curve that identifies the shape of a distribution for a variable.

**6.3** The percentage of all possible observations of a variable that lie within a specified range is approximately equal to the corresponding area under its density curve.

**6.5** The percentage of all possible observations of the variable that lie to the right of 4 equals the area under its density curve to the right of <u>4</u>, expressed as a percentage.

**6.7** (a) Since the area to the right of 15 is 0.324, 32.4% of all possible observations of the variable exceed 15.

(b) Since the total area under the density curve is one and the area to the right of 15 is 0.324, the area to the left of 15 must be 1 - 0.324 = 0.676. Therefore, 67.6% of all possible observations of the variable are at most 15.

**6.9** Since the area under the curve that lies between 15 and 20 is 0.414 and the total area under the density curve is one, the area <u>not</u> between 15 and 20 is 1 - 0.414 = 0.586. The area not between 15 and 20 would be the area to the left of 15 or the area to the right of 20. Therefore, 58.6% of all possible observations of the variable are either less than 15 or greater than 20.

**6.11** (a) Since 28.4% of all possible observations of the variable are less than 11, the area under the density curve that lies to the left of 11 is 0.284.

(b) Since the total area under the density curve is one and the area to the left of 11 is 0.284, the area to the right of 11 must be 1 - 0.284 = 0.716.

**6.13** No, because the total area to the right and left of 65 should be 1, not 0.9.

**6.15** The histogram will be roughly bell-shaped.

**6.17** Their distributions are identical. The mean and standard deviation completely determine the shape of a normal distribution. Thus if two normally distributed variables have the same mean and standard deviation, they also have the same distribution.

**6.19** (a) True. Both normal curves have the same spread. Spread is represented by $\sigma$. For each normal curve, $\sigma = 3$.

(b) False. The parameter $\mu$ affects where the normal curve is centered. Since this parameter is different for each normal curve, each normal curve is centered at a different place.

**6.21** True. The value of the parameter $\mu$ has no effect on the spread of a normal curve. The parameter $\mu$ affects only where the normal curve is centered. The spread of the normal curve is determined by the parameter $\sigma$.

**6.23**

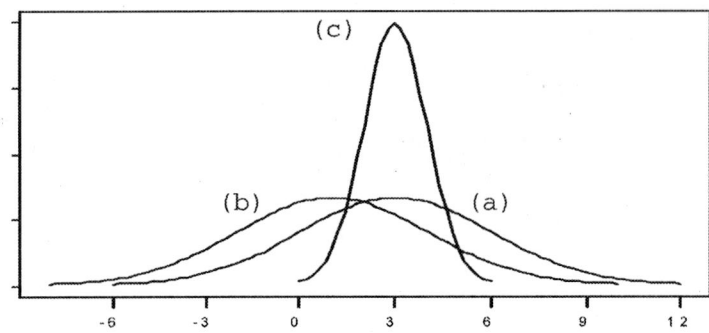

**6.25** The percentage of all possible observations of a normally distributed variable that lie between 2 and 3 is the same as the area under the associated normal curve between 2 and 3. If the variable is only approximately normally distributed, the percentage of all possible observations between 2 and 3 is only approximately the area under the associated normal curve between 2 and 3.

**6.27** If the area under a particular normal curve to the left of 105 is 0.6227, then 62.27% of all possible observations of the variable lie to the left of 105.

**6.29** (a) The following is a graph of the density curve with the equation $y = 2x$ for $0 < x < 1$.

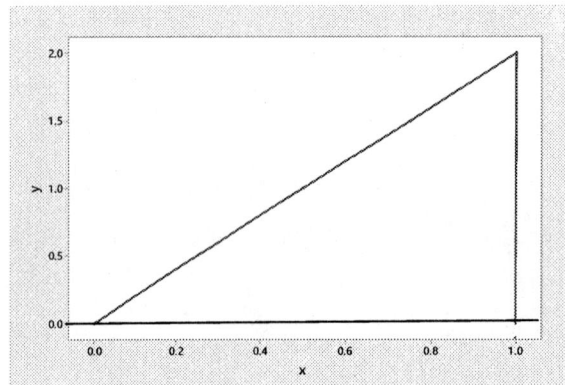

(b) The area to the left of any number $x$ between 0 and 1 would be a triangle with a base of $x$ and a height of $2x$. The area of a triangle is $1/2 \cdot base \cdot height = 1/2 \cdot x \cdot 2x = x^2$.

(c) The area between 1/2 and 3/4 can be calculated by finding the area to the left of 1/2 and subtracting that from the area to the left of 3/4. Using part (b), the area to the left of 1/2 is $(1/2)^2 = 1/4$ and the area to the left of 3/4 is $(3/4)^2 = 9/16$. Therefore, the area between 1/2 and 3/4 is $9/16 - 1/4 = 5/16$.

(d) The area to the left of 1/4, using part (b), is $(1/4)^2 = 1/16$. Therefore, the area to the right of 1/4 is one minus the area to the left, or $1 - 1/16 = 15/16$.

**6.31** (a) The following is a graph of the density curve with the equation $y = 1/30$ for $0 < x < 30$.

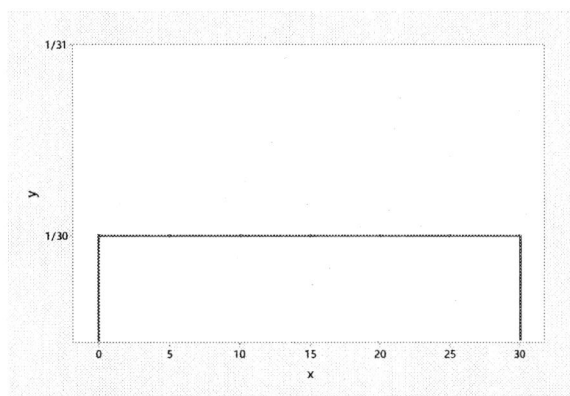

(b) The area to the left of any number x between 0 and 30 would be a rectangle with a base of x and a height of *1/30*. The area of a rectangle is *base·height* = x·1/30 = x/30.

(c) The probability that John waits for the train less than 5 minutes is the area under the density curve to the left of 5. Following part (b), the area to the left of 5 is 5/30 = 1/6. Therefore, John waits for the train less than 5 minutes about 16.7% of the time.

(d) The probability that John waits for the train between 10 and 15 minutes is the area under the density curve between 10 and 15. This area can be calculated by finding the area to the left of 10 and subtracting that from the area to the left of 15. Using part (b), the area to the left of 10 is 10/30 = 1/3 and the area to the left of 15 is 15/30 = 1/2. The area between 10 and 15 is 1/2 - 1/3 = 1/6. Therefore, John waits for the train between 10 and 15 minutes about 16.7% of the time.

(e) The probability that John waits for the train at least 20 minutes is the area under the density curve to the right of 20. The area to the left of 20, using part (b), is 20/30 = 2/3. The area to the right of 20 is one minus the area to the left, or 1 - 2/3 = 1/3. Therefore, John waits for the train at least 20 minutes about 33.3% of the time.

**6.33** (a) The following is a graph of the density curve with the equation $y = 1 - x/2$ for $0 < x < 2$.

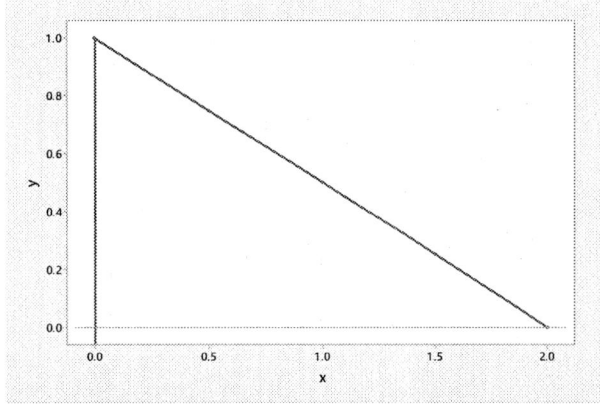

(b) The probability that the losses exceed $1.5 million is the area to the right of 1.5. The area to the left of 1.5, using the given information for area to the left, is $1.5 - 1.5^2/4 = 0.9375$. The area to the right of 1.5 is one minus the area to the left, or 1 − 0.9375 = 0.0625. Therefore, the losses exceed $1.5 million about 6.25% of the time.

**6.35** (a) The percentage of female students who are between 60 and 65 inches tall is 100(0.0450 + 0.0757 + 0.1170 + 0.1480 + 0.1713) = 55.70%.

**194** Chapter 6

(b) The area under the normal curve with parameters μ = 64.4 and σ = 2.4 between 60 and 65 is approximately 0.5570. This is only an estimate because the distribution of heights is only approximately normally distributed.

**6.37** (a)

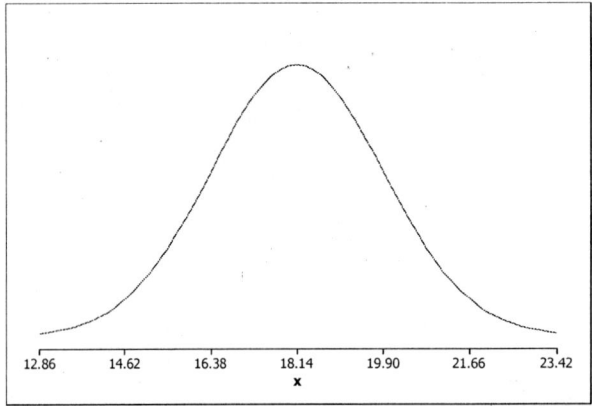

(b) z = (x − 18.14)/1.76

(c) z has a standard normal distribution (μ = 0 and σ = 1).

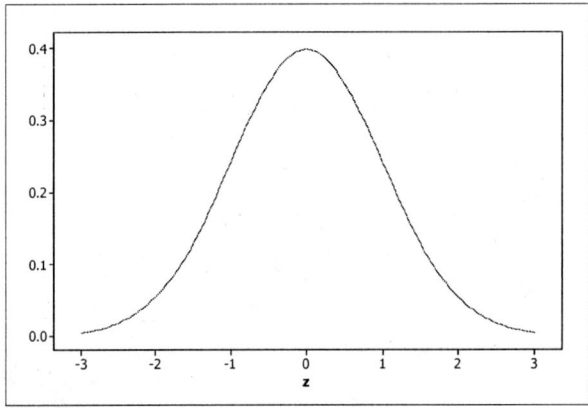

(d) The percentage of adult males *G. mollicoma* that have carapace lengths between 16 mm and 17 mm is equal to the area under the standard normal curve between −1.22 and −0.65.

(e) The percentage of adult males *G. mollicoma* that have carapace lengths exceeding 19 mm is equal to the area under the standard normal curve that lies to the right of 0.49.

**6.39** (a)

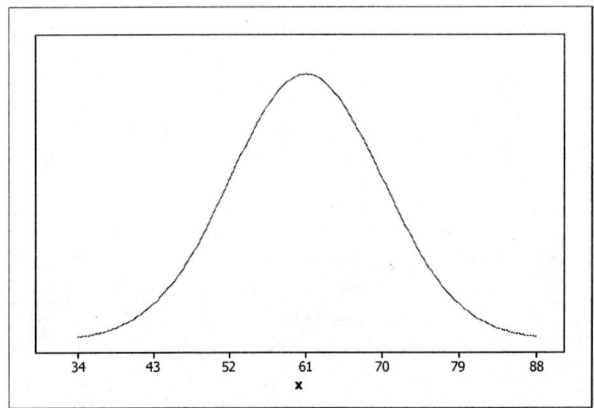

(b)  z = (x - 61)/9

(c)  z has a standard normal distribution (μ = 0 and σ = 1).

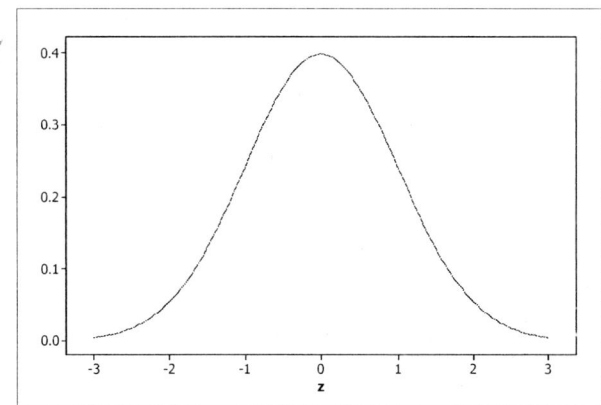

(d)  The percentage of finishers with times between 50 and 70 minutes is equal to the area under the standard normal curve between -1.22 and 1.00.

(e)  The percentage of finishers with times less than 75 minutes is equal to the area under the standard normal curve that lies to the left of 1.56.

**6.41** (a)

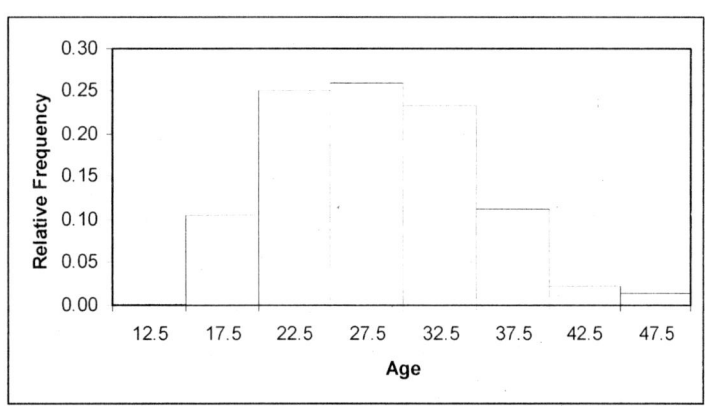

(b)  Yes. The distribution appears to be slightly right skewed, but could be approximated quite well by a normal distribution.

**6.43** (a)

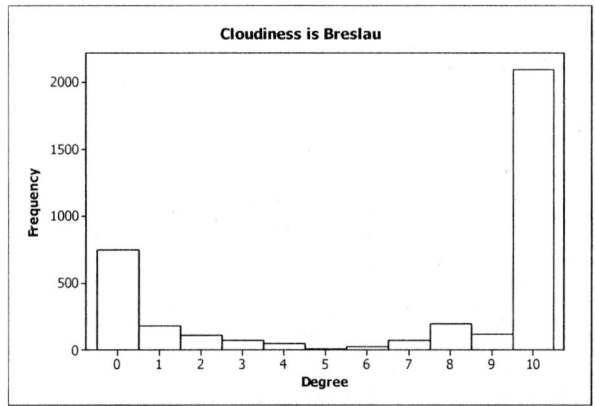

(b) The data is definitely not normal. The distribution instead looks bimodal with the most frequent values of 0 and 10 occurring at the far left and far right of the distribution. The rest of the distribution looks evenly distributed or uniform.

**6.45** (a) Using Minitab, we obtain the data from the WeissStats Resource Site and choose **Graph ▶ Histogram,** select **Simple** from the first row, and click **OK**. Then enter VERBAL and MATH in the Graphs variables text box and click OK. The resulting graphs follow.

 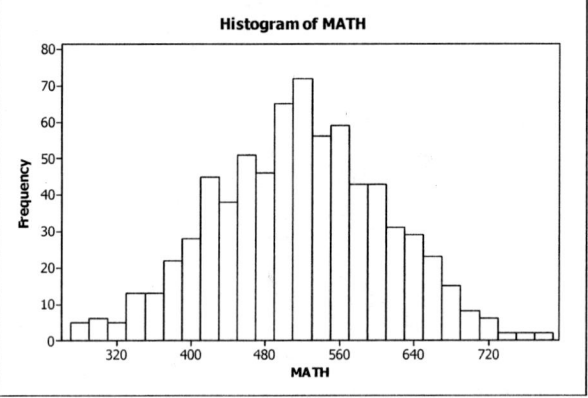

(a) Yes. The SAT verbal scores appearing in the left hand histogram have roughly a bell shaped distribution.
(b) Yes. The SAT verbal scores appearing in the right hand histogram have roughly a bell shaped distribution.

**6.47** The number of chips per bag could <u>not</u> be exactly normally distributed because that number is a discrete random variable, whereas any variable having a normal distribution is a continuous random variable.

**6.49** (a)

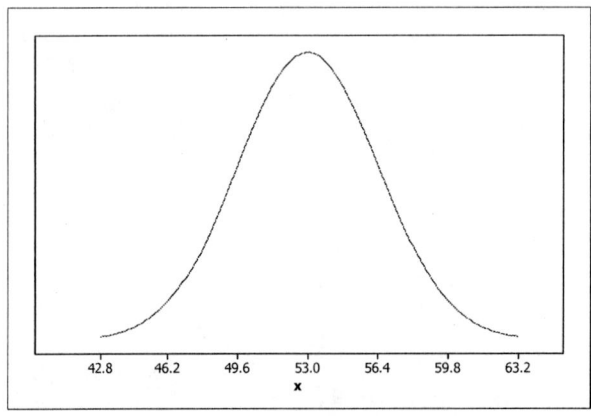

(b) Following the procedure in Example 6.4, we choose **Calc ▶ Random Data ▶ Normal...,** type 1500 in the **Generate rows of data** text box, click in the **Store in column(s)** text box and type DAYS, click in the **Mean** text box and type 53, click in the **Standard deviation** text box and type 3.4, and click **OK**.

(c) We would expect the sample mean and standard deviation to be near 53 and 3.4 respectively since these the corresponding values for the entire population.

(d) We choose **Calc ▶ Column Statistics...,** click on the **Mean** button, click in the **Input variable** text box and select DAYS, and click **OK**. Then repeat the process, selecting the **Standard deviation** button. The results are shown in the Session Window. The results we obtained were

Mean of DAYS = 52.9267 and Standard deviation of DAYS = 3.19167. Your results will vary from ours.

(e) We would expect the histogram for the sample to be bell-shaped since it should be similar to the normal distribution of the population.

(f) Using Minitab, we obtain the histogram below for the sample. Your histogram will likely be similar to ours, but not identical.

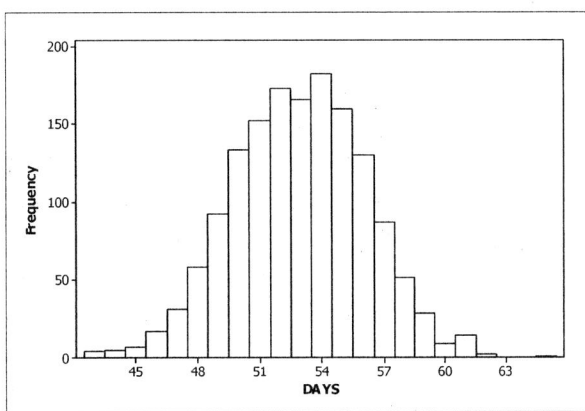

### Exercises 6.2

**6.51** The total area under the curve is 1, and the standard normal curve is symmetric about 0. Therefore, the area to the left of 0 is 0.5, and the area to the right of 0 is 0.5.

**6.53** The area under the standard normal curve to the right of 0.43 is 1 - the area to the left of 0.43. The area to the left of 0.43 is 0.6664. Therefore, the area to the right of 0.43 is 1 - 0.6664 = 0.3336.

**6.55** The area to the left of z = 3.00 is 0.9987 and the area to the left of -3.00 is 0.0013. Therefore the area between -3.00 and 3.00 is 0.9987 - 0.0013 = 0.9974 (99.74%).

**6.56** The standard normal curve is the associated normal curve for a standardized normally distributed variable. The standardized variable is labeled with the letter z, hence the name "z-curve."

**6.57** (a) Locate the row (tenths digit) and column (hundredths digit) of the specified z-score. The corresponding table entry is the area under the standard normal curve that lies to the left of the z-score.

(b) The area that lies under the standard normal curve to the right of a specified z-score is 1 - (area to the left of the z-score).

(c) The area that lies under the standard normal score between two specified z-scores, say $a$ and $b$, where $a < b$, is found by subtracting the area to the left of $a$ from the area to the left of $b$.

**6.59** (a)

Area = 0.9875

(b)

Area = 0.0594

(c)

Area = 0.5

(d)

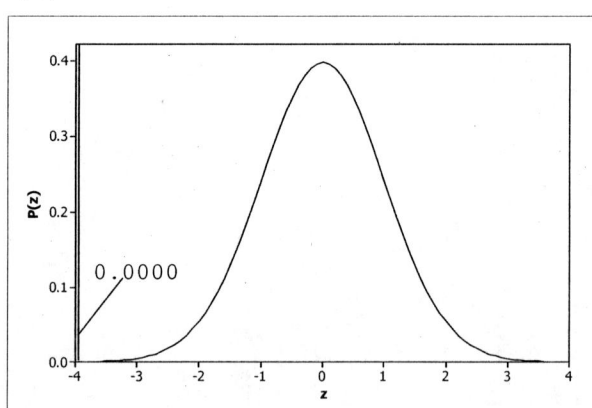

Area = 0.0000 (to 4 dp)

**6.61** (a)

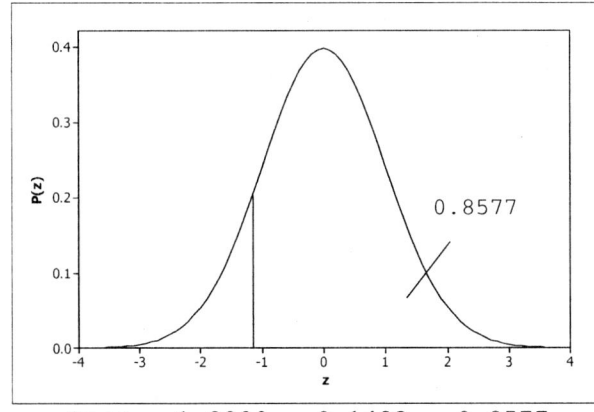

Area = 1.0000 − 0.1423 = 0.8577

(b)

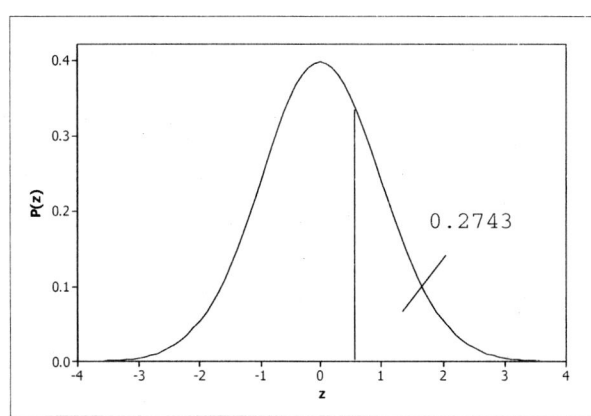

Area = 1.0000 − 0.7257 = 0.2743

(c)

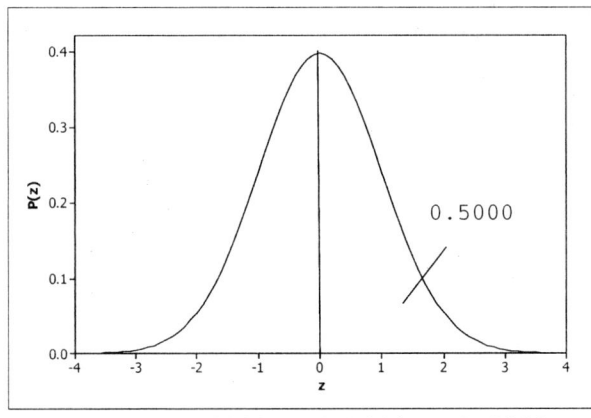

Area = 1.0000 − 0.5000 = 0.5000

(d)

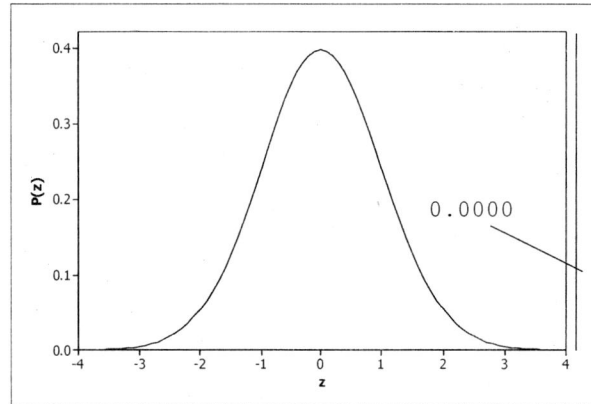

Area = 1.0000 − 1.0000 = 0.0000

**6.63** (a)

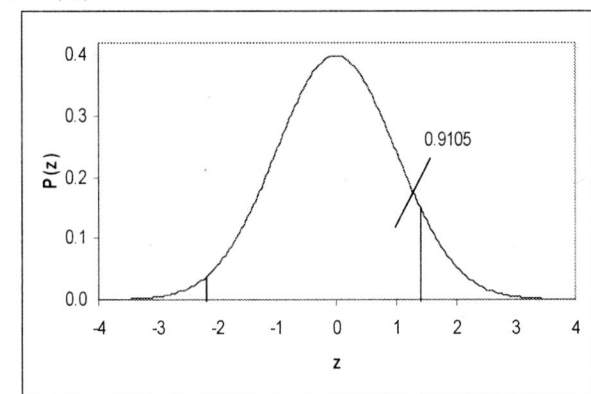

Area = 0.9251 − 0.0146 = 0.9105

(b)

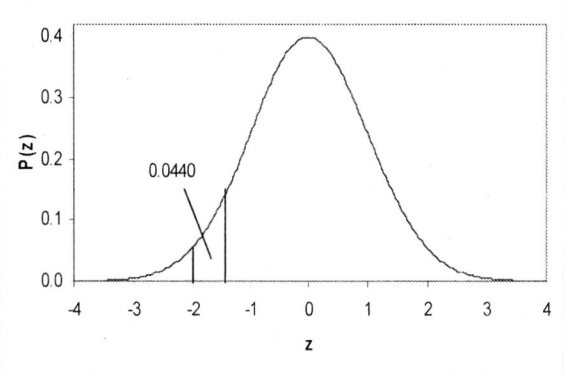

Area = 0.0668 − 0.0228 = 0.0440

(c)  (d)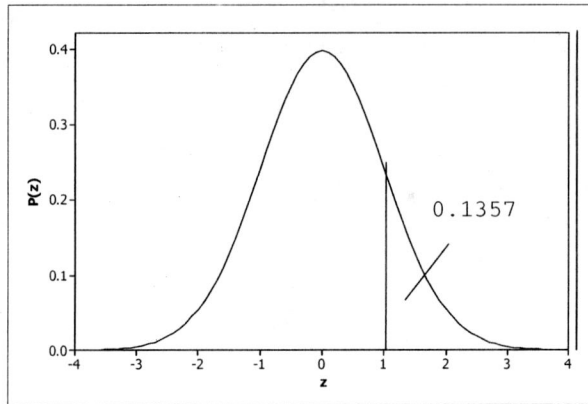

Area = 0.9345 − 0.7224 = 0.2121    Area = 1.0000 − 0.8643 = 0.1357

**6.65** (a)  (b)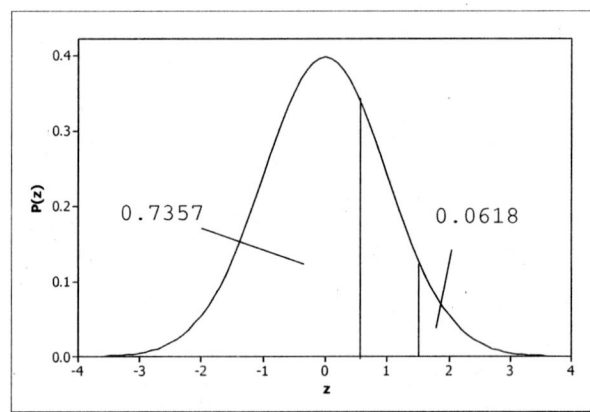

Area = 0.0170 + (1 − 0.9525) = 0.0645    Area = 0.7357 + (1 − 0.9382) = 0.7975

**6.67** (a) The area to the left of z = 1.28 is 0.8997. The area to the left of z = −1.28 is 0.1003. The area between z = −1.28 and z = 1.28 is 0.8997 − 0.1003 = 0.7994.
(b) The area to the left of z = 1.64 is 0.9495. The area to the left of z = −1.64 is 0.0505. The area between z = −1.64 and z = 1.64 is 0.9495 − 0.0505 = 0.8990.
(c) The area to the left of z = −1.96 is 0.0250. The area to the right of z = 1.96 is 1.0000 − 0.9750 = 0.0250. The area either to the left of z = −1.96 or to the right of z = 1.96 is 0.0250 + 0.0250 = 0.0500.
(d) The area to the left of z = −2.33 is 0.0099. The area to the right of z = 2.33 is 1.0000 − 0.9901 = 0.0099. The area either to the left of z = −2.33 or to the right of z = 2.33 is 0.0099 + 0.0099 = 0.0198.

**6.69** (a)

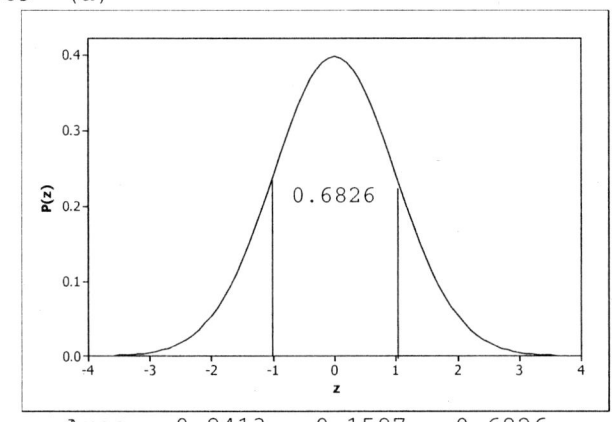

Area = 0.8413 − 0.1587 = 0.6826

(b)

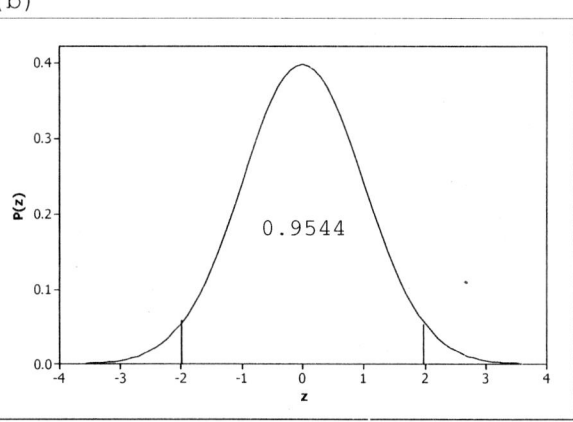

Area = 0.9772 − 0.0228 = 0.9544

(c) Area = 0.9987 − 0.0013 = 0.9974

**6.71**

z = −1.96

**6.73**

z = 0.67

**6.75**

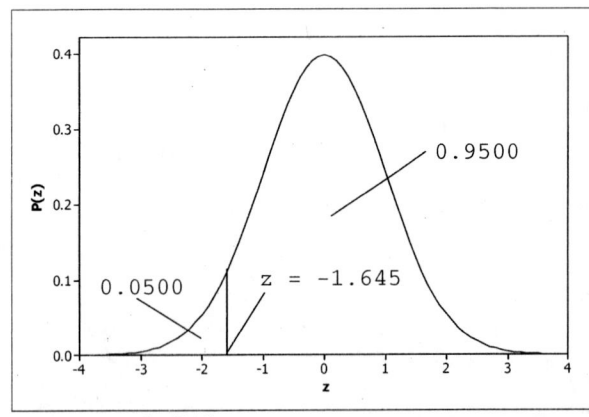

z = -1.645

Note: The area 0.0500 was exactly between the area of two z-scores on the table. Therefore, the two z-scores of -1.64 and -1.65 were averaged.

**6.77**

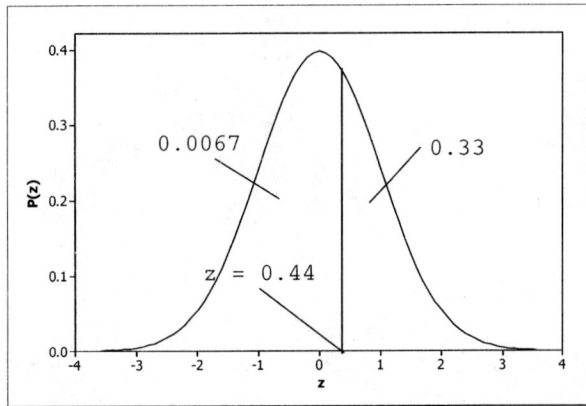

z = 0.44

**6.79** (a)

z = 1.88

(b)

z = 2.575

**6.81**

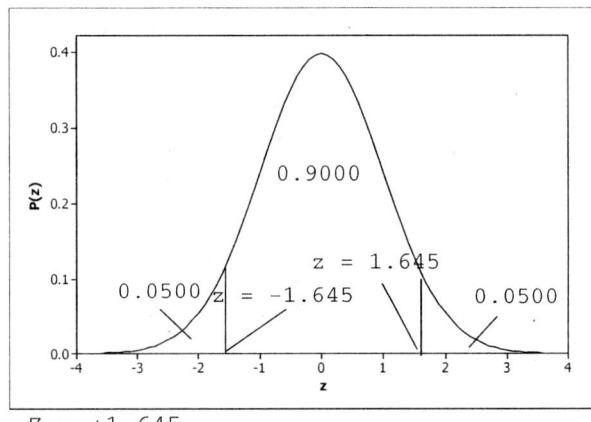

Z = ±1.645

**6.83**

| $z_{0.10}$ | $z_{0.05}$ | $z_{0.025}$ | $z_{0.01}$ | $z_{0.005}$ |
|---|---|---|---|---|
| 1.28 | 1.645 | 1.96 | 2.33 | 2.575 |

**6.85** (a) $z_\alpha$ (b) $-z_\alpha$ (c) $\pm z_{\alpha/2}$
(d)

For part (a):

For part (b):

For part (c):

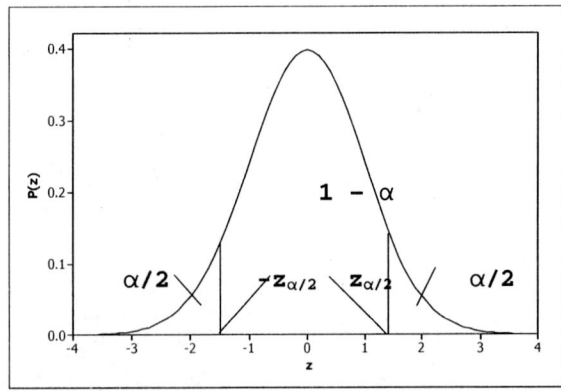

### Exercises 6.3

**6.87** The x values delimiting the interval within two standard deviations either side of the mean are $\mu - 2\sigma$ and $\mu + 2\sigma$. When these are standardized using $z = (x - \mu)/\sigma$, the former becomes $z = [(\mu - 2\sigma) - \mu]/\sigma = -2$ and the latter becomes $z = [(\mu + 2\sigma) - \mu]/\sigma = +2$.

**6.89** (a) For the variable values 1 and 7, the z-values are

$$z = \frac{1-6}{2} = -2.50 \text{ and } z = \frac{7-6}{2} = 0.50$$

The area to the left of z = -2.50 is 0.0062 and the area to the left of z = 0.50 is 0.6915. Therefore the area between z = -2.50 and z = 0.50 is 0.6915 - 0.0062 = 0.6853. Thus the percentage of all possible values of the variable that lie between 1 and 7 is 68.53%.

(b) For a variable value of 5, the z value is

$$z = \frac{5-6}{2} = -0.50$$

The area to the left of z = -0.50 is 0.3085. The area to right of z = -0.50 is 1 - 0.3085 = 0.6915. Thus the percentage of values of the variable exceeding 5 is 69.15%.

(c) For a variable value of 4, the z value is

$$z = \frac{4-6}{2} = -1.00$$

The area to the left of z = -1.00 is 0.1587. Thus the percentage of values of the variable that are less than 4 is 15.87%.

**6.91** (a) For the variable values 6 and 7, the z-values are

$$z = \frac{6-10}{3} = -1.33 \text{ and } z = \frac{7-10}{3} = -1.00$$

The area to the left of z = -1.33 is 0.0918 and the area to the left of z = -1.00 is 0.1587. Therefore the area between z = -1.33 and z = -1.00 is 0.1587 - 0.0918 = 0.0669. Thus the percentage of all possible values of the variable that lie between 6 and 7 is 6.69%.

(b) For a variable value of 10, the z value is

$$z = \frac{10-10}{3} = 0.00$$

The area to the left of z = 0.00 is 0.5000. The area to right of z = 0.00 is 1 - 0.5000 = 0.5000. Thus the percentage of values of the variable that are at least 10 is 50.00%.

(c) For a variable value of 17.5, the z value is

$$z = \frac{17.5-10}{3} = 2.50$$

The area to the left of z = 2.50 is 0.9938. Thus the percentage of values of the variable that are at most 17.5 is 99.38%.

**6.93** (a) Using Table II, we find that an area of 0.2500 lies to left of z = -0.67, an area of .5000 lies to the left of z = 0.00 and an area of 0.7500 lies to the left of 0.67. We convert each of these z-values to x-values using x = μ + zσ. Thus 25% of the values of the variable lie to the left of the first quartile, x = 6 + (-0.67)(2) = 4.66. Half (50%) of the values of the variable lie to the left of the second quartile (median), x = 6 + (0.00)(2) = 6. Finally, 75% of the values of the variable lie to the left of the third quartile, x = 6 + (0.67)(2) = 7.34.

(b) Using Table II, we find that an area of 0.8500 lies to the left of z = 1.04. We convert this z-value to an x-value using x = μ + zσ. Thus 85% of the values of the variable lie to the left of the 85th percentile, x = 6 + (1.04)(2) = 8.08.

(c) If 65% of the possible values exceed the value, then 35% of the possible values are less than the value. Using Table II, we find that an area of 0.3500 lies to the left of z = -0.39. We convert this z-value to an x-value using x = μ + zσ. Thus 65% of the values of the variable exceed x = 6 + (-0.39)(2) = 5.22.

(d) The two values that divide the area into a middle area of 0.95 and two outside areas of 0.025 have 2.5% of the curve and 97.5% of the curve to their left. Using Table II, we find that an area of 0.0250 lies to the left of z = -1.96 and an area of 0.9750 lies to the left of z = 1.96. We convert these z-values to x-values using x = μ + zσ. Thus 95% of the variable lie between x = 6 + (-1.96)(2) = 2.08 and x = 6 + (1.96)(2) = 9.92.

**6.95** (a) Using Table II, we find that an area of 0.2500 lies to left of z = -0.67, an area of .5000 lies to the left of z = 0.00 and an area of

0.7500 lies to the left of 0.67.  We convert each of these z-values to x-values using x = μ + zσ.  Thus 25% of the values of the variable lie to the left of the first quartile, x = 10 + (−0.67)(3) = 7.99.  Half (50%) of the values of the variable lie to the left of the second quartile (median), x = 10 + (0.00)(3) = 10.  Finally, 75% of the values of the variable lie to the left of the third quartile, x = 10 + (0.67)(3) = 12.01.

(b) The seventh decile is the same as the 70th percentile.  Using Table II, we find that an area of 0.7000 lies to the left of z = 0.52.  We convert this z-value to an x-value using x = μ + zσ.  Thus 70% of the values of the variable lie to the left of the seventh decile, x = 10 + (0.52)(3) = 11.56.

(c) If 35% of the possible values exceed the value, then 65% of the possible values are less than the value.  Using Table II, we find that an area of 0.6500 lies to the left of z = 0.39.  We convert this z-value to an x-value using x = μ + zσ.  Thus 35% of the values of the variable exceed x = 10 + (0.39)(3) = 11.17.

(d) The two values that divide the area into a middle area of 0.99 and two outside areas of 0.005 have 0.5% of the curve and 99.5% of the curve to their left.  Using Table II, we find that an area of 0.0050 lies to the left of z = −2.575 and an area of 0.9950 lies to the left of z = 2.575.  We convert these z-values to x-values using x = μ + zσ.  Thus 99% of the variable lie between x = 10 + (−2.575)(3) = 2.275 and x = 10 + (2.575)(3) = 17.725.

**6.97** (a) For carapace lengths 16 mm and 17 mm, the z-values are

$$z = \frac{16 - 18.14}{1.76} = -1.22 \text{ and } z = \frac{17 - 18.14}{1.76} = -0.65$$

The area to the left of z = −1.22 is 0.1112 and the area to the left of z = −0.65 is 0.2578.  Therefore the area between z = −1.22 and z = −0.65 is 0.2578 − 0.1112 = 0.1466.  Thus the percentage of carapace lengths that are between 16 mm and 17 mm is 14.66%.

(b) For a carapace length of 19 mm, the z value is

$$z = \frac{19 - 18.14}{1.76} = 0.49$$

The area to the left of z = 0.49 is 0.6879.  The area to right of 0.49 is  1 − 0.6879 = 0.3121.  Thus the percentage of adult G. mollicoma with carapace lengths exceeding 19 mm is 31.21%.

(c) Using Table II, we find that an area of 0.2500 lies to left of z = −0.67, an area of .5000 lies to the left of z = 0.00 and an area of 0.7500 lies to the left of 0.67.  We convert each of these z-values to x-values using x = μ + zσ.  Thus 25% of the carapace lengths of adult G. mollicoma lie to the left of the first quartile, x = 18.14 + (−0.67)(1.76) = 16.96 mm.  Half (50%) of the carapace lengths of adult G. mollicoma lie to the left of the second quartile (median), x = 18.14 + (0.00)(1.76) = 18.14 mm.  Finally. 75% of the carapace lengths of adult G. mollicoma lie to the left of the third quartile, x = 18.14 + (0.67)(1.76) = 19.32 mm.

(d) Using Table II, we find that an area of 0.9500 lies to the left of z = 1.645.  We convert this z-value to an x-value using x = μ + zσ.  Thus, the 95th percentile of the carapace lengths of adult G. mollicoma is x = 18.14 + (1.645)(1.76) = 21.04 mm.

**6.99** (a) For finishers with times of 50 and 70 minutes, the z-values are

$$z = \frac{50-61}{9} = -1.22 \text{ and } z = \frac{70-61}{9} = 1.00$$

The area to the left of z = -1.225 is 0.1112 and the area to the left of z = 1.00 is 0.8413. Therefore the area between z = -1.22 and z = 1.00 is 0.8413 - 0.1112 = 0.7301. Thus the percentage of finishers with times between 50 minutes and 70 minutes in the New York City 10 km run is 73.01%.

(b) For a finishing time of 75 minutes, the z value is

$$z = \frac{75-61}{9} = 1.56$$

The area to the left of z = 1.56 is 0.9406. Thus the percentage of finishers with times less than 765 minutes is 94.06%.

(c) Using Table II, we find that an area of 0.4000 lies to left of z = -0.25. We convert this z-value to an x-value using x = μ + zσ. Thus 40% of the finishers had times less than the 40th percentile, x = 61 + (-0.25)(9) = 58.75 minutes.

(d) The eighth decile is the same as the 80th percentile. Using Table II, we find that an area of 0.8000 lies to the left of z = 0.84. We convert this z-value to an x-value using x = μ + zσ. Thus, 80% of the finishing times were less than x = 61 + (0.84)(9) = 68.56 minutes.

**6.101** (a) For umbilical cord PH levels that are 7.00 and 7.50, the z values are

$$z = \frac{7-7.32}{0.1} = -3.20 \text{ and } z = \frac{7.50-7.32}{0.1} = 1.80$$

The area to the left of z = -3.20 is 0.0007 and the area to the left of z = 1.80 is 0.9641. The area between z = -3.20 and z = 1.80 is 0.9641 - 0.0007 = 0.9634. Therefore 96.34% of umbilical cord PH levels for preterm infants are between 7.00 and 7.50.

(b) For umbilical cord PH levels that are 7.40, the z value is

$$z = \frac{7.40-7.32}{0.1} = 0.80.$$ The area to the left of z = 0.80 is 0.7881.

Thus the area to the right of z = 0.80 is 1 - 0.7881 = 0.2119, implying that 21.19% of umbilical cord PH levels for preterm infants are over 7.40.

**6.103**

| Part | Standard deviations to either side of the mean | Area under normal curve | Percent |
| --- | --- | --- | --- |
| (a) | 1 | 0.3413 x 2 = 0.6826 | 68 |
| (b) | 2 | 0.4772 x 2 = 0.9544 | 95 |
| (c) | 3 | 0.4987 x 2 = 0.9974 | 99.7 |

**6.105** (a) Approximately 68% of Swedish men have brain weights between <u>1.29</u> and <u>1.51</u> kg.

(b) Approximately 95% of Swedish men have brain weights between 1.18 and 1.62 kg.

(c) Approximately 99.7% of Swedish men have brain weights between 1.07 and 1.73 kg.

(d)

**6.107** (a) From Table 6.1, exactly 11.70% of the heights of female students are between 62 and 63. For heights of 62 and 63 inches, the z values are

$$z = \frac{62-64.4}{2.4} = -1.00 \text{ and } z = \frac{63-64.4}{2.4} = -0.58$$

The area to the left of z = -1.00 is 0.1587 and the area to the left of z = -0.58 is 0.2810. Thus, the area between z = -1.00 and z = -0.58 is 0.2810 - 0.1587 = 0.1223 (12.23%). The two percentages are quite close.

(b) From Table 6.1, exactly = 0.1575 + 0.1100 + 0.0735 + 0.0374 + 0.0199 = 0.3983 (39.83%) of the heights of female students are between 65 and 70 inches. For heights of 65 and 70 inches, the z values are

$$z = \frac{65-64.4}{2.4} = 0.25 \text{ and } z = \frac{70-64.4}{2.4} = 2.33$$

The area to the left of z = 0.25 is 0.5987 and the area to the left of z = 2.33 is 0.9901. Thus, the area between z = 0.25 and a = 2.33 is 0.9901 - 0.5987 = 0.3914 (39.14%). The two percentages are quite close.

**6.109** (a) All of the times, 100%, should be at least 0 years because negative times do not make sense in the context of this problem.

(b) $P(X \geq 0) = 1 - P(X \leq 0) = 1 - P(Z \leq \frac{0-0.18}{0.624}) = 1 - P(Z \leq -0.29) = 1 - 0.3859 = 0.6141$

61.41% of times are at least 0 years.

(c) If x was normally distributed, the probability that the time is at least 0 is approximated to be 61.41%. However, really all of the times should be at least 0 because negative values do not make sense for the variable length of time since taking a math course. Therefore, the variable x is not approximately normally distributed.

**6.111** (a) $P(Y > 5) = P(Z > \frac{5-4.66}{0.75}) = P(Z > 0.45) = 1 - 0.6736 = 0.3264$. We interpret this to mean that about 32.64% of the nests of the booted eagle of western Europe are more than 5 km to the nearest marshland.

(b) $P(3 \leq Y \leq 6) = P(\frac{3-4.66}{0.75} \leq Z \leq \frac{6-4.66}{0.75}) = P(-2.21 \leq Z \leq 1.79)$

$= 0.9633 - 0.0136 = 0.9497$

We interpret this to mean that about 95% of the nests of the booted eagle of western Europe are between 3 and 6 km from the nearest marshland.

**6.113** $P(\mu - z_{\alpha/2}\sigma < x < \mu + z_{\alpha/2}\sigma) = P(-z_{\alpha/2}\sigma < x - \mu < z_{\alpha/2}\sigma)$

$= P(-z_{\alpha/2} < (x - \mu)/\sigma < z_{\alpha/2})$

$= P(-z_{\alpha/2} < z < z_{\alpha/2}) = (1 - \alpha/2) - \alpha/2$

$= 1 - \alpha$

**6.115** The $k$th percentile can be found by first finding the z-score with area of $k/100$ to its left or $1-k/100$ to its right. Then, multiplying this z-score by the standard deviation and adding it to the mean. Therefore, the $k$th percentile is $P_k = \mu + z_{(1-k/100)}\sigma$.

### Exercises 6.4

**6.117** Decisions about whether a variable is normally distributed often are important in subsequent analyses such as percentile calculations or in determining the type of statistical inference procedure to be used.

**6.119** If the normal probability plot is roughly linear except for a small number of points that lie well outside the overall pattern of the plot, it is possible that some or all of those points are outliers. If there is sufficient reason to remove the potential outliers from the sample and doing so results in a plot that is linear without outliers, the analysis can often be carried out with the remaining observations.

**6.121** The variable is approximately normally distributed because the normal probability plot is approximately linear.

**6.123** The variable is not normally distributed because the normal probability plot is not linear and is instead curved.

**6.125** The variable is not normally distributed because the normal probability plot is not linear and is instead curved.

**6.127** (a)

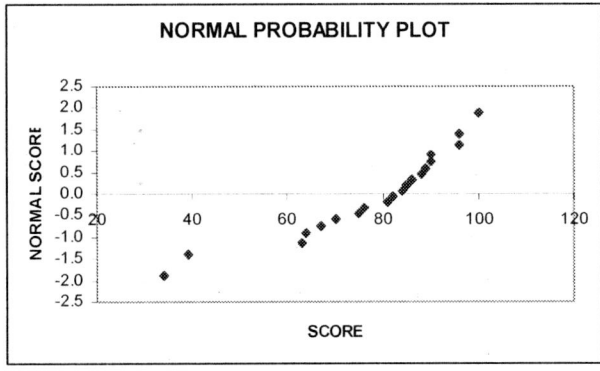

| Exam Score x | Normal Score y |
|---|---|
| 34 | -1.87 |
| 39 | -1.40 |
| 63 | -1.13 |
| 64 | -0.92 |
| 67 | -0.74 |
| 70 | -0.59 |
| 75 | -0.45 |
| 76 | -0.31 |
| 81 | -0.19 |
| 82 | -0.06 |
| 84 | 0.06 |
| 85 | 0.19 |
| 86 | 0.31 |
| 88 | 0.45 |
| 89 | 0.59 |
| 90 | 0.74 |
| 90.01 | 0.92 |
| 96 | 1.13 |
| 96.01 | 1.40 |
| 100 | 1.87 |

(b) Based on the probability plot, there appear to be two outliers in the sample: 34 and 39.

(c) Based on the probability plot, the sample does not appear to come from a normally distributed population.

**6.129** (a)

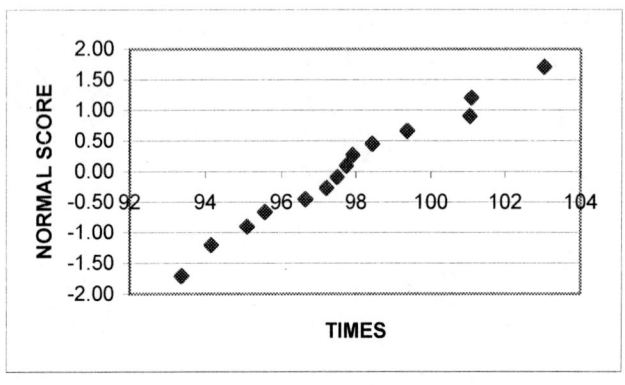

| Times x | Normal Score y |
|---|---|
| 93.37 | -1.71 |
| 94.15 | -1.20 |
| 95.10 | -0.90 |
| 95.57 | -0.66 |
| 96.63 | -0.45 |
| 97.19 | -0.27 |
| 97.47 | -0.09 |
| 97.73 | 0.09 |
| 97.91 | 0.27 |
| 98.44 | 0.45 |
| 99.38 | 0.66 |
| 101.05 | 0.90 |
| 101.09 | 1.20 |
| 103.02 | 1.71 |

(b) Based on the probability plot, there do not appear to be any outliers in the sample.

(c) Based on the probability plot, the sample appears to be from an approximately normally distributed population.

**6.131** (a) We enter the data in Excel in a column which we named "x" at the top. Highlight the name and the data with your mouse and then choose **Charts and Plots.** Select **Normal Probability Plot** from the drop down **Function**

**type** box and enter x in both the **Quantitative variable** and **Label variable** boxes. Click **OK**. The resulting plot is

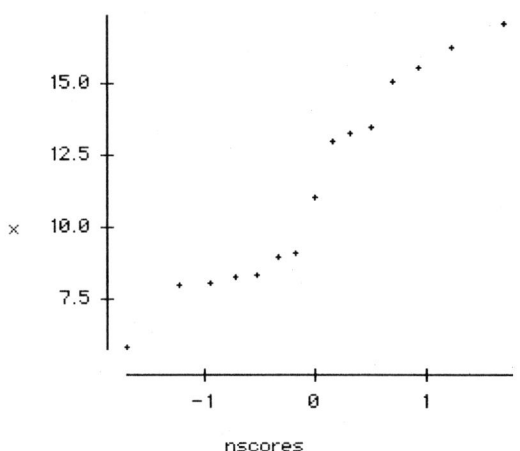

(b) From the plot, there do not appear to be any outliers.

(c) From the plot, normality seems to be a reasonable assumption.

**6.133** (a) We enter the data in Excel in a column which we named "DOU" at the top. Highlight the name and the data with your mouse and then choose **Charts and Plots**. Select **Normal Probability Plot** from the drop down **Function type** box and enter DOU in both the **Quantitative variable** and **Label variable** boxes. Click **OK**. The resulting plot is

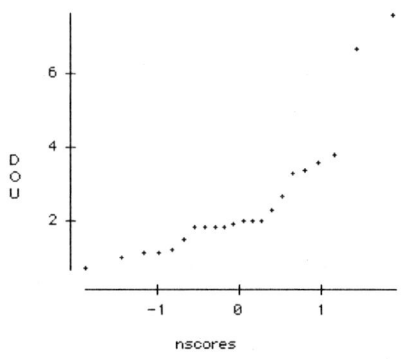

(b) From the plot, it appears that there may be two outliers, 7.6 and 6.7.
(c) From the plot, the data do not appear to be normally distributed. However, if the two outliers were excluded from the data, the remaining data may be normally distributed.

**6.135** (a) Using Minitab and assuming that the data are in a column named 'TEMP', choose **Graph ▶ Histogram...**, select the **Simple** plot and click **OK**. Specify TEMP in the **Graph Variables** text box and click **OK**. The resulting plot is

Aside from the unusual 'saw tooth' pattern of the left side of the graph, an assumption of normality for this data appears to be reasonable.

(b) Choose **Graph ▶ Probability Plot...**, select the **Single** plot and click **OK**. Specify TEMP in the Graph Variables text box. Click on the **Distribution** button, click the **Data Display** tab, select the **Symbols only** option button from the **Data Display** list and click **OK**. Now click the **Scale** button, click on the **Y-Scale Type** tab, select the **Score** option button from the **Y-Scale Type** list, and click **OK** twice. The resulting plot is

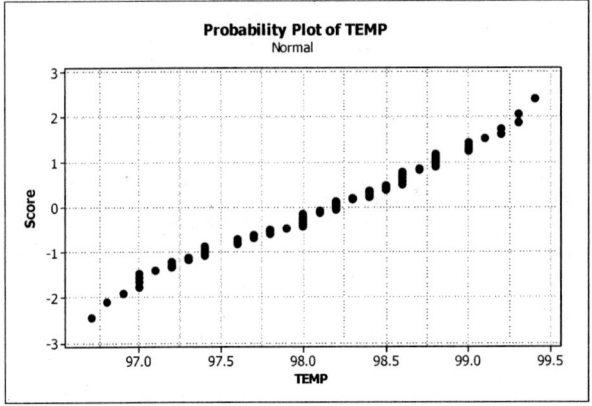

(b) There are no apparent outliers and the graph is roughly linear, so it seems reasonable to assume normality for these data.

(c) We arrived at the same conclusion from both graphs.

6.137 Choose **Graph ▶ Probability Plot...**, select the **Single** plot and click **OK**. Specify CHIPS in the **Graph Variables** text box. Click on the **Distribution** button, click the **Data Display** tab, select the **Symbols only** option button from the **Data Display** list and click **OK**. Now click the **Scale** button, click on the **Y-Scale Type** tab, select the **Score** option button from the **Y-Scale Type** list, and click **OK** twice. The resulting plot is

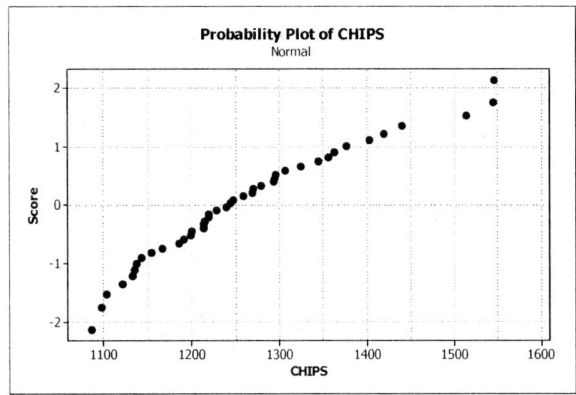

Yes. There do not appear to be any outliers in these data. Since the plot follows a fairly straight line, normality of the chips data seems to be a reasonable assumption.

**6.139** (a) Using Minitab, we will generate four columns of fifty observations each by choosing **Calc ▶ Random data ▶ Normal...**, entering 50 in the **Generate rows of data** text box, clicking in the **Store in column(s)** text box and typing C1-C4, clicking in the **Mean** text box and entering 266, clicking in the **Standard deviation** text box and entering 16, and clicking **OK**. Now click in the worksheet column title row and name the four columns GEST1, GEST2, GEST3, and GEST4.

Next we select **Graph ▶ Probability Plot**, select the **Single** version and click **OK**, then enter GEST1, GEST2, GEST3, and GEST4 in the **Graph Variables** text box. Click **OK**. The resulting four graphs are shown below.

(c) Yes. Since the data were generated from a normal distribution, we would expect the four plots to be roughly linear, as they are. Your simulation will likely result in data and graphs that differ from ours.

### Review Problems for Chapter 6

1. A density curve is a smooth curve that identifies the shape of a distribution for a variable. They are used to approximate percentages of all possible observations that lie within a range by finding corresponding area under its curve.

2. The percentage of all possible observations of a variable that lie between 25 and 50 equals the area under its density curve between <u>25 and 50</u>, expressed as a percentage.

3. (a) Since the area that lies to the left of 60 is 0.364, 36.4% of all possible observations of the variable are less than 60.

    (b) Since the area that lies to the left of 60 is 0.364, the area that lies to the right of 60 is 1.000 - 0.364 = 0.636. Thus, 63.6% of all possible observations of the variable are at least 60.

4. The area between 5 and 6 is 0.728. Therefore, the area outside of 5 and 6 (i.e. the area to the left of 5 or to the right of 6) is 1.000 - 0.728 = 0.272. Thus, 27.2% of the possible observations are either less than 5 or greater than 6.

5. (a) The following is a graph of the density curve with the equation
    $y = 1 - x/2$ for $0 < x < 2$.

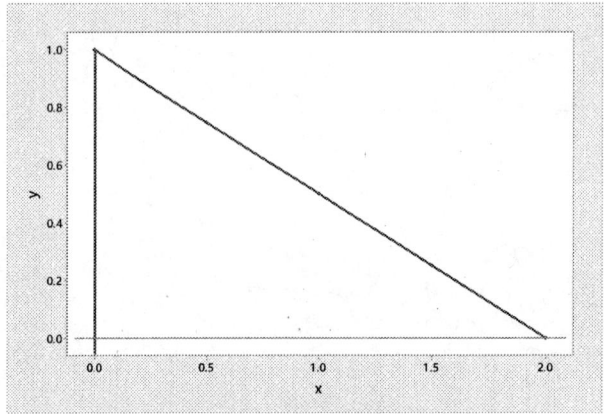

(b) The area to the right of any number $x$ between 0 and 2 would be a triangle with a base of $2 - x$ and a height of $1 - x/2$. The area of a triangle is $1/2 \cdot base \cdot height = 1/2 \cdot (2 - x) \cdot (1 - x/2) = 1 - x + x^2/4$. This is the area the right of $x$. The area to the left of $x$ would be $1 - (1 - x + x^2/4) = x - x^2/4$.

(c) The area between 1/2 and 1 can be calculated by finding the area to the left of 1/2 and subtracting that from the area to the left of 1. Using part (b), the area to the left of 1/2 is $1/2 - (1/2)^2/4 = 7/16$ and the area to the left of 1 is $1 - (1)^2/4 = 3/4$. Therefore, the area between 1/2 and 1 is $3/4 - 7/16 = 5/16$. 31.25% of the possible observations of the variable are between 1/2 and 1.

(d) The probability that the observation of the variable is at least 1.5 is the area to the right of 1.5. The area to the left of 1.5, using part (b), is $1.5 - 1.5^2/4 = 0.9375$. The area to the right of 1.5 is one minus the area to the left, or $1 - 0.9375 = 0.0625$. Therefore, 6.25% of the possible observations of the variable are at least 1.5.

6. Two primary reasons for studying the normal distribution are that:

(a) It is often appropriate to use the normal distribution as the distribution of a population or random variable.

(b) The normal distribution is frequently employed in inferential statistics.

7. (a) A variable is normally distributed if its distribution has the shape of a normal curve.

(b) A population is normally distributed if a variable of the population is normally distributed and it is the only variable under consideration.

(c) The parameters for a normal curve are the mean μ and the standard deviation σ.

8. (a) False. There are many types of distributions that could have the same mean and standard deviation.

(b) True. The mean and standard deviation completely determine a normal distribution, so if two normal distributions have the same mean and standard deviation, then those two distributions are identical.

9. The percentages for a normally distributed variable and areas under the corresponding normal curve (expressed as a percentage) are identical.

10. The distribution of the standardized version of a normally distributed variable is the standard normal distribution, that is, a normal distribution with a mean of 0 and standard deviation of 1.

11. (a) True. The mean completely determines the center of the distribution.

(b) True. The standard deviation completely determines the spread of the distribution.

12. (a) The (second) curve with σ = 6.2 has the largest spread.

(b) The first and second curves are centered at μ = 1.5.

(c) The first and third curves have the same spread because σ = 3 for both.

(d) The third curve is centered farthest to the left because it has the smallest value of μ = -2.7.

(e) The fourth curve is the standard normal curve because μ = 0 and σ = 1.

13. Key Fact 6.4.

**216** Chapter 6

14. (a) The table entry corresponding to the specified z-score is the area to the left of that z-score.
    (b) The area to the right of a specified z-score is found by subtracting the table entry from 1.
    (c) The area between two specified z-scores is found by subtracting the table entry for the smaller z-score from the table entry for the larger z-score.

15. (a) Find the table entry that is closest to the specified area. The z-score determined by locating the corresponding marginal values is the z-score that has the specified area to its left. If the specified area is exactly between two entries, then the two z-scores should be averaged.
    (b) Subtract the specified area from 1. Find the entry in the table that is closest to the result of the subtraction, averaging if the area is exactly between two table entries. The z-score determined by locating the corresponding marginal values is the z-score that has the specified area to its right.

16. The value $z_\alpha$ is the z-score that has area $\alpha$ to its right under the standard normal curve.

17. The empirical rule for variables states that for a normally distributed variable: 68% of all possible observations lie within one standard deviation to either side of the mean; 95% of all possible observations lie within two standard deviations to either side of the mean; 99.7% of all possible observations lie within three standard deviation to either side of the mean.

18. The normal scores for a sample of observations are the observations we would expect to get for a sample of the same size for a variable having the standard normal distribution.

19. If we observe the values of a normally distributed variable for a sample, then a normal probability plot should be roughly <u>linear</u>.

20.

21. (a) The area to the right of 1.05 is 1.0000 - 0.8531 = 0.1469.
    (b) The area to the left of -1.05 is 0.1469 (by symmetry).
    (c) The area between -1.05 and 1.05 is 0.8531 - 0.1469 = 0.7062.

**22.** (a) Area = 0.0013  (b) Area = 1 - 0.7291 = 0.2709

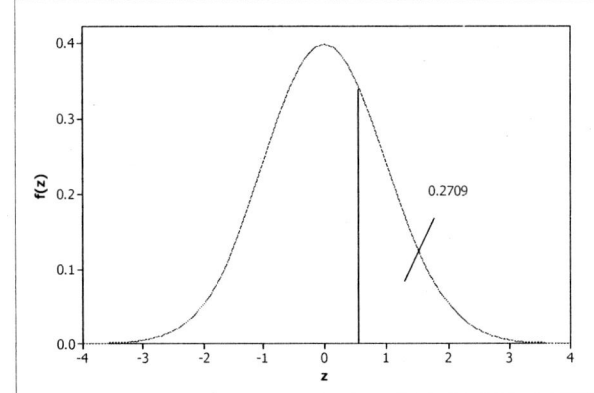

(c) Area = 0.9970 - 0.8665 = 0.1305  (d) Area = 1.000 - 0.0197 = 0.9803

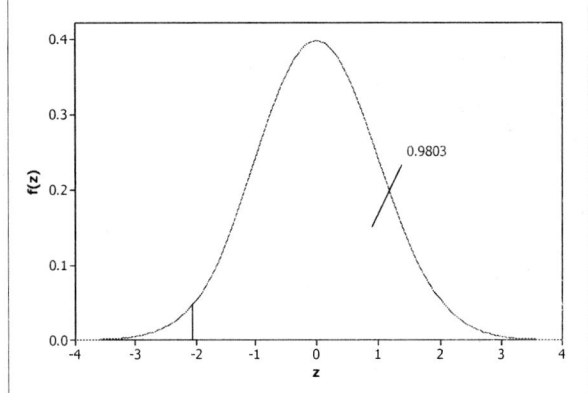

(e) Area = 0.0668 - 0.0000 = 0.0668  (f) Area = 0.8413 + (1 - 0.9987) = 0.8426

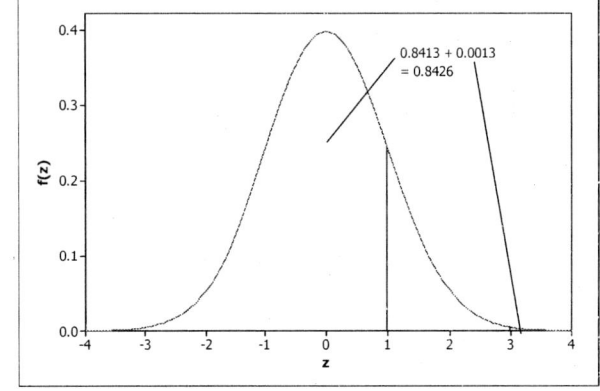

**23.** (a) $z = -0.52$

(b) z will have 0.9 to its left: $z = 1.28$

(c) $z_{0.025} = 1.96$; $z_{0.05} = 1.645$; $z_{0.01} = 2.33$; $z_{0.005} = 2.575$

(d) $-2.575$ and $+2.575$

**24.** (a) The following is a graph of the density curve with the equation $y = 2$ for $5.75 < x < 6.25$.

(b) The area to the left of any number $x$ between 5.75 and 6.25 would be a rectangle with a base of $x - 5.75$ and a height of 2. The area of a rectangle is base·height = $(x - 5.75) \cdot 2 = 2x - 11.5$.

(c) The probability that an observation is less than 6 is the area to the left of 6 under the density curve. Using part (b), the area to the left of 6 is $2(6) - 11.5 = 0.5$. Therefore, 50% of the cups dispensed have less than 6 fl oz.

(d) The probability that an observation is between 5.9 and 6.1 is the area between those two points under the density curve. Using part (b), the area to the left of 5.9 is $2(5.9) - 11.5 = 0.3$ and the area to the left of 6.1 is $2(6.1) - 11.5 = 0.7$. The area between 5.9 and 6.1 is then $0.7 - 0.3 = 0.4$. Therefore, 40% of the cups dispensed have between 5.9 and 6.1 fl oz.

(e) The probability that an observation is at least 5.8 is the area to the right of 5.8 under the density curve. Using part (b), the area to the left of 5.8 is $2(5.8) - 11.5 = 0.1$. The area to the right would be $1 - 0.1 = 0.9$. Therefore, 90% of the cups dispensed have at least 5.8 fl oz.

**25.** (a)

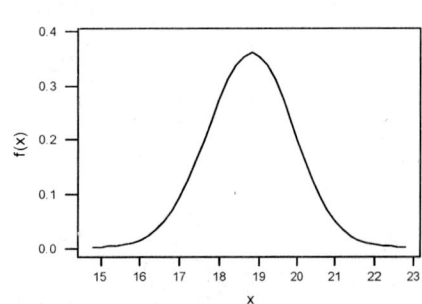

(b) $z = (x - 18.8)/1.1$

(c)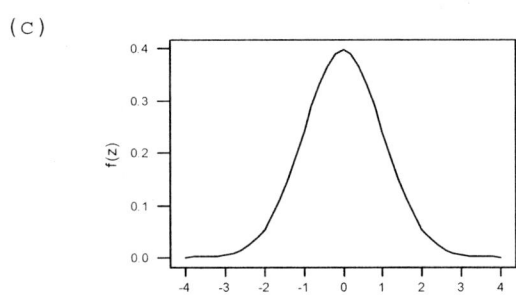

(d) P(17 ≤ x ≤ 20) = 0.8115

(e) The percentage of men who have forearm lengths less than 16 inches equals the area under the standard normal curve that lies to the <u>left</u> of <u>-2.55</u>.

26. (a)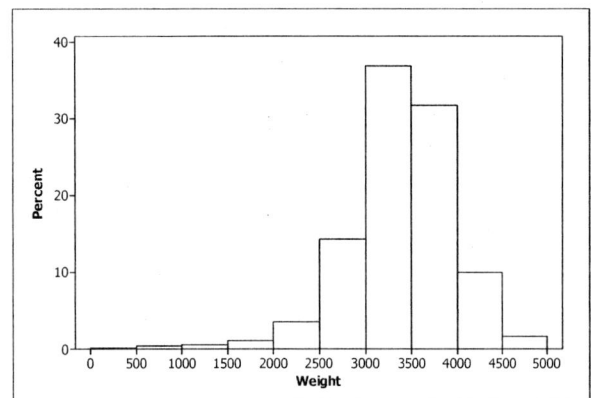

(b) No. The histogram indicates that the data are left skewed.

27. (a) The z-score for x = 304 is (304 − 283.3)/83.3 = 0.25. The area to the left of z = 0.25 is 0.5987, so the percentage of patients whose amount of blood loss during surgery is less than 304 ml is 59.87%.

(b) The z-score for x = 221 is (221 − 283.3)/83.3 = −0.75 and for x = 429, z = (429 − 283.3)/83.3 = 1.75. The area to the left of z = −0.75 is 0.2266 and the area to the left of z = 1.75 is 0.9599. Therefore, the percentage of patients whose amount of blood loss during surgery is between 221 and 429 ml is 0.9599 − 0.2266 = 0.7333, or 73.33%.

(c) The z-score for x = 450 is (450 − 283.3)/83.3 = 2.00. The area to the left of z = 2.00 is 0.9772, so the percentage of patients whose amount of blood loss during surgery is above 450 ml is 1.0000 − 0.9772 = 0.0228, or 2.28%.

28. (a) An area of 0.2500 lies to the left of z = −0.67; an area of 0.5000 lies to the left of z = 0.00; and an area of 0.7500 lies to the left of z = 0.67. To convert these z-scores to verbal GRE scores (x), we use x = μ + zσ. Thus $Q_1$ = 150 + (−0.67)(8.75) = 144.14; the median ($Q_2$) = 150 + (0.00)(8.75) = 150; and $Q_3$ = 150 + (0.67)(8.75) = 155.86. 25% of the verbal GRE scores will be less than 144.14; 50% of the scores will be less than 150; and 75% of the scores will be less than 155.86.

(b) An area of 0.9900 lies to the left of z = 2.33. To convert this z-scores to a verbal GRE score (x), we use x = μ + zσ. Thus the $99^{th}$ percentile is 150 + (2.33)(8.75) = 170.39. 99% of the verbal GRE scores will be less than 170.39.

29. (a) Approximately 68% of students who took the verbal portion of the GRE scored between <u>141.25</u> and <u>158.75</u>.

**220** Chapter 6

(b) Approximately 95% of students who took the verbal portion of the GRE scored between 132.50 and 167.50.

(c) Approximately 99.7% of students who took the verbal portion of the GRE scored between 123.75 and 176.25.

**30.** (a) Order the lengths from smallest to largest. Obtain the normal scores for 12 observations from Table III. The result is the table below. The normal probability plot is to the left of the table.

| Length | Normal score |
|--------|--------------|
| 47.6 | -1.64 |
| 48.1 | -1.11 |
| 48.5 | -0.79 |
| 52.4 | -0.53 |
| 54.6 | -0.31 |
| 55.4 | -0.10 |
| 55.8 | 0.10 |
| 57.1 | 0.31 |
| 58.6 | 0.53 |
| 62.5 | 0.79 |
| 65.1 | 1.11 |
| 67.9 | 1.64 |

(b) There do not appear to be any outliers.

(c) The plot does not appear linear, so it does not seem reasonable to assume that the variable is normally distributed.

**31.** (a) Using Minitab, enter the data in a column named EMPLOYEES. Select **Graph ▶ Probability Plot**, select **Single** and click **OK**. Enter EMPLOYEES in the **Graph variables,** click on the **Distributions** button and the **Data Display** tab. Check the **Symbols only** option and click **OK**. Then click on the **Scale** button and the **Y-Scale Type** tab and select the **Score** option. Click OK twice. The resulting graph is

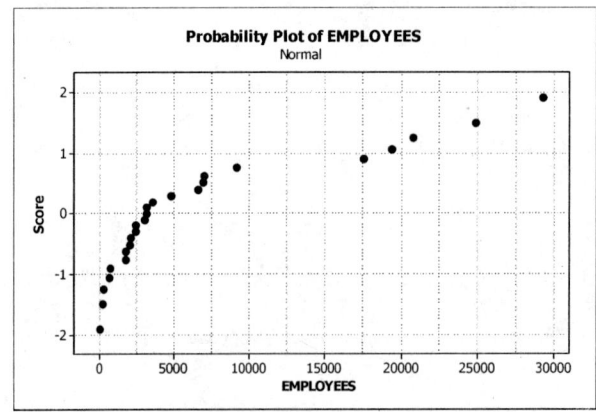

(b) There are five observations that are considerably greater than the other 20. If the distribution were otherwise normally distributed, these might be possible outliers.

(c) The graph is not close to being roughly linear. The data are very right skewed and hence not normally distributed, so the five possible outliers may be just part of the pattern of right skewness.

Copyright © 2016 Pearson Education, Inc.

# CHAPTER 7 SOLUTIONS

## Exercises 7.1

**7.1** Sampling is often preferable to conducting a census because it is quicker, less costly, and sometimes it is the only practical way to get information.

**7.3** (a) $\mu = \Sigma x/N = 6/3 = 2$

(b) for $n = 1$

| Sample | $\bar{x}$ |
|---|---|
| 1 | 1.0 |
| 2 | 2.0 |
| 3 | 3.0 |

for $n = 2$

| Sample | $\bar{x}$ |
|---|---|
| 1,2 | 1.2 |
| 1,3 | 2.0 |
| 2,3 | 2.5 |

for $n = 3$

| Sample | $\bar{x}$ |
|---|---|
| 1,2,3 | 3.0 |

See the dotplots in part (c).

(c)

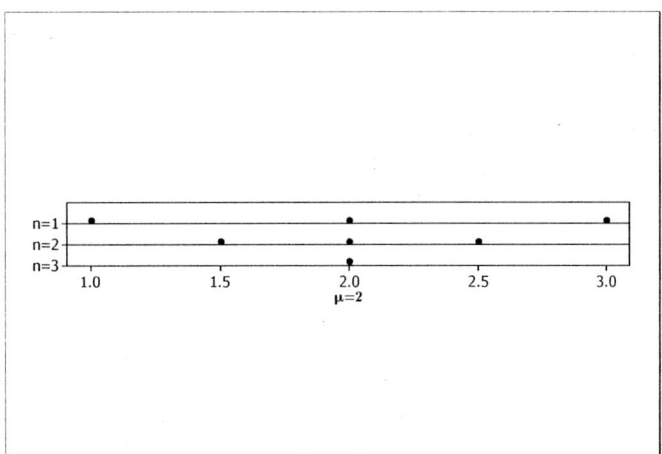

The sample means cluster more closely around the population mean as the sample size increases. Sampling error is smaller for large samples than for small samples.

(d) For $n = 1$, one out of the three sample means equaled the population mean. Thus, the probability that the sample mean equals the population mean for $n = 1$ is 1/3.

For $n = 2$, one out of the three sample means equaled the population mean. Thus, the probability that the sample mean equals the population mean for $n = 2$ is 1/3.

For $n = 3$, the one sample mean equals the population mean, Thus, the probability that the sample mean equals the population mean for $n = 3$ is 1.0.

(e) If the absolute value of the difference between the sample mean and the population mean is at most 0.5, that means that we are looking for sample means within the range 1.5 to 2.5, inclusive.

For $n = 1$, one out of the three sample means is in this range. Thus, the probability that the sampling error will be 0.5 or less for $n = 1$ is 1/3.

For $n = 2$, all of the three sample means are in this range. Thus, the probability that the sampling error will be 0.5 or less for $n = 2$ is 1.0.

For $n = 3$, the one sample mean is in this range. Thus, the probability that the sampling error will be 0.5 or less for $n = 3$ is 1.0.

**7.5** (a) $\mu = \Sigma x/N = 10/4 = 2.5$

(b) for $n = 1$

| Sample | $\bar{x}$ |
|---|---|
| 1 | 1.0 |
| 2 | 2.0 |
| 3 | 3.0 |
| 4 | 4.0 |

for $n = 2$

| Sample | $\bar{x}$ |
|---|---|
| 1,2 | 1.5 |
| 1,3 | 2.0 |
| 1,4 | 2.5 |
| 2,3 | 2.5 |
| 2,4 | 3.0 |
| 3,4 | 3.5 |

for $n = 3$

| Sample | $\bar{x}$ |
|---|---|
| 1,2,3 | 2.0 |
| 1,2,4 | 2.3 |
| 1,3,4 | 2.7 |
| 2,3,4 | 3.0 |

for n = 4

| Sample | $\bar{x}$ |
|---|---|
| 1,2,3,4 | 2.5 |

See the dotplots in part (c).

(c)

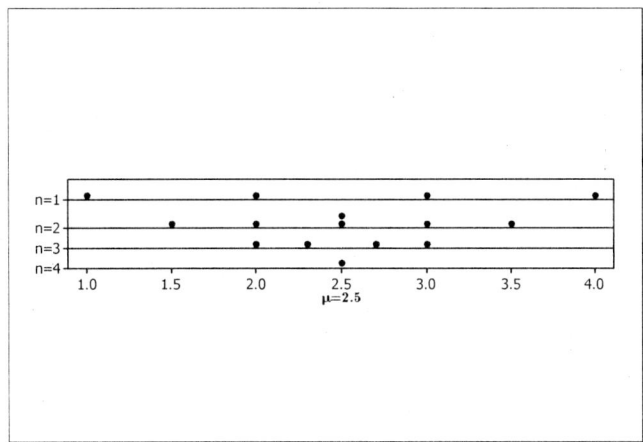

The sample means cluster more closely around the population mean as the sample size increases. Sampling error is smaller for large samples than for small samples.

(d) For n = 1, none of the four sample means equaled the population mean. Thus, the probability that the sample mean equals the population mean for n = 1 is 0.

For n = 2, two out of the six sample means equaled the population mean. Thus, the probability that the sample mean equals the population mean for n = 2 is 2/6 = 1/3.

For n = 3, none of the four sample means equaled the population mean. Thus, the probability that the sample mean equals the population mean for n = 3 is 0.

For n = 4, the one sample mean equals the population mean, Thus, the probability that the sample mean equals the population mean for n = 4 is 1.

(e) If the absolute value of the difference between the sample mean and the population mean is at most 0.5, that means that we are looking for sample means within the range 2.0 to 3.0, inclusive.

For n = 1, two out of the four sample means is in this range. Thus, the probability that the sampling error will be 0.5 or less for n = 1 is 2/4 = 1/2.

For n = 2, four of the six sample means are in this range. Thus, the probability that the sampling error will be 0.5 or less for n = 2 is 4/6 = 2/3.

For n = 3, all four of the sample means are in this range. Thus, the probability that the sampling error will be 0.5 or less for n = 3 is 1.

For n = 4, the one sample mean is in this range. Thus, the probability that the sampling error will be 0.5 or less for n = 4 is 1.0.

**7.7** (a) $\mu = \Sigma x/N = 15/5 = 3.0$

(b) for n = 1

| Sample | $\bar{x}$ |
|---|---|
| 1 | 1.0 |
| 2 | 2.0 |
| 3 | 3.0 |
| 4 | 4.0 |
| 5 | 5.0 |

for n = 2

| Sample | $\bar{x}$ |
|---|---|
| 1,2 | 1.5 |
| 1,3 | 2.0 |
| 1,4 | 2.5 |
| 1,5 | 3.0 |
| 2,3 | 2.5 |
| 2,4 | 3.0 |
| 2,5 | 3.5 |
| 3,4 | 3.5 |
| 3,5 | 4.0 |
| 4,5 | 4.5 |

for n = 3

| Sample | $\bar{x}$ |
|---|---|
| 1,2,3 | 2.0 |
| 1,2,4 | 2.3 |
| 1,2,5 | 2.7 |
| 1,3,4 | 2.7 |
| 1,3,5 | 3.0 |
| 1,4,5 | 3.3 |
| 2,3,4 | 3.0 |
| 2,3,5 | 3.3 |
| 2,4,5 | 3.7 |
| 3,4,5 | 4.0 |

for n = 4

| Sample | $\bar{x}$ |
|---|---|
| 1,2,3,4 | 2.50 |
| 1,2,3,5 | 2.75 |
| 1,2,4,5 | 3.00 |
| 1,3,4,5 | 3.25 |
| 2,3,4,5 | 3.50 |

for n = 5

| Sample | $\bar{x}$ |
|---|---|
| 1,2,3,4,5 | 3.0 |

See the dotplots in part (c).

(c)

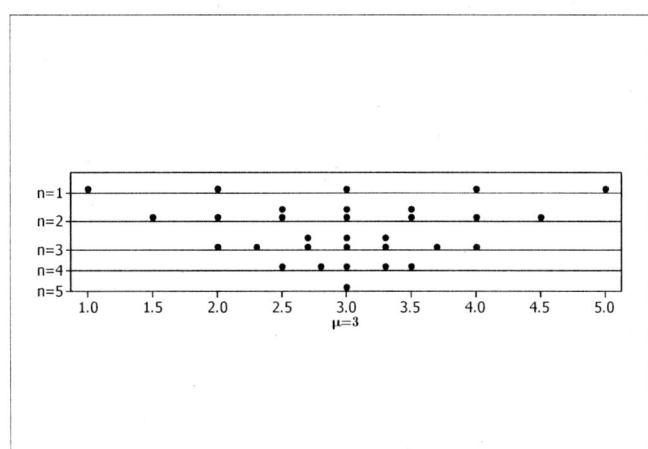

The sample means cluster more closely around the population mean as the sample size increases. Sampling error is smaller for large samples than for small samples.

(d) For $n = 1$, one of the five sample means equaled the population mean. Thus, the probability that the sample mean equals the population mean for $n = 1$ is 1/5.

For $n = 2$, two out of the ten sample means equaled the population mean. Thus, the probability that the sample mean equals the population mean for $n = 2$ is 2/10 = 1/5.

For $n = 3$, two out of the ten sample means equaled the population mean. Thus, the probability that the sample mean equals the population mean for $n = 3$ is 2/10 = 1/5.

For $n = 4$, one out of the five sample means equaled the population mean. Thus, the probability that the sample mean equals the population mean for $n = 4$ is 1/5.

For $n = 5$, the one sample mean equals the population mean, Thus, the probability that the sample mean equals the population mean for $n = 5$ is 1.

(e) If the absolute value of the difference between the sample mean and the population mean is at most 0.5, that means that we are looking for sample means within the range 2.5 to 3.5, inclusive.

For $n = 1$, one of the five sample means is in this range. Thus, the probability that the sampling error will be 0.5 or less for $n = 1$ is 1/5.

For $n = 2$, six of the ten sample means are in this range. Thus, the probability that the sampling error will be 0.5 or less for $n = 2$ is 6/10 = 3/5.

For $n = 3$, six of the ten sample means are in this range. Thus, the probability that the sampling error will be 0.5 or less for $n = 3$ is 6/10 = 3/5.

For $n = 4$, all of the five sample means are in this range. Thus, the probability that the sampling error will be 0.5 or less for $n = 4$ is 1.

For $n = 5$, the one sample mean is in this range. Thus, the probability that the sampling error will be 0.5 or less for $n = 5$ is 1.0.

**7.9** (a) $\mu = \Sigma x/N = 21/6 = 3.5$

(b) for $n = 1$

| Sample | $\bar{x}$ |
|---|---|
| 1 | 1.0 |
| 2 | 2.0 |
| 3 | 3.0 |
| 4 | 4.0 |
| 5 | 5.0 |
| 6 | 6.0 |

for $n = 2$

| Sample | $\bar{x}$ |
|---|---|
| 1,2 | 1.5 |
| 1,3 | 2.0 |
| 1,4 | 2.5 |
| 1,5 | 3.0 |
| 1,6 | 3.5 |
| 2,3 | 2.5 |
| 2,4 | 3.0 |
| 2,5 | 3.5 |
| 2,6 | 4.0 |
| 3,4 | 3.5 |
| 3,5 | 4.0 |
| 3,6 | 4.5 |
| 4,5 | 4.5 |
| 4,6 | 5.0 |
| 5,6 | 5.5 |

for $n = 3$

| Sample | $\bar{x}$ |
|---|---|
| 1,2,3 | 2.0 |
| 1,2,4 | 2.3 |
| 1,2,5 | 2.7 |
| 1,2,6 | 3.0 |
| 1,3,4 | 2.7 |
| 1,3,5 | 3.0 |
| 1,3,6 | 3.3 |
| 1,4,5 | 3.3 |
| 1,4,6 | 3.7 |
| 1,5,6 | 4.0 |
| 2,3,4 | 3.0 |
| 2,3,5 | 3.3 |
| 2,3,6 | 3.7 |
| 2,4,5 | 3.7 |

| Sample | $\bar{x}$ |
|---|---|
| 2,4,6 | 4.0 |
| 2,5,6 | 4.3 |
| 3,4,5 | 4.0 |
| 3,4,6 | 4.3 |
| 3,5,6 | 4.7 |
| 4,5,6 | 5.0 |

for $n = 4$

| Sample | $\bar{x}$ |
|---|---|
| 1,2,3,4 | 2.50 |
| 1,2,3,5 | 2.75 |
| 1,2,3,6 | 3.00 |
| 1,2,4,5 | 3.00 |
| 1,2,4,6 | 3.25 |
| 1,2,5,6 | 3.50 |
| 1,3,4,5 | 3.25 |
| 1,3,4,6 | 3.50 |
| 1,3,5,6 | 3.75 |
| 1,4,5,6 | 4.00 |
| 2,3,4,5 | 3.50 |
| 2,3,4,6 | 3.75 |
| 2,3,5,6 | 4.00 |
| 2,4,5,6 | 4.25 |
| 3,4,5,6 | 4.50 |

for $n = 5$

| Sample | $\bar{x}$ |
|---|---|
| 1,2,3,4,5 | 3.0 |
| 1,2,3,4,6 | 3.2 |
| 1,2,3,5,6 | 3.4 |
| 1,2,4,5,6 | 3.6 |
| 1,3,4,5,6 | 3.8 |
| 2,3,4,5,6 | 4.0 |

for $n = 6$

| Sample | $\bar{x}$ |
|---|---|
| 1,2,3,4,5,6 | 3.5 |

See the dotplots in part (c).

**228** Chapter 7

(c)

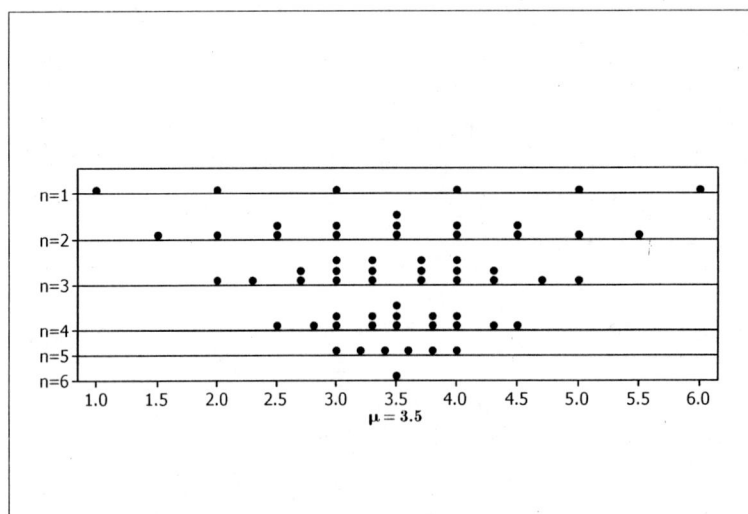

The sample means cluster more closely around the population mean as the sample size increases. Sampling error is smaller for large samples than for small samples.

(d) For $n = 1$, none of the six sample means equaled the population mean. Thus, the probability that the sample mean equals the population mean for $n = 1$ is 0.

For $n = 2$, three out of the fifteen sample means equaled the population mean. Thus, the probability that the sample mean equals the population mean for $n = 2$ is $3/15 = 1/5$.

For $n = 3$, none of the twenty sample means equaled the population mean. Thus, the probability that the sample mean equals the population mean for $n = 3$ is 0.

For $n = 4$, three out of the fifteen sample means equaled the population mean. Thus, the probability that the sample mean equals the population mean for $n = 4$ is $3/15 = 1/5$.

For $n = 5$, none of the six sample means equaled the population mean. Thus, the probability that the sample mean equals the population mean for $n = 5$ is 0.

For $n = 6$, the one sample mean equals the population mean, Thus, the probability that the sample mean equals the population mean for $n = 6$ is 1.

(e) If the absolute value of the difference between the sample mean and the population mean is at most 0.5, that means that we are looking for sample means within the range 3.0 to 4.0, inclusive.

For $n = 1$, two of the six sample means is in this range. Thus, the probability that the sampling error will be 0.5 or less for $n = 1$ is $2/6 = 1/3$.

For $n = 2$, seven of the fifteen sample means are in this range. Thus, the probability that the sampling error will be 0.5 or less for $n = 2$ is $7/15$.

For $n = 3$, twelve of the twenty sample means are in this range. Thus, the probability that the sampling error will be 0.5 or less for $n = 3$ is $12/20 = 3/5$.

For $n = 4$, eleven of the fifteen sample means are in this range. Thus, the probability that the sampling error will be 0.5 or less for n = 4 is 11/15.

For n = 5, all of the six sample means are in this range. Thus, the probability that the sampling error will be 0.5 or less for n = 5 is 1.

For n = 6, the one sample mean is in this range. Thus, the probability that the sampling error will be 0.5 or less for n = 6 is 1.0.

**7.11** (a) μ = Σx/N = 393/5 = 78.6 inches.

(b)

| Sample | Heights | $\bar{x}$ |
|---|---|---|
| B,W | 83, 76 | 79.5 |
| B,J | 83, 80 | 81.5 |
| B,C | 83, 74 | 78.5 |
| B,H | 83, 80 | 81.5 |
| W,J | 76, 80 | 78.0 |
| W,C | 76, 74 | 75.0 |
| W,H | 76, 80 | 78.0 |
| J,C | 80, 74 | 77.0 |
| J,H | 80, 80 | 80.0 |
| C,H | 74, 80 | 77.0 |

(c)

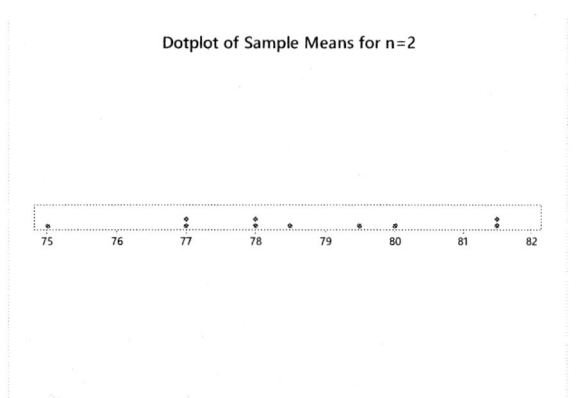

Dotplot of Sample Means for n=2

(d) $P(\bar{x} = \mu) = P(\bar{x} = 78.6) = 0.0$

(e) $P(78.6 - 1 \leq \bar{x} \leq 78.6 + 1) = P(77.6 \leq \bar{x} \leq 79.6)$
$= 4/10 = 0.4$

If we take a random sample of two heights, there is a 40% chance that the mean of the sample selected will be within one inch of the population mean.

**7.13** (b)

| Sample | Heights | $\bar{x}$ |
|---|---|---|
| B,W,J | 83,76,80 | 79.67 |
| B,W,C | 83,76,74 | 77.67 |
| B,W,H | 83,76,80 | 79.67 |
| B,J,C | 83,80,74 | 79.00 |
| B,J,H | 83,80,80 | 81.00 |
| B,C,H | 83,74,80 | 79.00 |
| W,J,C | 76,80,74 | 76.67 |
| W,J,H | 76,80,80 | 78.67 |
| W,C,H | 76,74,80 | 76.67 |
| J,C,H | 80,74,80 | 78.00 |

(c)

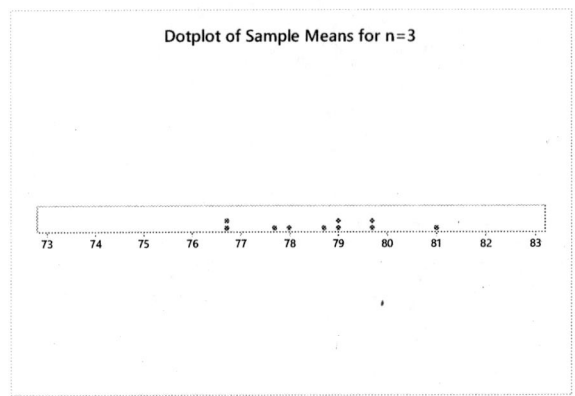

(d) $P(\bar{x} = \mu) = P(\bar{x} = 78.6) = 0.0$

(e) $P(78.6 - 1 \leq \bar{x} \leq 78.6 + 1) = P(77.6 \leq \bar{x} \leq 79.6) = 5/10 = 0.5$

If we take a random sample of three heights, there is a 50% chance that the mean of the sample selected will be within one inch of the population mean.

**7.15** (b)

| Sample | Heights | $\bar{x}$ |
|---|---|---|
| B,W,J,C,H | 83,76,80,74,80 | 78.60 |

(c)

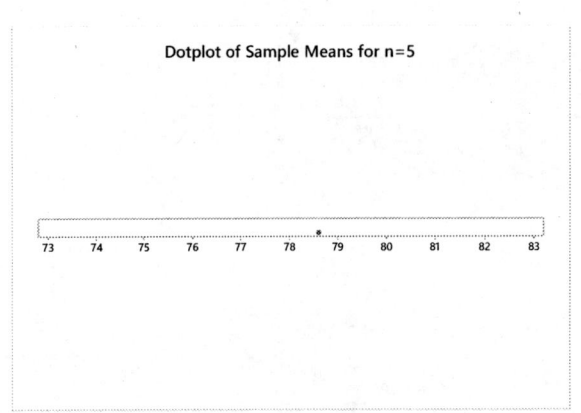

(d) $P(\bar{x} = \mu) = P(\bar{x} = 78.6) = 1.0$

(e) $P(78.6 - 1 \le \bar{x} \le 78.6 + 1) = P(77.6 \le \bar{x} \le 79.6) = 1.0$

If we take a random sample of five heights, there is a 100% chance that the mean of the sample selected will be within one inch of the population mean.

**7.17** (a) $\mu = \Sigma x/N = 279/6 = 46.5$ billion

(b)

| Sample | Wealth | $\bar{x}$ |
|--------|--------|-----------|
| G,B    | 72, 59 | 65.5      |
| G,E    | 72, 41 | 56.5      |
| G,C    | 72, 36 | 54.0      |
| G,D    | 72, 36 | 54.0      |
| G,W    | 72, 35 | 53.5      |
| B,E    | 59, 41 | 50.0      |
| B,C    | 59, 36 | 47.5      |
| B,D    | 59, 36 | 47.5      |
| B,W    | 59, 35 | 47.0      |
| E,C    | 41, 36 | 38.5      |
| E,D    | 41, 36 | 38.5      |
| E,W    | 41, 35 | 38.0      |
| C,D    | 36, 36 | 36.0      |
| C,W    | 36, 35 | 35.5      |
| D,W    | 36, 35 | 35.5      |

(c)

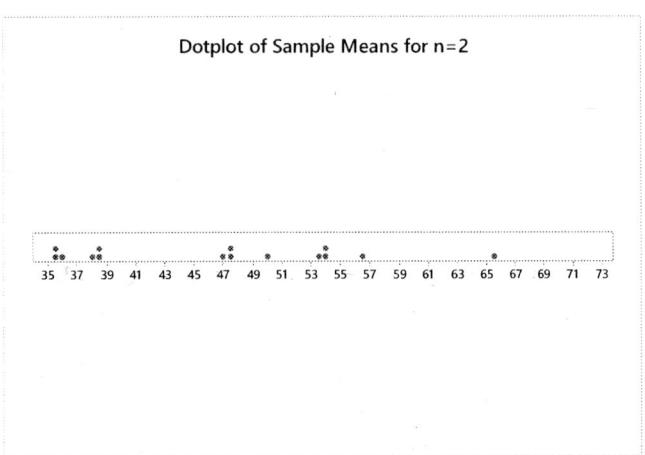

Dotplot of Sample Means for n=2

(d) $P(\bar{x} = \mu) = P(\bar{x} = 46.5) = 0/15 = 0$

(e) $P(46.5 - 3 \le \bar{x} \le 46.5 + 3) = P(43.5 \le \bar{x} \le 49.5) = 3/15 = 0.20$

If we take a random sample of two of these six rich Americans, there is a 20% chance that their mean wealth will be within three billion of the population mean wealth.

**7.19** (b)

| Sample | Wealth | $\bar{x}$ |
|---|---|---|
| G,B,E | 72, 59, 41 | 57.3 |
| G,B,C | 72, 59, 36 | 55.7 |
| G,B,D | 72, 59, 36 | 55.7 |
| G,B,W | 72, 59, 35 | 55.3 |
| G,E,C | 72, 41, 36 | 49.7 |
| G,E,D | 72, 41, 36 | 49.7 |
| G,E,W | 72, 41, 35 | 49.3 |
| G,C,D | 72, 36, 36 | 48.0 |
| G,C,W | 72, 36, 35 | 47.7 |
| G,D,W | 72, 36, 35 | 47.7 |
| B,E,C | 59, 41, 36 | 45.3 |
| B,E,D | 59, 41, 36 | 45.3 |
| B,E,W | 59, 41, 35 | 45.0 |
| B,C,D | 59, 36, 36 | 43.7 |
| B,C,W | 59, 36, 35 | 43.3 |
| B,D,W | 59, 36, 35 | 43.3 |
| E,C,D | 41, 36, 36 | 37.7 |
| E,C,W | 41, 36, 35 | 37.3 |
| E,D,W | 41, 36, 35 | 37.3 |
| C,D,W | 36, 36, 35 | 35.7 |

(c)

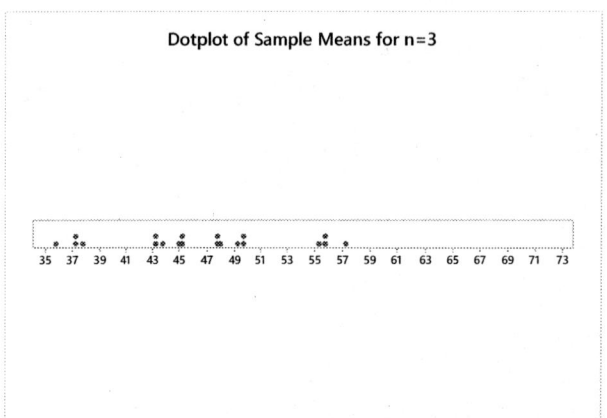

(d) $P(\bar{x} = \mu) = P(\bar{x} = 46.5) = 0/20 = 0.0$

(e) $P(46.5 - 3 \leq \bar{x} \leq 46.5 + 3) = P(43.5 \leq \bar{x} \leq 49.5)$
    $= 8/20 = 0.40$

If we take a random sample of three wealthy Americans, there is a 40% chance that their mean wealth will be within three billion of the population mean wealth.

**7.21** (b)

| Sample | Wealth | $\bar{x}$ |
|---|---|---|
| G,B,E,C,D | 72,59,41,36,36 | 48.8 |
| G,B,E,C,W | 72,59,41,36,35 | 48.6 |
| G,B,E,D,W | 72,59,41,36,35 | 48.6 |
| G,B,C,D,W | 72,59,36,36,35 | 47.6 |
| G,E,C,D,W | 72,41,36,36,35 | 44.0 |
| B,E,C,D,W | 59,41,36,36,35 | 41.4 |

(c)

Dotplot of Sample Means for n=5

(d) $P(\bar{x} = \mu) = P(\bar{x} = 46.5) = 0/6 = 0.000$

(e) $P(46.5 - 3 \leq \bar{x} \leq 46.5 + 3) = P(43.5 \leq \bar{x} \leq 49.5)$
$= 5/6 = 0.833$

If we take a random sample of five wealthy Americans, there is a 83.3% chance that their mean wealth will be within three billion of the population mean wealth.

**7.23** Increasing the sample size tends to reduce the sampling error.

**7.25** (a) If a sample of size n = 1 is taken from a population of size N, there are N possible samples.

(b) Since each sample mean is based upon a single observation, the possible values of the sample mean and the population values are the same.

(c) There is no difference between taking a random sample of size n = 1 from a population and selecting a member at random from the population.

## Exercises 7.2

**7.27** Obtaining the mean and standard deviation of $\bar{x}$ is a first step in approximating the sampling distribution of the mean by a normal distribution because the normal distribution is completely determined by its mean and standard deviation.

**7.29** Yes. The spread of the distribution of sample means becomes smaller as the sample size increases. Since that spread is measured by the standard deviation of all possible sample means, the standard deviation also gets smaller.

**7.31** Standard error of the sample mean. The standard deviation of $\bar{x}$ determines the amount of sampling error to be expected when a population mean is estimated by a sample mean.

**7.33** (a) For $n = 1$, the mean of the variable $\bar{x}$, $\mu_{\bar{x}}$, is found by summing up all of the possible values for $\bar{x}$ and dividing by the number of possible samples. We get $(1 + 2 + 3)/3 = 2.0$.

For $n = 2$, the mean of the variable $\bar{x}$, $\mu_{\bar{x}}$, is found by summing up all of the possible values for $\bar{x}$ and dividing by the number of possible samples. We get $(1.5 + 2 + 2.5)/3 = 2.0$.

For $n = 3$, the mean of the variable $\bar{x}$, $\mu_{\bar{x}}$, is found by summing up all of the possible values for $\bar{x}$ and dividing by the number of possible samples. We get $(2)/1 = 2.0$.

For each sample size, $\mu_{\bar{x}} = 2$.

(b) From part (a), in Exercise 7.3, we calculated $\mu = \Sigma x/N = 6/3 = 2$. Formula 7.1 states that $\mu_{\bar{x}} = \mu$. Thus, $\mu_{\bar{x}} = 2$.

**7.35** (a) For $n = 1$, the mean of the variable $\bar{x}$, $\mu_{\bar{x}}$, is found by summing up all of the possible values for $\bar{x}$ and dividing by the number of possible samples. We get $(1 + 2 + 3 + 4)/4 = 2.5$.

For $n = 2$, the mean of the variable $\bar{x}$, $\mu_{\bar{x}}$, is found by summing up all of the possible values for $\bar{x}$ and dividing by the number of possible samples. We get $(1.5 + 2 + 2.5 + 2.5 + 3 + 3.5)/6 = 2.5$.

For $n = 3$, the mean of the variable $\bar{x}$, $\mu_{\bar{x}}$, is found by summing up all of the possible values for $\bar{x}$ and dividing by the number of possible samples. We get $(2 + 2.3 + 2.7 + 3)/4 = 2.5$.

For $n = 4$, the mean of the variable $\bar{x}$, $\mu_{\bar{x}}$, is found by summing up all of the possible values for $\bar{x}$ and dividing by the number of possible samples. We get $(2.5)/1 = 2.5$.

For each sample size, $\mu_{\bar{x}} = 2.5$.

(b) From part (a), in Exercise 7.5, we calculated $\mu = \Sigma x/N = 10/4 = 2.5$. Formula 7.1 states that $\mu_{\bar{x}} = \mu$. Thus, $\mu_{\bar{x}} = 2.5$.

**7.37** (a) For $n = 1$, the mean of the variable $\bar{x}$, $\mu_{\bar{x}}$, is found by summing up all of the possible values for $\bar{x}$ and dividing by the number of possible samples. We get $(1 + 2 + 3 + 4 + 5)/5 = 3.0$.

For $n = 2$, the mean of the variable $\bar{x}$, $\mu_{\bar{x}}$, is found by summing up all of the possible values for $\bar{x}$ and dividing by the number of possible samples. We get $(1.5 + 2 + 2.5 + 3 + 2.5 + 3 + 3.5 + 3.5 + 4 + 4.5)/10 = 3.0$.

For $n = 3$, the mean of the variable $\bar{x}$, $\mu_{\bar{x}}$, is found by summing up all of the possible values for $\bar{x}$ and dividing by the number of possible samples. We get $(2 + 2.3 + 2.7 + 2.7 + 3 + 3.3 + 3 + 3.3 + 3.7 + 4)/10 = 3.0$.

For $n = 4$, the mean of the variable $\bar{x}$, $\mu_{\bar{x}}$, is found by summing up all of the possible values for $\bar{x}$ and dividing by the number of possible samples. We get $(2.5 + 2.75 + 3 + 3.25 + 3.5)/5 = 3.0$.

For $n = 5$, the mean of the variable $\bar{x}$, $\mu_{\bar{x}}$, is found by summing up all of the possible values for $\bar{x}$ and dividing by the number of possible samples. We get $(3)/1 = 3.0$. For each sample size, $\mu_{\bar{x}} = 3.0$.

(b) From part (a), in Exercise 7.7, we calculated $\mu = \Sigma x/N = 15/5 = 3$. Formula 7.1 states that $\mu_{\bar{x}} = \mu$. Thus, $\mu_{\bar{x}} = 3$.

**7.39** (a) For $n = 1$, the mean of the variable $\bar{x}$, $\mu_{\bar{x}}$, is found by summing up all of the possible values for $\bar{x}$ and dividing by the number of possible samples. We get $(1 + 2 + 3 + 4 + 5 + 6)/6 = 3.5$.

For $n = 2$, the mean of the variable $\bar{x}$, $\mu_{\bar{x}}$, is found by summing up all of the possible values for $\bar{x}$ and dividing by the number of possible samples. We get $(1.5 + 2 + 2.5 + 3 + 3.5 + 2.5 + 3 + 3.5 + 4 + 3.5 + 4 + 4.5 + 4.5 + 5 + 5.5)/15 = 3.5$.

For $n = 3$, the mean of the variable $\bar{x}$, $\mu_{\bar{x}}$, is found by summing up all of the possible values for $\bar{x}$ and dividing by the number of possible samples. We get $(2 + 2.3 + 2.7 + 3 + 2.7 + 3 + 3.3 + 3.3 + 3.7 + 4 + 3 + 3.3 + 3.7 + 3.7 + 4 + 4.3 + 4 + 4.3 + 4.7 + 5)/20 = 3.5$.

For $n = 4$, the mean of the variable $\bar{x}$, $\mu_{\bar{x}}$, is found by summing up all of the possible values for $\bar{x}$ and dividing by the number of possible samples. We get $(2.5 + 2.75 + 3 + 3 + 3.25 + 3.5 + 3.25 + 3.5 + 3.75 + 4 + 3.5 + 3.75 + 4 + 4.25 + 4.5)/15 = 3.5$.

For $n = 5$, the mean of the variable $\bar{x}$, $\mu_{\bar{x}}$, is found by summing up all of the possible values for $\bar{x}$ and dividing by the number of possible samples. We get $(3 + 3.2 + 3.4 + 3.6 + 3.8 + 4)/6 = 3.5$.

For each sample size, $\mu_{\bar{x}} = 3.5$.

(b) From part (a), in Exercise 7.9, we calculated $\mu = \Sigma x/N = 21/6 = 3.5$. Formula 7.1 states that $\mu_{\bar{x}} = \mu$. Thus, $\mu_{\bar{x}} = 3.5$.

**7.41** (a) $\mu_{\bar{x}} = \dfrac{\Sigma x}{N} = \dfrac{393}{5} = 78.6$

(b)
$$\mu_{\bar{x}} = \dfrac{\Sigma \bar{x}}{N} = \dfrac{79.5+81.5+78.5+81.5+78.0+75.0+78.0+77.0+80.0+77.0}{10}$$
$$= \dfrac{786}{10} = 78.6$$

(c) $\mu_{\bar{x}} = \mu = 78.6$

**7.43** (b)
$$\mu_{\bar{x}} = \dfrac{\Sigma \bar{x}}{N} = \dfrac{79.7+77.7+79.7+79.0+81.0+79.0+76.7+78.7+76.7+78.0}{10}$$
$$= \dfrac{786}{10} = 78.6$$

(c) $\mu_{\bar{x}} = \mu = 78.6$

**7.45** (b)
$$\mu_{\bar{x}} = \dfrac{\Sigma \bar{x}}{N} = \dfrac{78.6}{1} = 78.6$$

(c) $\mu_{\bar{x}} = \mu = 78.6$

**7.47** (a) The population consists of all babies born in 1991. The variable is the birth weight of the baby.

(b) $\mu_{\bar{x}} = \mu = 3369$ grams; $\sigma_{\bar{x}} = \sigma/\sqrt{n} = 581/\sqrt{200} = 41.08$ grams

(c) $\mu_{\bar{x}} = \mu = 3369$ grams; $\sigma_{\bar{x}} = \sigma/\sqrt{n} = 10/\sqrt{400} = 29.05$ grams

**7.49** (a) $\mu_{\bar{x}} = \mu = \$65,100$; $\sigma_{\bar{x}} = \sigma/\sqrt{n} = 7200/\sqrt{50} = \$1018.23$

Thus, for samples of size 50, the mean and standard deviation of all possible sample means are respectively, $65,100 and $1018.23.

(b) $\mu_{\bar{x}} = \mu = \$65,100$; $\sigma_{\bar{x}} = \sigma/\sqrt{n} = 7200/\sqrt{100} = \$720.00$

Thus, for samples of size 100, the mean and standard deviation of all possible sample means are respectively, $65,100 and $720.

**7.51** (a) On the average, we would expect the sample mean of the four times to be the same as the population mean of 437 days.

(b) The standard deviation of the sample mean is
$\sigma_{\bar{x}} = \sigma/\sqrt{n} = 399/\sqrt{4} = 199.5$ days, so 99.74% of the time we would expect the sample mean to fall within 3 standard deviations of 437 days, i.e., between 437 - 3(199.5) = -161.5 and 437 + 3(199.5) = 1035.5. Since the lower limit is clearly below zero and the time between earthquakes cannot be negative, we would expect the sample mean to fall between 0 and 1035.5 days 99.74% of the time.

**7.53** (a) Enter the data into cells A1 to A487 of Excel. In cell B1, enter the expression =STDEVP(A1:A487) to obtain the population standard deviation of the data in B1. The result is 94.334 Note: Using procedures in Excel will yield a SAMPLE standard deviation of 94.431, which you could then convert to a POPULATION standard deviation by multiplying by the square root of 486/487 to obtain the same result.

(b) In Excel, in cells C1 to C487, enter the numbers 1 to 487. In cell D1, enter the expression =SQRT((487-C1)/(487-1))*94.334/SQRT(C1). In Cell E1, enter the expression =94.334/SQRT(C1). Copy cells D1 and E1 to the clipboard and then paste them into cells D2 through D487. This will produce your table in three columns. To save space, we show the table below using six columns and every tenth value of n after n = 30.

| n | Eq. 7.1 | Eq. 7.2 | n | Eq. 7.1 | Eq. 7.2 |
|---|---|---|---|---|---|
| 1 | 94.33 | 94.33 | 120 | 7.48 | 8.61 |
| 2 | 66.64 | 66.70 | 130 | 7.09 | 8.27 |
| 3 | 54.35 | 54.46 | 140 | 6.74 | 7.97 |
| 4 | 47.02 | 47.17 | 150 | 6.41 | 7.70 |
| 5 | 42.01 | 42.19 | 160 | 6.12 | 7.46 |
| 6 | 38.31 | 38.51 | 170 | 5.84 | 7.24 |
| 7 | 35.43 | 35.65 | 180 | 5.59 | 7.03 |
| 8 | 33.11 | 33.35 | 190 | 5.35 | 6.84 |
| 9 | 31.18 | 31.44 | 200 | 5.13 | 6.67 |
| 10 | 29.55 | 29.83 | 210 | 4.91 | 6.51 |
| 11 | 28.15 | 28.44 | 220 | 4.71 | 6.36 |
| 12 | 26.92 | 27.23 | 230 | 4.52 | 6.22 |
| 13 | 25.84 | 26.16 | 240 | 4.34 | 6.09 |
| 14 | 24.87 | 25.21 | 250 | 4.17 | 5.97 |
| 15 | 24.00 | 24.36 | 260 | 4.00 | 5.85 |
| 16 | 23.22 | 23.58 | 270 | 3.84 | 5.74 |
| 17 | 22.50 | 22.88 | 280 | 3.68 | 5.64 |
| 18 | 21.84 | 22.23 | 290 | 3.53 | 5.54 |
| 19 | 21.24 | 21.64 | 300 | 3.38 | 5.45 |
| 20 | 20.68 | 21.09 | 310 | 3.23 | 5.36 |
| 21 | 20.16 | 20.59 | 320 | 3.09 | 5.27 |
| 22 | 19.67 | 20.11 | 330 | 2.95 | 5.19 |
| 23 | 19.22 | 19.67 | 340 | 2.81 | 5.12 |
| 24 | 18.79 | 19.26 | 350 | 2.68 | 5.04 |
| 25 | 18.40 | 18.87 | 360 | 2.54 | 4.97 |
| 26 | 18.02 | 18.50 | 370 | 2.41 | 4.90 |
| 27 | 17.66 | 18.15 | 380 | 2.27 | 4.84 |
| 28 | 17.33 | 17.83 | 390 | 2.13 | 4.78 |
| 29 | 17.01 | 17.52 | 400 | 2.00 | 4.72 |
| 30 | 16.70 | 17.22 | 410 | 1.85 | 4.66 |

| | | | | | | |
|---|---|---|---|---|---|---|
| | 40 | 14.30 | 14.92 | 420 | 1.71 | 4.60 |
| | 50 | 12.65 | 13.34 | 430 | 1.56 | 4.55 |
| | 60 | 11.42 | 12.18 | 440 | 1.40 | 4.50 |
| | 70 | 10.44 | 11.28 | 450 | 1.23 | 4.45 |
| | 80 | 9.65 | 10.55 | 460 | 1.04 | 4.40 |
| | 90 | 8.99 | 9.94 | 470 | 0.81 | 4.35 |
| | 100 | 8.42 | 9.43 | 480 | 0.52 | 4.31 |
| | 110 | 7.92 | 8.99 | 487 | 0.00 | 4.27 |

The results obtained with Equation 7.2 are reasonably accurate when n is 5% of N or less (n < 25). As n gets larger, the discrepancy between the results of the two equations continues to grow.

**7.55** (a) We start with Equation 7.1 and substitute 1 for the sample size n.

$$\sigma_{\bar{x}} = \sqrt{\frac{N-n}{N-1}} \frac{\sigma}{\sqrt{n}} = \sqrt{\frac{N-1}{N-1}} \frac{\sigma}{\sqrt{n}} = \frac{\sigma}{\sqrt{n}}.$$

The result on the right side of this expression is the same as Equation 7.2.

(b) The two equations should be the same when n = 1 because it makes no difference whether you are sampling with (Equation 7.2) or without (Equation 7.1) replacement when there is no second observation to be made.

(c) If a sample of size N is drawn from a population of size N without replacement, all observations will have been drawn and there can be only one result for the sample mean and that is the population mean. There can be no variation from the population mean and therefore the standard deviation of the distribution of the sample mean must be zero.

(d) $\sigma_{\bar{x}} = \sqrt{\frac{N-n}{N-1}} \frac{\sigma}{\sqrt{n}} = \sqrt{\frac{N-N}{N-1}} \frac{\sigma}{\sqrt{n}} = \sqrt{\frac{0}{N-1}} \frac{\sigma}{\sqrt{n}} = 0 \cdot \frac{\sigma}{\sqrt{n}} = 0$.

**7.57** (a) When n = 1, $\sqrt{\frac{N-n}{N-1}} = \sqrt{\frac{N-1}{N-1}} = 1$. When n > 1, N − n will be smaller than N − 1, and the square root will be less than 1.

If n ≤ 0.05N, then N − n ≥ N − 0.05N = 0.95N and

$$\sqrt{\frac{N-n}{N-1}} \geq \sqrt{\frac{0.95N}{N-1}} = \sqrt{\frac{0.95}{1-1/N}} \geq \sqrt{\frac{0.95}{1}} = 0.975 > 0.97.$$

In the third expression, we divided the numerator and denominator by N. In the fourth expression, we used the fact that 1 − 1/N is smaller than 1 and therefore the ratio under the square root sign in the third expression is larger than 0.95/1. These two results demonstrate that when

n ≤ 0.05N, $0.97 \leq \sqrt{\frac{N-n}{N-1}} \leq 1$.

(b) Since Equation 7.1 is the same as Equation 7.2 multiplied by the factor in part (a), our result in part (a) shows that whenever n ≤ 0.05N, Equation 7.1 will yield a result that is at least as large as 97% of that yielded by Equation 7.2 and never any larger than that yielded by Equation 7.2. There will never be more than a 3% discrepancy between Equation 7.1 and Equation 7.2 when n ≤ 0.05N.

(c) Equation 7.2 is used when the population is infinite or can be considered infinite due to sampling **with** replacement. If sampling is from a finite population **without** replacement, we must account for that

**7.59** (a) Theoretically, the mean of possible sample means is 266 and the standard deviation is $16/\sqrt{9} = 5.33$.

We have Minitab take 2000 random samples of size n = 9 from a normally distributed population with mean 266 and standard deviation 16 by choosing **Calc ▶ Random Data ▶ Normal...**, entering 2000 in the **Generate rows of data** text box, entering C1-C9 in the **Store in column(s)** text box, entering 266 in the **Mean** text box and 16 in the **Standard deviation** text box, and clicking **OK**. These commands tell Minitab to place a total of 18000 observations from the population into columns C1-C9, with 2000 observations in each column. Then, our first random sample of size n = 9 is the first row of columns C1-C9, our second random sample of size n = 9 is the second row of columns C1-C9, and so on.

(c) We compute the sample mean of each of the 2000 samples by choosing **Calc ▶ Row statistics...**, clicking on the **Mean** button, entering C1-C9 in the **Input variables** text box and XBAR in the **Store result in** text box, and clicking **OK**. This command instructs Minitab to compute the means of the 2000 rows of C1-C9 and to place those means in a column named XBAR. Thus, the 2000 -values are now in XBAR. (We will not print these values.)

(d) We would expect the mean of the 2000 sample means to be roughly 266 and the standard deviation of the sample means to be about 5.33 since we are taking a sample of 2000 means from a theoretical sampling distribution with mean and standard deviation given in part (a).

(e) The mean is obtained by choosing **Calc ▶ Column statistics...**, clicking on the **Mean** button, selecting XBAR in the **Input variable** text box, and clicking **OK**. The result is

Mean of XBAR = 266.03

Similarly, the standard deviation is obtained by choosing **Calc ▶ Column statistics...**, clicking on the **Standard deviation** button, selecting XBAR in the **Input variable** text box, and clicking **OK**. The result is

Standard deviation of XBAR = 5.2753

Thus, the mean of our 2000 means was 266.03 and the standard deviation was 5.2753.

(b) The answers in part (e) differ from the theoretical values given in part (d) as a result of random error or sampling variability.

**Exercises 7.3**

**7.61** (a) The sampling distribution of the mean is approximately normally distributed with mean

µ = 100 and standard deviation $\sigma_{\bar{x}} = 28/\sqrt{49} = 4$.

(b) No assumptions were made about the distribution of the population.

(c) Part (a) cannot be answered if the sample size is n = 16. Since the

**240** Chapter 7

distribution of the population is not specified, we need a sample size of at least 30 to apply Key Fact 7.6.

**7.63** (a) The probability distribution of $\bar{x}$ is normal with mean μ and standard deviation $\sigma_{\bar{x}} = \sigma/\sqrt{n}$.

(b) The answer to part (a) does not depend on how large the sample size is because the population being sampled is normally distributed.

(c) The mean of $\bar{x}$ is μ; its standard deviation is $\sigma_{\bar{x}} = \sigma/\sqrt{n}$.

(d) No. The mean and standard deviation of the sampling distribution of the sample mean are always as given in part (c) regardless of the distribution of the variable under consideration.

**7.65** (a) All four graphs are centered at the same place because the population mean of $\bar{x}$ is μ and because normal curves are centered at their μ-parameter.

(b) Since $\sigma_{\bar{x}} = \sigma/\sqrt{n}$, we see that $\sigma_{\bar{x}}$ decreases as *n* increases. This results in a diminishing of the spread because the spread of a distribution is determined by its σ-parameter. As a consequence, we see that the larger the sample size, the greater the likelihood for small sampling error.

(c) The graphs in Figure 7.6(a) are bell-shaped because, for normally distributed populations, the random variable $\bar{x}$ is always normally distributed regardless of the sample size.

(d) The graphs in Figures 7.6(b) and 7.6(c) become bell-shaped as the sample size increases because of the central limit theorem; the probability distribution of $\bar{x}$ tends to a normal distribution as the sample size increases.

**7.67** (a) Because the weights themselves are normally distributed, the sampling distribution for means of samples of size 3 will also be normal and will have mean $\mu_{\bar{x}} = \mu = 1.40 \, kg$ and standard deviation $\sigma_{\bar{x}} = 0.11/\sqrt{3} = 0.064$. Thus, the distribution of the possible sample means for samples of three brain weights will be normal with mean 1.40 kg and standard deviation 0.064 kg.

(b) Because the weights themselves are normally distributed, the sampling distribution for means of samples of size 12 will also be normal and will have mean $\mu_{\bar{x}} = \mu = 1.40 \, kg$ and standard deviation $\sigma_{\bar{x}} = 0.11/\sqrt{12} = 0.032$. Thus, the distribution of the possible sample means for samples of twelve brain weights will be normal with mean 1.40 kg and standard deviation 0.032 kg.

(c) To facilitate the comparison of the three graphs, we have overlaid them on one set of axes.

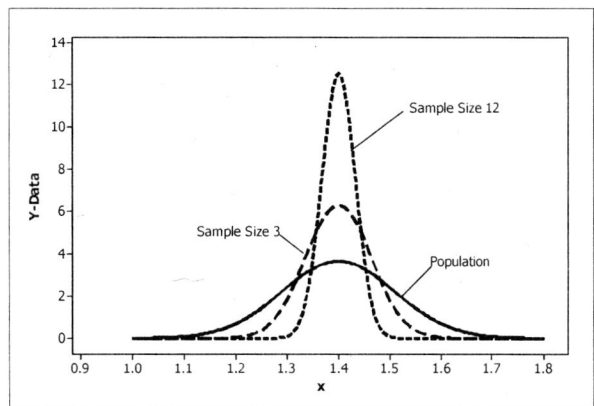

(d) $\sigma_{\bar{x}} = 0.11/\sqrt{3} = 0.064$; we want $P(1.30 \le \bar{x} \le 1.50)$.

```
z-score computations:                    Area less than z:
```

$$\bar{x} = 1.30 \to z = \frac{1.30 - 1.40}{0.064} = -1.56 \qquad 0.0594$$

$$\bar{x} = 1.50 \to z = \frac{1.50 - 1.40}{0.064} = 1.56 \qquad 0.9406$$

Total area = 0.9406 - 0.0594 = 0.8812

Note: If you round the standard deviation of the sample means differently, you may reach a slightly different answer.

Thus, 88.12% of samples of three Swedish men will have mean brain weights within 0.1 kg of the population mean brain weight of 1.40 kg.

(e) $\sigma_{\bar{x}} = 0.11/\sqrt{12} = 0.032$; we want $P(1.30 \le \bar{x} \le 1.50)$.

```
z-score computations:                    Area less than z:
```

$$\bar{x} = 1.30 \to z = \frac{1.30 - 1.40}{0.032} = -3.13 \qquad 0.0009$$

$$\bar{x} = 1.50 \to z = \frac{1.50 - 1.40}{0.032} = 3.13 \qquad 0.9991$$

Total area = 0.9991 - 0.0009 = 0.9982

Note: If you round the standard deviation of the sample means differently, you may reach a slightly different answer.

Thus, 99.82% of samples of twelve Swedish men will have mean brain weights within 0.1 kg of the population mean brain weight of 1.40 kg.

**7.69** (a) The sampling distribution of the sample mean for samples of size 64 will be approximately normal with a mean of $55.4 thousand and a standard deviation of $\sigma_{\bar{x}} = 9.2/\sqrt{64} = 1.15$. Thus, if all possible sample means for samples of size 64 were found, their distribution would be approximately normal with mean $55.4 thousand and standard deviation $1,150.

(b) The sampling distribution of the sample mean for samples of size 256 will be approximately normal with a mean of $55.4 thousand and a standard deviation of $\sigma_{\bar{x}} = 9.2/\sqrt{256} = 0.575$. Thus, if all possible sample means for samples of size 256 were found, their distribution would be approximately normal with mean $55.4 thousand and standard deviation $575.

(c) No. These results follow from the central limit theorem, which implies that for large samples (n ≥ 30), the distribution of the sample mean will be approximately normal regardless of the distribution of the original variable.

(d) $\sigma_{\bar{x}} = 9.2/\sqrt{64} = 1.15$; we want P(54.4 ≤ $\bar{x}$ ≤ 56.4).

z-score computations:              Area less than z:

$\bar{x} = 54.4 \to z = \dfrac{54.4 - 55.4}{1.15} = -0.87$              0.1922

$\bar{x} = 56.4 \to z = \dfrac{56.4 - 55.4}{1.15} = 0.87$              0.8078

Total area = 0.8078 - 0.1922 = 0.6156

Thus, 61.56% of samples of size 64 will have a mean that is within $1000 of the population mean, i.e., 61.56% of the samples will have a sampling error less than $1000 for samples of size 64.

(e) $\sigma_{\bar{x}} = 9.2/\sqrt{256} = 0.575$; we want P(54.4 ≤ $\bar{x}$ ≤ 56.4).

z-score computations:              Area less than z:

$\bar{x} = 54.4 \to z = \dfrac{54.4 - 55.4}{0.575} = -1.74$              0.0409

$\bar{x} = 56.4 \to z = \dfrac{56.4 - 55.4}{0.575} = 1.74$              0.9591

Total area = 0.9591 - 0.0409 = 0.9182

Thus, 91.82% of samples of size 256 will have a mean that is within $1000 of the population mean, i.e., 91.82% of the samples will have a sampling error less than $1000 for samples of size 256.

**7.71** (a) The sampling distribution of the sample mean for n = 80 will be approximately normal with some mean μ and standard deviation $\sigma_{\bar{x}} = 8.3/\sqrt{80} = 0.928$.

(b) No. The sample size of 80 is large enough so that the Central Limit Theorem applies and the sampling distribution will be approximately normal.

(c) We want P(μ - 2 ≤ $\bar{x}$ ≤ μ + 2)
z-score computations:              Area less than z:

$\bar{x} = \mu - 2 \to z = \dfrac{(\mu - 2) - \mu}{0.928} = -2.16$              0.0154

$\bar{x} = \mu + 2 \to z = \dfrac{(\mu + 2) - \mu}{0.928} = 2.16$              0.9846

Total area = 0.9846 - 0.0154 = 0.9692

Thus, the probability is approximately 0.9692 that the sampling error made in estimating the population mean length of stay on the intervention ward by the mean length of stay of a sample of 80 patients will be at most 2 days.

**7.73** $\sigma_{\bar{x}} = 1150/\sqrt{500} = 51.43; P(|\bar{x} - \mu| \leq 100)$

z-score computations:                                     Area less than z:

$\bar{x} = \mu - 100 \rightarrow z = \dfrac{(\mu - 100) - \mu}{51.43} = -1.94$            0.0262

$\bar{x} = \mu + 100 \rightarrow z = \dfrac{(\mu + 100) - \mu}{51.43} = 1.94$             0.9738

Required Area = 0.9738 - 0.0262 = 0.9476

Thus, there is a 0.9476 probability that the mean tariff rate of a sample of 500 randomly selected ethanol shipments used to estimate the mean tariff rate of all ethanol shipments will have a sampling error less than $100.

**7.75** In solving this problem, we have to assume that the distribution of calcium intakes of adults with incomes below the poverty level is approximately normally distributed. Then, the distribution of the sample means will be approximately normally distributed. The distribution of the sample means have a mean 1000 mg and standard deviation of $\sigma_{\bar{x}} = \sigma/\sqrt{n} = 188/\sqrt{18} = 44.312$.

We want $P(\bar{x} \leq 947.4)$:

z-score computation:                                       Area less than z:

$\bar{x} = 947.4 \rightarrow z = \dfrac{947.4 - 1000}{44.312} = -1.19$             0.1170

Total area = 0.1170

11.70% of all samples of size 18 of adults with incomes below the poverty level will have mean calcium intakes of at most 947.4 mg.

**7.77** In solving this problem, we have to assume that the distribution of post-work hear rate for casting workers is approximately normally distributed. Then, the distribution of the sample means will be approximately normally distributed. The distribution of the sample means have a mean 72 bpm and standard deviation of $\sigma_{\bar{x}} = \sigma/\sqrt{n} = 11.2/\sqrt{29} = 2.080$. We want $P(\bar{x} > 78.3)$:

z-score computation:                                       Area less than z:

$\bar{x} = 78.3 \rightarrow z = \dfrac{78.3 - 72}{2.080} = 3.03$                 0.9988

Total area = 1 - 0.9988 = 0.0012.

The probability is 0.0012 that samples of 29 casting workers will have a mean post=work hear rate exceeding 78.3 bpm.

**7.79** (a) 68.26      (b) 95.44      (c) 99.74      (d) 100(1 - α)

**7.81** (a) No. The original population is quite skewed. A sample size of 4 is quite small, so we would expect that a histogram of the means of such a sample would still exhibit some degree of skewness.

(b) We'll use Excel to carry out the simulation. First enter the values 1 through 7 in cells A1 to A7, then enter the frequencies 19.4 through 1.6 from the table in Example 7.9 into cells C1 through C7. In C8, enter the expression =Sum(C1:C7). Then in cell B1, enter the expression =C1/$C$8, copy this cell to the clipboard and then paste it into cells B2 through B7. Columns A and B now contain the probability distribution of x, the household size. Now from the Tools menu, select **Data Analysis ▶ Random Number Generation**. Enter 4 in the **Number of Variables** text box, enter 1000 in the **Number of Random Numbers** text box, choose **Discrete** in the **Distribution** box, enter the range A1:B7 in the **Value and Probability Input Range** text box, and enter A11 in the **Output Range** box. Click on **OK**. You will get 4 columns of 100 numbers between 1 and 7 in columns A, B, C, and D, starting in row 11 and ending in row 1010. In cell E11, enter the expression =AVERAGE(A11:D11). Copy this cell to the clipboard and then paste it into the cells E12 through E1010. Type the label **Mean** in cell E10. Use the mouse to highlight cells E10 though E1010. Now from the menu, select **Charts and Plots**. In the **Function Type** box, select **Histogram**, Click on **Mean** in the **Names and Columns** box and enter it in the **Quantitative Variables** box. Then click **OK**. A histogram of the sampling distribution of $\bar{x}$ for our samples of size 4 is shown below. Yours is very likely to be different. Our histogram is slightly skewed to the right.

(c) We would expect this histogram to be closer to bell-shaped than the one in part (a), but 10 is still a relatively small sample size, so there may still be a trace of skewness. Using the same procedure as in part (b), but with 10 variables, our result below is more symmetrical, but we definitely did not expect the one tall spike. Note that the scale is different from the first histogram. These means are more closely concentrated.

(d) Since 100 is definitely a large sample size, we would expect the histogram of the sample means to be bell-shaped. Using the same procedure as in part (b), but with 100 variables, our result below is close to bell-shaped and again is more concentrated than the first two.

**7.83** (a) We will describe a procedure using Minitab to sketch the exponential curve for this population, although the sketch could be done using any spreadsheet program as well. We begin by selecting typical values of x that will eventually be substituted into the exponential function itself to trace out values for y. Values of x selected are the integers 0 through 15. Name two columns in the Data window X and Y and enter the numbers 0, 1, 2, ...,15 in the X column. Then choose **Calc** ▶

**Probability Distributions** ▶ **Exponential...**, click on the **Probability density** button, click in the **Mean** text box and type 8.7. Then enter **X** in the **Input column** text box and **Y** in the **Optional storage** text box.

Click **OK**. Now choose **Graph** ▶ **Scatterplot...**, select the **With connect line** version and click **OK**. Select Y for row 1 of the **Y** column and X for row 1 of the **X** column. To plot a continuous curve, click on the **Data view** button and check only the **Connect line** box. Click **OK** twice. The result is shown below. The exponential distribution is reverse J-shaped.

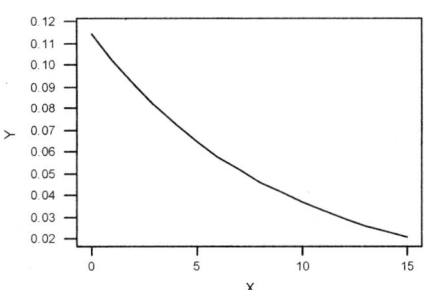

(b) To have Minitab take 1000 random samples of size n = 4 from an exponentially distributed population with mean μ = 8.7, we choose **Calc**

▶ **Random Data** ▶ **Exponential...**, enter 1000 in the **Generate rows of data** text box, enter C1-C4 in the **Store in column(s)** text box, enter 8.7 in the **Mean** text box, and click **OK**. These commands tell Minitab to place a total of 4000 observations from the population into columns C1-C4, with 1000 observations in each column. Then, our first random sample of size n = 4 is the first row of columns C1-C4, our second random sample of size n = 4 is the second row of columns C1-C4, and so on.

(c) We compute the sample mean of each of the 1000 samples by choosing **Calc ▶ Row statistics...**, clicking on the **Mean** button, entering C1-C4 in the **Input variables** text box and XBAR in the **Store result in** text box, and clicking **OK**. This command instructs Minitab to compute the means of the 1000 rows of C1-C4 and to place those means in XBAR. Thus, the 1000 -values are now in XBAR. (We will not print these values.) Your results will differ from ours.

(d) The mean of the 1000 sample means is obtained by choosing **Calc ▶ Column statistics...**, clicking on the **Mean** button, selecting XBAR in the **Input variable** text box, and clicking **OK**. The result is

Mean of XBAR = 8.4743

Similarly, the standard deviation of the 1000 sample means is obtained by choosing **Calc ▶ Column statistics...**, clicking on the **Standard deviation** button, selecting XBAR in the **Input variable** text box, and clicking **OK**. The result is

Standard deviation of XBAR = 4.2498

(e) Theoretically, for all possible sample means for samples of size 4, the mean is the same as the population mean, 8.7, and the standard deviation is $\sigma/\sqrt{n} = 8.7/\sqrt{4} = 4.35$. Both of the simulated values obtained in part (d) are close to these values.

(f) To get a histogram of the 1000 sample means stored in XBAR, choose **Graph ▶ Histogram...**, selecting the **Simple** version and click **OK**. Enter XBAR in the **Graph variables** text box, and click **OK**. The result is shown below at the right. The histogram is not bell-shaped; given the right-skewness of the population, we would expect for samples of only size four, that the distribution of means would be somewhat skewed to the right.

(g) Repeating the process above for samples of size 40 (use columns C1-C40 for the samples), we find that

Mean of XBAR = 8.7367 and   Standard deviation of XBAR = 1.4204

In theory, the mean of all possible samples of size 40 is 8.7 and the standard deviation is $\sigma/\sqrt{n} = 8.7/\sqrt{40} = 1.38$. The simulated values are very close to these.

A histogram of the means of samples of size 40 is shown following, obtained as above. The simulated distribution of sample means is close to bell-shaped as we would expect for samples over size 30.

## Review Problems for Chapter 7

1. Sampling errors are errors resulting from using a sample to estimate a population characteristic.

2. The sampling distribution of a statistic is the set of all possible observations of the statistic for a sample of a given size. It is important to know the sampling distribution in order to answer questions about the accuracy of estimating a population parameter using the sample statistic.

3. Two other terms are 'sampling distribution of the sample mean' and 'the distribution of the variable $\bar{x}$.'

4. The set of possible sample means exhibits less and less variability as the sample size increases, that is, the set becomes more and more clustered about the population mean. This means that as the sample size increases, there is a greater chance that the value of the sample mean from any sample is close to the value of the population mean.

5. (a) $\mu = \$108/6 = \$18$ (thousands)

(b)

| Sample | Salaries | |
|---|---|---|
| A,B,C,D | 8, 12, 16, 20 | 14 |
| A,B,C,E | 8, 12, 16, 24 | 15 |
| A,B,C,F | 8, 12, 16, 28 | 16 |
| A,B,D,E | 8, 12, 20, 24 | 16 |
| A,B,D,F | 8, 12, 20, 28 | 17 |
| A,B,E,F | 8, 12, 24, 28 | 18 |
| A,C,D,E | 8, 16, 20, 24 | 17 |
| A,C,D,F | 8, 16, 20, 28 | 18 |
| A,C,E,F | 8, 16, 24, 28 | 19 |
| A,D,E,F | 8, 20, 24, 28 | 20 |
| B,C,D,E | 12, 16, 20, 24 | 18 |
| B,C,D,F | 12, 16, 20, 28 | 19 |
| B,C,E,F | 12, 16, 24, 28 | 20 |
| B,D,E,F | 12, 20, 24, 28 | 21 |
| C,D,E,F | 16, 20, 24, 28 | 22 |

(c)
```
                           •
              •     •   •     •
      •   •   •   •   •   •   •   •   •
      +----+----+----+----+----+----+----+----+
      14   15   16   17   18   19   20   21   22
```

(d) $P(|\bar{x} - \mu| \leq 1) = P(\bar{x} = 17) + P(\bar{x} = 18) + P(\bar{x} = 19)$
$= 2/15 + 3/15 + 2/15 = 7/15 = 0.4667$

6. (a) $\mu_{\bar{x}} = \dfrac{\sum \bar{x}}{N} = \dfrac{270}{15} = 18.0$. The mean of the means of all of the possible samples of size 4 is $18.0 thousand.

(b) Yes. The mean of the sampling distribution of $\bar{x}$ is always the same as the mean of the population, which is $18 thousand in this case.

7. (a) For a normally distributed population, the random variable $\bar{x}$ is normally distributed, regardless of the sample size. Also, we know that $\mu_{\bar{x}} = \mu$. Consequently, since the normal curve for a normally distributed population or random variable is centered at its μ-parameter, all three curves are centered at the same place.

(b) Curve B corresponds to the larger sample size. Since $\sigma_{\bar{x}} = \sigma/\sqrt{n}$, the larger the sample size, the smaller the value of σ and, hence, the smaller the spread of the normal curve for $\bar{x}$. Thus, Curve B, which has the smaller spread, corresponds to the larger sample size.

(c) The spread of each curve is different because $\sigma_{\bar{x}} = \sigma/\sqrt{n}$ and the spread of a normal curve is determined by $\sigma_{\bar{x}}$. Thus, different sample sizes result in normal curves with different spreads.

(d) Curve B corresponds to the sample size that will tend to produce less sampling error. The smaller the value of $\sigma_{\bar{x}}$, the smaller the sampling error tends to be.

(e) When x is normally distributed, $\bar{x}$ always has a normal distribution as well.

**8.** (a) The sampling error results from using the sample mean income tax, $\bar{x}$, of the 308,946 tax returns as an estimate of the mean income tax, μ, of all 2010 tax returns.

(b) The sampling error is $11,266 - $11,354 = -$88.

(c) No, not necessarily. However, increasing the sample size from 308,946 to 400,000 would increase the likelihood for a smaller sampling error.

(d) Increase the sample size.

**9.** (a) The population consists of all new car sales in the U.S. in the given year. The variable under consideration is the amount spent for a new car.

(b) $\mu_{\bar{x}} = \$30,803; \sigma_{\bar{x}} = \$10200/\sqrt{50} = \$1442.50$

(c) $\mu_{\bar{x}} = \$30,803; \sigma_{\bar{x}} = \$10200/\sqrt{100} = \$1020.00$

(d) The value of $\sigma_{\bar{x}}$ will be smaller than $1020 because $\sigma_{\bar{x}} = \sigma/\sqrt{n}$. Thus, the larger the sample size, the smaller the value of $\sigma_{\bar{x}}$.

**10.** (a) False. By the central limit theorem, the random variable $\bar{x}$ is approximately normally distributed. Furthermore, $\mu_{\bar{x}} = \mu = 45$, and $\sigma_{\bar{X}} = \sigma/\sqrt{n} = 7/\sqrt{196} = 0.5$. Thus, P(31 ≤ $\bar{x}$ ≤ 59) equals the area under the normal curve with parameters $\mu_{\bar{x}} = 45$ and $\sigma_{\bar{x}} = 0.5$ that lies between 31 and 59. Applying the usual techniques, we find that area to be 1.0000 to four decimal places. Hence, there is almost a 100% chance that the mean of the sample will be between 31 and 59.

(b) This is not possible to tell, since we do not know the distribution of the population.

(c) True. Referring to part (a), we see that P(44 ≤ $\bar{x}$ ≤ 46) equals the area under the normal curve with parameters $\mu_{\bar{x}} = 45$ and $\sigma_{\bar{x}} = 0.5$ that lies between 44 and 46. Applying the usual techniques, we find that area to be 0.9544. Hence, there is approximately a 95.44% chance that the mean of the sample will be between 44 and 46.

**11.** (a) False. Since the population is normally distributed, so is the random variable $\bar{x}$. Furthermore, $\mu_{\bar{x}} = \mu = 45$ and $\sigma_{\bar{X}} = \sigma/\sqrt{n} = 7/\sqrt{196} = 0.5$.

Thus, P(31 ≤ $\bar{x}$ ≤ 59) equals the area under the normal curve with parameters $\mu_{\bar{x}} = 45$ and $\sigma_{\bar{x}} = 0.5$ that lies between 31 and 59. Applying the usual techniques, we find that area to be 1.0000 to four decimal places. Hence, there is almost a 100% chance that the mean of the sample will be between 31 and 59.

(b) True. Since the population is normally distributed, percentages for the population are equal to areas under the normal curve with parameters μ = 45 and σ = 7. Applying the usual techniques, we find that the area under that normal curve between 31 and 59 is 0.9544.

(c) True. From part (a), we see that the random variable $\bar{x}$ is normally distributed with $\mu_{\bar{x}} = 45$ and $\sigma_{\bar{x}} = 0.5$. Applying the usual techniques, we find that area to be 0.9544. Hence, we find that there is a 95.44% chance that the mean of the sample will be between 44 and 46.

12. (a) $\mu = 8.5g \text{ and } \sigma = 0.3g$

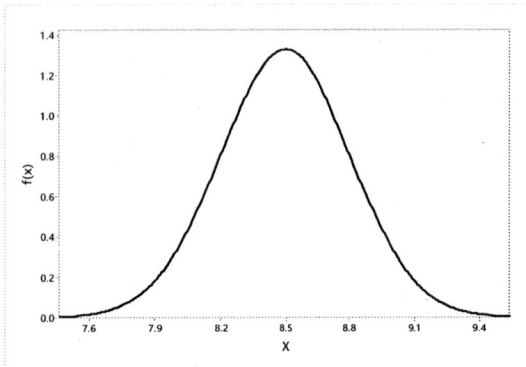

(b) normal distribution with $\mu_{\bar{x}} = 8.5g \text{ and } \sigma_{\bar{x}} = 0.3/\sqrt{4} = 0.15g$

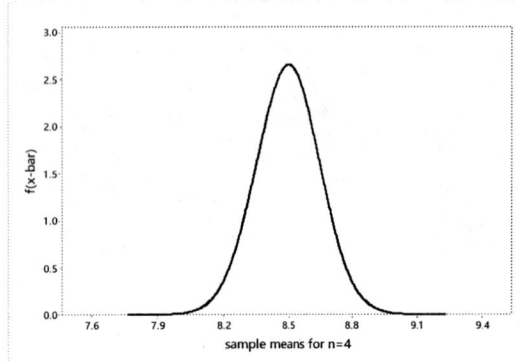

(c) normal distribution with $\mu_{\bar{x}} = 8.5g$ and $\sigma_{\bar{x}} = 0.3/\sqrt{9} = 0.10g$

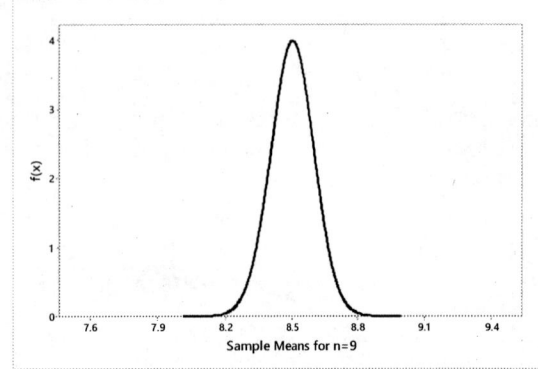

**13.** $n = 4: \mu_{\bar{x}} = 8.5$ and $\sigma_{\bar{x}} = 0.3/\sqrt{4} = 0.15$

(a) P(8.275 ≤ x̄ ≤ 8.725):

z-score computations:            Area less than z:

$\bar{x} = 8.275 \to z = \dfrac{8.275 - 8.5}{0.15} = -1.50$      0.0668

$\bar{x} = 8.725 \to z = \dfrac{8.725 - 8.5}{0.15} = 1.50$      0.9332

Total area = 0.9332 - 0.0668 = 0.8664, or 86.64%.

(b) The probability is 0.8664 that a sample of size four will have a mean within 0.225g of the population mean of 8.5g.

(c) There is an 86.64% chance that the sampling error when using the mean weights, x̄, of the four possums will be less than 0.225g.

(d) $n = 9: \mu_{\bar{x}} = 8.5$ and $\sigma_{\bar{x}} = 0.3/\sqrt{9} = 0.10$

z-score computations:            Area less than z:

$\bar{x} = 8.275 \to z = \dfrac{8.275 - 8.5}{0.10} = -2.25$      0.0122

$\bar{x} = 8.725 \to z = \dfrac{8.725 - 8.5}{0.1} = 2.25$      0.9878

Total area = 0.9878 - 0.0122 = 0.9756 or 97.56%

The probability is 0.9756 that a sample of size 9 will have a mean within 0.225g of the population mean of 8.5g.

There is an 97.56% chance that the sampling error when using the mean weights, x̄, of the nine possums will be less than 0.225g

**14.** (a) Since 60 is a fairly large sample size, the distribution of the sample mean will be approximately normal with a mean of 4.60 mmol/l and a standard deviation of $\sigma_{\bar{x}} = \sigma/\sqrt{n} = 0.16/\sqrt{60} = 0.0207$.

(b) Since 120 is a large sample size, the distribution of the sample mean will be approximately normal with a mean of 4.60 mmol/l and a standard deviation of $\sigma_{\bar{x}} = \sigma/\sqrt{n} = 0.16/\sqrt{120} = 0.0146$.

(c) No. The Central Limit Theorem ensures that when n is large, the distribution of the sample mean will be approximately normal regardless of whether the blood glucose levels themselves are normally distributed.

**15.** (a) $n = 500: \mu_{\bar{x}} = \mu$ and $\sigma_{\bar{x}} = 50{,}900/\sqrt{500} = 2276.32$

P(μ - 2,000 ≤ x̄ ≤ μ + 2,000):

**252** Chapter 7

z-score computations:                                                Area less than z:

$$\bar{x} = \mu - 2000 \rightarrow z = \frac{(\mu - 2000) - \mu}{2276.32} = -0.88$$                    0.1894

$$\bar{x} = \mu + 2000 \rightarrow z = \frac{(\mu + 2000) - \mu}{2276.32} = 0.88$$                    0.8106

Total area = 0.8106 - 0.1894 = 0.6212

(b) To answer part (a), it is not necessary to assume that the population is normally distributed because the sample size is large and, therefore, $\bar{x}$ is approximately normally distributed, regardless of the distribution of the population of life-insurance amounts.

If the sample size were 20 instead of 500, it would be necessary to assume normality because the sample size is small.

(c) $n = 5000: \mu_{\bar{x}} = \mu$ and $\sigma_{\bar{x}} = 50{,}900/\sqrt{5000} = 719.83$

z-score computation:                                                 Area less than z:

$$\bar{x} = \mu - 2000 \rightarrow \frac{(\mu - 2000) - \mu}{719.83} = -2.78$$                    0.0027

$$\bar{x} = \mu + 2000 \rightarrow \frac{(\mu + 2000) - \mu}{719.83} = 2.78$$                    0.9973

Total area = 0.9973 - 0.0027 = 0.9946

**16.** (a) $P(\bar{x} \leq 4.5)$:

z-score computation:                                                 Area less than z:

$$\bar{x} = 4.5 \rightarrow z = \frac{4.5 - 5}{0.5} = -1.00$$                                      0.1587

Total area = 0.1587

If the paint lasts 4.5 years, one would not consider this to be substantial evidence against the manufacturer's claim that the paint will last an average of five years.

Assuming the manufacturer's claim is correct, the probability is 0.1587 that the paint will last 4.5 years or less on a (randomly selected) house painted with the paint. In other words, there is a (fairly high) 15.87% chance that the paint would last 4.5 years or less, even if the manufacturer's claim is correct.

(b) $P(\bar{x} \leq 4.5)$:

$n = 10: \mu_{\bar{x}} = 5$ and $\sigma_{\bar{x}} = 0.5/\sqrt{10} = 0.158$

z-score computation:                                                 Area less than z:

$$\bar{x} = 4.5 \rightarrow z = \frac{4.5 - 5}{0.158} = -3.16$$                                    0.0008

Total area = 0.0008

For 10 houses, if the paint lasts an average of 4.5 years, I would consider this to be substantial evidence against the manufacturer's claim that the paint will last an average of five years.

Assuming the manufacturer's claim is correct, the probability is 0.0008 that the paint will last an average of 4.5 years or less for 10 (randomly selected) houses painted with the paint. In other words, there is less than a 0.1% chance that that would occur, if the manufacturer's claim is correct.

(c) $P(\bar{x} \leq 4.9)$:

z-score computation:  Area less than z:

$$\bar{x} = 4.9 \rightarrow z = \frac{4.9 - 5}{0.158} = -0.63 \qquad 0.2643$$

Total area = 0.2643

For 10 houses, if the paint lasts an average of 4.9 years, I would not consider this to be substantial evidence against the manufacturer's claim that the paint will last an average of five years.
Assuming the manufacturer's claim is correct, the probability is 0.2643 that the paint will last an average of 4.9 years or less for 10 (randomly selected) houses painted with the paint. In other words, there is a (fairly high) 26.43% chance that that would occur, even if the manufacturer's claim is correct.

**17.** (a) The sample size is large enough that the distribution of the sample means will be approximately normal with mean 6.83 and standard deviation of $\sigma_{\bar{x}} = \sigma / \sqrt{n} = 4.28 / \sqrt{100} = 0.428$.

We want $P(\bar{x} > 7.5)$:

z-score computation:  Area less than z:

$$\bar{x} = 7.5 \rightarrow z = \frac{7.5 - 6.83}{0.428} = 1.57 \qquad 0.9418$$

Total area = 1 - 0.9418 = 0.0582.

5.82% of all samples of size 100 days during the decade in question will have a mean degree of cloudiness exceeding 7.5.

(b) For samples of size 5, it is not reasonable to use a normal distribution to obtain the percentage required in part (a). From the frequency distribution, we can see that the distribution of degree of cloudiness is definitely not normally distributed. Therefore a sample of at least 30 is required before we can assume that the distribution of the sample means is approximately normally distributed.

**18.** (a) We have Minitab take 1000 random samples of size n = 4 from a normally distributed population with mean 49.34 and standard deviation 10.14 by choosing **Calc ▶ Random Data ▶ Normal...**, entering 1000 in the **Generate rows of data** text box, entering C1-C4 in the **Store in column(s)** text box, entering 49.34 in the **Mean** text box and 10.14 in the **Standard deviation** text box, and clicking **OK**.

(b) We compute the sample mean of each of the 1000 samples by choosing **Calc ▶ Row statistics...**, clicking on the **Mean** button, typing C1-C4 in the **Input variables** text box and XBAR in the **Store result in** text box, and clicking **OK**. (We will not print these values.)

(c) The mean is obtained by choosing **Calc ▶ Column statistics...**, clicking on the **Mean** button, selecting XBAR in the **Input variable** text box, and clicking **OK**. Our result is

   Mean of XBAR = 49.1875

Similarly, the standard deviation is obtained by choosing **Calc ▶ Column statistics...**, clicking on the **Standard deviation** button, selecting XBAR in the **Input variable** text box, and clicking **OK**. Our result is

   Standard deviation of XBAR = 5.18840

To get a histogram of the 1000 sample means stored in XBAR, choose **Graph ▶ Histogram...**, select the **Simple** version and click **OK**. Enter XBAR in the **Graph variables** text box, and click **OK**. Our result follows. Individual results will vary.

(d) Theoretically, the distribution of all possible sample means for samples of size four from this normal population should have mean 49.34, standard deviation $\sigma_{\bar{x}} = \sigma/\sqrt{n} = 10.14/\sqrt{4} = 5.07$, and a normal distribution.

(e) The histogram is close to bell-shaped, is centered near 49.34, and most of the data lies within three standard deviations (3 x 5.07 = 15.21) of 49.34, i.e. between 34.13 and 64.55. The mean of the 1000 sample means is 49.19, very close to 49.34, and the standard deviation of the sample means is 5.19, very close to 5.07.

**19.** (a) A uniform distribution between 0 and 1:

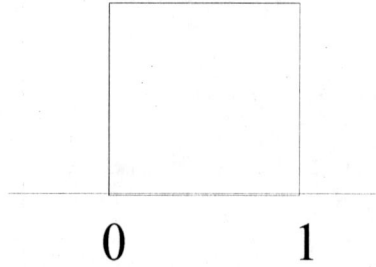

(b) We have Minitab take 2000 random samples of size n = 2 from a uniformly distributed population between 0 and 1 by choosing **Calc ▶ Random Data ▶ Uniform...**, entering 2000 in the **Generate rows of data** text box, entering C1-C2 in the **Store in column(s)** text box, entering 0.0 in the **Lower endpoint** text box and 1.0 in the **Upper endpoint** text box, and clicking **OK**.

(c) We compute the sample mean of each of the 1000 samples by choosing **Calc ▶ Row statistics...**, clicking on the **Mean** button, typing C1-C2 in the **Input variables** text box and XBAR in the **Store result in** text box, and clicking **OK**. (We will not print these values.)

(d) The mean of the sample means is obtained by choosing **Calc ▶ Column statistics...**, clicking on the **Mean** button, selecting XBAR in the **Input variable** text box, and clicking **OK**. Our result is

Mean of XBAR = 0.50454

Similarly, the standard deviation is obtained by choosing **Calc ▶ Column statistics...**, clicking on the **Standard deviation** button, selecting XBAR in the **Input variable** text box, and clicking **OK**. Our result is

Standard deviation of XBAR = 0.20545

(e) Theoretically, the distribution of all possible sample means for samples of size two from this uniform normal population should have mean

$(0 + 1)/2 = .5$ and standard deviation $\sigma/\sqrt{n} = ((1-0)/\sqrt{12})/\sqrt{2} = 0.2041$. Minitab simulation results are very close to these theoretical values.

(f) To get a histogram of the 1000 sample means stored in XBAR, choose **Graph ▶ Histogram...**, select the **Simple** version and click **OK**. Enter XBAR in the **Graph variables** text box, and click **OK**. The result is shown following.

The histogram is more triangle-shaped than bell-shaped. Since the population is far from normal and the sample size is only two, we would not expect the sampling distribution to be bell-shaped. In fact, it can be shown using advanced methods that the sum of two uniformly distributed variables has a triangular distribution.

(g) Repeating the process above, but using C1-C35 for the data in each sample, we obtained

Mean of XBAR = 0.49932

Standard deviation of XBAR = 0.049240

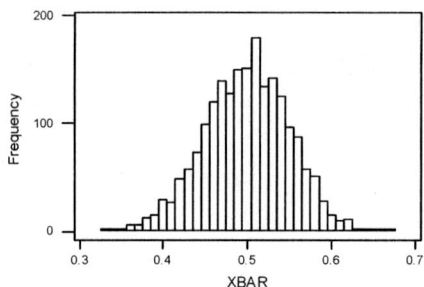

The theoretical distribution of the means has mean 0.5 and standard deviation $\sigma/\sqrt{n} = ((1-0)/\sqrt{12})/\sqrt{35} = 0.0488$.

The Minitab simulation values are very close to these. The histogram for the means of the samples of size 35 is shown. It is much more bell-shaped than for samples of size two, as we would expect since n > 30.

## CHAPTER 8 SOLUTIONS

### Exercises 8.1

**8.1** The value of a statistic that is used to estimate a parameter is called a <u>point estimate</u> of the parameter.

**8.3** Margin of error indicates how accurate our estimate ($\bar{x}$) is an estimate for the value of the unknown parameter ($\mu$).

**8.5** Approximately 95%, or 950, of the confidence intervals will contain the value of the unknown parameter.

**8.7** $\bar{x} = 230/10 = 23.0$

**8.9** $s = \sqrt{\dfrac{\sum x^2 - (\sum x)^2 / n}{n-1}} = \sqrt{\dfrac{6140 - (230)^2 / 10}{9}} = \sqrt{94.44} = 9.72$

**8.11** (a) The approximate confidence interval will be

$$\bar{x} - 2\sigma/\sqrt{n} \text{ to } \bar{x} + 2\sigma/\sqrt{n}$$
$$20 - 2(3)/\sqrt{36} \text{ to } 20 + 2(3)/\sqrt{36}$$
$$19 \text{ to } 21$$

(b) The margin of error is 1. We can be approximately 95% confident that the unknown parameter, $\mu$, will be within 1 of our sample mean, $\bar{x} = 20$.

(c) The endpoints of a confidence interval are found by point estimate ± margin of error. For the interval in part (a), we have 20 ± 1.

**8.13** (a) The approximate confidence interval will be

$$\bar{x} - 2\sigma/\sqrt{n} \text{ to } \bar{x} + 2\sigma/\sqrt{n}$$
$$30 - 2(4)/\sqrt{25} \text{ to } 30 + 2(4)/\sqrt{25}$$
$$28.4 \text{ to } 31.6$$

(b) The margin of error is 1.6. We can be approximately 95% confident that the unknown parameter, $\mu$, will be within 1.6 of our sample mean, $\bar{x} = 30$.

(c) The endpoints of a confidence interval are found by point estimate ± margin of error. For the interval in part (a), we have 30 ± 1.6.

**8.15** (a) The approximate confidence interval will be

$$\bar{x} - 2\sigma/\sqrt{n} \text{ to } \bar{x} + 2\sigma/\sqrt{n}$$
$$50 - 2(5)/\sqrt{16} \text{ to } 50 + 2(5)/\sqrt{16}$$
$$47.5 \text{ to } 52.5$$

(b) The margin of error is 2.5. We can be approximately 95% confident that the unknown parameter, $\mu$, will be within 2.5 of our sample mean, $\bar{x} = 50$.

**258** Chapter 8

(c) The endpoints of a confidence interval are found by point estimate ± margin of error. For the interval in part (a), we have 50 ± 2.5.

**8.17** (a) $\bar{x} = \$526538/20 = \$26326.90$

(b) Since some sampling error is expected, it is unlikely that the sample mean will exactly equal the population mean $\mu$.

**8.19** (a) The approximate confidence interval will be

$$\bar{x} - 2\sigma/\sqrt{n} \text{ to } \bar{x} + 2\sigma/\sqrt{n}$$
$$26326.90 - 2(8100)/\sqrt{20} \text{ to } 26326.90 + 2(8100)/\sqrt{20}$$
$$\$22,704.47 \text{ to } \$29,949.33$$

(b) Since we know that approximately 95% of all samples of 20 wedding costs have the property that the interval from $\bar{x}$ − $3622.43 to $\bar{x}$ + $3622.43 contains $\mu$, we can be 95% confident that the interval from $22707.47 to $29952.33 contains $\mu$.

(c) We can't be certain that the population mean lies in the interval, but we are 95% confident that it does.

**8.21** (a) n = 35; a point estimate for the mean fuel tank capacity is

$\bar{x}$ = 664.9/35 = 19.00 gallons. This number is called a point estimate because it consists of a single value.

(b) The approximate confidence interval will be

$$\bar{x} - 2\sigma/\sqrt{n} \text{ to } \bar{x} + 2\sigma/\sqrt{n}$$
$$19.00 - 2(3.50)/\sqrt{35} \text{ to } 19.00 + 2(3.50)/\sqrt{35}$$
$$17.82 \text{ to } 20.18$$

(c) We could see if a histogram looked bell-shaped or if a normal probability plot produced a relatively straight line.

(d) It is not necessary that the fuel tank capacities be exactly normally distributed because for samples of size 35 by the Central Limit Theorem, the distribution of the sample means will be approximately normally distributed, regardless of the population distribution. Therefore, the confidence interval will be approximately correct.

**8.23** (a) Using Minitab to retrieve the data from the WeissStats Resource Site, we obtained the following normal probability plot by choosing **Stat ▶ Basic Statistics ▶ Normality Test,** entering LENGTHS in the **Variable** text box, and clicking **OK**.

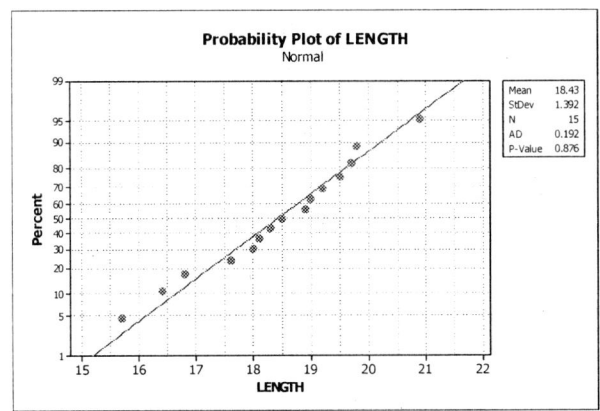

(b) Yes. There do not appear to be any outliers, and the data points fall very close to a straight line.

(c) The sample mean is 18.43 mm. Since $P(\bar{x} - 2\sigma/\sqrt{n} < \mu < \bar{x} + 2\sigma/\sqrt{n}) = 0.9544$, we can be approximately 95% confident that the mean μ is somewhere between $\bar{x} - 2\sigma/\sqrt{n}$ and $\bar{x} + 2\sigma/\sqrt{n}$. Since n = 15, $\bar{x}$ = 18.43, and $\sigma$ = 1.76, the 95% confidence interval is

$$\bar{x} - 2\sigma/\sqrt{n} \text{ to } \bar{x} + 2\sigma/\sqrt{n}$$
$$18.43 - 2(1.76)/\sqrt{15} \text{ to } 18.43 + 2(1.76)/\sqrt{15}$$
$$17.52 \text{ to } 19.34$$

We can be approximately 95% confident that the interval 17.52 mm to 19.34 mm contains the value of the mean $\mu$.

(d) Yes. 18.14 lies between 17.52 and 19.34, so the interval we obtained in part (c) does contain μ. It is possible that our confidence interval will not contain μ. This should happen only about 5% of the time.

**8.25** Since $P(\bar{x} - 3\sigma/\sqrt{n} < \mu < \bar{x} + 3\sigma/\sqrt{n}) = 0.9974$, we can be 99.7% confident that the mean μ is somewhere between $\bar{x} - 3\sigma/\sqrt{n}$ and $\bar{x} + 3\sigma/\sqrt{n}$. Since n = 36, $\bar{x}$ = 63.28, and $\sigma$ = 7.2, the 99.7% confidence interval is (in thousands)

$$\bar{x} - 3\sigma/\sqrt{n} \text{ to } \bar{x} + 3\sigma/\sqrt{n}$$
$$63.28 - 3(7.2)/\sqrt{36} \text{ to } 63.28 + 3(7.2)/\sqrt{36}$$
$$59.68 \text{ to } 66.88$$

We can be 99.7% confident that the interval $59,680 to $66,880 contains the value of the mean $\mu$.

### Exercises 8.2

**8.27** (a) Confidence level = 0.90; $\alpha$ = 0.10

(b) Confidence level = 0.99; $\alpha$ = 0.01

**8.29** (a) By saying that a 1-$\alpha$ confidence interval is exact, we mean that the true confidence level is equal to 1-$\alpha$.

**260** Chapter 8

- (b) By saying that a $1-\alpha$ confidence interval is approximately correct, we mean that the true confidence level is only approximately equal to $1-\alpha$.

**8.31** When we use the abbreviation "normal population," we mean that the variable under consideration is normally distributed on the population of interest.

**8.33** A statistical procedure is robust if it is insensitive to departures from the assumptions on which it is based.

**8.35** The z-interval procedure is reasonable since the sample size is very large.

**8.37** The z-interval procedure is reasonable since the population is roughly normal and the sample size is over 15.

**8.39** The z-interval procedure is reasonable since the sampling distribution of $\bar{x}$ will be very close to normal for a sample size of 250 even though the population itself is far from normal.

**8.41** A 95% confidence interval will give a more precise (shorter interval) estimate of $\mu$ than will a 99% confidence interval.

**8.43** The margin of error is the standard error multiplied by $z_{\alpha/2}$.

**8.45** Increasing the sample size while keeping the same confidence level will decrease the margin of error and hence increase the accuracy of estimating a population mean by a sample mean.

**8.47** Increasing the confidence level while keeping the same sample size will increase the margin of error and hence decrease the accuracy of estimating a population mean by a sample mean.

**8.49** (a) The length of the confidence interval is twice the margin of error; i.e., 2 x 3.4 = 6.8.

   (b) The confidence interval is 52.8 ± 3.4 = (49.4, 56.2).

   (c)

   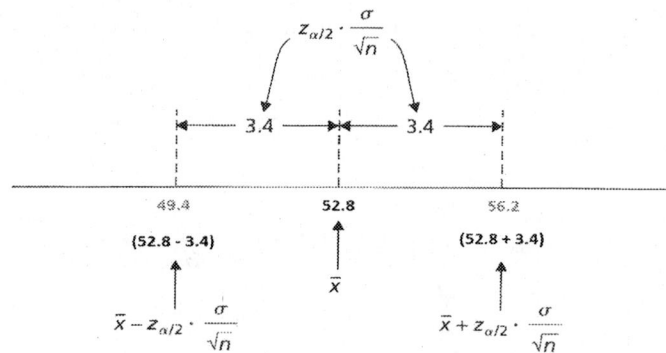

**8.51** (a) The margin of error is 1/2 the length of the confidence interval; i.e., (1/2) x 20 = 10.

   (b) The confidence interval is 60 ± 10 = (50, 70).

(c)

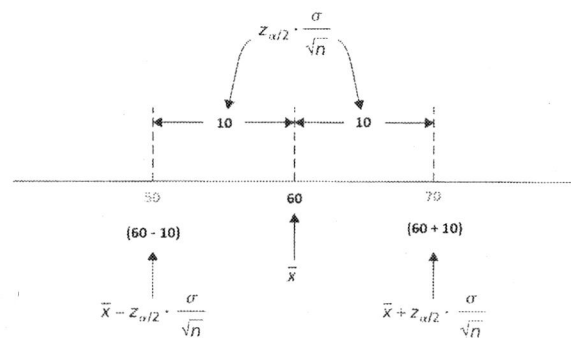

**8.53** True. The length of the confidence interval is twice the margin of error.

**8.55** False. One must also know the sample mean.

**8.57** False. One must also know n and $\sigma$.

**8.59** True. $E = z_{\alpha/2}(\sigma/\sqrt{n})$.

**8.61** (a) We want a whole number because the number of observations to be taken cannot be fractional. It must be an integer because we cannot sample a partial member of a population.

(b) The original number computed is the smallest value of n that will provide the required margin of error. If we were to round down, the actual sample size would be slightly too small.

**8.63** (a) The approximate 95% confidence interval for μ is

$$\bar{x} - z_{\alpha/2}\sigma/\sqrt{n} \text{ to } \bar{x} + z_{\alpha/2}\sigma/\sqrt{n}$$
$$20 - 1.96(3)/\sqrt{36} \text{ to } 20 + 1.96(3)/\sqrt{36}$$
$$19.02 \text{ to } 20.98$$

(b) The length of the confidence interval is 20.98 − 19.03 = 1.96. The margin of error is half the length of the interval and equal to 1.96/2 = 0.98.

(c) Formula 8.1 states $E = z_{\alpha/2}\sigma/\sqrt{n} = 1.96(3)/\sqrt{36} = 0.98$.

**8.65** (a) The 90% confidence interval for μ is

$$\bar{x} - z_{\alpha/2}\sigma/\sqrt{n} \text{ to } \bar{x} + z_{\alpha/2}\sigma/\sqrt{n}$$
$$30 - 1.645(4)/\sqrt{25} \text{ to } 30 + 1.645(4)/\sqrt{25}$$
$$28.684 \text{ to } 31.316$$

(b) The length of the confidence interval is 31.316 − 28.684 = 2.632. The margin of error is half the length of the interval and equal to 2.632/2 = 1.316.

(c) Formula 8.1 states $E = z_{\alpha/2}\sigma/\sqrt{n} = 1.645(4)/\sqrt{25} = 1.316$.

**8.67** (a) The 99% confidence interval for μ is

$$\bar{x} - z_{\alpha/2}\sigma/\sqrt{n} \text{ to } \bar{x} + z_{\alpha/2}\sigma/\sqrt{n}$$
$$50 - 2.575(5)/\sqrt{16} \text{ to } 50 + 2.575(5)/\sqrt{16}$$
$$46.781 \text{ to } 53.219$$

(b) The length of the confidence interval is 53.219 − 46.781 = 6.438. The margin of error is half the length of the interval and equal to 6.438/2 = 3.219.

(c) Formula 8.1 states $E = z_{\alpha/2}\sigma/\sqrt{n} = 2.575(5)/\sqrt{16} = 3.219$.

**8.69** The sample mean is $113.97 million / 18 = $6.332 million. The 95% confidence interval for μ is

$$\bar{x} - z_{\alpha/2}\sigma/\sqrt{n} \text{ to } \bar{x} + z_{\alpha/2}\sigma/\sqrt{n}$$
$$6.332 - 1.96(2.04)/\sqrt{18} \text{ to } 6.332 + 1.96(2.04)/\sqrt{18}$$
$$\$5.389 \text{ to } \$7.274 \text{ (million)}$$

We can be 95% confident that the interval from $5.389 million to $7.274 million contains the population mean venture capital investment in the fiber optics business sector.

**8.71** The sample mean is 6.31 / 12 = 0.526 ppm. The 95% confidence interval for μ is

$$\bar{x} - z_{\alpha/2}\sigma/\sqrt{n} \text{ to } \bar{x} + z_{\alpha/2}\sigma/\sqrt{n}$$
$$0.526 - 2.575(0.37)/\sqrt{12} \text{ to } 0.526 + 2.575(0.37)/\sqrt{12}$$
$$0.251 \text{ to } 0.801 \text{ ppm}$$

We can be 99% confident that the interval from 0.251 to 0.801 ppm contains the population mean cadmium level in *Boletus pinicola* mushrooms.

**8.73** n = 32; $\bar{x}$ = 33.4; σ = 42

Step 1: $\alpha$ = 0.05; $z_{\alpha/2}$ = $z_{0.025}$ = 1.96

Step 2:

$$\bar{x} - z_{\alpha/2}\sigma/\sqrt{n} \text{ to } \bar{x} + z_{\alpha/2}\sigma/\sqrt{n}$$
$$33.4 - 1.96(42)/\sqrt{32} \text{ to } 33.4 + 1.96(42)/\sqrt{32}$$
$$18.8 \text{ to } 48.0$$

We can be 95% confident that the mean duration of imprisonment, $\mu$, of all East German political prisoners with chronic PTSD is somewhere between 18.8 and 48.0 months.

**8.75** $n = 18$; $\bar{x} = 6.33$; $\sigma = 2.04$

(a) Step 1: $\alpha = 0.01$; $z_{\alpha/2} = z_{0.005} = 2.575$

Step 2:
$$\bar{x} - z_{\alpha/2}\sigma/\sqrt{n} \text{ to } \bar{x} + z_{\alpha/2}\sigma/\sqrt{n}$$
$$6.33 - 2.575(2.04)/\sqrt{18} \text{ to } 6.33 + 2.575(2.04)/\sqrt{18}$$
$$5.09 \text{ to } 7.57$$

(b) The confidence interval in part (a) is longer than the one in Exercise 8.69 because we have changed the confidence level from 95% in Exercise 8.69 to 99% in this exercise. Notice that increasing the confidence level from 95% to 99% increases the $z_{\alpha/2}$-value from 1.96 to 2.575. The larger z-value, in turn, results in a longer interval. In order to achieve a higher level of confidence that the interval contains the population mean, we need a longer interval.

(c)

We can be 95% confident that μ lies in here

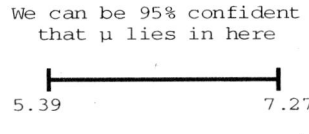
5.39  7.27

We can be 99% confident that μ lies in here

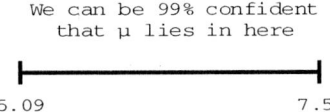
5.09  7.57

(d) The 95% confidence interval is shorter and therefore provides a more accurate estimate of μ.

**8.77** $n = 120$; $\bar{x} = 102.72$ days; $\sigma = 32$ days

(a) Step 1: $\alpha = 0.05$; $z_{\alpha/2} = z_{0.025} = 1.96$

Step 2:
$$\bar{x} - z_{\alpha/2}\sigma/\sqrt{n} \text{ to } \bar{x} + z_{\alpha/2}\sigma/\sqrt{n}$$
$$102.72 - 1.96(32)/\sqrt{120} \text{ to } 102.72 + 1.96(32)/\sqrt{120}$$
$$96.994 \text{ to } 108.446$$

(b) Step 1: $\alpha = 0.10$; $z_{\alpha/2} = z_{0.05} = 1.645$

Step 2:
$$\bar{x} - z_{\alpha/2}\sigma/\sqrt{n} \text{ to } \bar{x} + z_{\alpha/2}\sigma/\sqrt{n}$$
$$102.72 - 1.645(32)/\sqrt{120} \text{ to } 102.72 + 1.645(32)/\sqrt{120}$$
$$97.915 \text{ to } 107.525$$

**264** Chapter 8

(c)

> We can be 90% confident that μ lies within
> 
> ├─────────────┤
> 97.915     107.525
> 
> We can be 95% confident that μ lies within
> 
> ├─────────────┤
> 96.994     108.446

(d) The 90% interval is narrower and therefore provides a more accurate estimate of μ.

**8.79** (a) The margin of error for the 95% confidence interval is (108.446 − 96.994)/2 = 5.726 days.

(b) The margin of error for the 90% confidence interval is (107.525 − 97.915)/2 = 4.805 days.

(c) The margin of error for the 90% confidence interval is smaller than the margin of error for the 95% confidence interval.

(d) This is illustrating the principle that decreasing the confidence level, while keeping the sample size the same, decreases the margin of error.

**8.81** $n = 30$; $\bar{x} = 105.43$ days; $\sigma = 32$ days

(a) Step 1: $\alpha = 0.05$; $z_{\alpha/2} = z_{0.025} = 1.96$

Step 2:

$$\bar{x} - z_{\alpha/2}\sigma/\sqrt{n} \text{ to } \bar{x} + z_{\alpha/2}\sigma/\sqrt{n}$$

$$105.43 - 1.96(32)/\sqrt{30} \text{ to } 105.43 + 1.96(32)/\sqrt{30}$$

$$93.979 \text{ to } 116.881$$

(b)

> Ex 8.77
> 
> We can be 95% confident that μ lies within
> 
> ├─────────────┤
> 96.994     108.446
> 
> Ex 8.81
> 
> We can be 95% confident that μ lies within
> 
> ├─────────────┤
> 93.979     116.881

(c) The margin of error for this exercise is (116.881 − 93.979)/2 = 11.451 days while the margin of error for the 95% confidence interval in

Exercise 8.77 is 5.726 days. The margin of error for this exercise is larger than the margin of error for the 95% confidence interval in Exercise 8.77.

(d) This is illustrating the principle that decreasing the sample size, while keeping the confidence level fixed, increases the margin of error.

**8.83** $n = 36$; $\bar{x} = \$63.28$ thousand; $\sigma = \$7.2$ thousand

(a) Step 1: $\alpha = 0.05$; $z_{\alpha/2} = z_{0.025} = 1.96$

Step 2:
$$\bar{x} - z_{\alpha/2}\sigma/\sqrt{n} \text{ to } \bar{x} + z_{\alpha/2}\sigma/\sqrt{n}$$
$$63.28 - 1.96(7.2)/\sqrt{36} \text{ to } 63.28 + 1.96(7.2)/\sqrt{36}$$
$$60.928 \text{ to } 65.632$$

(b) The interval in Example 8.2 is 60.88 to 65.68. Example 8.2 used 2 (From Property 2 of the empirical rule) as an approximation of $z_{\alpha/2}$. This exercise used 1.96. Example 8.2 gave an approximate 95% confidence interval.

**8.85** $n = 10$; $\bar{x} = 34.2$ cm; $\sigma = 2.1$ cm

(a) $\alpha = 0.10$; $z_{\alpha/2} = z_{0.05} = 1.645$, so the 90% confidence interval for $\mu$ is

$$\bar{x} - 1.645(2.1)/\sqrt{10} \text{ to } \bar{x} + 1.645(2.1)/\sqrt{10}$$
$$34.2 - 1.645(2.1)/\sqrt{10} \text{ to } 34.2 + 1.645(2.1)/\sqrt{10}$$
$$33.1 \text{ to } 35.3 \text{ cm}$$

(b) The margin of error is $E = 1.645\sigma/\sqrt{n} = 1.645(2.1)/\sqrt{10} = 1.1$ cm.

(c) We are 90% confident that our error in estimating $\mu$ by $\bar{x}$ is at most 1.1 cm.

(d) $\alpha = 0.05$; $z_{\alpha/2} = z_{0.025} = 1.96$, so $n = \left(\dfrac{z_{\alpha/2} \cdot \sigma}{E}\right)^2 = \left(\dfrac{1.96 \cdot 2.1}{0.5}\right)^2 = 67.76$ or 68.

**8.87** (a) $E = 0.5(\$7.247$ million $- \$5.389$ million$) = 0.943$ million

(b) $E = 1.96\sigma/\sqrt{n} = 1.96(2.04)/\sqrt{18} = \$0.942$ million. Note: there is a small difference in the answers do to rounding.

**8.89** (a) $E = (48.0 - 18.8)/2 = 14.6$ months

(b) We are 95% confident that the maximum error made in using $\bar{x}$ to estimate $\mu$ is 14.6 months.

(c) The margin of error of the estimate is specified to be $E = 12.0$ months. Also, for a 99% confidence interval, $z_{\alpha/2} = z_{0.005} = 2.575$.

$$n = \left[\dfrac{z_{\alpha/2}\sigma}{E}\right]^2 = \left[\dfrac{2.575(42)}{12}\right]^2 = 81.2 \to 82$$

**266** Chapter 8

(d)
$$\bar{x} - z_{\alpha/2}\sigma/\sqrt{n} \text{ to } \bar{x} + z_{\alpha/2}\sigma/\sqrt{n}$$
$$36.2 - 2.575(12)/\sqrt{82} \text{ to } 36.2 + 2.575(12)/\sqrt{82}$$
$$24.3 \text{ to } 48.1$$

**8.91** Using Minitab, retrieve the data from the WeissStats Resource Site. Click on the **Graphs** button and check the box for **Histogram** then click **OK**. Enter BRIGHTNESS in the **Variable** text box and click **OK**. To get the probability plot and stem-and leaf diagram, choose **Stat, Basic Statistics,** then **Normality Test.** Enter BRIGHTNESS in the **Variable** text box and click **OK**. The resulting graphs are as follows.

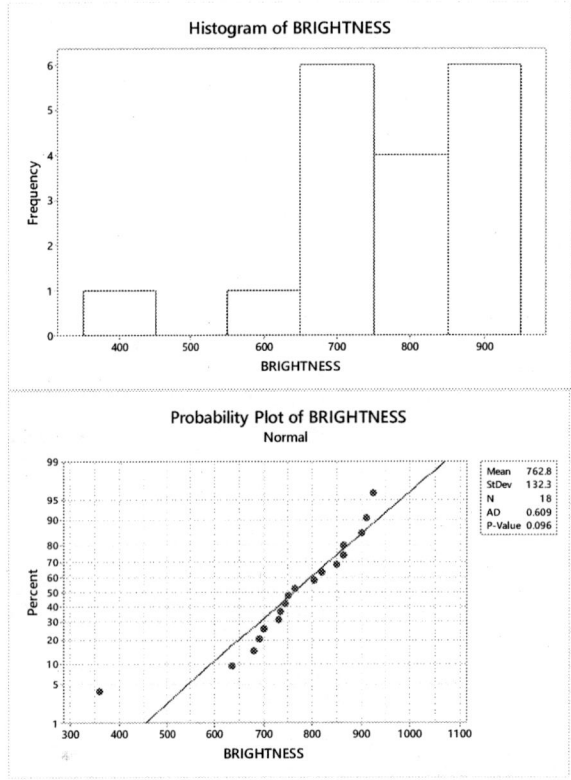

From the graphs, we determine that the variable is not normally distributed and contains an outlier. Also the sample size is small. Therefore it is not reasonable to apply the z-interval procedure.

**8.93** (a) Using Minitab, retrieve the data from the WeissStats Resource Site. Then choose **Stat ▸ Basic statistics ▸ 1-Sample z.** Enter TIME in the **Samples in Columns** text box and enter 30 in the **Standard deviation** text box. Click on the **Graphs** button and check the boxes for **Histogram of data** and **Boxplot of data**, then click **OK** twice. The confidence interval appears in the Sessions Window s

```
Variable   N     Mean    StDev   SE Mean      95% CI
TIME      20   289.950   30.741    6.708   (276.802, 303.098)
```

The histogram and boxplot are shown following.

To get the probability plot and stem-and leaf diagram, choose **Stat ▶ Basic statistics ▶ Normality test**, enter TIME in the **Variable** text box and click **OK**. Then choose **Graph ▶ Stem-and-Leaf**, and enter TIME in the **Graph Variables** text box and click **OK**. The results are

```
Stem-and-leaf of TIME   N  = 20
Leaf Unit = 1.0

  2   25   45
  6   26   0779
  9   27   235
 10   28   7
 10   29   01257
  5   30   22
  3   31
  3   32
  3   33   3
  2   34   4
 HI 374
```

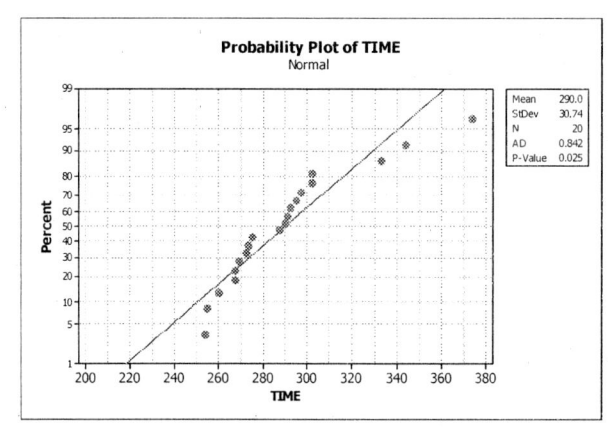

(c) The value 374 is a potential outlier. Remove it from the data and repeat the process in part (a). The result is

```
Variable   N    Mean    StDev   SE Mean       95% CI
TIME      19  285.526  24.174    6.882   (272.037, 299.016)
```

(d) The graphs (not shown here) that were produced with the confidence interval now show that 344 is an outlier in the sample of 19 observations. The data are also skewed to the right both before and after deleting the first outlier. The relatively small sample size makes it unwise to use the z-interval procedure with these data.

**8.95** $n = 900$; $\alpha = 0.05$; $z_{\alpha/2} = z_{0.025} = 1.96$; $\sigma = 12.1$

$$E = 1.96(12.1)/\sqrt{900} = 0.791$$

**268** Chapter 8

**8.97** (a) The margin of error of the estimate is specified to be E = 2 years.

$$n = \left[\frac{z_{\alpha/2}\sigma}{E}\right]^2 = \left[\frac{1.96(13.36)}{2.0}\right]^2 = 171.4 \rightarrow 172$$

(b) We used s in place of $\sigma$ because $\sigma$ was unknown. We can do this because the sample of size 36 is large enough to provide an estimate of $\sigma$ and the variation is not likely to change much from one year to the next.

**8.99** (a) We will get the confidence interval for part (c) along with the histogram and boxplot. Using Minitab, retrieve the data from the WeissStats Resource Site. Then choose **Stat ▶ Basic statistics ▶ 1-Sample z**. Enter TEMP in the **Samples in Columns** text box and enter 0.63 in the **Standard deviation** text box. Click on the Options button and enter 99.0 in the **Confidence level** text box and click **OK**. Click on the **Graphs** button and check the boxes for **Histogram of data** and **Boxplot of data**, then click **OK** twice. To get the probability plot and stem-and-leaf diagram, choose **Stat ▶ Basic statistics ▶ Normality test**, enter TEMP in the **Variable** text box and click **OK**. Then choose **Graph ▶ Stem-and-Leaf**, and enter TEMP in the **Graph Variables** text box and click **OK**. The results are

```
Stem-and-leaf of TEMP    N = 93
Leaf Unit = 0.10

   1    96  7
   3    96  89
   8    97  00001
  13    97  22233
  19    97  444444
  26    97  6666777
  31    97  88889
  45    98  00000000000111
 (10)   98  2222222233
  38    98  4444445555
  28    98  66666666677
  17    98  8888888
  10    99  00001
   5    99  2233
   1    99  4
```

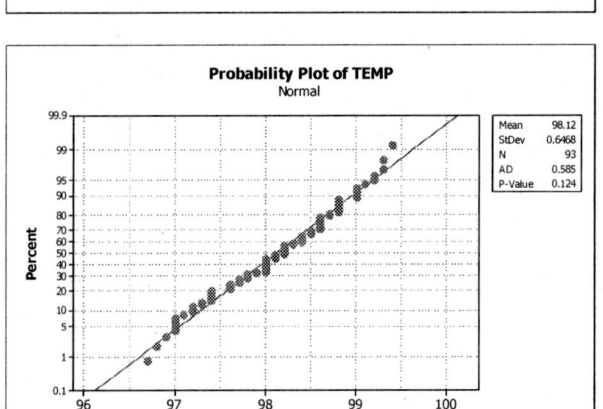

(b) Yes. There are no outliers, the sample size is large (93), and the distribution of the data appears to be approximately normal.

(c) The output from part (a) in the Sessions Window is

```
Variable    N      Mean    StDev   SE Mean         99% CI
TEMP       93    98.1237   0.6468   0.0653   (97.9554, 98.2919)
```

We can be 99% confident that the interval (97.9554, 98.2919) contains the mean body temperature μ of healthy humans. The output is surprising since the confidence interval does not inclue the generally accepted normal temperature of 98.6° F.

**8.101** (a) We will get the confidence interval for part (c) along with the histogram and boxplot. Using Minitab, retrieve the data from the WeissStats Resource Site. Then choose **Stat ▶ Basic statistics ▶ 1-Sample z.** Enter SPEED in the **Samples in Columns** text box and enter 3.2 in the **Standard deviation** text box. Click on the **Options** button and enter 95.0 in the **Confidence level** text box and click **OK**. Click on the **Graphs** button and check the boxes for **Histogram of data** and **Boxplot of data**, then click **OK** twice. To get the probability plot and stem-and-leaf diagram, choose **Stat ▶ Basic statistics ▶ Normality test**, enter SPEED in the **Variable** text box and click **OK**. Then choose **Graph ▶ Stem-and-Leaf**, and enter SPEED in the **Graph Variables** text box and click **OK**. The results are

```
Variable    N      Mean    StDev   SE Mean         95% CI
SPEED      35    59.5257   4.2739   0.5409   (58.4656, 60.5859)
```

## 270 Chapter 8

```
Stem-and-leaf of SPEED   N = 35
Leaf Unit = 0.10

  2    52  46
  2    53
  4    54  78
  7    55  459
  8    56  5
 13    57  35688
 16    58  137
 (5)   59  02678
 14    60  12679
  9    61  36
  7    62  36
  5    63  4
  4    64
  4    65  02
  2    66  0
HI 753
```

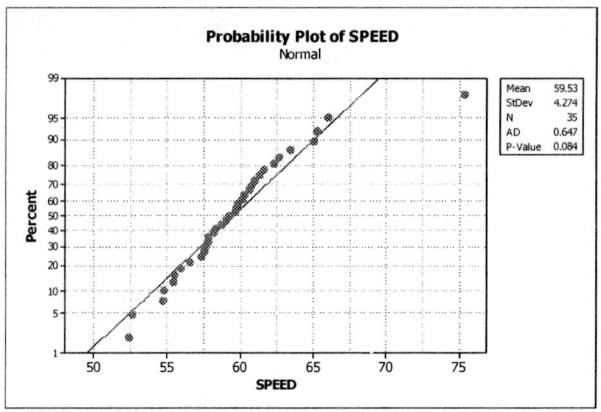

(c) After removing 75.3 from the data, the 95% confidence interval is

```
Variable   N    Mean    StDev   SE Mean      95% CI
SPEED      34   59.0618  3.3253  0.5488   (57.9861, 60.1374)
```

(d) The outlier had an effect on the mean of about 0.46 mph and, therefore, about the same effect on the endpoints of the confidence interval. The movement of the interval is about one-fourth the length of the interval and is significant in that sense. On the other hand, the change in the mean represents less than a 1% error and is probably not very important, although this is a matter of judgment. The sample size of 35 appears to be large enough that one outlier of the size observed does not have a great effect.

**8.103** (a) $\alpha = 0.05$; $z_{\alpha/2} = z_{0.025} = 1.96$; $\sigma = 10$

$$E = 1.96(10) / \sqrt{4} = 9.80$$

(b) $\alpha = 0.05$; $z_{\alpha/2} = z_{0.025} = 1.96$; $\sigma = 10$

$$E = 1.96(10) / \sqrt{16} = 4.90$$

(c) It appears that quadrupling the sample size will halve the margin of error. Therefore, increasing $n$ from 16 to 64 will decrease the margin of error from 4.90 to 2.45.

**8.105** We can be $100(1 - \alpha)\%$ confident that the mean $\mu$ is greater than the lower confidence bound and $100(1 - \alpha)\%$ confident that the mean $\mu$ is less than the upper confidence bound.

**8.107** (a) The sample mean is $6.31 / 12 = 0.526$ ppm. The 99% lower confidence bound for $\mu$ is

$$\bar{x} - z_\alpha \sigma / \sqrt{n}$$

$$0.526 - 2.33(0.37) / \sqrt{12}$$

$$0.277 \text{ ppm}$$

We can be 99% confident that the population mean cadmium level in *Boletus pinicola* mushrooms is greater than 0.277 ppm.

(b) This lower confidence bound is greater than the lower confidence limit of 0.251 found in Exercise 8.71. This is because the z-value used for a one-sided 99% confidence bound is 2.33, whereas, the z-value used for a two-sided 99% confidence interval is 2.575.

### Exercises 8.3

**8.109** (a) The standardized version of $\bar{x}$ is

$$z = (\bar{x} - \mu)/(\sigma/\sqrt{n}) = (108-100)/(16/\sqrt{4}) = 1.00.$$

(b) The studentized version of $\bar{x}$ is

$$t = (\bar{x} - \mu)/(s/\sqrt{n}) = (108-100)/(12/\sqrt{4}) = 1.333.$$

**8.111** (a) Standard normal

(b) t-distribution with 11 degrees of freedom

**8.113** The variation in the possible values of the standardized version of $\bar{x}$ is due only to the variation in $\bar{x}$ while the variation in the studentized version results not only from the variation in $\bar{x}$, but also from the variation in the sample standard deviation.

**8.115** For df = 6:

(a) $t_{0.10} = 1.440$  (b) $t_{0.025} = 2.447$  (c) $t_{0.01} = 3.143$

**8.117** (a) $t_{0.10} = 1.323$  (b) $t_{0.01} = 2.518$

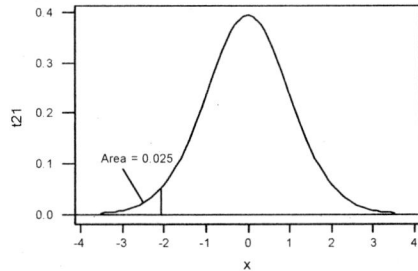

(c) $-t_{0.025} = -2.080$  (d) $\pm t_{0.05} = \pm 1.721$

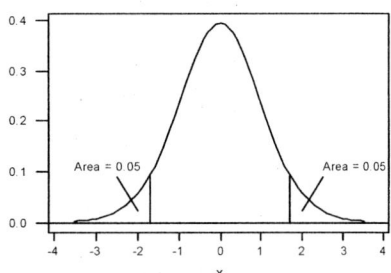

**8.119** It is reasonable to use the t-interval procedure since the sample size is large and for large degrees of freedom (99), the t-distribution is very

**272** Chapter 8

similar to the standard normal distribution. Another way of expressing this is that the sampling distribution of $\bar{x}$ is approximately normal when n is large, so the standardized and studentized versions of $\bar{x}$ are essentially the same.

**8.121** $E = t_{\alpha/2} s / \sqrt{n}$

**8.123** (a) df = 35 $\alpha$ = 0.05; $t_{\alpha/2}$ = $t_{0.025}$ = 2.030

The 95% confidence interval for μ is

$$\bar{x} - t_{\alpha/2} s / \sqrt{n} \text{ to } \bar{x} + t_{\alpha/2} s / \sqrt{n}$$
$$20 - 2.030(3) / \sqrt{36} \text{ to } 20 + 2.030(3) / \sqrt{36}$$
$$18.985 \text{ to } 21.015$$

(b) The length of the confidence interval is 21.015 − 18.985 = 2.030. The margin of error is half the length of the interval and equal to 2.030/2 = 1.015.

(c) The formula states $E = t_{\alpha/2} s / \sqrt{n} = 2.030(3) / \sqrt{36} = 1.015$.

**8.125** (a) df = 24 $\alpha$ = 0.10; $t_{\alpha/2}$ = $t_{0.05}$ = 1.711

The 90% confidence interval for μ is

$$\bar{x} - t_{\alpha/2} s / \sqrt{n} \text{ to } \bar{x} + t_{\alpha/2} s / \sqrt{n}$$
$$30 - 1.711(4) / \sqrt{25} \text{ to } 30 + 1.711(4) / \sqrt{25}$$
$$28.6312 \text{ to } 31.3688$$

(b) The length of the confidence interval is 31.3688 − 28.6312 = 2.7376. The margin of error is half the length of the interval and equal to 2.7376/2 = 1.3688.

(c) The formula states $E = t_{\alpha/2} s / \sqrt{n} = 1.711(4) / \sqrt{25} = 1.3688$.

**8.127** (a) df = 15 $\alpha$ = 0.01; $t_{\alpha/2}$ = $t_{0.005}$ = 2.947

The 99% confidence interval for μ is

$$\bar{x} - t_{\alpha/2} s / \sqrt{n} \text{ to } \bar{x} + t_{\alpha/2} s / \sqrt{n}$$
$$50 - 2.947(5) / \sqrt{16} \text{ to } 50 + 2.947(5) / \sqrt{16}$$
$$46.3163 \text{ to } 53.6838$$

(b) The length of the confidence interval is 53.6838 − 46.3163 = 7.3675. The margin of error is half the length of the interval and equal to 7.3675/2 = 3.6838.

(c) The formula states $E = t_{\alpha/2} s / \sqrt{n} = 2.947(5) / \sqrt{16} = 3.6838$.

**8.129** n = 30; df = 29; $t_{\alpha/2}$ = $t_{0.05}$ = 1.699; $\bar{x}$ = 27.97 minutes; s = 10.04 minutes

$$\bar{x} - t_{\alpha/2} s/\sqrt{n} \text{ to } \bar{x} + t_{\alpha/2} s/\sqrt{n}$$
$$27.97 - 1.699(10.04)/\sqrt{30} \text{ to } 27.97 + 1.699(10.04)/\sqrt{30}$$
$$24.86 \text{ to } 31.08$$

We are 90% confident that the interval 24.86 to 31.08 minutes contains the mean commute times for all working adults in the Washington, D.C. area.

**8.131** n = 10; df = 9; $t_{\alpha/2}$ = $t_{0.025}$ = 2.262; $\bar{x}$ = 2.33 hours; s = 2.002 hours

(a)
$$\bar{x} - t_{\alpha/2} s/\sqrt{n} \text{ to } \bar{x} + t_{\alpha/2} s/\sqrt{n}$$
$$2.33 - 2.262(2.002)/\sqrt{10} \text{ to } 2.33 + 2.262(2.002)/\sqrt{10}$$
$$0.90 \text{ to } 3.76$$

(b) We are 95% confident that the interval 0.90 to 3.76 hours contains the mean of additional sleep obtained by all patients using laevohysocyamine hydrobromide. Since this interval was entirely above zero, we could conclude that the treatment was effective.

**8.133** n = 42; df = 41; $t_{\alpha/2}$ = $t_{0.025}$ = 2.020; $\bar{x}$ = 1261.6 chips; s = 117.6

(a)
$$\bar{x} - t_{\alpha/2} s/\sqrt{n} \text{ to } \bar{x} + t_{\alpha/2} s/\sqrt{n}$$
$$1261.6 - 2.020(117.6)/\sqrt{42} \text{ to } 1261.6 + 2.020(117.6)/\sqrt{42}$$
$$1224.94 \text{ to } 1298.26$$

(b) We are 95% confident that the interval 1224.94 to 1298.26 chips contains the mean number of chips per bag for all 18oz bags of Chips Ahoy! cookies. Since this interval was entirely above 1000, we could conclude that the average 18oz bag of Chips Ahoy! cookies contains at least 1000 chips.

**8.135** We used Minitab to obtain a boxplot and a normal probability plot of the data, which are shown below. The plots show that the value 1643 is a potential outlier and the distribution is skewed right. Since the sample size is only 10, it is not reasonable to use the t-interval procedure to obtain a confidence interval for the population mean.

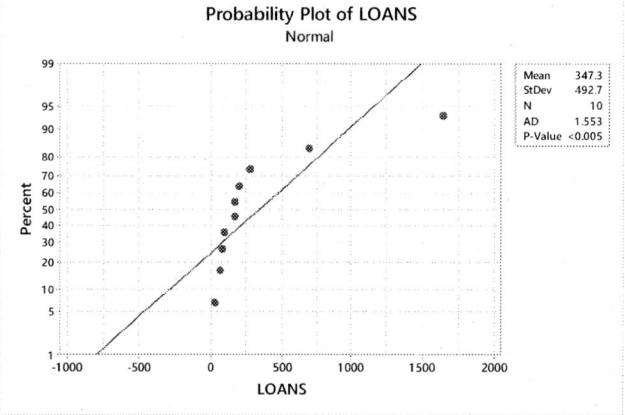

**8.137** We used Minitab to obtain a boxplot and a normal probability plot of the data, which are shown below. The plots show no outliers and the distribution is close to symmetrical. Since the sample size is 20, it is reasonable to use the t-interval procedure to obtain a confidence interval for the population mean.

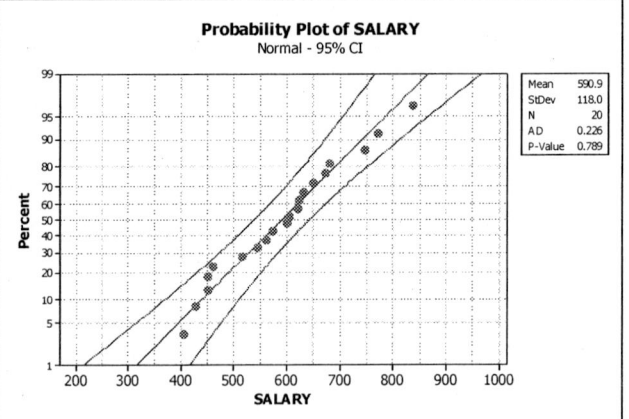

**8.139** (a) We will get the confidence interval for part (c) along with the histogram and boxplot. Using Minitab, retrieve the data from the WeissStats Resource Site. Then choose **Stat ▶ Basic statistics ▶ 1-Sample t.** Enter DEPTH in the **Samples in Columns** text box. Click on the **Options** button and enter 90.0 in the **Confidence level** text box and click **OK**. Click on the **Graphs** button and check the boxes for **Histogram of data** and **Boxplot of data**, then click **OK** twice. To get the probability plot and stem-and leaf diagram, choose **Stat ▶ Basic statistics ▶ Normality test**, enter DEPTH in the **Variable** text box and click **OK**. Then choose **Graph ▶ Stem-and-Leaf**, and enter DEPTH in the **Graph Variables** text box and click **OK**. The results are

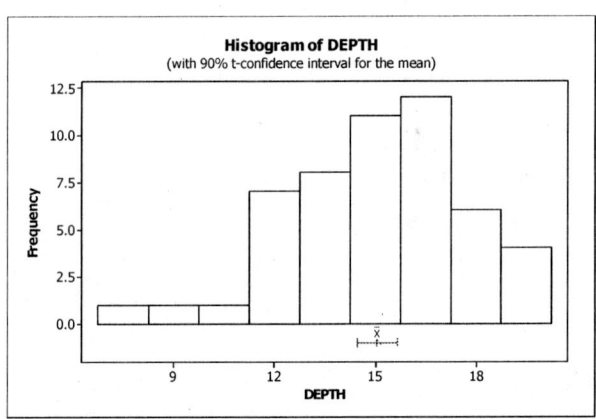

```
Stem-and-leaf of DEPTH   N = 51
Leaf Unit = 0.10

    1    8   2
    2    9   7
    2   10
    4   11   08
   12   12   01123588
   16   13   3399
   23   14   0245799
   (8)  15   00134689
   20   16   002567789
   11   17   24459
    6   18   2389
    2   19   37
```

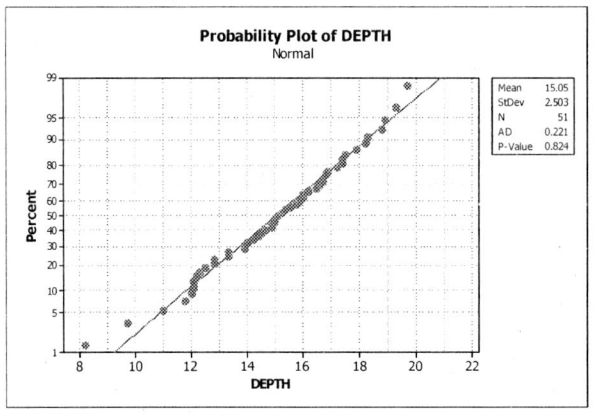

(b) Yes. The plots indicate no outliers, the sample size is quite large at 51, and the distribution of the data is only a little left skewed.

(c) The output for the confidence interval obtained in part (a) is

```
Variable   N    Mean    StDev   SE Mean       90% CI
DEPTH     51  15.0510  2.5028   0.3505   (14.4636, 15.6383)
```

We can be 90% confident that the mean, $\mu$, for all burrow depths is somewhere between 14.46 cm and 15.64 cm.

**8.141** (a) We will get the confidence interval for part (c) along with the histogram and boxplot. Using Minitab, retrieve the data from the WeissStats Resource Site. Then choose **Stat ▶ Basic statistics ▶ 1-Sample t.** Enter WITHOUT in the **Samples in Columns** text box. Click on the **Options** button and enter 95.0 in the **Confidence level** text box and click **OK**. Click on the **Graphs** button and check the boxes for **Histogram of data** and **Boxplot of data**, then click **OK** twice. To get the probability plot and stem-and leaf diagram, choose **Stat ▶ Basic statistics ▶ Normality test**, enter WITHOUT in the **Variable** text box and click **OK**. Then choose **Graph ▶ Stem-and-Leaf**, and enter WITHOUT in the **Graph Variables** text box and click **OK**. The results are

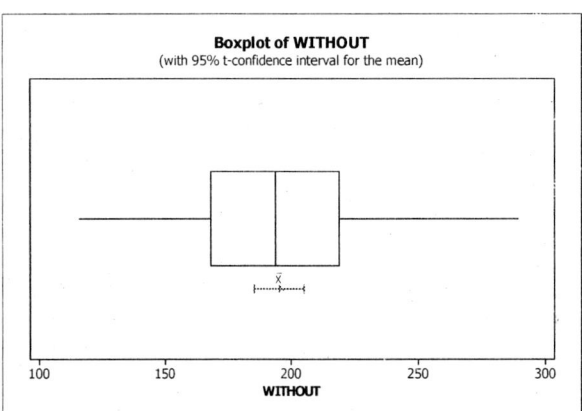

```
Stem-and-leaf of WITHOUT  N = 51
Leaf Unit = 10

  1    1  1
  2    1  3
  9    1  4455555
 19    1  6666667777
 (8)   1  88999999
 24    2  000000000111
 12    2  2223333
  5    2  45
  3    2  66
  1    2  8
```

(b) Yes. The plots indicate no outliers, the sample size is quite large at 51, and the distribution of the data is quite symmetric.

(c) The output for the confidence interval obtained in part (a) is

```
Variable    N     Mean    StDev   SE Mean      95% CI
WITHOUT    51   195.275   36.110    5.056   (185.118, 205.431)
```

We can be 95% confident that the mean, $\mu$, for all plasma cholesterol concentrations of patients without evidence of heart disease is somewhere between 185.118 mg/dl and 203.431 mg/dl.

(d) Repeating the process in part (a) using the WITH data, we obtain

```
Stem-and-leaf of WITH  N = 320
Leaf Unit = 10

   2    1  01
   8    1  333333
  19    1  44444455555
  63    1  6666666666666667777777777777777777777777777777
 126    1  888888888888888888888888999999999999999999999999999999999
 (50)   2  00000000000000000000000111111111111111111111111111
 144    2  22222222222222222222222233333333333333333333333333333333
  87    2  44444444444444444444444444455555555555555555
  44    2  66666666666666677777777
  21    2  8888889999
  11    3  00011
   6    3  2

HI 33, 33, 34, 38, 41
```

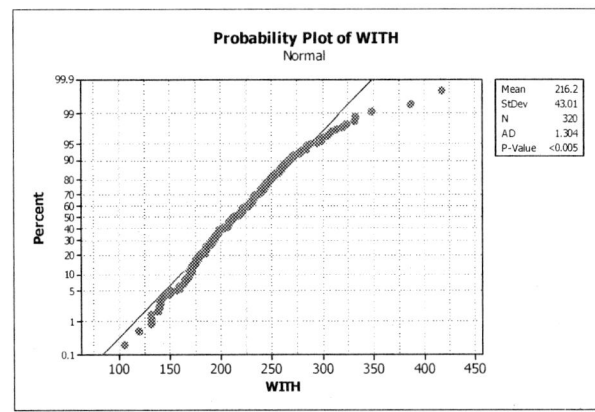

The data are quite symmetric except for five outliers on the high side. Nevertheless, the large sample size of 320 makes it reasonable to use the t-interval procedure with these data. The results of that procedure are

```
Variable    N     Mean    StDev   SE Mean      95% CI
WITH      320  216.191   43.015    2.405   (211.460, 220.921)
```

We can be 95% confident that the mean, $\mu$, for all plasma cholesterol concentrations of patients with evidence of heart disease is somewhere between 211.460 mg/dl and 220.921 mg/dl.

**8.143** (a) We could not provide entries for every possible degrees of freedom because the number of possibilities is infinite.

(b) As the degrees of freedom increase, the difference in consecutive entries becomes very small, making it unnecessary to list every possibility.

(c) Anytime the actual degrees of freedom lies between two consecutive entries in the table, we should use the table value associated with the lower number of degrees of freedom. Thus for $t_{0.05}$ with 87 df, we should use the value associated with 80 df, 1.664; for 125 df, use 1.660; for 650 df, use 1.660; and for 3000 df, use 1.645. This is a conservative approach, resulting in margins of error that are never smaller than we are entitled to have.

**8.145** (a) Your results will vary. To obtain the 2000 samples using Minitab, Choose **Calc ▶ Random Data ▶ Normal...**, enter 2000 in the **Generate rows of data** text box, enter C1-C5 in the **Store in Column(s):** text box, enter .270 in the **Mean** text box, and enter .031 in the **Standard deviation:** text box. Click **OK**.

(b) To find the mean and sample standard deviation in each row (sample), choose **Calc ▶ Row statistics...** and click on **Mean**. Enter C1-C5 in the **Input variable(s):** text box and enter C6 in the **Store result in:** text box. Repeat this last process, selecting **Standard Deviation** instead of **Mean** and put the results in C7.

(c) To obtain the Standardized version of each mean in the sample, choose **Calc ▶ Calculator...**, enter STANDARD in the **Store results in variable:** text box, enter (C6-.270)/(.031/SQRT(5)) in the **Expression:** text box and click **OK**.

**278** Chapter 8

(d) To facilitate a comparison in part (k),., do part (h) now. Choose **Graph ▶ Histogram...**, select the **Simple** version, and click **OK**. Enter STANDARD STUDENT in the **Graph variables** text box. Click the **Multiple graphs** button, click the **On separate graphs** button and check the box for **Same X, including same bins**. Click **OK** and click **OK**. Our first graph is shown below, but yours will differ, yet look similar.

(e) In theory, the distribution of the standardized version of $\bar{x}$ is standard normal.

(f) The histogram in (d) appears to be very close to standard normal. Recall that the distribution is centered at zero and 99.7% of the data should be within 3 standard deviations of the mean. It does appear that this is so for this simulated data.

(g) To obtain the Studentized version of each $\bar{x}$ in the sample, choose **Calc ▶ Calculator...**, enter STUDENT in the **Store results in variable:** text box, enter (C6-.270)/(C7/SQRT(5)) in the **Expression:** text box and click **OK**.

(h) This graph was produced in produced (d). Your graph should be similar to ours.

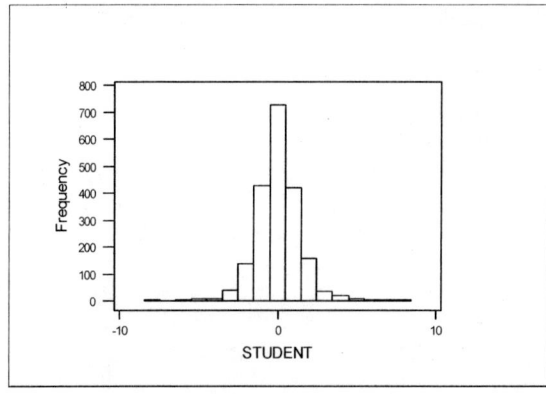

(i) In theory, the distribution of the studentized version of $\bar{x}$ is a t-distribution with 4 degrees of freedom.

(j) The distribution shown is symmetric about zero as is a t-distribution.

(k) The histogram of the studentized version in (h) is more spread out than that of the standardized version in (d). The reason is that there is more variability in the t-distribution due to the extra uncertainty arising from the use of s instead of $\sigma$.

**8.147** (a)  n=30; $\bar{x}$ = 27.97; s = 10.04; df = 29; $t_\alpha$ = $t_{0.10}$ = 1.311

$$\bar{x} + t_\alpha s / \sqrt{n}$$
$$27.97 + 1.311(10.04)/\sqrt{30}$$
$$30.37 \text{ minutes}$$

We can be 90% confident that the population mean commute time of all commuters in Washington, D.C. is less than 30.37 minutes.

(b) This upper confidence bound is less than the upper confidence limit of 31.08 found in Exercise 8.129. This is because the t-value used for a one-sided 90% confidence bound is 1.311, whereas, the t-value used for a two-sided 90% confidence interval is 1.699.

**8.149** (a) n=30; $\bar{x}$ = 52.040; s = 2.807; df = 29; $t_\alpha$ = $t_{0.05}$ = 1.699

$$\bar{x} - t_\alpha s / \sqrt{n}$$
$$52.040 - 1.699(2.807)/\sqrt{30}$$
$$51.169 \text{ g}$$

(b) We can be 95% confident that the population mean weight of all small bags of peanut M&Ms is greater than 51.169 grams.

(c) According to our confidence interval in part (b), we are 95% confident that on average the small bags of peanut M&Ms will be greater than the advertised weight on the package of 49.3 grams. It seems like the bags are slightly heavier than advertised. This is good news for the consumer, rather than slightly lighter than advertised.

**8.151** Using Minitab, we will generate 1000 samples of size 25 for the bootstrap confidence interval. First open the data file for Table 8.6 from Example 8.10 in Minitab from the WeissStats Resource Site. Generate all the data at once for the 1000 samples of size 25 (25,000 observations). Click **Calc**, **Random Data,** then **Sample from Columns.** The number of rows to sample is 25000, from column LOSS, and stored in column C2. Check Sample with replacement and click OK. Then separate the 25,000 observations into 1000 samples of size 25 by clicking **Calc, Make Patterned Data,** then **Simple Set of Numbers.** Store patterned data in C3, from first value 1 to last value 25, in steps of 1. Choose Number of times to list each value equal to 1 and Number of times to list the sequence 1000. Click OK. Separate the samples into 1000 rows by cicking **Data** then **Unstack Columns.** Unstack the data in C2, Using subscripts in C3. Click store unstacked data after last column in use, and check name the columns containing the unstacked data. Click OK. Now, calculate the mean of each of the 1000 samples by clicking **Calc** then **Row Statistics.** Choose Mean, and double click each of the 25 columns for the samples (columns C4-C28) to move them to the Input variables. Store the results in C29. Click OK. Title the column C29 as MEANS. Calculate the 2.5[th] and 97.5[th] percentile of the MEANS column by finding the 25th and 975th observation of the MEANS column after ordering them from smallest to largest. We calculated the 2.5[th] percentile as 418.92 and the 97.5[th] percentile as 613.72. Therefore, our 95% bootstrap confidence interval is 418.92 to 613.72. The 95% t-interval calculated in Example 8.10 was 405.07 to 621.57. The two intervals were close. However, the bootstrapping method resulted in a narrower and more accurate estimate of the population mean.

## Review Problems for Chapter 8

1. A point estimate of a parameter consists of a single value with no indication of the accuracy of the estimate. A confidence interval consists of an interval of numbers obtained from a point estimate of the parameter together with a percentage that specifies how confident we are that the parameter lies in the interval.

2. False. We are 95% confident that the mean lies in the interval from 33.8 to 39.0, but about 5% of the time, the procedure will produce an interval that does not contain the population mean. Therefore, we cannot say that the mean must lie in the interval.

3. No. The z-interval procedure can be used almost anytime with large samples because the sampling distribution of $\bar{x}$ is approximately normal for large n. The same is true for the t-interval procedure because when n is large, the t distribution is very similar to the normal distribution. However, when n is small, especially when n is 15 or less, the z-interval and t-interval procedures will not provide reliable estimates if the distribution of the underlying variable is not normal. For sample sizes in the range of 15 to 30, both procedures can be used if the data is roughly normal and has no outliers.

4. Approximately 950 of 1000 95% confidence intervals for a population mean would actually contain the true value of the mean.

5. Before applying a particular statistical inference procedure, we should look at graphical displays of the sample data to see if there appear to be any violations of the conditions required for the use of the procedure.

6. (a) Reducing the sample size from 100 to 50 will reduce the accuracy of the estimate (result in a longer confidence interval).

   (b) Reducing the confidence level from .95 to .90 while maintaining the sample size will increase the accuracy of the estimate (result in a shorter confidence interval).

7. (a) The length of the confidence interval is twice the margin of error or 2 x 10.7 = 21.4.

   (b) The confidence interval will be 75.2 ± 10.7 = (64.5, 85.9)

   (c) The confidence interval will be 75.2 ± 10.7.

8. (a) $E = z_{\alpha/2}(\sigma/\sqrt{n}) = 1.645(12/\sqrt{9}) = 6.58$

   (b) To obtain the confidence interval, you also need to know $\bar{x}$.

9. (a) The standardized value of $\bar{x}$ is $z = \dfrac{\bar{x} - \mu}{\sigma/\sqrt{n}} = \dfrac{262.1 - 266}{16/\sqrt{10}} = -0.77$

   (b) The studentized value of $\bar{x}$ is $t = \dfrac{\bar{x} - \mu}{s/\sqrt{n}} = \dfrac{262.1 - 266}{20.4/\sqrt{10}} = -0.61$

10. (a) standard normal distribution

    (b) t distribution with 14 degrees of freedom

11. The curve that looks more like the standard normal curve has the larger degrees of freedom because, as the number of degrees of freedom gets larger, t-curves look increasingly like the standard normal curve.

12. The t-interval procedure should be used.

13. The z-interval procedure should be used.

14. The z-interval procedure should be used.
15. Neither procedure should be used.
16. The z-interval procedure should be used.
17. Neither procedure should be used.
18. (a) $t_{0.025} = 2.101$           (b) $t_{0.05} = 1.734$

 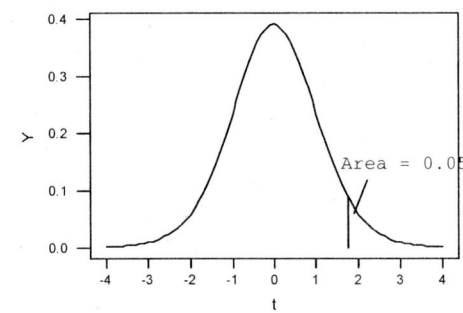

(c) $-t_{0.10} = -1.330$           (d) $\pm t_{0.005} = \pm 2.878$

 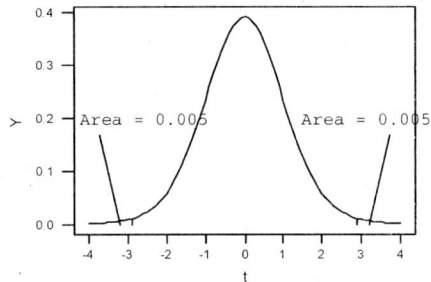

19. $n = 36$, $\bar{x} = 58.53$, $\sigma = 13.0$, $z_{\alpha/2} = z_{0.025} = 1.96$

$$\bar{x} - z_{\alpha/2}(\sigma/\sqrt{n}) \text{ to } \bar{x} + z_{\alpha/2}(\sigma/\sqrt{n})$$
$$58.53 - 1.96 \cdot (13.0/\sqrt{36}) \text{ to } 58.53 + 1.96 \cdot (13.0/\sqrt{36})$$
$$54.3 \text{ to } 62.8 \text{ years}$$

20. A confidence-interval estimate specifies how confident we are that a (unknown) parameter lies in the interval. This interpretation is presented correctly by (c). A *specific* confidence interval either will or will not contain the true value of the population mean μ; the *specific* interval is either sure to contain μ or sure not to contain μ. This interpretation is *not* presented correctly by (a), (b), and (d).

**21.** $n = 712$, $\bar{x} = 9.15$ years, $\sigma = 17.2$ years, $z_{\alpha/2} = z_{0.05} = 1.645$

(a)
$$\bar{x} - z_{\alpha/2}\sigma/\sqrt{n} \text{ to } \bar{x} + z_{\alpha/2}\sigma/\sqrt{n}$$
$$9.15 - 1.645(17.2)/\sqrt{712} \text{ to } 9.15 + 1.645(17.2)/\sqrt{712}$$
$$8.090 \text{ to } 10.210$$

We can be 90% confident that the mean sentence length, $\mu$, of all federally sentenced adult male prisoners is between 8.09 and 10.21 years.

(b) Since the sample size is very large, the distribution of sample means will be approximately normal regardless of the shape of the original distribution. It would be nice if the normal probability plot were roughly linear and did not indicate the presence of any extreme outliers, but some non-linearity and a few moderate outliers will not likely invalidate the use of the z-interval procedure.

**22.** (a) $E = z_{\alpha/2}\sigma/\sqrt{n} = 1.645(17.2)/\sqrt{712} = 1.060$

(b) We can be 90% confident that the maximum error made in using $\bar{x}$ to estimate μ is 1.06 years.

(c) The margin of error of the estimate is specified to be $E = 0.1$ mm.

$$n = \left[\frac{z_{\alpha/2}\,\sigma}{E}\right]^2 = \left[\frac{1.645\,(17.2)}{0.5}\right]^2 = 3202.2 \rightarrow 3203$$

(d)
$$\bar{x} - z_{\alpha/2}\sigma/\sqrt{n} \text{ to } \bar{x} + z_{\alpha/2}\sigma/\sqrt{n}$$
$$10.1 - 1.645(17.2)/\sqrt{3203} \text{ to } 10.1 + 1.645(17.2)/\sqrt{3203}$$
$$9.60 \text{ to } 10.60$$

**23.** $n = 16$; $df = 15$; $t_{\alpha/2} = t_{0.025} = 2.131$; $\bar{x} = 85.99$ mm Hg; $s = 8.08$ mm Hg

(a)
$$\bar{x} - t_{\alpha/2}s/\sqrt{n} \text{ to } \bar{x} + t_{\alpha/2}s/\sqrt{n}$$
$$85.99 - 2.131(8.08)/\sqrt{16} \text{ to } 85.99 + 2.131(8.08)/\sqrt{16}$$
$$81.69 \text{ to } 90.29$$

We can be 95% confident that the mean arterial blood pressure $\mu$ of all children of diabetic mothers is somewhere between 81.69 and 90.29 mm Hg.

(b) Using Minitab, choose **Stat ▶ Basic statistics ▶ 1-Sample t...**, enter PRESSURE in the **Samples in columns** text box. Click the **options…** button, type 95 in the **Confidence interval** text box, click the **Graphs** and check the boxes for **Histogram of data** and **Boxplot of data**, and click **OK** twice. Then choose **Stat ▶ Basic statistics ▶ Normality test**, enter PRESSURE in the **Variable** text box and click **OK**. Finally, choose **Graph ▶ Stem-and-Leaf**, enter PRESSURE in the **Graph Variables** text box and click **OK**. The results are

```
Stem-and-leaf of PRESSURE   N = 16
    Leaf Unit = 1.0

    1    6   9
    1    7
    3    7   58
    8    8   12244
    8    8   678
    5    9   014
    2    9   6
    1   10   1
```

(c) Yes. There are no outliers and the distribution of the data is approximately normal.

24. (a) Using Minitab, choose **Stat ▶ Basic statistics ▶ 1-Sample t...**, enter PRICE in the **Samples in columns** text box. Click the **options...** button, enter 90 in the **Confidence interval** text box and click **OK**, click the **Graphs** button and check the boxes for **Histogram of data** and **Boxplot of data**, and click **OK** twice. Then choose **Stat ▶ Basic statistics ▶ Normality test**, enter PRICE in the **Variable** text box and click **OK**. Finally, choose **Graph ▶ Stem-and-Leaf**, enter PRICE in the **Graph Variables** text box and click **OK**. The confidence interval is

```
Variable   N    Mean    StDev   SE Mean       90% CI
PRICE     18  1964.72  206.45    48.66  (1880.07, 2049.37)
```

We can be 90% confident that the interval ($1880.07, $2049.37) contains the mean diamond price for one-half carat diamonds.

**284** Chapter 8

(b) The graphs obtained by the procedures in part (a) are

```
Stem-and-leaf of PRICE   N  = 18
Leaf Unit = 10

LO  144


  2    16   7
  3    17   1
  5    18   27
 (6)   19   448899
  7    20   377
  4    21   04
  2    22   3
  1    23   1
```

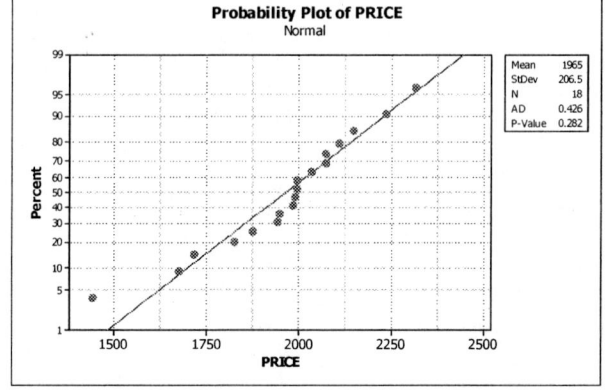

(c) No. Since the sample size is small (18), and there is an outlier (1442), it is not reasonable to use the t-interval procedure.

25. We used Minitab to obtain a boxplot and a normal probability plot of the data, which are shown below. The plots show that the value 907 is a potential outlier and the distribution is skewed right. Since the sample size is only 14, it is not reasonable to use the t-interval procedure to obtain a confidence interval for the population mean.

26. (a) We will import the data into Minitab.

(b) Choose **Stat ▶ Basic statistics ▶ 1-Sample t..**, enter DURATION in the **Samples in columns** text box. Click the **options...** button, enter 99 in the **Confidence interval** text box and click **OK**, click the **Graphs** button and check the boxes for **Histogram of data** and **Boxplot of data**, and click **OK** twice. Then choose **Stat ▶ Basic statistics ▶ Normality test**, enter DURATION in the **Variable** text box and click **OK**. Finally, choose **Graph ▶ Stem-and-Leaf**, enter DURATION in the **Graph Variables** text box and click **OK**. The graphs are

```
Stem-and-leaf of DURATION   N = 90
Leaf Unit = 0.10

    2   44   33
    3   45   0
    3   46
    4   47   3
    9   48   00036
   16   49   0000033
   22   50   003366
   34   51   000000003333
   45   52   00000003666
   45   53   000
   42   54   000000033666
   30   55   0000336666
   20   56   000003336
   11   57   0
   10   58   00036
    5   59   00356
```

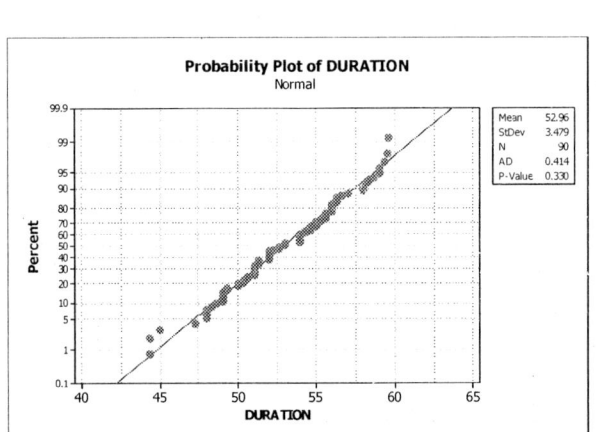

(c) Yes. The sample size is large (90) and there are no outliers.

(d) The procedure in part (a) resulted in the following confidence interval.

```
Variable    N    Mean    StDev   SE Mean        99% CI
DURATION   90  52.9633  3.4794   0.3668   (51.9979, 53.9287)
```
We can be 99% confident that the interval (51.9979, 53.9287) contains the population mean µ of the larval duration of convict surgeonfish in days.

27. (a) Using Minitab, choose **Calc ▶ Random data ▶ Sample from columns,** enter 35 in the **Sample ___ rows from column(s)** text box, enter MILEAGE in the

**286** Chapter 8

first text box and enter SAMPLE in the **Store samples in** text box. Click **OK**.

(b) Choose **Stat ▶ Basic statistics ▶ 1-Sample t..**, enter SAMPLE in the **Samples in columns** text box. Click the **options…** button, enter 95 in the **Confidence interval** text box and click **OK**. The result is

```
Variable    N     Mean    StDev   SE Mean       95% CI
SAMPLE     35   24.4857  4.0174   0.6791   (23.1057, 25.8657)
```

(c) Choose **Calc ▶ Column statistics** and check the box for the mean. Enter MILEAGE in the **Input variable** text box and click **OK**. The population mean is 24.8234. This value lies in the 95% confidence interval found in part (b). It would not necessarily have to lie in the confidence interval. Since the sample was randomly selected, it is possible to select a sample for which the confidence interval does not contain the population mean.

**28.** (a) The population consists of all eruptions of the Old Faithful Geyser. The variable under consideration is the time between eruptions (in minutes).

(b) Using Minitab, choose **Stat ▶ Basic statistics ▶ 1-Sample t..**, enter TIME in the **Samples in columns** text box. Click the **options…** button, enter 99 in the **Confidence interval** text box and click **OK**. The resulting confidence interval is

```
Variable    N     Mean    StDev   SE Mean       99% CI
TIME      500   91.5780  9.2022   0.4115   (90.5139, 92.6421)
```

We can be 99% confident that the population mean time between eruptions of the Old Faithful Geyser lies in the interval from 90.5139 to 92.6421 minutes.

(c) Strictly speaking, the confidence interval applies to the population from which the sample was drawn. There were no observations in the sample from the future, so the interval is not relevant to the future. In geologic terms, however, five years is insignificant, so the confidence interval is probably still somewhat meaningful. One should consider that unforeseen events, such as earthquakes or volcanic eruptions in the region could have a substantial effect on the time between eruptions. In that case, the current sample would not be representative of the population in five years.

**29.** (a) Your results will vary. To obtain the 3000 samples using Minitab, Choose **Calc ▶ Random Data ▶ Normal...**, enter 3000 in the **Generate rows of data** text box, type C1-C4 in the **Store in Column(s):** text box, enter 4.66 in the **Mean** text box, and enter 0.75 in the **Standard deviation:** text box. Click **OK**.

(b) To find the mean and sample standard deviation in each row (sample), choose **Calc ▶ Row statistics...** and click on **Mean**. Enter C1-C4 in the **Input variable(s):** text box and enter XBAR in the **Store result in:** text box. Repeat this last process, selecting **Standard Deviation** instead of **Mean** and put the results in SD.

(c) To obtain the Standardized version of each $\bar{x}$ in the sample, choose **Calc ▶ Calculator...**, enter STANDARD in the **Store results in variable:** text box, enter ('XBAR'-4.66)/(0.75/SQRT(4)) in the **Expression:** text box and click **OK**.

(d) To facilitate a comparison in part (h), **do part (g) now**. Then choose **Graph ▶ Histogram...**, select the **Simple** version, and click **OK**. Enter STANDARD and STUDENT in the **Graph variables** text box. Click on the **Multiple graphs** button, then click the **On separate graphs** button and check the boxes for **Same X, including same bins** and **Same Y**. Click **OK** twice. The first graph is shown below. Yours will differ, yet look similar.

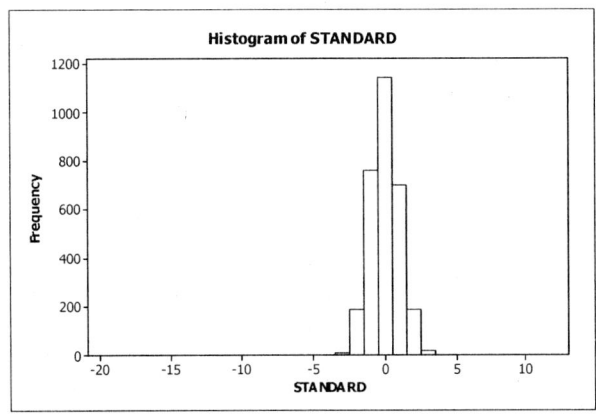

(e) In theory, the distribution of the standardized version of $\bar{x}$ is standard normal.

(f) The histogram in (e) appears to be very close to standard normal. Recall that the distribution is centered at zero and 99.7% of the data should be within 3 standard deviations of the mean. It does appear that this is so for this simulated data. (g) To obtain the Studentized version of each $\bar{x}$ in the sample, choose **Calc ▶ Calculator...**, enter STUDENT in the **Store results in variable** text box, enter ('SBAR'-4.66)/('SD'/SQRT(4) in the **Expression:** text box and click **OK**.

(h) This graph was produced by the procedure in part (d). Our second graph follows, but yours will differ, yet look similar.

(i) In theory, the distribution of the studentized version of $\bar{x}$ is a t-distribution with 3 degrees of freedom.

(j) The distribution shown is symmetric about zero as is a t-distribution.

(k) The histogram of the studentized version in (h) is much more spread out than that of the standardized version in (d). The reason is that there is more variability in the t-distribution due to the extra uncertainty arising from the use of s instead of σ.

# CHAPTER 9 SOLUTIONS

### Exercises 9.1

**9.1** A hypothesis is a statement that something is true.

**9.3** The decision criterion specifies whether or not the null hypothesis should be rejected in favor of the alternative hypothesis.

**9.5** (a) $H_a: \mu \neq \mu_0$; two-tailed

(b) $H_a: \mu < \mu_0$; left-tailed

(c) $H_a: \mu > \mu_0$; right-tailed

**9.7** (a) If the null hypothesis is in fact false, you cannot possibly make a Type I error. You would either reject the null hypothesis when it is false, which is a correct decision, or you would not reject the null hypothesis when it is false, which is a Type II error.

(b) If the null hypothesis is in fact false, you could possibly make a Type II error. You would either reject the null hypothesis when it is false, which is a correct decision, or you would not reject the null hypothesis when it is false, which is a Type II error.

**9.9** True. The significance level is equal to the probability of a Type I error, which is the probability of rejecting a true null hypothesis. Therefore, the smaller the significance level, the smaller the probability of rejecting a true null hypothesis.

**9.11** A Type I error is made when a true null hypothesis is rejected. The probability of making this error is denoted by $\alpha$. A Type II error is made when a false null hypothesis is not rejected. We denote the probability of a Type II error by $\beta$.

**9.13** In this exercise, we are told that failing to reject the null hypothesis corresponds to approving the nuclear reactor for use. This action — approving the nuclear reactor suggests that the null hypothesis must be something like: "The nuclear reactor is safe." This further suggests that the alternative hypothesis is something like: "The nuclear reactor is unsafe." Putting things together, the Type II error in this situation is: "Approving the nuclear reactor for use when, in fact, it is unsafe." This type of error has consequences that are catastrophic. Thus, the property that we want the Type II error probability to exhibit is that it be small.

**9.15** Let $\mu$ denote the mean cadmium level in *Boletus pinicola* mushrooms.

(a) $H_0: \mu = 0.5$ ppm   (b) $H_a: \mu > 0.5$ ppm   (c) right-tailed test

**9.17** Let $\mu$ denote the mean daily intake of iron by adult females under age 51.

(a) $H_0: \mu = 18$ mg/day   (b) $H_a: \mu < 18$ mg/day   (c) left-tailed

**9.19** Let $\mu$ denote the mean length of imprisonment for motor-vehicle theft offenders in Australia.

(a) $H_0: \mu = 16.7$ months   (b) $H_a: \mu \neq 26.7$ months

(c) two-tailed

**9.21** Let $\mu$ denote the mean body temperature of healthy humans.

(a) $H_0: \mu = 98.6°F$   (b) $H_a: \mu \neq 98.6°F$   (c) two-tailed

**9.23** (a) A Type I error would occur if, in fact, $\mu = 0.5$ ppm, but the results of the sampling lead to the conclusion that $\mu > 0.5$ ppm.

(b) A Type II error would occur if, in fact, $\mu > 0.5$ ppm, but the results of the sampling fail to lead to that conclusion.

(c) A correct decision would occur if, in fact, $\mu = 0.5$ ppm and the results of the sampling do not lead to the rejection of that fact; or if, in fact, $\mu > 0.5$ ppm and the results of the sampling lead to that conclusion.

(d) If, in fact, the mean cadmium level in *Boletus pinicola* mushrooms is equal to 0.5 ppm, and we do not reject the null hypothesis that $\mu = 0.5$ ppm, we made a correct decision.

(e) If, in fact, the mean cadmium level in *Boletus pinicola* mushrooms is greater than to 0.5 ppm, and we do not reject the null hypothesis that $\mu = 0.5$ ppm, we made a Type II error.

**9.25** (a) A Type I error would occur if, in fact, $\mu = 18$ mg, but the results of the sampling lead to the conclusion that $\mu < 18$ mg.

(b) A Type II error would occur if, in fact, $\mu < 18$ mg, but the results of the sampling fail to lead to that conclusion.

(c) A correct decision would occur if, in fact, $\mu = 18$ mg and the results of the sampling do not lead to the rejection of that fact; or if, in fact, $\mu < 18$ mg and the results of the sampling lead to that conclusion.

(d) If the mean iron intake equals the RDA of 18 mg, and we reject the null hypothesis that $\mu = 18$ mg, we made a Type I error.

(e) If, in fact, the mean iron intake is less than the RDA of 18 mg, and we reject the null hypothesis that $\mu = 18$ mg, we made a correct decision.

**9.27** (a) A Type I error would occur if, in fact, $\mu = 16.7$ months, but the results of the sampling lead to the conclusion that $\mu \neq 16.7$ months.

(b) A Type II error would occur if, in fact, $\mu \neq 16.7$ months, but the results of the sampling fail to lead to that conclusion.

(c) A correct decision would occur if, in fact, $\mu = 16.7$ months and the results of the sampling do not lead to the rejection of that fact; or if, in fact, $\mu \neq 16.7$ months and the results of the sampling lead to that conclusion.

(d) If, in fact, the mean length of imprisonment equals 16.7 months, and we do not reject the null hypothesis that $\mu = 16.7$ months, we made a correct decision.

(e) If, in fact, the mean length of imprisonment does not equal 16.7 months, and we do not reject the null hypothesis that $\mu = 16.7$ months, we made a Type II error.

**9.29** (a) A Type I error would occur if, in fact, $\mu = 98.6°$ F, but the results of the sampling lead to the conclusion that $\mu \neq 98.6°$ F.

(b) A Type II error would occur if, in fact, $\mu \neq 98.6°$ F, but the results of the sampling fail to lead to that conclusion.

(c) A correct decision would occur if, in fact, $\mu = 98.6°$ F and the results of the sampling do not lead to the rejection of that fact; or if, in fact, $\mu \neq 98.6°$ F and the results of the sampling lead to that conclusion.

(d) If the mean temperature of all healthy humans equals 98.6° F, and we reject the null hypothesis that $\mu = 98.6°$ F, we made a Type I error.

(e) If, in fact, the temperature of all healthy humans is not equal to 98.6° F, and we reject the null hypothesis that $\mu = 98.6°$ F, we made a correct decision.

**Exercises 9.2**

**9.31** A test statistic is a statistic used to decide whether to reject the null hypothesis.

**9.33** The nonrejection region is a set of values of the test statistic that do not lead to rejection of the null hypothesis.

**9.35** (a) Rejection region: $z \geq 1.645$

(b) Nonrejection region: $z < 1.645$

(c) Critical value: $z = 1.645$

(d) Significance level: $\alpha = 0.05$

(e)

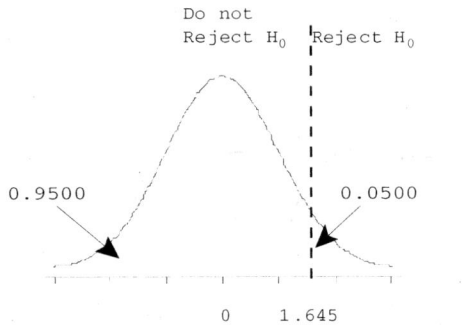

(f) Right-tailed test

**9.37** (a) Rejection region: $z \leq -2.33$

(b) Nonrejection region: $z > -2.33$

(c) Critical value: $z = -2.33$

(d) Significance level: $\alpha = 0.01$

(e)

(f) Left-tailed test

**9.39** (a) Rejection region: $z \leq -1.645$ or $z \geq -1.645$

(b) Nonrejection region: $-1.645 < z < -1.645$

(c) Critical values: $z = -1.645$ and $z = 1.645$

(d) Significance level: $\alpha = 0.10$

(e)

(f) Two-tailed test

**9.41** Critical values: $\pm z_{0.05} = \pm 1.645$

**9.43** Critical value: $-z_{0.01} = -2.33$

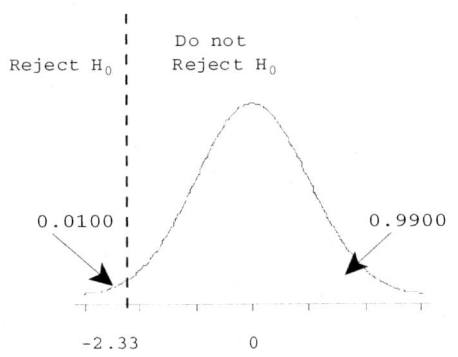

**9.45** Critical value: $z_{0.01} = 2.33$

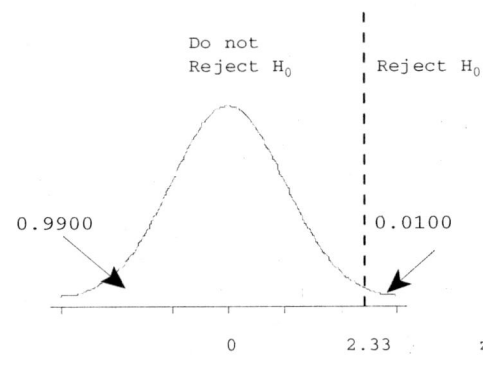

**294** Chapter 9

**Exercises 9.3**

**9.47** (1) It allows the reader to assess significance at any desired level, and (2) it permits the reader to evaluate the strength of the evidence against the null hypothesis.

**9.49** The P-value for a one-sample z-test is obtained as:

(a) $P(z \leq$ observed z value) for a left-tailed test

(b) $P(z \geq$ observed z value) for a right-tailed test

(c) $2P(z \geq$ absolute value of the observed z value) for a two-tailed test.

**9.51** (a) Do not reject the null hypothesis.

(b) Reject the null hypothesis.

(c) Reject the null hypothesis.

**9.53** A P-value of 0.02 provides stronger evidence against the null hypothesis than does a value of 0.03. It says that if the null hypothesis is true, the data are less likely than they are when the P-value is 0.03.

**9.55** Strength of the evidence against the null hypothesis is moderate.

**9.57** Strength of the evidence against the null hypothesis is strong.

**9.59** Strength of the evidence against the null hypothesis is weak or none.

**9.61** Strength of the evidence against the null hypothesis is very strong.

**9.63** (a) z = 2.03, P-value = 1.000 − 0.9788 = 0.0212. At a 5% significance level, we would reject the null hypothesis in favor of the alternative hypothesis.

(b) z = −0.31, P-value = 1.000 − 0.3783 = 0.6217. At a 5% significance level, we would not reject the null hypothesis in favor of the alternative hypothesis.

**9.65** (a) z = −0.74, P-value = 0.2296. At a 5% significance level, we would not reject the null hypothesis in favor of the alternative hypothesis.

(b) z = 1.16, P-value = 0.8770. At a 5% significance level, we would not reject the null hypothesis in favor of the alternative hypothesis.

**9.67** (a) z = −1.66,  Left-tail probability = 0.0485

P-value = 0.0485 x 2 = 0.0970

At a 5% significance level, we would not reject the null hypothesis in favor of the alternative hypothesis.

(b) z = 0.52,  Right-tail probability = 1.0000 − 0.6985 = 0.3015

P-value = 0.3015 x 2 = 0.6030

At a 5% significance level, we would not reject the null hypothesis in favor of the alternative hypothesis.

**9.69** (a) The P-value is expressed as $P(z \leq z_0)$ if the hypothesis test is left-tailed.

(b) The P-value is expressed as $2 \cdot P(|z| \geq |z_0|)$ if the test is two-tailed.

**9.71** Given that x can be transformed to z (and $x_0$ to $z_0$), we have:

1. $P(x \geq x_0) = P(z \geq z_0)$, for a right-tailed test

2. $P(x \leq x_0) = P(z \leq z_0)$, for a left-tailed test

3. $2 \cdot \min \{P(x \leq x_0), P(x \geq x_0)\} = 2 \cdot \min \{P(z \leq z_0), P(z \geq z_0)\}$

$$= \begin{cases} 2 \cdot P(z \leq z_0) \text{ if } z_0 < 0 \\ 2 \cdot P(z \geq z_0) \text{ if } z_0 \geq 0 \end{cases}$$

$$\text{By symmetry} = \begin{cases} 2 \cdot P(z \geq -z_0) \text{ if } z_0 < 0 \\ 2 \cdot P(z \leq z_0) \text{ if } z_0 \geq 0 \end{cases} = 2 \cdot P(z \geq |z_0|)$$

$$\text{By symmetry} = P(z \leq -|z_0|) + P(z \geq |z_0|) = P(|z| \geq |z_0|)$$

### Exercises 9.4

**9.73** The z-test is not an appropriate method for highly skewed data when the sample size is less than 30.

**9.75** The z-test is appropriate for large samples with no outliers even if the data are mildly skewed.

**9.77** For the critical value approach, reject $H_0$ if z < -1.645;
$z = (20-22)/(4/\sqrt{32}) = -2.83$; therefore, reject $H_0$ and conclude that $\mu < 22$.
For the P-value approach, the P-value is 0.0023. Since the P-value is less than the significance level, reject $H_0$ and conclude that $\mu < 22$.

**9.79** For the critical value approach, reject $H_0$ if z > 1.645;
$z = (24-22)/(4/\sqrt{15}) = 1.94$; therefore, reject $H_0$ and conclude that $\mu > 22$.
For the P-value approach, the P-value is 1.000 - 0.9738 = 0.0262. Since the P-value is less than the significance level, reject $H_0$ and conclude that $\mu > 22$.

**9.81** For the critical value approach, reject $H_0$ if z < -1.96 or z > 1.96;
$z = (23-22)/(4/\sqrt{24}) = 1.22$; therefore, do not reject $H_0$. The data do not provide sufficient evidence to support $H_a$: $\mu \neq 22$.
For the P-value approach, the right tailed probability is 0.1112 and the P-value would be 0.2224. Since the P-value is greater than the significance level, do not reject $H_0$. The data do not provide sufficient evidence to support $H_a$: $\mu \neq 22$.

**9.83** n = 12, $\sigma$ = 0.37 ppm, $\bar{x}$ = 6.31/12 = 0.526 ppm

Step 1: $H_0$: $\mu$ = 0.5 ppm, $H_a$: $\mu$ > 0.5 ppm

Step 2: $\alpha$ = 0.05

Step 3: $z = (0.526 - 0.5)/(0.37/\sqrt{12}) = 0.24$

Step 4: Critical-Value Approach: Critical value = $z_\alpha = z_{0.05} = 1.645$.

P-Value Approach: P-value is $P(Z > 0.24) = 1 - 0.5948 = 0.4052$.

Step 5: Critical-Value Approach: Since 0.24 < 1.645, do not reject $H_0$.

P-value Approach: Since 0.4052 > 0.05, do not reject $H_0$.

Step 6: At the 5% significance level, the data do not provide sufficient evidence to conclude that the mean cadmium level $\mu$ of *Boletus pinicola* mushrooms is greater than the safety limit of 0.5 ppm.

**9.85** n = 45, $\bar{x}$ = 14.68, $\sigma$ = 4.2

Step 1: $H_0$: $\mu$ = 18 mg, $H_a$: $\mu$ < 18 mg

Step 2: $\alpha$ = 0.01

Step 3: $z = (14.68 - 18)/(4.2/\sqrt{45}) = -5.30$

Step 4: Critical-Value Approach: Critical value $= -z_\alpha = -z_{0.01} = -2.33$.

P-Value Approach: P-value is $P(Z < -5.30) \approx 0.0000$.

Step 5: Critical-Value Approach: Since -5.30 < -2.33, reject $H_0$.

P-value Approach: Since 0.0000 < 0.01, reject $H_0$.

Step 6: At the 1% significance level, the data provide sufficient evidence to conclude that adult females under the age of 51 are, on the average, getting less than the RDA of 18 mg of iron. Considering that iron deficiency causes anemia and that iron is required for transporting oxygen in the blood, this result could have practical significance as well.

**9.87** $n = 100$, $\bar{x} = 17.8$ months, $\sigma = 6.0$ months

Step 1: $H_0$: $\mu = 16.7$ months, $H_a$: $\mu \ne 16.7$ months

Step 2: $\alpha = 0.05$

Step 3: $z = (17.8 - 16.7)/(6.0/\sqrt{100}) = 1.83$

Step 4: Critical-Value Approach: Critical value $= \pm z_{\alpha/2} = \pm z_{0.025} = \pm 1.96$.

P-Value Approach: P-value is $2 \cdot P(Z > 1.83) = 2 \cdot (1 - 0.9664) = 0.0672$.

Step 5: Critical-Value Approach: Since 1.83 < 1.96, do not reject $H_0$.

P-value Approach: Since 0.0672 > 0.05, do not reject $H_0$.

Step 6: At the 5% significance level, the data do not provide sufficient evidence to conclude that the mean length of imprisonment $\mu$ of motor-vehicle theft offenders in Sydney differs from the national mean in Australia.

**9.89** (a) Using Minitab, with the data in a column named GAIN, we choose **Stat ▶ Basic Statistics ▶ 1-Sample z...**, click in the **Samples in columns** text box and specify GAIN, click in the **Standard deviation** text box and type 0.42, and click in the **Test mean** text box and type 0.2. Click the **Options...** button, enter 95 in the **Confidence level** text box, click the arrow button at the right of the **Alternative** drop-down list box and select **greater than** and click **OK**. Click on the **Graphs** button and check the boxes for **Histogram of Data** and **Boxplot of data**. Then click **OK** twice. The result of the test is

```
Test of mu = 0.2 vs > 0.2
The assumed standard deviation = 0.42
                                                95%
                                               Lower
Variable   N     Mean      StDev    SE Mean    Bound     Z      P
GAIN      20   0.295000   0.499974  0.093915  0.140524  1.01  0.156
```

(b) The histogram and boxplot were produced by the procedure in part (a). Now choose **Stat ▶ Basic Statistics ▶ Normality test** and enter GAIN in the **Variable** text box. Click **OK**. Then choose **Graph ▶ Stem-and-Leaf** , enter GAIN in the **Graph variables** text box and click **OK**. The results are

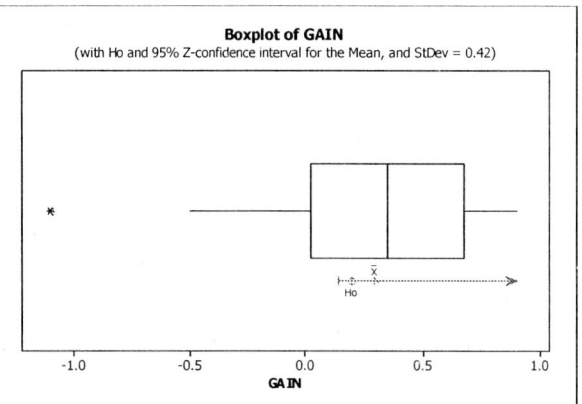

```
Stem-and-leaf of GAIN   N = 20
Leaf Unit = 0.10

 LO -11

  2   -0   5
  3   -0   2
  4   -0   1
  6    0   01
 10    0   2233
 10    0   45
  8    0   6667
  4    0   8889
```

(c) Repeating the procedure of part (a), we obtain

```
Test of mu = 0.2 vs > 0.2
The assumed standard deviation = 0.42

                                          95%
                                         Lower
Variable   N    Mean     StDev   SE Mean  Bound     Z      P
GAIN      19  0.368421  0.387374 0.096355 0.209932  1.75  0.040
```

(d) The original sample size is only 20. The plots in part (b) indicate that the value -1.1 is a potential outlier. The z-test should not be used with the original data. This is further confirmed by the fact that using all of the data leads to z = 1.01, whereas, deleting the outlier leads to z = 1.75. This is enough of a change to alter our conclusion from not rejecting the null hypothesis to rejecting it. If there is no good reason for deleting the outlier, then the z-test is inappropriate for these data.

9.91 (a) Using Minitab, with the data in a column named TEMP, we choose **Stat ▶ Basic Statistics ▶ 1-Sample z...**, click in the **Samples in columns** text box and specify TEMP, click in the **Standard deviation** text box and enter 0.63, and click in the **Hypothesized** text box and enter 98.6. Click the **Options...** button, enter 95 in the **Confidence level** text box, click the arrow button at the right of the **Alternative** drop-down list box and select **not equal** and click **OK**. Click on the **Graphs** button and check the boxes for **Histogram of Data** and **Boxplot of data**. Then click **OK** twice. Now choose **Stat ▶ Basic Statistics ▶ Normality test** and enter CHARGE in the **Variable** text box. Click **OK**. Then choose **Graph ▶ Stem-and-Leaf**, enter CHARGE in the **Graph variables** text box and click **OK**. The graphs are

```
Stem-and-leaf of TEMP   N = 93
Leaf Unit = 0.10

   1    96  7
   3    96  89
   8    97  00001
  13    97  22233
  19    97  444444
  26    97  6666777
  31    97  88889
  45    98  00000000000111
 (10)   98  2222222233
  38    98  4444445555
  28    98  66666666677
  17    98  8888888
  10    99  00001
   5    99  2233
   1    99  4
```

(b) Yes. The sample size 93 is large and the distribution of the data is quite symmetric.

(c) Yes. The procedure in part (a) also produced the results of the test which are

Test of mu = 98.6 vs not = 98.6
The assumed standard deviation = 0.63

| Variable | N | Mean | StDev | SE Mean | 99% CI | Z | P |
|---|---|---|---|---|---|---|---|
| TEMP | 93 | 98.1237 | 0.6468 | 0.0653 | (97.9554, 98.2919) | -7.29 | 0.000 |

The P-value is 0.000. Since the P-value is less than 0.01, we reject the null hypothesis and conclude that the mean body temperature of healthy humans is different from the generally accepted value of 98.6°F.

**9.93** (a) Using Minitab, with the data in a column named BILL, we choose **Stat ▶ Basic Statistics ▶ 1-Sample z...**, click in the **Samples in columns** text box and specify BILL, click in the **Standard deviation** text box and enter 25, and click in the **Hypothesized mean** text box and enter 48.16. Click the **Options...** button, enter 95 in the **Confidence level** text box, click the arrow button at the right of the **Alternative** drop-down list box and select **less than** and click **OK**. Click on the **Graphs** button and check the boxes for **Histogram of Data** and **Boxplot of data**. Then click **OK** twice. Now choose **Stat** ▶ **Basic Statistics** ▶ **Normality test** and enter BILL in the **Variable** text box. Click **OK**. Then choose **Graph** ▶ **Stem-and-Leaf**, enter BILL in the **Graph variables** text box and click **OK**. The graphs are

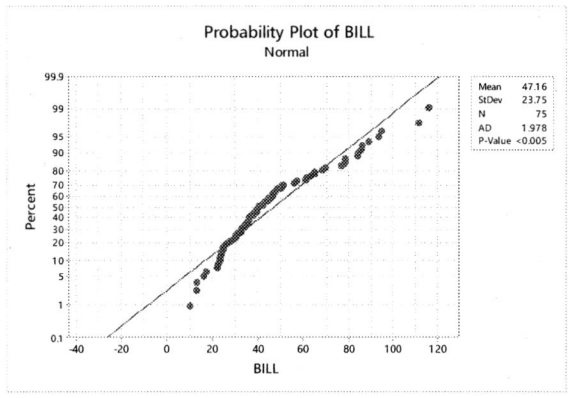

```
Stem-and-leaf of BILL   N = 75
Leaf Unit = 1.0

    5    1   03367
   17    2   223334445679
   36    3   0012223445566678999
  (14)   4   00223444666778
   25    5   00167
   20    6   0135589
   13    7   6888
    9    8   44569
    4    9   35
    2   10
    2   11   16
```

(b) The results of the test carried out by the procedure in part (a) are

```
Test of μ = 48.16 vs < 48.16
The assumed standard deviation = 25
```

| Variable | N | Mean | StDev | SE Mean | 95% Upper Bound | Z | P |
|---|---|---|---|---|---|---|---|
| BILL | 75 | 47.16 | 23.75 | 2.89 | 51.91 | -0.35 | 0.365 |

The P-value for the test is 0.365. Since the P-value is greater than 0.05, we do not reject the null hypothesis. The data do not provide sufficient evidence that the mean local monthly cell phone bill has decreased from the 2009 mean of $48.16.

(c) After deleting the two outliers 116.13 and 111.50, the graphs and test results are

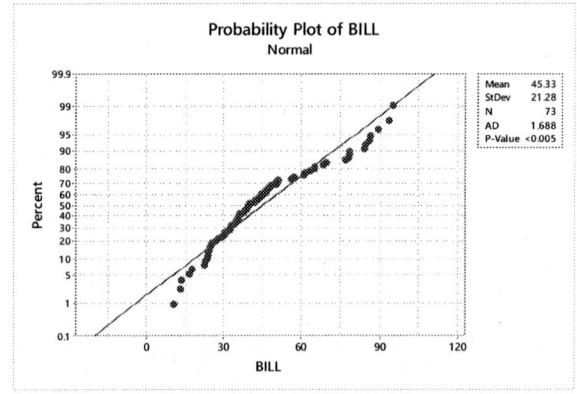

```
Stem-and-leaf of BILL   N = 73
Leaf Unit = 1.0

    3   1  033
    5   1  67
   13   2  22333444
   17   2  5679
   26   3  001222344
   36   3  5566678999
   (8)  4  00223444
   29   4  666778
   23   5  001
   20   5  67
   18   6  013
   15   6  5589
   11   7
   11   7  6888
    7   8  44
    5   8  569
    2   9  3
    1   9  5
```

Test of μ = 48.16 vs < 48.16
The assumed standard deviation = 25

| Variable | N | Mean | StDev | SE Mean | 95% Upper Bound | Z | P |
|---|---|---|---|---|---|---|---|
| BILL | 73 | 45.33 | 21.28 | 2.93 | 50.15 | -0.97 | 0.167 |

(d) Although the value of z has been changed from -0.35 to -0.97 by deleting the two outliers, the conclusion remains the same. We do not reject the null hypothesis.

**9.95** (a) n = 28, $\sigma$ = $8.45, $\bar{x}$ = $1788.62/28 = $63.88

The 90% confidence interval is $63.88 \pm 1.645(8.45)/\sqrt{28} = (61.25, 66.51)$.

The hypothesized mean ($66.52) lies outside of the confidence interval, so we should reject the null hypothesis. Since the test statistic is $z = (63.88 - 66.52)/(8.45/\sqrt{28}) = -1.653$, which is less than the lower critical value of -1.645, the hypothesis test also leads to the conclusion that we should reject the null hypothesis.

(b) n = 100, $\bar{x}$ = 17.8 months, $\sigma$ = 6.0 months

The 90% confidence interval is $17.8 \pm 1.96(6.0)/\sqrt{100} = (16.6, 19.0)$.

The hypothesized mean (16.7) lies inside of the confidence interval, so we should not reject the null hypothesis. Since the test statistic is $z = (17.8 - 16.7)/(6.0/\sqrt{100}) = 1.83$, which is less than the upper critical value of 1.96, the hypothesis test also leads to the conclusion that we should not reject the null hypothesis.

**9.97** (a) n = 12, $\sigma$ = 0.37 ppm, $\bar{x}$ = 6.31/12 = 0.526 ppm

The 95% lower level confidence bound is $0.526 - 1.645(0.37)/\sqrt{12} = 0.350$.

The hypothesized mean (0.5) lies above the lower confidence bound, so we should not reject the null hypothesis. Since the test statistic is $z = (0.526 - 0.5)/(0.37/\sqrt{12}) = 0.24$, which is less than the critical value of 1.645, the hypothesis test also leads to the conclusion that we should not reject the null hypothesis.

(b) n = 30, $\bar{x}$ = 78.3, $\sigma$ = 11.2

The 95% lower level confidence bound is $78.3 - 1.645(11.2)/\sqrt{30} = 74.94$.

The hypothesized mean (72) lies below the lower confidence bound, so we should reject the null hypothesis. Since the test statistic is $z = (78.3 - 72)/(11.2/\sqrt{30}) = 3.08$, which is greater than the critical value of 1.645, the hypothesis test also leads to the conclusion that we should reject the null hypothesis.

**Exercises 9.5**

**9.99** (a) Neither the z-test nor the t-test is appropriate for small samples containing outliers or exhibiting extremely non-normal distributional shapes.

(b) A non-parametric test that is neither affected by a few outliers nor by non-normality might be appropriate.

**9.101** (a) 0.01 < P < 0.025

(b) Reject $H_0$ for $\alpha \geq 0.025$; Do not reject $H_0$ for $\alpha \leq 0.01$; Undecided for $0.01 < \alpha < 0.025$.

**9.103** (a) P < 0.005

(b) Reject $H_0$ for $\alpha \geq 0.005$; Undecided for $\alpha < 0.005$.

**9.105** (a) 0.01 < P < 0.02

(b) Reject $H_0$ for $\alpha \geq 0.02$; Do not reject $H_0$ for $\alpha \leq 0.01$; Undecided for $0.01 < \alpha < 0.02$.

**9.107** (a) df = 31; t = −2.828

(b) P < 0.005; Reject $H_0$; Evidence against $H_0$ is very strong.

**9.109** (a) df = 14; t = 1.936

(b) 0.025 < P < 0.05; Reject $H_0$; Evidence against $H_0$ is strong.

**9.111** (a) df = 23; t = 1.225

(b) P > 0.20; Do not reject $H_0$; Evidence against $H_0$ is weak or none.

**9.113** $H_0$: μ = 4.55, $H_a$: μ ≠ 4.55, α = 0.10; Critical values: ±1.729

$$t = (\bar{x} - \mu)/(s/\sqrt{n}) = (4.760 - 4.55)/(2.297/\sqrt{20}) = 0.409.$$

Since −1.729 < 0.409 < 1.729, we do not reject $H_0$. The P-value is P > 0.20. The data does not provide sufficient evidence to conclude that the amount of television watched per day last year by the average person differed from that in 2005.

**9.115** n = 20, df = 19, $\bar{x}$ = 49.0, s = 10.0

Step 1: $H_0$: $\mu$ = 45, $H_a$: $\mu$ > 45

Step 2: $\alpha$ = 0.05

Step 3: $t = \dfrac{49.0 - 45}{10.0/\sqrt{20}} = 1.789$

Step 4: Critical Value Approach: Critical value = 1.729

P-Value Approach: 0.025 < P-value < 0.05

Step 5: Since 1.789 > 1.729, reject $H_0$. Since the P-value < $\alpha$, reject $H_0$.

Step 6: At the 5% significance level, the data do provide sufficient evidence to conclude that the mean angle between the body and head of an alligator during a death roll is greater than 45 degrees.

**9.117** n = 187, df = 186, $\bar{x}$ = 0.64, s = 0.15

Step 1: $H_0$: $\mu$ = 0.9, $H_a$: $\mu$ < 0.9

Step 2: $\alpha$ = 0.01

Step 3: $t = \dfrac{0.64 - 0.90}{0.15/\sqrt{187}} = -23.703$

Step 4: Critical Value Approach: Critical value = −2.345

P-Value Approach: P < 0.005

Step 5: Since −23.703 < −2.345, reject $H_0$. Since the p-value < $\alpha$, reject $H_0$.

Step 6: At the 1% significance level, the data provide sufficient evidence to conclude that the mean ABI for women with peripheral arterial disease is less than the healthy ABI of 0.9. Thus, we conclude that such women do have an unhealthy ABI. The practical significance of this result is that the ABI may be a good tool for determining the possibility of peripheral arterial disease in women.  There could also be other causes of a low ABI, so this test by itself may not be able to determine the precise ailment.

**9.119** We used Minitab to produce the following histogram.

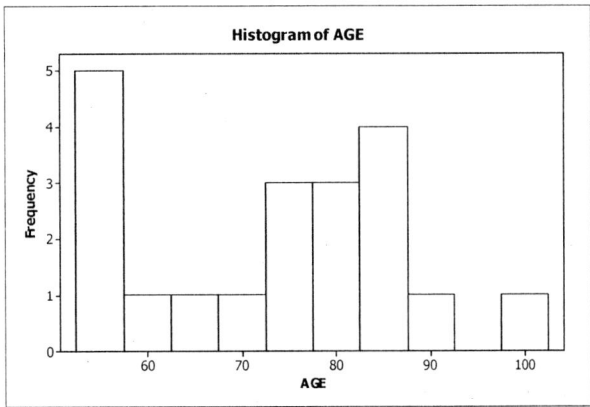

The sample size is only 20.  Although there are no outliers, the distribution is not very close to being normally distributed.  It does not appear to be reasonable to use a t-test with these data.

**9.121** We used Minitab to produce the following normal probability plot.

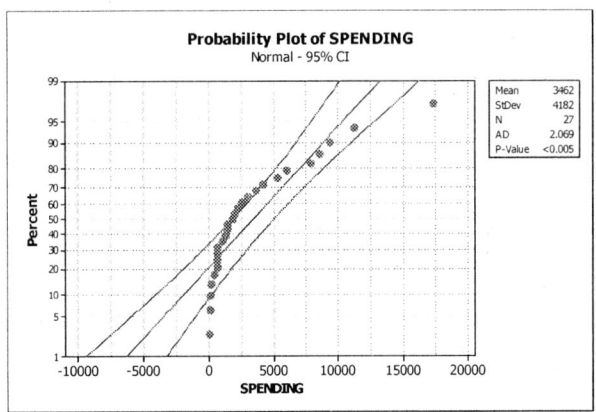

**9.123** (a) Using Minitab, with the data in a column named PRESSURE, we choose **Stat ▶ Basic Statistics ▶ 1-Sample t...**, click in the **Samples in columns** text box and specify PRESSURE, click in the **Test mean** text box and enter 80. Click the **Options...** button, enter 90 in the **Confidence level** text box, click the arrow button at the right of the **Alternative** drop-down list box and select **greater than** and click **OK**.  Click on the **Graphs** button and check the boxes for **Histogram of Data** and **Boxplot of data**. Then click **OK** twice.  Now choose **Stat ▶ Basic Statistics ▶ Normality test** and enter PRESSURE in the **Variable** text box.  Click **OK**.  Then choose **Graph ▶ Stem-and-Leaf**, enter PRESSURE in the **Graph variables** text box and click **OK**.  The results are

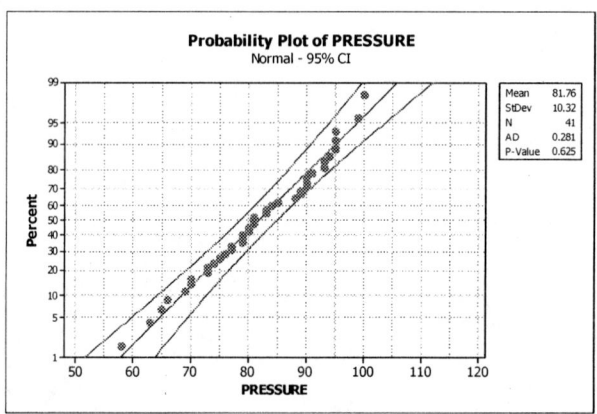

```
Stem-and-leaf of PRESSURE   N = 41
Leaf Unit = 1.0

    1    5    8
    2    6    3
    5    6    569
   10    7    00334
   17    7    5677999
   (8)    8    00111334
   16    8    5899
   12    9    0001334
    5    9    5559
    1   10    0
```

(b) The large sample size, lack of potential outliers, and the nearly linear probability plot all indicate that a t-test is reasonable for these data.

(c) The first procedure in part (a) also yielded the following test results:

Test of mu = 80 vs > 80

```
                                    90%
                                   Lower
Variable    N    Mean    StDev  SE Mean   Bound      T      P
PRESSURE   41  81.7561  10.3242  1.6124  79.6551   1.09  0.141
```

We see that t = 1.09 and the P-value is 0.141. Since the P-value is greater than the significance level 0.10, we do not reject the null hypothesis and conclude that the data do not provide evidence that the mean diastolic blood pressure of bus drivers in Stockholm exceeds the normal pressure of 88 mm Hg.

9.125 (a) Using Minitab, with the data in a column named RENT, we choose **Stat ▶ Basic Statistics ▶ 1-Sample t...**, click in the **Samples in columns** text box and specify RENT, click in the **Test mean** text box and enter 949. Click the **Options...** button, enter 95 in the **Confidence level** text box, click the arrow button at the right of the **Alternative** drop-down list box and select **greater than** and click **OK**.

```
Test of μ = 949 vs > 949

Variable    N    Mean   StDev   SE Mean   95% Lower Bound    T      P
  RENT     100  960.96  89.85    8.98         946.04        1.33  0.093
```
The P-value for the test is 0.093, larger than the significance lever 0.05, so we do not reject the null hypothesis. The data do not provide sufficient evidence to conclude that the mean rent for a two-bedroom unit is greater than the FMR of $949.

(b) After removing the outlier 662 and following the procedure in part (a), the results are

```
Test of μ = 949 vs > 949

Variable    N    Mean   StDev   SE Mean   95% Lower Bound    T      P
RENT       99   963.98  85.05    8.55         949.79        1.75  0.041
```
(c) Now the P-value is 0.041, leading to rejection of the null hypothesis. The sample mean increased by $3.02 and the standard deviation decreased by about $4.8.

(d) The sample size is large in both cases, yet the effect of the outlier is considerable. Caution should be used and perhaps the data should be analyzed using a method that is not influenced by outliers.

**9.127** (a) Yes, it is permissible to use the *t*-test because the assumptions have been met since the sample size is large.

(b) Yes, it is permissible to use the Wilcoxon signed-rank test because the assumptions have been met since the distribution is symmetric.

(c) The *t*-test should be used instead of the Wilcoxon signed-rank test because the *t*-test will tend to give more accurate results.

**9.129** (a) n = 187, df = 186, $-t_\alpha$ = -2.345, s = 0.15, $\bar{x}$ = 0.64

The 95% upper confidence bound is $0.64 + 2.345(0.15)/\sqrt{187} = 0.666$.

The hypothesized mean (0.9) lies above the upper confidence bound, so we should reject the null hypothesis. Since the test statistic is $t = (0.64 - 0.9)/(0.15/\sqrt{187}) = -23.703$, which is less than the critical value of -2.345, the hypothesis test also leads to the conclusion that we should reject the null hypothesis.

(b) n = 30, df = 29, $-t_\alpha$ = -1.311, s = 0.74, $\bar{x}$ = 1.91

The 90% upper confidence bound is $1.91 + 1.311(0.74)/\sqrt{30} = 2.087$.

The hypothesized mean (2) lies below the upper confidence bound, so we should not reject the null hypothesis. Since the test statistic is $t = (1.91 - 2)/(0.74/\sqrt{30}) = -0.67$, which is greater than the critical value of -1.311, the hypothesis test also leads to the conclusion that we should not reject the null hypothesis.

### Review Problems for Chapter 9

**1.** (a) A null hypothesis always specifies a single value for the parameter of a population which is of interest.

(b) The alternative hypothesis reflects the purpose of the hypothesis test, which can be to determine that the parameter of interest is greater than, less than, or different from the single value specified in the null hypothesis.

(c) The test statistic is a quantity calculated from the sample, under the assumption that the null hypothesis is true, which is used as a basis for deciding whether or not to reject the null hypothesis.

(d) The significance level is the probability of making a Type I error. That is, it is the probability of rejecting a true null hypothesis.

2. (a) The statement is expressing the fact that there is variability in the net weights of the boxes' content and some boxes may actually contain less than the printed weight on the box. However, the net weights for each day's production will average a bit more than the printed weight.

(b) To test the truth of this statement, we would use a null hypothesis that stated that the population mean net weight of the boxes was <u>equal</u> to the printed weight and an alternative hypothesis that stated that the population mean net weight of the boxes was <u>greater than</u> the printed weight.

(c) Null hypothesis: Population mean net weight = 76 oz

Alternative hypothesis: Population mean net weight > 76 oz

or

$H_0: \mu = 76$    $H_a: \mu > 76$

3. (a) Roughly speaking, there is a range of values of the test statistic which one could reasonably expect to occur if the null hypothesis were true. If the value of the test statistic is one that would not be expected to occur when the null hypothesis is true, then we reject the null hypothesis.

(b) To make this procedure objective and precise, we specify the probability with which we are willing to reject the null hypothesis when it is actually true. This is called the significance level of the test and is usually some small number like 0.05 or 0.01. Specifying the significance level allows us to determine the range of values of the test statistic that will lead to rejection of the null hypothesis. If the computed value of the test statistic falls in this "rejection region," then the null hypothesis is rejected. If it does not fall in the rejection region, then the null hypothesis is not rejected.

4. We would use the alternative hypothesis $\mu \neq \mu_0$ if we wanted to determine whether the population mean were <u>different from</u> the value $\mu_0$ specified in the null hypothesis. We would use the alternative hypothesis $\mu > \mu_0$ if we wanted to determine whether the population mean were <u>greater than</u> from the value $\mu_0$ specified in the null hypothesis. We would use the alternative hypothesis $\mu < \mu_0$ if we wanted to determine whether the population mean were <u>less than</u> the value $\mu_0$ specified in the null hypothesis.

5. (a) A Type I error is made whenever the null hypothesis is true, but the value of the test statistic leads us to reject the null hypothesis. A Type II error is made whenever the null hypothesis is false, but the value of the test statistic leads us to not reject the null hypothesis.

(b) The probability of a Type I error is represented by $\alpha$ and that of a Type II error by $\beta$.

(c) If the null hypothesis is true, the test statistic can lead us to either reject or not reject the null hypothesis. The first is the correct decision, while the latter constitutes a Type I error. Thus a Type I error is the only type of error possible when the null hypothesis is true.

(d) If the null hypothesis is not rejected, a correct decision has been made if the null hypothesis is, in fact, true. But if the null hypothesis is false, we have made a Type II error. Thus a Type II error is the only type of error possible when the null hypothesis is not rejected.

6. The probability of a Type II error is increased when the significance level is decreased for a fixed sample size.

7. (a) The rejection region is a set of values of the test statistic that lead to rejection of the null hypothesis.

   (b) The nonrejection region is a set of values of the test statistic that lead to not rejecting the null hypothesis.

   (c) The critical values are values of the test statistic that separate the rejection region from the nonrejection region.

8. True. You would reject the null hypothesis if your test statistic is within the rejection region or equal to the critical value.

9. Assuming that the null hypothesis is true, find the value of the test statistic for which the probability of obtaining a value less than this value (the critical value) is 0.05.

10. (a) For a right tailed one-mean z-test at 1% significance, the critical value is $z_\alpha = z_{0.01} = 2.33$.

    (b) For a left tailed one-mean z-test at 1% significance, the critical value is $-z_\alpha = -z_{0.01} = -2.33$.

    (c) For a two tailed one-mean z-test at 1% significance, the critical value is $\pm z_{\alpha/2} = \pm z_{0.005} = \pm 2.575$.

11. (a) Rejection region:  $z \geq 1.28$

    (b) Nonrejection region:  $z < 1.28$

    (c) Critical value:  $z = 1.28$

    (d) Significance level:  $\alpha = 0.10$

    (e)

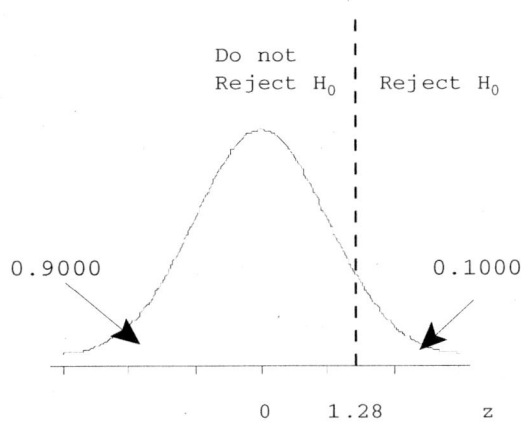

Critical Value
Nonrejection region | Rejection region

**308** Chapter 9

 (f) Right-tailed test

12. Step 1: State the null and alternative hypothesis.
 Step 2: Choose $\alpha$.
 Step 3: Calculate the test statistic.
 Step 4: Calculate the critical value(s).
 Step 5: If the test statistic falls within the rejection region, reject the null hypothesis. Otherwise, do not reject the null hypothesis.
 Step 6: Interpret your conclusion in context of the problem.

13. The P-value of a hypothesis test is the probability, assuming that the null hypothesis is true, of getting a value of the test statistic that is as extreme or more extreme than the one actually obtained.

14. True. If the null hypothesis were true, a value of the test statistic with a P-value of 0.02 would be more extreme than one with a P-value of 0.03.

15. If the P-value is less than or equal to the significance level, reject the null hypothesis. Otherwise, do not reject the null hypothesis.

16. The P-value of a hypothesis test is also called the observed significance level since it represents the smallest possible significance level at which the null hypothesis could have been rejected.

17. The P-value is the probability, assuming that the null hypothesis is true, of seeing a test statistic at least as extreme than that observed.

18. (a) The P-value for a left-tailed test with a test statistic of $z = -1.25$ is $P(Z < -1.25) = 0.1056$. This P-value of 0.1056 is greater than the significance level of 0.05. We would not reject the null hypothesis.

 (b) The P-value for a right-tailed test with a test statistic of $z = 2.36$ is $P(Z > 2.36) = 0.0091$. This P-value of 0.0091 is less than the significance level of 0.05. We would reject the null hypothesis.

 (c) The P-value for a two-tailed test with a test statistic of $z = 1.83$ is $2 \cdot P(Z > 1.83) = 2 \cdot (0.0336) = 0.0672$. This P-value of 0.0672 is greater than the significance level of 0.05. We would not reject the null hypothesis.

19. Step 1: State the null and alternative hypothesis.
 Step 2: Choose $\alpha$.
 Step 3: Calculate the test statistic.
 Step 4: Calculate the P-value.
 Step 5: If the P-value is less than or equal to the significance level, reject the null hypothesis. Otherwise, do not reject the null hypothesis.
 Step 6: Interpret your conclusion in context of the problem.

20. The P-value of 0.062 is within the range of 0.05 < P-value < 0.10. Therefore, there is moderate evidence against the null hypothesis.

21. (a) A hypothesis test is exact if the actual significance level is the same as the one that is stated.

 (b) A hypothesis test is approximately correct if the actual significance level only approximately equals $\alpha$.

22. A statistically significant result occurs when the value of the test statistic falls in the rejection region. A result has practical significance when it is statistically significant and the result also is

different enough from results expected under the null hypothesis to be important to the consumer of the results. By taking large enough sample sizes, almost any result can be made statistically significant due to the increased ability of the test to detect a false null hypothesis, but small differences from the conditions expressed by the null hypothesis may not be important, that is, they may not have practical significance.

23. (a) If the population standard deviation is unknown, and the population is normal or the sample size is large, we can use the one-mean t-statistic, $t = (\bar{x} - \mu_0)/(s/\sqrt{n})$.

   (b) If the population standard deviation is known, and the population is normal or the sample size is large, we can use the one-mean z-statistic, $z = (\bar{x} - \mu_0)/(\sigma/\sqrt{n})$.

24. Let μ denote last year's mean cheese consumption by Americans.

   (a) H₀: $\mu$ = 33.0 lb

   (b) Hₐ: $\mu$ > 33.0 lb

   (c) This is a right-tailed test.

25. (a) A Type I error would occur if, in fact, $\mu$ = 33.0 lb, but the results of the sampling lead to the conclusion that $\mu$ > 33.0 lb.

   (b) A Type II error would occur if, in fact, $\mu$ > 33.0 lb, but the results of the sampling fail to lead to that conclusion.

   (c) A correct decision would occur if, in fact, $\mu$ = 33.0 lb and the results of the sampling do not lead to the rejection of that fact; or if, in fact, $\mu$ > 33.0 lb and the results of the sampling lead to that conclusion.

   (d) If, in fact, last year's mean consumption of cheese for all Americans has not increased over the 2010 mean of 33.0 lb, and we reject the null hypothesis that $\mu$ = 33.0 lb, we made a Type I error.

   (e) If, in fact, last year's mean consumption of cheese for all Americans has increased over the 2010 mean of 33.0 lb, and we reject the null hypothesis that $\mu$ = 33.0 lb, we made a correct decision.

26. (a) n = 35, $\bar{x}$ = 1218/35 = 34.8, $\sigma$ = 6.9

   Step 1: H₀: $\mu$ = 33.0 lb, Hₐ: $\mu$ > 33.0 lb

   Step 2: $\alpha$ = 0.10

   Step 3: z = (34.8 - 33.0)/(6.9/$\sqrt{35}$) = 1.54

   Step 4: Critical Value Approach: Critical value = 1.28

   P-value Approach: P-value is $P(Z > 1.54) = 0.0618$

   Step 5: Since 1.54 > 1.28, reject H₀. Since the P-value of 0.0618 is less than the significance level, reject H₀.

   Step 6: At the 10% significance level, the data provide sufficient evidence to conclude that last year's mean cheese consumption $\mu$ for all Americans has increased over the 2010 mean of 33.0 lb.

(b) Given the conclusion in part (a), if an error has been made, it must be a Type I error. This is because, given that the null hypothesis was rejected, the only error that could be made is the error of rejecting a true null hypothesis.

27. $n = 12$, $\bar{x} = \$455.0$, $s = \$86.8$

   Step 1: $H_0$: $\mu = \$468$, $H_a$: $\mu < \$468$

   Step 2: $\alpha = 0.05$

   Step 3: $t = (455.0 - 468)/(86.8/\sqrt{12}) = -0.52$

   Step 4: Critical Value Approach: Critical value = $-1.796$

   P-Value Approach: $P > 0.10$

   Step 5: Since $-0.52 > -1.796$, do not reject $H_0$. Since the P-value is greater than the significance level, do not reject $H_0$.

   Step 6: At the 5% significance level, the data do not provide sufficient evidence to conclude that the mean value lost because of purse snatching has decreased from the 2012 mean of $468.

28. (a) If the odds-makers are estimating correctly, the mean point-spread error is zero.

   (b) It seems reasonable to assume that the distribution of point spread errors is approximately normal. In any case, the sample size of 2109 is very large, so the t-test of $H_0$: $\mu = 0$ vs. $H_a$: $\mu \neq 0$ is appropriate. At the 5% significance level, the critical values are $\pm 1.960$. Since $t = (-0.2 - 0.0)/(10.9/\sqrt{2109}) = 0.843$, we do not reject $H_0$.

   (c) There is not sufficient evidence to conclude that the mean point-spread is different from zero.

29. The z-test should be used because the population standard deviation is known, the sample is large, and the distribution is approximately normal.

30. The t-test should be used because the population standard deviation is unknown and the sample size is greater than 30.

31. The z-test should be used because the population standard deviation is known and the distribution is approximately normal.

32. The z-test should be used because the population standard deviation is known and the sample is large.

33. The t-test should be used because the population standard deviation is unknown and the sample size is greater than 30.

34. Neither test should be used because the sample size is small and the distribution is not normally distributed.

35. Neither test should be used because the sample size is small and the distribution is not normally distributed.

36. Neither test should be used because the sample size is small and the distribution is not normally distributed.

37. Neither test should be used because the sample size is small and the distribution is not normally distributed.

38. Neither test should be used because the sample size is small and the distribution is not normally distributed.

**39.** (a,b) Using Minitab, choose **Stat ▶ Basic Statistics ▶ 1-Sample t,** enter CONSUMPTION in the **Samples in columns** text box and 57.5 in the **Test mean** text box. Click on the **Graphs** button and check the boxes for **Histogram of data** and **Boxplot of data**. Click **OK**. Click on the **Options** button, enter 95 in the **Confidence level** text box and select **Less than** from the **Alternative** drop down box. Click **OK** twice. Then choose **Stat ▶ Basic Statistics ▶ Normality test** and enter CONSUMPTION in the **Variable** text box and click **OK**. Finally, choose **Graph ▶ Stem-and-Leaf** and enter CONSUMPTION in the **Graph Variables** text box and click **OK**. The results are

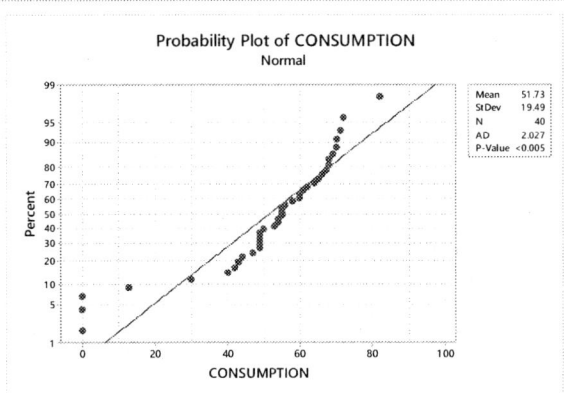

```
Stem-and-leaf of CONSUMPTION   N = 40
Leaf Unit = 1.0

    3   0   000
    4   1   3
    4   2
    5   3   0
   15   4   0234799999
   (9)  5   034455568
   16   6   00124567889
    5   7   0012
    1   8   2
```

Test of μ = 57.5 vs < 57.5

```
Variable       N    Mean   StDev   SE Mean   95% Upper Bound    T       P
CONSUMPTION   40   51.73   19.49     3.08          56.92      -1.87   0.034
```

The P-value of the test is 0.034, which is less than the 0.05

significance level. We reject $H_0$. The data provide sufficient evidence to conclude that the mean beef consumption last year is less than the 2011 mean of 57.5 lbs.

(c) After removing the four outliers, the test results are

Test of μ = 57.5 vs < 57.5

```
Variable       N    Mean   StDev   SE Mean   95% Upper Bound      T      P
CONSUMPTION   36   57.11   11.02    1.84              60.21   -0.21  0.417
```

Now the P-value of the test is 0.417, which is greater than the 0.05 significance level. We do not reject $H_0$. The data do not provide sufficient evidence to conclude that the mean beef consumption last year is less than the 2011 mean of 57.5 lbs.

(d) The outliers had a very large effect on the test results. Although the sample size was 40, four outliers are too many for the t-test to be appropriate. If the outliers are not recording errors, they represent legitimate observations from the population and should not be deleted. If they were deleted, the test results would not yield valid conclusions about the population. With so many outliers, it is just possible that the population itself is quite left skewed as is the sample.

(e) This is more appropriately done with a non-parametric test that is insensitive to outliers.

40. (a) Using Minitab, choose **Stat ▶ Basic Statistics ▶ 1-Sample z,** enter BMI in the **Samples in columns** text box and 25 in the **Test mean** text box. Click on the **Graphs** button and check the boxes for **Histogram of data** and **Boxplot of data**. Click **OK**. Click on the **Options** button, enter 95 in the **Confidence level** text box and select **Greater than** from the **Alternative** drop down box. Click **OK** twice. Then choose **Stat ▶ Basic Statistics ▶ Normality test** and enter BMI in the **Variable** text box and click **OK**. Finally, choose **Graph ▶ Stem-and-Leaf** and enter BMI in the **Graph Variables** text box and click **OK**. The graphic results are

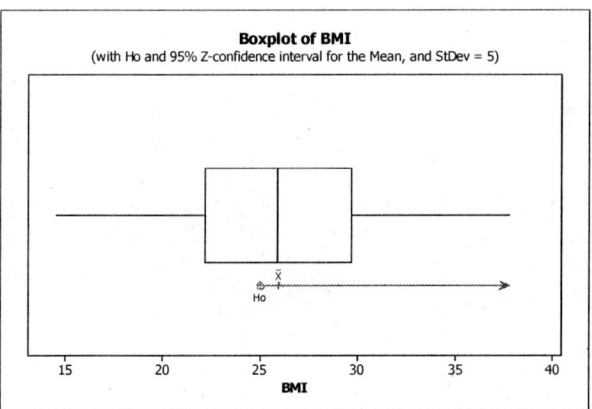

```
Stem-and-leaf of BMI   N = 75
Leaf Unit = 1.0

    1   1  4
    4   1  677
    6   1  88
   16   2  0000111111
   29   2  2222222233333
   (9)  2  444444455
   37   2  666666777
   28   2  88888899999
   17   3  0000111
   10   3  22233
    5   3  4555
    1   3  7
```

(b) Yes. There are no outliers, the sample size is large, and the data is reasonably normally distributed.

(c) The process in part (a) yielded the following test results:

```
Test of mu = 25 vs > 25
The assumed standard deviation = 5

                                          95%
                                        Lower
Variable   N     Mean    StDev  SE Mean  Bound    Z      P
BMI       75   25.9867  5.0293  0.5774  25.0370  1.71  0.044
```

Since the P-value of 0.044 is less than the significance level 0.05, we reject the null hypothesis. The data do provide sufficient evidence to conclude that the mean BMI of U.S. adults is greater than that for a healthy weight.

**41.** (a) Using Minitab, we choose **Graph ▶ Histogram,** choose the **Simple** version, enter BEER in the **Graph variables** text box, and click **OK**. The result is

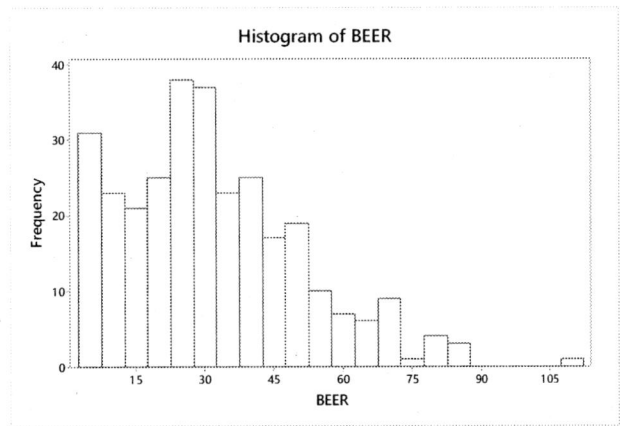

(b) There may be an outlier at about 110.

(c) Since the sample size is very large (300), we can use the z-test or the t-test. Since the standard deviation is unknown, we use the t-test. Using Minitab, we obtained the following result using all of the data.

Test of μ = 28.2 vs ≠ 28.2

```
Variable    N    Mean   StDev   SE Mean      99% CI          T      P
BEER       300   31.52  19.43    1.12     (28.61, 34.43)   2.96   0.003
```

After eliminating the potential outlier (112), we repeated the process and obtained

Test of μ = 28.2 vs ≠ 28.2

```
Variable    N    Mean   StDev   SE Mean      99% CI          T      P
BEER       299   31.26  18.92    1.09     (28.42, 34.09)   2.79   0.006
```

Clearly, elimination of the value 110 had little effect on the value of t or on the P-value. In either case, we reject the null hypothesis. The data do provide sufficient evidence to claim that the mean annual consumption of beer in Missouri is different from the national mean.

# CHAPTER 10 SOLUTIONS

## Exercises 10.1

**10.1** Answers will vary.

**10.3** (a) $\mu_1$, $\sigma_1$, $\mu_2$, and $\sigma_2$ are parameters; $\bar{x}_1$, $s_1$, $\bar{x}_2$, and $s_2$ are statistics.

(b) $\mu_1$, $\sigma_1$, $\mu_2$, and $\sigma_2$ are fixed numbers; $\bar{x}_1$, $s_1$, $\bar{x}_2$, and $s_2$ are random variables.

**10.5** (a) $H_0: \mu_1 = \mu_2$  $H_a: \mu_1 \neq \mu_2$  (b) two tailed

**10.7** (a) $H_0: \mu_1 = \mu_2$  $H_a: \mu_1 > \mu_2$  (b) right tailed

**10.9** (a) $H_0: \mu_1 = \mu_2$  $H_a: \mu_1 < \mu_2$  (b) left tailed

**10.11** It is the sampling distribution of the difference of the two sample means that allows us to determine whether the difference of the sample means can reasonably be attributed to sampling error or whether it is large enough for us to conclude that the population means are different.

**10.13** We can be 95% confident that the mean for population one is between 15 to 20 units greater than the mean for population two.

**10.15** We can be 90% confident that the mean for population one is between 5 to 10 units less than the mean for population two.

**10.17** We can be 99% confident that the mean for population one is between 20 units less to 15 units greater than the mean for population two.

**10.19** (a) The mean of $\bar{x}_1 - \bar{x}_2$ is $\mu_1 - \mu_2 = 40 - 40 = 0$

The standard deviation of $\bar{x}_1 - \bar{x}_2$ is $\sigma_{\bar{x}_1 - \bar{x}_2} = \sqrt{\dfrac{\sigma_1^2}{n_1} + \dfrac{\sigma_2^2}{n_2}} = \sqrt{\dfrac{12^2}{9} + \dfrac{6^2}{4}} = 5$

(b) No. The determination of the mean of $\bar{x}_1 - \bar{x}_2$ is the same for all populations. The formula for the standard deviation of $\bar{x}_1 - \bar{x}_2$ is dependent on the samples being independent, but not on the populations from which they came.

(c) No. It is not known that the two populations are normally distributed (which would lead to $\bar{x}_1 - \bar{x}_2$ being normally distributed), and the sample sizes of 9 and 4 are too small to claim that $\bar{x}_1$ and $\bar{x}_2$ are normally distributed (which would also lead to $\bar{x}_1 - \bar{x}_2$ being normally distributed).

**10.21** (a) The mean of $\bar{x}_1 - \bar{x}_2$ is $\mu_1 - \mu_2 = 40 - 40 = 0$

The standard deviation of $\bar{x}_1 - \bar{x}_2$ is $\sigma_{\bar{x}_1 - \bar{x}_2} = \sqrt{\dfrac{\sigma_1^2}{n_1} + \dfrac{\sigma_2^2}{n_2}} = \sqrt{\dfrac{12^2}{9} + \dfrac{6^2}{4}} = 5$

(b) Yes. Since $x_1$ and $x_2$ are each normally distributed, then so are $\bar{x}_1$ and $\bar{x}_2$. Since the difference of two normally distributed random variables is also normally distributed, so is $\bar{x}_1 - \bar{x}_2$. See also Key Fact 10.1.

(c) Since $\bar{x}_1 - \bar{x}_2$ is normally distributed with mean 0 and standard deviation 5, P(-10< $\bar{x}_1 - \bar{x}_2$ < 10) = P([-10 - 0]/5 < z < [10 - 0]/5)
=P(-2 < z < 2) = 0.9772 - 0.0228 = 0.9544, so 95.44% of the differences in sample means from samples of size 9 and 4 from the two populations will lie between -10 and 10.

**10.23** $H_0$: $\mu_1 = \mu_2$, $H_a$: $\mu_1 > \mu_2$

**10.25** (a) The variable is systolic blood pressure (SBP).

(b) The two populations are the adolescent offspring of diabetic mothers

(ODM) and of nondiabetic mothers (ODN).

(c) $H_0: \mu_1 = \mu_2$, $H_a: \mu_1 > \mu_2$ where $\mu_1$ is the mean SBP for ODM and $\mu_2$ is the mean SBP for ODN.

(d) Right tailed

**10.27** (a) The variable is household vehicle miles traveled (VMT).

(b) The two populations are households in the Midwest and in the South.

(c) $H_0: \mu_1 = \mu_2$, $H_a: \mu_1 \ne \mu_2$ where $\mu_1$ is the mean VMT for households in the Midwest and $\mu_2$ is the mean VMT for households in the South.

(d) two-tailed

**10.29** (a) The variable is operative times in minutes.

(b) The two populations are operations performed using a dynamic system (Z-plate) and operations performed using a static system (ALPS plate).

(c) $H_0: \mu_1 = \mu_2$, $H_a: \mu_1 < \mu_2$ where $\mu_1$ is the mean time of operations performed using the dynamic system and $\mu_2$ is the mean time of operations performed using the static system.

(d) left-tailed

**10.31** (a) We have Minitab take 2000 observations from a normally distributed population having mean 100 and standard deviation 16 by choosing **Calc ▶ Random Data ▶ Normal...**, typing 2000 in the **Generate rows of data** text box, typing X1 in the **Store in columns** text box, typing 100 in the **Mean** text box and 16 in the **Standard deviation** text box, and clicking **OK**.

(b) We repeat part (a) for a normally distributed population having mean 120 and standard deviation 12 by choosing **Calc ▶ Random Data ▶ Normal...**, typing 2000 in the **Generate rows of data** text box, typing X2 in the **Store in columns** text box, typing 120 in the **Mean** text box and 12 in the **Standard deviation** text box, and clicking **OK**.

(c) To determine the difference between each pair of observations in parts (a) and (b), we choose **Calc ▶ Calculator...**, type DIFF in the **Store results in variable** text box, type X1 - X2 in the **Expression** text box, and click **OK**. DIFF will be stored in C3.

(d) To obtain a histogram of the 2000 differences found in part (c), we choose **Graph ▶ Histogram...**, select the **Simple** version and click **OK**. Enter DIFF in the **Graph variables** text box and click **OK**.

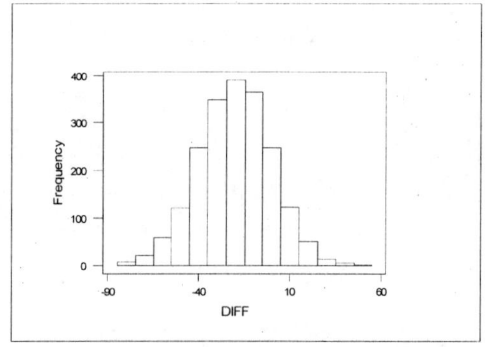

The histogram is bell-shaped because the random variable named DIFF is the difference of two random variables which themselves are normally distributed. A random variable which is the difference of two normally

## Section 10.2

**10.33** (a) The four conditions required for using the pooled t-procedure are (i) both samples are simple random samples, (ii) independence of the two samples, (iii) the variable under consideration is normally distributed in each of the two populations or both samples are large, and (iv) the population standard deviations are equal.

(b) Independence and randomness are absolutely essential. Moderate violations of the normality assumption are not serious problems even for small or moderate size samples. Moderate violations of the equal standard deviation requirement are not serious if the two sample sizes are roughly equal.

**10.35** Not Reasonable. One sample standard deviation is more than twice the other one. This is a good indication that the population standard deviations are not equal. Since the two sample sizes are not roughly equal (6 and 14), nor are both of them large, there appears to be a serious violation of the equal standard deviations requirement of the pooled t-test.

**10.37** Reasonable. The standard deviations are roughly equal and both sample sizes are large.

**10.39** (a) $s_p = \sqrt{\dfrac{14(2.1)^2 + 14(2.3)^2}{15+15-2}} = \sqrt{4.85} = 2.2023$

Step 1: $H_0: \mu_1 = \mu_2$, $H_a: \mu_1 \neq \mu_2$
Step 2: $\alpha = 0.05$
Step 3: $t = \dfrac{10-12}{2.2023\sqrt{(1/15)+(1/15)}} = -2.487$
Step 4: df = 28, Critical values = ±2.048
0.01 < P-value < 0.02.
Step 5: Since −2.487 < −2.048, reject $H_0$.

(b) 95% CI = $(10-12) \pm 2.048(2.2023)\sqrt{1/15 + 1/15} = (-3.647, -0.353)$

**10.41** (a) $s_p = \sqrt{\dfrac{9(4)^2 + 14(5)^2}{10+15-2}} = \sqrt{21.47826} = 4.6345$

Step 1: $H_0: \mu_1 = \mu_2$, $H_a: \mu_1 > \mu_2$
Step 2: $\alpha = 0.05$
Step 3: $t = \dfrac{20-18}{4.6345\sqrt{(1/10)+(1/15)}} = 1.057$
Step 4: df = 23, Critical value = 1.714
P-value > 0.10.
Step 5: Since 1.057 < 1.714, do not reject $H_0$.

(b) 90% CI = $(20-18) \pm 1.714(4.6345)\sqrt{1/10 + 1/15} = (-1.243, 5.243)$

**10.43** (a) $s_p = \sqrt{\dfrac{19(4)^2 + 14(5)^2}{20+15-2}} = \sqrt{19.81818} = 4.4518$

Step 1: $H_0: \mu_1 = \mu_2$, $H_a: \mu_1 < \mu_2$
Step 2: $\alpha = 0.05$
Step 3: $t = \dfrac{20-24}{4.4518\sqrt{(1/20)+(1/15)}} = -2.631$

**318** Chapter 10

Step 4:      df = 33, Critical value = -1.692
0.005 < P-value < 0.010.
Step 5:      Since -2.631 < -1.697, reject $H_0$.

(b)    90% CI = $(20-24) \pm 1.697(4.4518)\sqrt{(1/20+1/15)} = (-6.580, -1.420)$

**10.45** Population 1: Fraud, $n_1 = 10$, $\bar{x}_1 = 10.12$, $s_1 = 4.90$

Population 2: Firearms, $n_2 = 10$, $\bar{x}_2 = 18.78$, $s_2 = 4.64$

$$s_p = \sqrt{\frac{9(4.90)^2 + 9(4.64)^2}{10+10-2}} = \sqrt{22.7698} = 4.7718$$

Step 1:    $H_0: \mu_1 = \mu_2$,    $H_a: \mu_1 < \mu_2$
Step 2:    $\alpha = 0.05$
Step 3:    $t = \dfrac{10.12 - 18.78}{4.7718\sqrt{(1/10)+(1/10)}} = -4.058$

Step 4:    df = 18, Critical value = -1.734
P-value < 0.005.
Step 5:    Since -4.06 < -1.734, reject $H_0$.

Step 6:    At the 5% significance level, the data do provide sufficient evidence to conclude that, on the average, the mean time served for fraud is less than that served for firearms offenses.

**10.47** Population 1: Fortified, $n_1 = 14$, $\bar{x}_1 = 9.0$, $s_1 = 37.4$

Population 2: Unfortified, $n_2 = 12$, $\bar{x}_2 = 1.6$, $s_2 = 34.6$

$$s_p = \sqrt{\frac{13(37.4)^2 + 11(34.6)^2}{24}} = \sqrt{1306.36} = 36.1436$$

Step 1:    $H_0: \mu_1 = \mu_2$,    $H_a: \mu_1 > \mu_2$
Step 2:    $\alpha = 0.05$
Step 3:    $t = \dfrac{9.0 - 1.6}{36.1436\sqrt{1/14 + 1/12}} = 0.520$

Step 4:    Critical value = 1.711
P-value > 0.10
Step 5:    Since 0.520 < 1.711, do not reject $H_0$.
Step 6:    At the 5% significance level, the data do not provide sufficient evidence that drinking fortified orange juice reduces PTH level more than drinking unfortified orange juice.

**10.49** Population 1: Males, $n_1 = 91$, $\bar{x}_1 = 11.1$, $s_1 = 0.9$

Population 2: Females, $n_2 = 109$, $\bar{x}_2 = 10.1$, $s_2 = 0.7$

$$s_p = \sqrt{\frac{90(0.9)^2 + 108(0.7)^2}{91+109-2}} = \sqrt{0.6355} = 0.7972$$

Step 1:    $H_0: \mu_1 = \mu_2$,    $H_a: \mu_1 \neq \mu_2$
Step 2:    $\alpha = 0.01$
Step 3:    $t = \dfrac{11.1 - 10.1}{0.7972\sqrt{1/91 + 1/109}} = 8.83$

Step 4: df = 198, Critical values = ±2.601 (using technology)
For the p-value approach, 2{P(T > 8.883)} < 0.01.

Step 5: Since 8.83 > 2.601, reject $H_0$. Because the P-value is smaller than the significance level, reject $H_0$.

Step 6: At the 1% significance level, the data provide sufficient evidence that there is a difference in the mean spleen lengths of male and female Nigerians.

In each of Exercises 10.51-10.56, the value of $s_p$ has been obtained in Exercise 10.45-10.50, respectively.

**10.51** $t_{\alpha/2} = 1.734$, 90%CI $= (10.12 - 18.78) \pm 1.734(4.7718)\sqrt{1/10 + 1/10} = (-12.36, -4.96)$

**10.53** $t_{\alpha/2} = 1.711$, 90%CI $= (9.0 - 1.6) \pm 1.711(36.1436)\sqrt{1/14 + 1/12} = (-16.93, 31.73)$

**10.55** $t_{\alpha/2} = 2.601$, 99%CI $= (11.1 - 10.1) \pm 2.601(0.7972)\sqrt{1/91 + 1/109} = (0.71, 1.29)$

**10.57** (a) Using Minitab, we choose **Graph ▶ Probability Plot**, select **Single** and click **OK**. Enter VEGETARIAN and OMNIVORE in the **Graph Variables** text box. Click on the **Multiple Graphs** button and select **On separate graphs**. Click **OK** twice. Note that the standard deviations are shown on the plots. Then choose **Graph ▶ Boxplot**, click on the **Simple** plot in the **Multiple Y's** row and click **OK**..Enter VEGETARIAN and OMNIVORE in the **Graph Variables** text box and click **OK**. The results are

The standard deviations are 18.82 for Vegetarians and 18.97 for Omnivores.

(b) Choose **Stat ▶ Basic Statistics ▶ 2-Sample t**, click on **Samples in different columns** and enter VEGETARIANS in the **First** textbox and OMNIVORES in the **Second** text box. Check the box for **Assume equal**

**variances** and click on the **Options** button. Enter 99 in the **Confidence level** textbox, 0 in the **Test difference** text box and select **not equal** in the **Alternative drop down box**. Click **OK** twice. The results are

```
Two-sample T for VEGETARIAN vs OMNIVORE

             N   Mean   StDev   SE Mean
VEGETARIAN   51  39.0   18.8    2.6
OMNIVORE     53  49.9   19.0    2.6

Difference = mu (VEGETARIAN) - mu (OMNIVORE)
Estimate for difference:  -10.8853
99% CI for difference:  (-20.6147, -1.1559)
T-Test of difference = 0 (vs not =): T-Value = -2.94  P-Value = 0.004
DF = 102
Both use Pooled StDev = 18.8964
```

Since the P-value of 0.004 is less than the significance level of 0.01, the data do provide sufficient evidence that there is a difference between the mean daily protein intakes of female vegetarians and female omnivores.

(c) The 99% confidence interval is (-20.6147, -1.1559).

(d) Yes. The probability plots indicate that normality is reasonable, the sample sizes are large, and there are no outliers. Furthermore, the two sample standard deviations are almost equal, making the pooled t procedure a good choice.

**10.59** (a) Using Minitab, we choose **Graph ▶ Probability Plot**, select **Single** and click **OK**. Enter STINGER and REGULAR in the **Graph Variables** text box. Click on the **Multiple Graphs** button and select **On separate graphs**. Click **OK** twice. Note that the standard deviations are shown on the plots. Then choose **Graph ▶ Boxplot**, click on the **Simple** plot in the **Multiple Y's** row and click **OK**...Enter STINGER and REGULAR in the **Graph Variables** text box and click **OK**. The results are

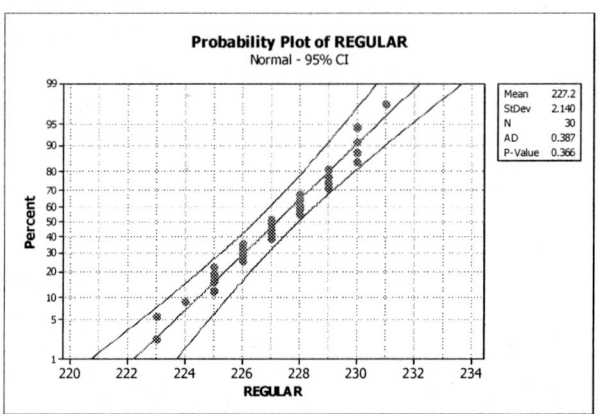

The standard deviations are
2.778 for STINGER and
2.140 for REGULAR.

(b) Choose **Stat ▶ Basic Statistics ▶ 2-Sample t**, click on **Samples in different columns** and enter STINGER in the **First** text box and REGULAR in the **Second** text box. Check the box for **Assume equal variances** and click on the **Options** button. Enter 99 in the **Confidence level** textbox, 0 in the **Test difference** text box and select **greater than** in the **Alternative drop down** box. Click **OK** twice. The results are

```
Two-sample T for STINGER vs REGULAR

          N    Mean   StDev  SE Mean
STINGER  30  240.93    2.78     0.51
REGULAR  30  227.20    2.14     0.39

Difference = mu (STINGER) - mu (REGULAR)
Estimate for difference:  13.7333
99% lower bound for difference:  12.2015
T-Test of difference = 0 (vs >): T-Value = 21.45  P-Value = 0.000  DF = 58
Both use Pooled StDev = 2.4798
```
Since the P-value of 0.000 is less than the significance level of 0.01, the data do provide sufficient evidence that the Stinger tee improves total distance traveled.

(c) Repeat the process in part (b), but select **not equal** in the **Alternative drop down** box. The confidence interval is
99% CI for difference: (12.0281, 15.4386)

(d) Yes. The probability plots indicate that normality is reasonable, the sample sizes are large, and there are no outliers. Furthermore, the two sample standard deviations are almost equal, making the pooled t procedure a good choice.

**10.61** If $n_1 = n_2$, say both equal n, then

$$s_p^2 = \frac{(n_1-1)s_1^2 + (n_2-1)s_2^2}{n_1+n_2-2} = \frac{(n-1)s_1^2 + (n-1)s_2^2}{n+n-2}$$

$$= \frac{(n-1)(s_1^2+s_2^2)}{2n-2} = \frac{(n-1)(s_1^2+s_2^2)}{2(n-1)} = \frac{(s_1^2+s_2^2)}{2}$$

Thus, if $n_1 = n_2$, then $s_p^2$ is just the mean of $s_1^2$ and $s_2^2$.

**10.63** (a) The test value of t = -0.927 fell between the critical values ±2.052, leading to nonrejection of the null hypothesis. The 95% confidence interval was (-4.69, 1.77). This includes zero, so the null hypothesis is not rejected.

(b) The test value of t = −1.977 fell between the critical values ±1.982, leading to nonrejection of the null hypothesis. The 95% confidence interval was (−2.603, 0.003). This includes zero, so the null hypothesis is not rejected. Note: These results are so close to rejection that if intermediate rounding is done during the calculations, it is possible to obtain results rejecting the null hypothesis and getting a confidence interval that does not include zero.

**10.65** (a) The test value of t = 0.520 fell to the left of the critical value 1.711, leading to nonrejection of the null hypothesis. The 95% lower confidence bound is −16.928. This is not positive, so the null hypothesis is not rejected.

(b) The test value of t = = 12.88 fell to the right of the critical value 2.328, leading to rejection of the null hypothesis. The 99% lower confidence bound is 54.48. This is positive, so the null hypothesis is rejected.

**10.67** (a) Yes, it is permissible to use the pooled *t*-test because the assumptions have been met.

(b) Yes, it is permissible to use the Mann-Whitney test because the assumptions have been met.

(c) The pooled *t*-test should be used instead of the Mann-Whitney test because the *t*-test will tend to give more accurate results.

### Exercises 10.3

**10.69** Pooled t-test. When the population standard deviations are equal, the pooled t-test is more powerful than the non-pooled t-test.

**10.71** Nonpooled t-test. The sample standard deviations are enough different to raise doubt about the equality of the population standard deviations.

**10.73** (a) Pooled t-test. Both populations are normally distributed and the population standard deviations are equal.

(b) Nonpooled t-test. Both populations are normally distributed, but the population standard deviations are unequal.

(c) Neither. Both populations are skewed and it is given that both sample sizes are small.

(d) Neither. Only one population is normally distributed; the other is skewed. Since the sample sizes are small, the nonnormality of one population rules out the use of either t-test.

**10.75** (a) Step 1: $H_0: \mu_1 = \mu_2$, $H_a: \mu_1 \neq \mu_2$

Step 2: $\alpha = 0.05$

Step 3: $t = \dfrac{10-12}{\sqrt{(2^2/15)+(5^2/15)}} = -1.438$

Step 4: $\Delta = \dfrac{\left[\left(\dfrac{s_1^2}{n_1}\right)+\left(\dfrac{s_2^2}{n_2}\right)\right]^2}{\dfrac{\left(\dfrac{s_1^2}{n_1}\right)^2}{n_1-1}+\dfrac{\left(\dfrac{s_2^2}{n_2}\right)^2}{n_2-1}} = \dfrac{\left[\left(\dfrac{2^2}{15}\right)+\left(\dfrac{5^2}{15}\right)\right]^2}{\dfrac{\left(\dfrac{2^2}{15}\right)^2}{15-1}+\dfrac{\left(\dfrac{5^2}{15}\right)^2}{15-1}} = 18.37; df = 18$

Critical values = ±2.101
0.100 < P-value < 0.200.

Step 5: Since −2.101 < −1.438 < 2.101, do not reject $H_0$.

(b) 95% CI = $(10-12) \pm 2.101\sqrt{(2^2/15 + 5^2/15)} = (-4.921, 0.921)$

**10.77** (a) Step 1: $H_0: \mu_1 = \mu_2$, $H_a: \mu_1 > \mu_2$

Step 2: $\alpha = 0.05$

Step 3: $t = \dfrac{20-18}{\sqrt{(4^2/10)+(5^2/15)}} = 1.107$

Step 4: $\Delta = \dfrac{\left[\left(\dfrac{s_1^2}{n_1}\right)+\left(\dfrac{s_2^2}{n_2}\right)\right]^2}{\dfrac{\left(\dfrac{s_1^2}{n_1}\right)^2}{n_1-1}+\dfrac{\left(\dfrac{s_2^2}{n_2}\right)^2}{n_2-1}} = \dfrac{\left[\left(\dfrac{4^2}{10}\right)+\left(\dfrac{5^2}{15}\right)\right]^2}{\dfrac{\left(\dfrac{4^2}{10}\right)^2}{10-1}+\dfrac{\left(\dfrac{5^2}{15}\right)^2}{15-1}} = 22.10; df = 22$

Critical value = 1.717
P-value > 0.100.

Step 5: Since 1.107 < 1.717, do not reject $H_0$.

(b) 90% CI = $(20-18) \pm 1.717\sqrt{(4^2/10 + 5^2/15)} = (-1.103, 5.103)$

**10.79** (a) Step 1: $H_0: \mu_1 = \mu_2$, $H_a: \mu_1 < \mu_2$

Step 2: $\alpha = 0.05$

Step 3: $t = \dfrac{20-24}{\sqrt{(6^2/20)+(2^2/15)}} = -2.782$

Step 4: $\Delta = \dfrac{\left[\left(\dfrac{s_1^2}{n_1}\right)+\left(\dfrac{s_2^2}{n_2}\right)\right]^2}{\dfrac{\left(\dfrac{s_1^2}{n_1}\right)^2}{n_1-1}+\dfrac{\left(\dfrac{s_2^2}{n_2}\right)^2}{n_2-1}} = \dfrac{\left[\left(\dfrac{6^2}{20}\right)+\left(\dfrac{2^2}{15}\right)\right]^2}{\dfrac{\left(\dfrac{6^2}{20}\right)^2}{20-1}+\dfrac{\left(\dfrac{2^2}{15}\right)^2}{15-1}} = 24.322; df = 24$

Critical value = -1.711
0.005 < P-value < 0.010.

Step 5: Since -2.782 < -1.711, reject $H_0$.

(b) 90% CI = $(20-24) \pm 1.711\sqrt{(6^2/20 + 2^2/15)} = (-6.460, -1.540)$

**10.81** Population 1: Chronic, $n_1 = 32$, $\bar{x}_1 = 28.8$, $s_1 = 9.2$

Population 2: Remitted, $n_2 = 20$, $\bar{x}_2 = 22.1$, $s_2 = 5.7$

Step 1: $H_0: \mu_1 = \mu_2$, $H_a: \mu_1 \neq \mu_2$

Step 2: $\alpha = 0.10$

Step 3: $t = \dfrac{25.8-22.1}{\sqrt{(9.2^2/32)+(5.7^2/20)}} = 1.791$

Step 4: $\Delta = \dfrac{\left[\left(\dfrac{s_1^2}{n_1}\right)+\left(\dfrac{s_2^2}{n_2}\right)\right]^2}{\dfrac{\left(\dfrac{s_1^2}{n_1}\right)^2}{n_1-1}+\dfrac{\left(\dfrac{s_2^2}{n_2}\right)^2}{n_2-1}} = \dfrac{\left[\left(\dfrac{9.2^2}{32}\right)+\left(\dfrac{5.7^2}{20}\right)\right]^2}{\dfrac{\left(\dfrac{9.2^2}{32}\right)^2}{32-1}+\dfrac{\left(\dfrac{5.7^2}{20}\right)^2}{20-1}} = 49.99995; df = 49$

                    Critical values = ±1.677
                    0.050 < P-value < 0.100.
        Step 5:     Since 1.791 > 1.677, reject H₀.
                    For the P-value approach, 0.05 <2{P(t > 1.791)} < 0.10.
                    Therefore, because the P-value is smaller than the significance
                    level, reject H₀.
        Step 6:     At the 10% significance level, the data do provide sufficient
                    evidence to conclude that there is a difference between the mean
                    age at arrest.

**10.83** Population 1: Dynamic, $n_1 = 14$, $\bar{x}_1 = 7.36$, $s_1 = 1.22$
Population 2: Static, $n_2 = 6$, $\bar{x}_2 = 10.50$, $s_2 = 4.59$

Step 1: $H_0: \mu_1 = \mu_2$, $H_a: \mu_1 < \mu_2$
Step 2: $\alpha = 0.05$
Step 3: $t = \dfrac{7.36 - 10.50}{\sqrt{(1.22^2/14) + (4.59^2/6)}} = -1.651$

Step 4: $\Delta = \dfrac{\left[\left(\dfrac{s_1^2}{n_1}\right) + \left(\dfrac{s_2^2}{n_2}\right)\right]^2}{\dfrac{\left(\dfrac{s_1^2}{n_1}\right)^2}{n_1-1} + \dfrac{\left(\dfrac{s_2^2}{n_2}\right)^2}{n_2-1}} = \dfrac{\left[\left(\dfrac{1.22^2}{14}\right) + \left(\dfrac{4.59^2}{6}\right)\right]^2}{\dfrac{\left(\dfrac{1.22^2}{14}\right)^2}{14-1} + \dfrac{\left(\dfrac{4.59^2}{6}\right)^2}{6-1}} = 5.305; df = 5$

Critical value = −2.015
0.050 < P-value < 0.100.

Step 5: Since −1.651 > −2.015, do not reject H₀.
For the P-value approach, 0.05 < P(t<−1.651) < 0.10. Therefore, since the P-value is larger than the significance level, do not reject H₀.

Step 6: At the 5% significance level, the data do not provide sufficient evidence to conclude that the mean number of acute postoperative days in the hospital is smaller with the dynamic system than with the static system.

**10.85** Population 1: Psychotic patients, $n_1 = 10$, $\bar{x}_1 = 0.02426$, $s_1 = 0.00514$
Population 2: Non-psychotic patients, $n_2 = 15$, $\bar{x}_2 = 0.01643$, $s_2 = 0.00470$

(a) Step 1: $H_0: \mu_1 = \mu_2$, $H_a: \mu_1 > \mu_2$
Step 2: $\alpha = 0.01$
Step 3: $t = \dfrac{0.02426 - 0.01643}{\sqrt{(0.00514^2/10) + (0.00470^2/15)}} = 3.860$

Step 4: $\Delta = \dfrac{\left[\left(\dfrac{s_1^2}{n_1}\right) + \left(\dfrac{s_2^2}{n_2}\right)\right]^2}{\dfrac{\left(\dfrac{s_1^2}{n_1}\right)^2}{n_1-1} + \dfrac{\left(\dfrac{s_2^2}{n_2}\right)^2}{n_2-1}} = \dfrac{\left[\left(\dfrac{0.00514^2}{10}\right) + \left(\dfrac{0.01643^2}{15}\right)\right]^2}{\dfrac{\left(\dfrac{1.00514^2}{10}\right)^2}{10-1} + \dfrac{\left(\dfrac{0.01643^2}{15}\right)^2}{15-1}} = 18.20; df = 18$

Critical value = 2.552 (approximately)
P-value < 0.005.

Step 5: Since 3.860 > 2.552, reject H₀.
For the P-value approach, P(t > 3.860) < 0.005.   Therefore,

since the P-value is smaller than the significance level, reject $H_0$.

Step 6: At the 1% significance level, the data do provide sufficient evidence to conclude that dopamine activity is higher, on average, in psychotic patients than in non-psychotic patients.

**10.87** $t_{\alpha/2} = 1.677$; 90% CI $= (25.8-22.1) \pm 1.677\sqrt{9.2^2/32 + 5.7^2/20} = (0.235, 7.165)$

**10.89** $t_{\alpha/2} = 2.015$; 90% CI $= (7.36-10.50) \pm 2.015\sqrt{1.22^2/14 + 4.59^2/6} = (-6.973, 0.693)$

**10.91** $t_{\alpha/2} = 2.552$; 98% CI $= (0.02426-0.01643) \pm 2.552\sqrt{0.00514^2/10 + 0.00470^2/15}$
$= (0.00265, 0.01301)$

**10.93** (a) Choose nonpooled t-prodedures. There is clearly much more variability in the data for males than in the data for females.

(b) No. The two sample sizes are small and moderate. The data for the males is not normally distributed or has a large outlier at 39.2. Neither t-procedure is reasonable under these circumstances.

**10.95** (a) Using Minitab, we choose **Graph ▶ Probability Plot**, select **Single** and click **OK**. Enter HEALTHCARE and TECHNOLOGY in the **Graph Variables** text box. Click on the **Multiple Graphs** button and select **On separate graphs**. Click **OK** twice. Note that the standard deviations are shown on the plots. Then choose **Graph ▶ Boxplot**, click on the **Simple** plot in the **Multiple Y's** row and click **OK**...Enter HEALTHCARE and TECHNOLOGY in the **Graph Variables** text box and click **OK**. The resulting standard deviations are 4.684% for HEALTHCARE and 1.504% for TECHNOLOGY.

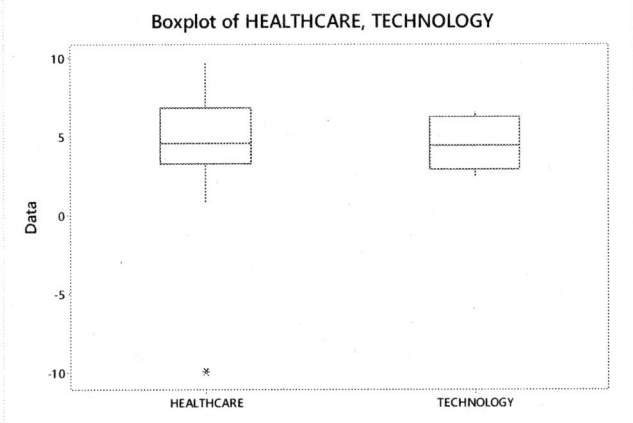

The sample standard deviations for Healthcare and Technology are very different from each other. If we had to choose between pooled and nonpooled, we would choose the nonpooled *t*-procedure.

(b) It is not reasonable to use the nonpooled t-procedure because the healthcare data is not normally distributed and contains an outlier.

**10.97** Population 1: Dynamic, $n_1 = 14$, $\bar{x}_1 = 394.6$, $s_1 = 84.7$

Population 2: Static, $n_2 = 6$, $\bar{x}_2 = 468.3$, $s_2 = 38.2$

(a) $s_p = \sqrt{\dfrac{13(84.7)^2 + 5(38.2)^2}{18}} = \sqrt{5586.63} = 74.7438$

Step 1: $H_0: \mu_1 = \mu_2$, $H_a: \mu_1 < \mu_2$
Step 2: $\alpha = 0.05$
Step 3: $t = \dfrac{394.6 - 468.3}{74.7438\sqrt{1/14 + 1/6}} = -2.021$
Step 4: Critical value = $-1.734$
$0.025 < $ P-value $< 0.05$
Step 5: Since $-2.021 < -1.734$, reject $H_0$.
Step 6: At the 5% significance level, the data do provide sufficient evidence to conclude that the mean operative time is less with the dynamic system than with the static system.

(b) Both tests ended up rejecting the null hypothesis test. However, the P-value for the non-pooled approach was smaller than the P-value for the pooled approach.

(c) The non-pooled test at 1% significance:

Step 1: $H_0: \mu_1 = \mu_2$, $H_a: \mu_1 < \mu_2$
Step 2: $\alpha = 0.01$
Step 3: $t = \dfrac{394.6 - 468.3}{\sqrt{(84.7^2/14) + (38.2^2/6)}} = -2.681$
Step 4: Critical value = $-2.567$
$0.005 < $ P-value $< 0.01$
Step 5: Since $-2.681 < -2.567$, reject $H_0$.
Step 6: At the 1% significance level, the data do provide sufficient evidence to conclude that the mean operative time is less with the dynamic system than with the static system.

The pooled test at 1% significance:

$s_p = \sqrt{\dfrac{13(84.7)^2 + 5(38.2)^2}{18}} = \sqrt{5586.63} = 74.7438$

Step 1: $H_0: \mu_1 = \mu_2$, $H_a: \mu_1 < \mu_2$
Step 2: $\alpha = 0.01$
Step 3: $t = \dfrac{394.6 - 468.3}{74.7438\sqrt{1/14 + 1/6}} = -2.021$
Step 4: Critical value = $-2.552$
$0.025 < P(t < -2.021) < 0.05$
Step 5: Since $-2.021 > -2.552$, do not reject $H_0$.
Step 6: At the 1% significance level, the data do not provide sufficient evidence to conclude that the mean operative time is less with the dynamic system than with the static system.

The pooled t-test resulted in not rejecting the null hypothesis while the nonpooled t-test resulted in rejecting the null hypothesis.

(d) The nonpooled t-test is more appropriate. One sample standard deviation is more than two times as large as the other, making it unlikely that the two population standard deviations are equal. The fact that the two sample sizes are also quite different makes it essential that the pooled t-test not be used.

**10.99** (a) Using Minitab, we choose **Graph ▶ Probability Plot**, select **Single** and click **OK**. Enter REGULAR and STINGER in the **Graph Variables** text box. Click on the **Multiple Graphs** button and select **On separate graphs**. Click **OK** twice. Note that the standard deviations are shown on the plots. Then choose **Graph ▶ Boxplot**, click on the **Simple** plot in the **Multiple Y's** row and click **OK**...Enter REGULAR and STINGER in the **Graph Variables** text box and click **OK**. The results are

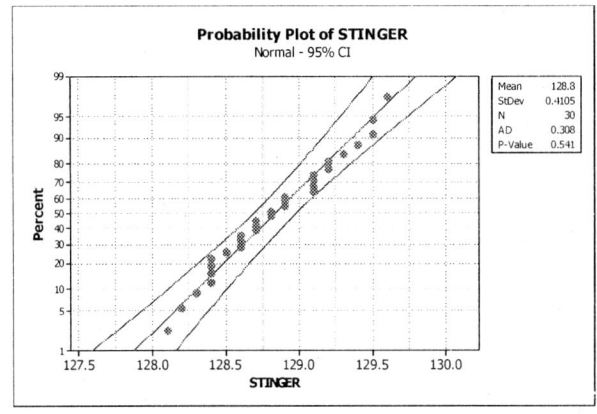

Standard deviations are 0.8944 for REGULAR and 0.4105 for STINGER.

(b) Nonpooled t-procedure. The samples are both large enough (30), normality is reasonable, there are no potential outliers, but one standard deviation is more than twice the other, making it unlikely that the population standard deviations are equal.

(c) Choose **Stat ▶ Basic Statistics ▶ 2-Sample t**, click on **Samples in different columns** and enter REGULAR in the **First** text box and STINGER in the **Second** text box. Check the box for **Assume equal variances** and click on the **Options** button. Enter 95 in the **Confidence level** textbox, 0 in the **Test difference** text box and select **Less than** in the **Alternative** drop down box. Click **OK** twice. Then repeat this procedure with the **Assume equal variances** box unchecked. The results are

Two-sample T for REGULAR vs STINGER

```
             N      Mean   StDev  SE Mean
REGULAR     30    127.007  0.894   0.16
STINGER     30    128.833  0.410   0.075

Difference = mu (REGULAR) - mu (STINGER)
Estimate for difference:  -1.82667
95% upper bound for difference:  -1.52634
T-Test of difference = 0 (vs <): T-Value = -10.17  P-Value = 0.000  DF = 58
Both use Pooled StDev = 0.6959

Two-sample T for REGULAR vs STINGER

             N      Mean   StDev  SE Mean
REGULAR     30    127.007  0.894   0.16
STINGER     30    128.833  0.410   0.075

Difference = mu (REGULAR) - mu (STINGER)
Estimate for difference:  -1.82667
95% upper bound for difference:  -1.52413
T-Test of difference = 0 (vs <): T-Value = -10.17  P-Value = 0.000  DF = 40
```
Both procedures result in a P-value of 0.000, which is less than the significance level 0.05. The data do provide sufficient evidence that the ball velocity is less with the regular tee than with the Stinger tee.

(d) To get the confidence intervals, repeat the process in part (b), entering 90 in the **Confidence level** text box and selecting **not equal** in the **Alternative** drop down box. The confidence interval portions of the results are

```
Pooled      90% CI for difference:  (-2.12700, -1.52634)
Nonpooled   90% CI for difference:  (-2.12921, -1.52413)
```

The results of the two procedures are almost identical.

**10.101** (a) Use the pooled t-test because the distributions are normal and the population standard deviations are the same.
(b) Use the non-pooled t-test because the distributions are normal and the population standard deviations are not the same.
(c) Use the Mann-Whitney test because the distributions are not normal but they have the same shape.
(d) Use none of the above tests because the distributions are both not normal and they do not have the same shape.

**10.103** (a) No. The highest value (24.6) and the lowest value (9.3) are both in the "South" data.
(b) Not applicable.
(c) Yes. The highest value (87.8) is in the non-migratory group and he lowest value (79.7) is in the migratory group.
(d) There are 7 values in the high (non-migratory) group higher than the highest value (84.5) in the low group. There are 6 values in the low (migratory) group lower than the lowest value (82.1) in the high group. There is one tie. Thus c = 7 + 6 + 1/2 = 13.5. Since c $\geq$ 10, reject the null hypothesis at the 1% significance level. This is the same result as obtained by the nonpooled and pooled t-tests.

**10.105** (a) Population 1: Dynamic, $n_1 = 14$, $\bar{x}_1 = 7.36$, $s_1 = 1.22$

Population 2: Static, $n_2 = 6$, $\bar{x}_2 = 10.50$, $s_2 = 4.59$
From Exercise 10.83, df = 5 and $\alpha = 0.05$

$$t_\alpha = t_{0.05} = 2.015;\ (7.36-10.50)+2.015\sqrt{1.22^2/14+4.59^2/6} = 0.693$$

The 95% upper confidence bound for $\mu_1 - \mu_2$ is greater than zero.

Therefore, the left-tailed hypothesis test for $H_0$: $\mu_1 = \mu_2$ vs.

$H_a$: $\mu_1 < \mu_2$ will not be rejected at 5% significance. This compares to the result of the hypothesis test in Exercise 10.83.

(b) Population 1: Intervention, $n_1$ = 10, $\bar{x}_1$ = 67.9, $s_1$ = 5.49
Population 2: Control, $n_2$ = 31, $\bar{x}_2$ = 66.81, $s_2$ = 9.04
From Exercise 10.84, df = 25 and $\alpha$ = 0.05

$$t_\alpha = 1.708;\ (67.90-66.81)+1.708\sqrt{5.49^2/10+9.04^2/31} = 5.150$$

The 95% upper confidence bound for $\mu_1 - \mu_2$ is greater than zero.

Therefore, the left-tailed hypothesis test for $H_0$: $\mu_1 = \mu_2$ vs.

$H_a$: $\mu_1 < \mu_2$ will not be rejected at 5% significance. This compares to the result of the hypothesis test in Exercise 10.84.

## Exercises 10.4

**10.107** Paired sampling helps to remove extraneous sources of variation, resulting in a smaller sampling error for the estimated difference between the sample means. This, in turn, makes it more likely that differences between the means will be detected when those differences actually exist.

**10.109** The samples must be paired (this is essential), and the population of all paired differences must be normally distributed or the sample must be large (the procedure works reasonably well even for small or moderate size samples if the paired-difference variable is not normally distributed provided that the deviation from normality is small).

**10.111** (a) TV viewing time     (b) Married men and married women
(c) Married couples
(d) Paired difference variable = TV viewing time of Man - TV viewing time of Woman
(e) $H_0$: $\mu_M = \mu_W$ ; $H_a$: $\mu_M \neq \mu_W$     (f) Two-tailed

**10.113** (a) VAS sensory ratings (b) People under and not under hypnosis
(c) pair is a person (1) under and (2) not under hypnosis (same person)
(d) Paired difference variable = VAS sensory rating under hypnosis - VAS sensory rating not under hypnosis.
(e) $H_0$: $\mu_1 = \mu_2$ ; $H_a$: $\mu_1 < \mu_2$     (f) Left-tailed

**10.115** (a) total antioxidant capacity     (b) fresh and stored breastmilk
(c) pair is a woman (1) stored breastmilk and (2) fresh breastmilk
(d) Paired difference variable = antioxidant capacity of stored breastmilk - antioxidant capacity of fresh breastmilk
(e) $H_0$: $\mu_1 = \mu_2$ ; $H_a$: $\mu_1 < \mu_2$     (f) Left-tailed

**10.117** (a) Step 1: $H_0$: $\mu_1 = \mu_2$, $H_a$: $\mu_1 \neq \mu_2$
Step 2: $\alpha$ = 0.10
Step 3: The paired differences, d = $x_1 - x_2$, are

| 2 | 1 | 3 | 6 | 4 | -1 | 3 |
|---|---|---|---|---|----|---|

$$\bar{d} = \frac{\sum d_i}{n} = \frac{18}{7} = 2.571 \qquad s_d = \sqrt{\frac{\sum d^2 - (\sum d)^2/n}{n-1}} = \sqrt{\frac{76-324/7}{6}} = 2.225$$

$$t = \frac{\bar{d}}{s_d/\sqrt{n}} = \frac{2.571}{2.225/\sqrt{7}} = 3.057$$

**330** Chapter 10

Step 4: df = 6; critical values = ±1.943
0.02 < P-value < 0.05
Step 5: Since 3.057 > 1.943 and the P-value is less than 10%, reject $H_0$.

**10.119**(a) Step 1: $H_0$: $\mu_1 = \mu_2$, $H_a$: $\mu_1 > \mu_2$
Step 2: $\alpha = 0.10$
Step 3: The paired differences, $d = x_1 - x_2$, are

| 4 | -1 | 1 | 5 | 3 | 0 | -5 | -6 |

$$\bar{d} = \frac{\sum d_i}{n} = \frac{1}{8} = 0.125 \quad s_d = \sqrt{\frac{\sum d^2 - (\sum d)^2 / n}{n-1}} = \sqrt{\frac{113 - 1/8}{7}} = 4.016$$

$$t = \frac{\bar{d}}{s_d / \sqrt{n}} = \frac{0.125}{4.016 / \sqrt{8}} = 0.088$$

Step 4: df = 7; critical values = 1.415
P-value > 0.10
Step 5: Since 0.088 < 1.415 and the P-value is greater than 10%, do not reject $H_0$.

**10.121**(a) Step 1: $H_0$: $\mu_1 = \mu_2$, $H_a$: $\mu_1 < \mu_2$
Step 2: $\alpha = 0.10$
Step 3: The paired differences, $d = x_1 - x_2$, are

| -3 | -3 | -2 | 3 | -6 | 0 | -2 | -7 | -1 |

$$\bar{d} = \frac{\sum d_i}{n} = \frac{-21}{9} = -2.333 \quad s_d = \sqrt{\frac{\sum d^2 - (\sum d)^2 / n}{n-1}} = \sqrt{\frac{121 - 441/9}{8}} = 3.000$$

$$t = \frac{\bar{d}}{s_d / \sqrt{n}} = \frac{-2.333}{3.000 / \sqrt{9}} = -2.333$$

Step 4: df = 8; critical values = -1.397
0.01 < P-value < 0.025
Step 5: Since -2.333 < -1.397 and the P-value is less than 10%, reject $H_0$.

**10.123**(a) The variable is the height of the plant.
(b) The two populations are the cross-fertilized plants and the self-fertilized plants.
(c) The paired-difference variable is the difference in heights of the cross-fertilized plants and the self-fertilized plants grown in the same pot.
(d) Yes. They represent the difference in heights of two plants grown under the same conditions (same pot), one from each category of plants.
(e) Step 1: $H_0$: $\mu_1 = \mu_2$, $H_a$: $\mu_1 \neq \mu_2$
Step 2: $\alpha = 0.05$
Step 3: The paired differences are given.

$$t = \frac{\bar{d}}{s_d / \sqrt{n}} = \frac{20.93}{37.74 / \sqrt{15}} = 2.148$$

Step 4: df = 14; critical values = ±2.145
0.02 < P-Value < 0.05

Step 5: Since 2.148 > 2.145, reject $H_0$. The data do provide sufficient evidence at the 5% significance level that

Copyright © 2016 Pearson Education, Inc.

there is a difference between the mean heights of cross-fertilized and self-fertilized plants.

(f) For the 0.01 significance level, the critical values are ±2.977 and the P-value is the same as in part (e). Since t = 2.148 is not in the rejection region, do not reject the null hypothesis at the 0.01 significance level.

**10.125** Population 1: Weights after treatment for anorexia nervosa
Population 2: Weights before treatment for anorexia nervosa

Step 1: $H_0: \mu_1 = \mu_2$, $H_a: \mu_1 > \mu_2$
Step 2: $\alpha = 0.05$
Step 3: The paired differences 'Weight after - Weight before' are
11.0, 5.5, 9.4, 13.6, -2.9, 10.7, -0.1, 7.4, 21.5, -5.3, -3.8, 11.4, 13.4, 13.1, 9.0, 3.9, 5.7

For these differences, $\sum d = 123.5$ and $\sum d^2 = 1716.9$

$$\bar{d} = \sum d / n = 123.5 / 17 = 7.26$$

$$s_d = \sqrt{\frac{\sum d^2 - (\sum d)^2 / n}{n-1}} = \sqrt{\frac{1716.9 - 123.5^2 / 17}{17-1}} = 7.16$$

$$t = \frac{\bar{d}}{s_d / \sqrt{n}} = \frac{7.26}{7.16 / \sqrt{17}} = 4.181$$

Step 4: df = 16; critical values = 1.746
P-Value < 0.005
Step 5: Since 4.181 > 1.746, reject $H_0$. The data do provide sufficient evidence at the 5% significance level that family therapy is effective in helping anorexic young women gain weight.
For the P-value approach, since the P-value is smaller than the significance level, reject $H_0$.

**10.127** Population 1: Corneal thickness in normal eyes
Population 2: Corneal thickness in glaucoma eyes

Step 1: $H_0: \mu_1 = \mu_2$, $H_a: \mu_1 > \mu_2$
Step 2: $\alpha = 0.10$
Step 3: The paired differences, $d = x_1 - x_2$, are
-4, 0 12, 18, -4, -12, 6, 16

For these differences, $\sum d = 32$ and $\sum d^2 = 936$

$$\bar{d} = \sum d / n = 32 / 8 = 4.0$$

$$s_d = \sqrt{\frac{\sum d^2 - (\sum d)^2 / n}{n-1}} = \sqrt{\frac{936 - 32^2 / 8}{8-1}} = 10.7438$$

$$t = \frac{\bar{d}}{s_d / \sqrt{n}} = \frac{4.0}{10.7438 / \sqrt{8}} = 1.053$$

**332** Chapter 10

Step 4: df = 7, Critical value = 1.415
P-Value > 0.10

Step 5: Since 1.053 < 1415, do not reject $H_0$.
For the P-value approach, since the P-value is larger than the significance level, do not reject $H_0$.

Step 6: The data do not provide sufficient evidence at the 10% significance level that the mean corneal thickness is greater in normal eyes than in eyes with glaucoma.

**10.129** From Exercise 10.123, df = 14.

(a) $$20.93 \pm 2.145 \cdot \frac{37.74}{\sqrt{15}} = 20.93 \pm 20.90 = 0.03 \text{ to } 41.83 \text{ eighths of an inch}$$

We can be 95% confident that the difference, $\mu_1 - \mu_2$, between the mean heights of cross-fertilized and self-fertilized Zea mays is somewhere between 0.03 and 41.83 eighths of an inch.

(b) $$20.93 \pm 2.977 \cdot \frac{37.74}{\sqrt{15}} = 20.93 \pm 29.01 = -8.08 \text{ to } 49.94 \text{ eighths of an inch}$$

We can be 99% confident that the difference, $\mu_1 - \mu_2$, between the mean heights of cross-fertilized and self-fertilized Zea mays is somewhere between -8.08 and 49.94 eighths of an inch.

**10.131** From Exercise 10.125, df = 16.

$$7.26 \pm 1.746 \cdot \frac{7.16}{\sqrt{17}} = 7.26 \pm 3.03 = 4.23 \text{ to } 10.29 \text{ pounds}$$

We can be 90% confident that the mean weight gain, $\mu_1 - \mu_2$, resulting from family-therapy treatment by anorexic young women is somewhere between 4.23 and 10.29 pounds.

**10.133** From Exercise 10.127, df = 7.

$$4.0 \pm 1.415 \cdot \frac{10.7434}{\sqrt{8}} = 4.0 \pm 5.37 = (-1.37, 9.37) \text{ microns}$$

We can be 80% confident that the difference, $\mu_1 - \mu_2$, in corneal thickness of patients who had glaucoma in one eye but not the other is between -1.37 and 9.37 microns.

**10.135** Using Minitab, with the differences in a column named DIFF, choose **Graph ▶ Probability plot...**, select the **Single** version and click **OK**. Enter DIFF in the **Graph variables text box** and click **OK**. Then choose **Graph ▶ Boxplot**, select the **One Y Simple** version and click **OK**. Enter DIFF in **Graph variables** text box and click **OK**.

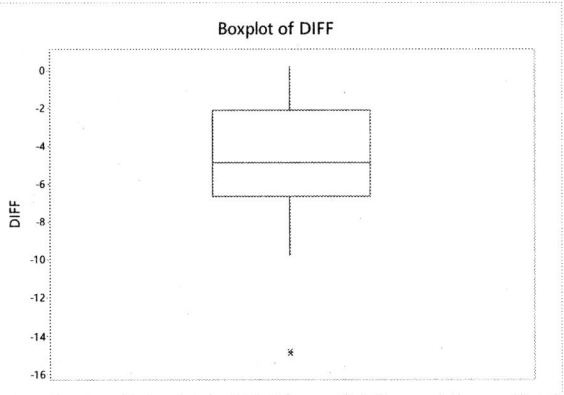

The sample size is moderate (25), and there is an outliers in the differences. Since the t-test is based on an assumption of normality, the t-test is not appropriate for this data. However, since the sample size is close to 30, perhaps an analysis of the differences with and without the outlier can be explored to see the effect of the outlier on the conclusion.

10.137  Using Minitab, enter DIFF at the top of column C3, then choose **Calc ▶ Calculator,** enter DIFF in the **Store result in variable** box, enter 'RESOLUTION'-'ONSET' in the Expression box, and click **OK.** Choose **Graph ▶ Probability plot...,** select the **Single** version and click **OK.** Enter ONSET RESOLUTION DIFF in the **Graph variables** text box and click **OK.** Then choose **Graph ▶ Boxplot,** select the **Multiple Y's Simple** version and click **OK.** Enter ONSET and RESOLUTION in the **Graph Variables** box and click **OK.** Finally, choose **Graph ▶ Boxplot,** select the **One Y Simple** version and click **OK.** Enter DIFF in **Graph variables** text box and click **OK.** The results are

**334** Chapter 10

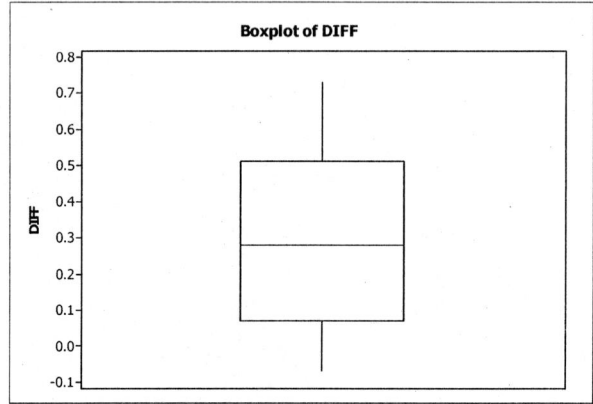

(b) No. The sample size is small, there is an outlier, and the normal probability plot is not linear.
(c) No. The sample size is small and there is an outlier. The probability plot is closer to linear than for the ONSET data, but overall, t-procedures are not advisable.
(d) Yes. There are no outliers and the probability plot is reasonably linear.
(e) They imply that the assumptions for using a paired t-procedure may be met even though the assumptions for using a t-procedure may not be met for the original variables.

10.139 (a) Using Minitab, choose **Stat ▸ Basic Statistics ▸ Paired t...**, enter MEN in the **First sample** box and WOMEN in the **Second Sample** box. Click on the **Options** button and enter 99 in the **Confidence** level box, 0.0 in the **Test mean** box, and **greater than** in the **Alternative** drop down box. Click **OK** twice. The results are

Paired T for MEN - WOMEN

```
                 N      Mean    StDev   SE Mean
MEN             75    35.4800   9.7695   1.1281
WOMEN           75    32.0800   7.0265   0.8114
Difference      75    3.40000   7.23542  0.83547
```

99% lower bound for mean difference: 1.41341
T-Test of mean difference = 0 (vs > 0): T-Value = 4.07   P-Value = 0.000

The P-value is 0.000, which is less than the significance level 0.01. Therefore, the data do provide sufficient evidence that the mean age of Norwegian men at the time of marriage exceeds that of Norwegian women.

(b) The 99% confidence interval is found by repeating the procedure of part (a) using **not equal** in the **Alternative** droop down box.  The result is
99% CI for mean difference: (1.19108, 5.60892)
We can be 99% confident that the mean difference in ages of Norwegian men and women at the time of marriage lies somewhere between 1.19 and 5.61 years.

(c) We remove the pairs (18,32) and (53,29) from the data and repeat parts (a) and (b).  The results are
Paired T for MEN - WOMEN

|  | N | Mean | StDev | SE Mean |
|---|---|---|---|---|
| MEN | 73 | 35.4795 | 9.4650 | 1.1078 |
| WOMEN | 73 | 32.1233 | 7.1140 | 0.8326 |
| Difference | 73 | 3.35616 | 6.61095 | 0.77375 |

99% lower bound for mean difference: 1.51520
T-Test of mean difference = 0 (vs > 0): T-Value = 4.34   P-Value = 0.000

99% CI for mean difference: (1.30893, 5.40340)
By removing the outliers, the t-value increases from 4.07 to 4.34, the P-value remains at 0.000, and the conclusion remains the same.  The 99% confidence interval is shortened by about 0.3 years and the mean difference changes from 3.40 to 3.36.

**10.141** Data is obtained from two populations whose members can be naturally paired.  By letting d represent the difference between the values in each pair, we can reduce the data set from pairs of numbers to a single number representing each pair.  Since the mean of the paired differences is the same as the difference of the two population means, we can test the equality of the two population means by testing to see if the mean of the paired differences is zero (or some other value if appropriate).  If the paired differences are normally distributed, then the one-sample t-test can be used to carry out the test.

**10.143** Answers will vary.

**10.145** (a) The paired *t*-test
(b) Neither test
(c) The paired Wilcoxn signed-rank test


**Review Problems for Chapter 10**

1. Randomly sample independently from both populations, compute the means of both samples and reject the null hypothesis if the sample means differ by too much.  Otherwise, do not reject the null hypothesis.
2. Sample independent pairs of observations from the two populations, compute the difference of the two observations in each pair, compute the mean of the differences, and reject the null hypothesis if the sample mean of the differences differs from zero by too much.  Otherwise, do not reject the null hypothesis.
3. (a) The pooled t-test requires that the population standard deviations be equal whereas the nonpooled t-test does not.
   (b) It is absolutely essential that the assumption of independence be satisfied.
   (c) The normality assumption is especially important for both t-tests for small samples.  With large samples, the Central Limit Theorem applies and the normality assumption is less important.
4. Unless you are quite sure that the population standard deviations are equal, the nonpooled *t*-procedures should be used instead of the pooled *t*-procedures.

**336** Chapter 10

**5.** A paired sample may reduce the estimate of the standard error of the mean of the differences, making it more likely that a difference of a given size will be judged significant.

**6.** (a) Population 1: Male; $n_1 = 13$, $\bar{x}_1 = 2127$, $s_1 = 513$
Population 2: Female; $n_2 = 14$, $\bar{x}_2 = 1843$, $s_2 = 446$
Step 1: $H_0$: $\mu_1 = \mu_2$, $H_a$: $\mu_1 > \mu_2$
Step 2: $\alpha = 0.05$,
Step 3:

$$s_p = \sqrt{\frac{12(513)^2 + 13(446)^2}{25}} = 479.33$$

$$t = \frac{2127 - 1843}{479.33\sqrt{\frac{1}{13} + \frac{1}{14}}} = 1.538$$

Step 4: df = 25, Critical value = 1.708
$0.05 < P(t > 1.538) < 0.10$
Step 5: Since 1.538 < 1.708 and since the P-value is larger than the significance level, do not reject $H_0$.
Step 6: At the 5% significance level, the data do not provide enough evidence to conclude that the mean right-leg strength of males exceeds that of females.

**7.**

$$2127 - 1843 \pm 1.708(479.33)\sqrt{\frac{1}{13} + \frac{1}{14}}$$

$$284 \pm 315.33$$

$$-31.33 \text{ to } 599.33$$

We can be 90% confident that the difference, $\mu_1 - \mu_2$, between the mean right-leg strengths of males and females is between -31.3 and 599.3 newtons.

**8.** Population 1: Florida, $n_1 = 24$, $\bar{x}_1 = 5.46$, $s_1 = 1.59$
Population 2: Virginia, $n_2 = 44$, $\bar{x}_2 = 7.59$, $s_2 = 2.68$

$$\Delta = \frac{\left[\left(\frac{s_1^2}{n_1}\right) + \left(\frac{s_2^2}{n_2}\right)\right]^2}{\frac{\left(\frac{s_1^2}{n_1}\right)^2}{n_1 - 1} + \frac{\left(\frac{s_2^2}{n_2}\right)^2}{n_2 - 1}} = \frac{\left[\left(\frac{5.46^2}{24}\right) + \left(\frac{7.59^2}{44}\right)\right]^2}{\frac{\left(\frac{5.46^2}{24}\right)^2}{24 - 1} + \frac{\left(\frac{7.59^2}{44}\right)^2}{44 - 1}} = 65.45; \text{ df} = 65$$

Step 1: $H_0$: $\mu_1 = \mu_2$, $H_a$: $\mu_1 < \mu_2$
Step 2: $\alpha = 0.01$
Step 3:

$$t = \frac{5.4583 - 7.5909}{\sqrt{\frac{1.5874^2}{24} + \frac{2.6791^2}{44}}} = -4.119$$

Step 4: Critical value = -2.385
P(t < -4.119) < 0.005
Step 5: Since -4.119 < -2.385 and also the P-value is smaller than the significance level, reject $H_0$.
Step 6: At the 1% significance level, the data provide sufficient evidence to conclude that the average litter size of cottonmouths in Florida is less than that in Virginia.

9.
$$(5.46 - 7.49) \pm 2.385 \sqrt{\frac{1.59^2}{24} + \frac{2.68^2}{44}}$$

$$-2.13 \pm 1.24$$

$$-3.37 \text{ to } -0.89$$

We can be 98% confident that the average difference, $\mu_1 - \mu_2$, between cottonmouth litter sizes in Florida and Virginia is somewhere between -3.37 and -0.89.

10. (a) We used Minitab to produce the following normal probability plot and boxplot of the differences.

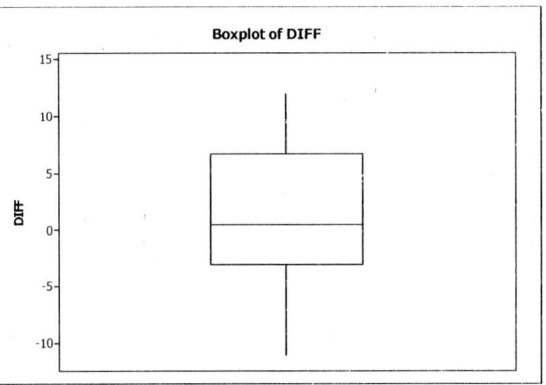

(b) Yes. The normal probability plot is reasonably linear and the boxplot shows no outliers.

(c) Step 1: $H_0: \mu_1 = \mu_2$, $H_a: \mu_1 \neq \mu_2$
Step 2: $\alpha = 0.10$
Step 3: Paired differences, d = $x_1 - x_2$:

| 12 | 7 | 1 | 0 |
|---|---|---|---|
| -11 | 0 | 6 | -4 |

From the probability plot, we see that the sample mean of the differences is 1.375 and the standard deviation is 7.090. Therefore,

$$t = \frac{1.375}{7.090/\sqrt{8}} = 0.549$$

Step 4: Critical values = ±1.895
2{P(t > 0.549)} > 0.20
Step 5: Since -1.895 < 0.549 < 1.895 and also since the P-value is greater than the significance level, do not reject $H_0$.
Step 6: At the 10% significance level, the data do not provide sufficient evidence to conclude that there is a difference

in the mean length of time that ice stays on the two lakes.

**11.**

$$1.375 \pm 1.860 \cdot 7.090/\sqrt{8} = 1.375 \pm 4.749 = -3.374 \text{ to } 6.124 \text{ days}$$

We can be 90% confident that the difference, $\mu_1 - \mu_2$, between the mean numbers of days ice stays on Lakes Mendota and Monona is somewhere between -3.374 and 6.124 days.

**12.** The paired $t$-test
**13.** The non-pooled $t$-test
**14.** None of the procedures
**15.** The non-pooled $t$-test
**16.** None of the procedures
**17.** None of the procedures
**18.** (a) Using Minitab, choose **Graph ▶ Probability plot...**, select the **Single** version and click **OK**. Enter MALE FEMALE in the **Graph Variables** text box, and click **OK**. Then choose **Graph ▶ Boxplot**, select the **Multiple Y's Simple** version and click **OK**. Enter MALE FEMALE **in the Graph variables** text box, and click **OK**. The standard deviations will be produced on the probability plot legend. The results are

 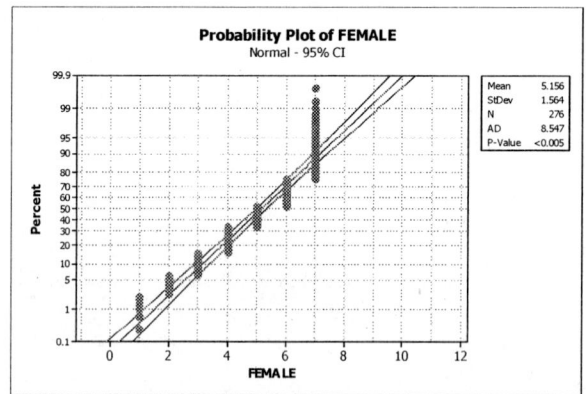

The standard deviations are 1.513 for MALE and 1.564 for FEMALE.

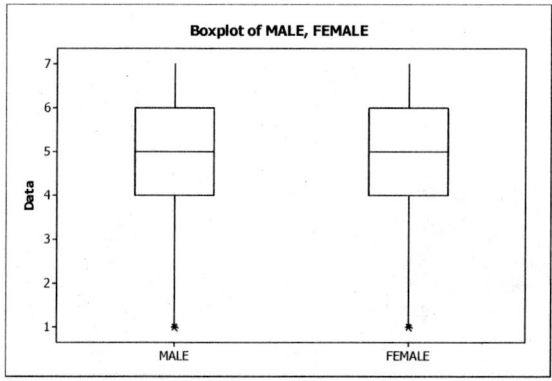

(b) Pooled t-procedures. The sample sizes are both large, there are no potential outliers in either sample, and the sample standard deviations are nearly equal.

(c) To carry out the pooled t-test, choose **Stat ▶ Basic Statistics ▶ 2-Sample t**, click on **Samples in different columns**, enter MALE in the **First** text box and FEMALE in the **Second** text box, select **not equal** in the **Alternative** box, enter 90 in the **Confidence level** text box, click on **Assume equal variances** to make certain that there is an **X** in the box, and click **OK**. The results are

```
Two-sample T for MALE vs FEMALE

          N    Mean   StDev   SE Mean
MALE     77    5.00   1.51    0.17
FEMALE  276    5.16   1.56    0.094

Difference = mu (MALE) - mu (FEMALE)
Estimate for difference:  -0.155797
95% CI for difference:  (-0.549387, 0.237793)
T-Test of difference = 0 (vs not =): T-Value = -0.78  P-Value = 0.437  DF = 351
Both use Pooled StDev = 1.5528
```

The P-value is 0.437, which is greater than the significance level 0.05. Do not reject the null hypothesis. The data do not provide sufficient evidence to conclude that there is a difference in mean dating satisfaction of male and female college students.

(d) A 95% confidence interval for the difference between mean dating satisfaction of male and female college students was produced as part of the process in part (b). The interval is (-0.55, 0.24).

(e) Yes. Although the data for both genders are left-skewed and discrete (possible values include only the whole numbers from 1 to 7), the large sample sizes justify the use of the pooled t-procedures.

19. (a) Using Minitab, choose **Graph ▶ Probability plot...**, select the **Single** version and click **OK**. Enter OVER 65 and CONTROL in the **Graph Variables** text box, and click **OK**. Then choose **Graph ▶ Boxplot**, select the **Multiple Y's Simple** version and click **OK**. Enter OVER 65 and CONTROL in **the Graph variables** text box, and click **OK**. The standard deviations will be produced on the probability plot legend. The results are

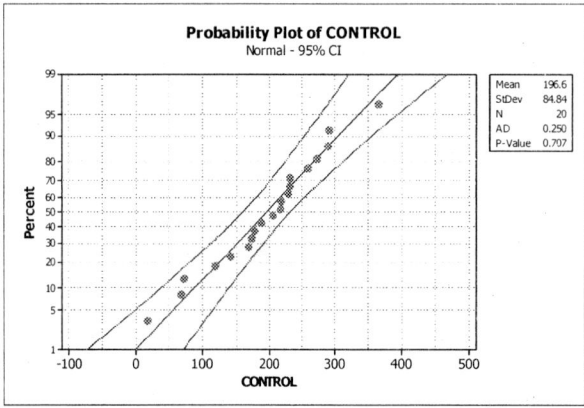

The standard deviations are
36.63 for OVER 65 and 84.84
for CONTROL,

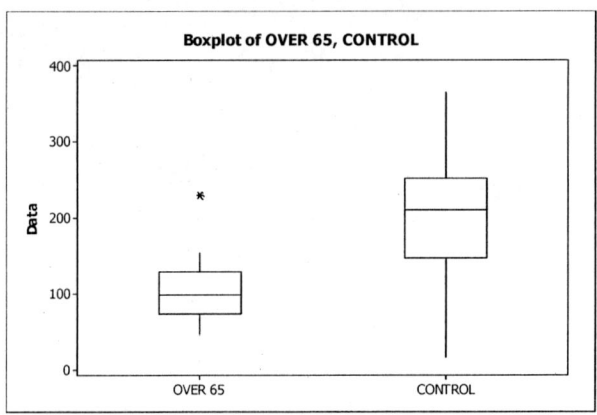

(b) The nonpooled t-procedure is preferable since the ratio of the standard deviations is greater than 2.

(c) To carry out the nonpooled t-test, choose **Stat ▶ Basic Statistics ▶ 2-Sample t**, click on **Samples in different columns**, enter 'OVER 65' in the **First** text box and CONTROL in the **Second** text box, and leave the **Assume equal variances** box unchecked. Click on the **Options** button, enter 99 in the **Confidence level** text box, select **less than** in the **Alternative** box, and click **OK** twice. The results are

```
Two-sample T for OVER 65 vs CONTROL

          N    Mean   StDev   SE Mean
OVER 65  41   102.1    36.6      5.7
CONTROL  20   196.6    84.8       19

Difference = mu (OVER 65) - mu (CONTROL)
Estimate for difference:  -94.4768
99% upper bound for difference:  -44.7729
T-Test of difference = 0 (vs <): T-Value = -4.77   P-Value = 0.000   DF = 22
```

Since the P-value = 0.000, which is less than the significance level 0.01, reject the mull hypothesis. The data do provide sufficient evidence to conclude that, on average, men over 65 have a lower IGF-1 level than younger men.

(d) To obtain the 99% confidence interval, follow the procedure in part (c), but select not equal in the Alternative drop down box. The confidence interval is given by
   99% CI for difference:  (-150.3322, -38.6215)

We can be 99% confident that the difference in IGF-1 levels, on average, between men over 65 and younger men is somewhere between -150.3 and -38.6.

(e) Yes. The CONTROL sample of size 20 is reasonably normally distributed with no outliers. The OVER 65 sample of size 41 is also close to normal with one outlier. The effect of the outlier is not great enough to alter the conclusion (Deleting it changes the value of t from -4.77 to -4.99).

20. (a) Using Minitab, choose **Stat ▶ Basic Statistics ▶ Paired t**, click on **Samples in columns**, enter MEN in the **First** text box and WOMEN in the **Second** text box. Click on the **Options** button, enter 95 in the

**Confidence level** text box, select **greater than** in the **Alternative** box, and click **OK**. Click on the **Graphs** button, check the box for **Boxplot of differences**, and click **OK** twice. The results are

Paired T for MEN - WOMEN

```
              N    Mean   StDev   SE Mean
MEN          50    1030    820      116
WOMEN        50     899    818      116
Difference   50   131.1   93.8     13.3
```

95% lower bound for mean difference: 108.9
T-Test of mean difference = 0 (vs > 0): T-Value = 9.89  P-Value = 0.000

Since the P-value = 0.000, which is less than the significance level 0.05, reject the null hypothesis. The data do provide sufficient evidence to conclude that, on average, the weekly earnings of male full-time and salary workers exceed that of women.

(b) Follow the procedure in part (a), but enter 90 in the **Confidence level** text box and select **not equal** from the **Alternative** drop down box. The resulting interval is

90% CI for mean difference: (108.9, 153.3)

We can be 90% confident that the mean difference in weekly earnings of male and female full-time wage and salary workers is somewhere between $108.9 and $153.3.

(c) To create a column of differences, click in the **Sessions Window** and then click on the **Editor** tab and on **Enable Commands**. An MTB> prompt will appear at the bottom of the Sessions Window. After the prompt, enter Let c3=c1-c2 and press ENTER. Click at the top of Column 3 and name the column DIFF. Choose **Graph ▶ Probability plot...**, select the **Single** version and click **OK**. Enter DIFF in the **Graph Variables** text box, and click **OK**. The standard deviations will be produced on the probability plot legend. A boxplot was produced by the procedure in part (a). Now choose **Graph ▶ Stem-and-Leaf...**, enter DIFF in the **Graph Variables** text box, and click **OK**. The results are

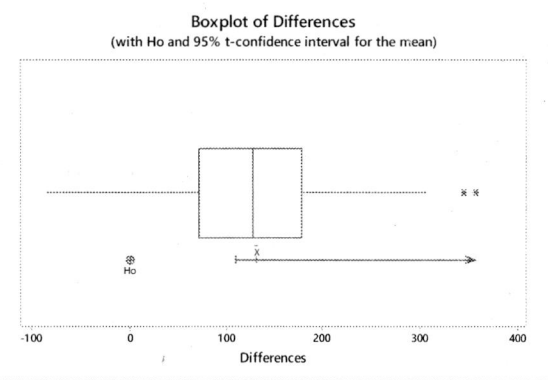

```
Stem-and-leaf of DIFF   N  = 50
Leaf Unit = 10

    2   -0  85
    3   -0  3
    9    0  222333
   19    0  5577788899
   (8)   1  00011234
   23    1  555666666778
   11    2  000114
    5    2  67
    3    3  04
```

(d)  Yes.  The sample size is large (50), there are only two mild outliers on the right, and the normal probability plot is close to linear.  The outliers may cause some concern since they are in direction that helps to reject the null hypothesis.  If so, a nonparametric method may be used.

## CHAPTER 11 SOLUTIONS

### Exercises 11.1

**11.1** Answers will vary.

**11.3** A population proportion $p$ is a parameter since it is a descriptive measure for a population. A sample proportion $\hat{p}$ is a statistic since it is a descriptive measure for a sample.

**11.5** (a) A sample proportion is the proportion (percentage) of a sample from the population that has the specified attribute.

(b) $\hat{p}$

**11.7** The "number of failures" is an abbreviation for the number of members of the sample that do not have the specified attribute, and it is denoted by n - x, where n is the size of the sample.

**11.9** (a) p = 2/5 = 0.4

(b)

| Sample | Number of females x | Sample proportion $\hat{p}$ |
|---|---|---|
| J,G | 1 | 0.5 |
| J,P | 0 | 0.0 |
| J,C | 0 | 0.0 |
| J,F | 1 | 0.5 |
| G,P | 1 | 0.5 |
| G,C | 1 | 0.5 |
| G,F | 2 | 1.0 |
| P,C | 0 | 0.0 |
| P,F | 1 | 0.5 |
| C,F | 1 | 0.5 |

(c) The population proportion is marked by the vertical line below.

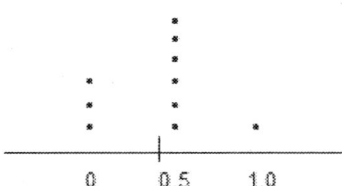

(d) $\mu_{\hat{p}} = \sum \hat{p}/10 = 4.0/10 = 0.4$

(e) The answers to (a) and (d) are the same. $\hat{p}$ is a sample mean. The mean of the sampling distribution of $\hat{p}$ is the same as the mean of the population, which is p.

**344** Chapter 11

**11.11** (b)

| Sample | Number of females x | Sample proportion $\hat{p}$ |
|---|---|---|
| J,P,C | 0 | 0 |
| J,P,G | 1 | 1/3 |
| J,P,F | 1 | 1/3 |
| J,C,G | 1 | 1/3 |
| J,C,F | 1 | 1/3 |
| J,G,F | 2 | 2/3 |
| P,C,G | 1 | 1/3 |
| P,C,F | 1 | 1/3 |
| P,G,F | 2 | 2/3 |
| C,G,F | 2 | 2/3 |
|  |  |  |

(c) The population proportion is marked by the vertical line below.

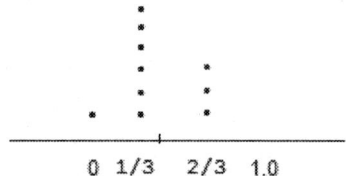

(d) $\mu_{\hat{p}} = \sum \hat{p}/10 = (12/3)/10 = 0.4$

(e) The answers to (a) and (d) are the same. $\hat{p}$ is a sample mean. The mean of the sampling distribution of $\hat{p}$ is the same as the mean of the population, which is p.

**11.13** (b)

| Sample | Number of females X | Sample proportion $\hat{p}$ |
|---|---|---|
| J,P,C,G,F | 2 | 0.4 |

(c) The population proportion is marked by the vertical line below.

(d) $\mu_{\hat{p}} = \sum \hat{p}/1 = 0.4/1 = 0.4$

(e) The answers to (a) and (d) are the same. $\hat{p}$ is a sample mean. The mean of the sampling distribution of $\hat{p}$ is the same as the mean of the population, which is p.

Copyright © 2016 Pearson Education, Inc.

**11.15** (a) The population consists of all #1 NBA draft picks since 1947.
   (b) The specified attribute is being other than a U.S. national.
   (c) The 10.4% is a population proportion since all of the #1 picks are known. There is no need to sample this group.

**11.17** (a) $\hat{p} = 0.79$; $z_{0.005} = 2.575$; the margin of error is

$$E = z_{0.005} \cdot \sqrt{\hat{p}(1-\hat{p})/n} = 2.575\sqrt{0.79(0.21)/21355} = 0.00718$$

   (b) The margin of error will be smaller for a 90% confidence interval. Specifically, 2.575 will be replaced by 1.645 in the formula for E and everything else stays the same. More generally speaking, in order to have a higher level of confidence in an interval, one needs to have a wider interval.

**11.19** (a) 0.4  (b) $0.4 < \hat{p} < 0.6$

**11.21** (a) 0.2  (b) $0.2 < \hat{p} < 0.8$

**11.23** (a) 0.5  (b) none

**11.25** (a) $\hat{p} = 8/40 = 0.2$
   (b) x = 8 and n - x = 32. Both are at least 5, so the one-proportion z-interval procedure is appropriate.
   (c) The 95% confidence interval is

$$\hat{p} \pm z_{\alpha/2}\sqrt{\hat{p}(1-\hat{p})/n} = 0.2 \pm 1.96\sqrt{0.2(1-0.2)/40} = (0.076, 0.324)$$

   (d) The margin of error is $E = z_{\alpha/2}\sqrt{\hat{p}(1-\hat{p})/n} = 1.96\sqrt{0.2(1-0.2)/40} = 0.1240$. The confidence interval in terms of point estimate ± margin of error is 0.2 ± 0.1240.

**11.27** (a) $\hat{p} = 35/50 = 0.70$
   (b) x = 35 and n - x = 15. Both are at least 5, so the one-proportion z-interval procedure is appropriate.
   (c) The 99% confidence interval is

$$\hat{p} \pm z_{\alpha/2}\sqrt{\hat{p}(1-\hat{p})/n} = 0.70 \pm 2.575\sqrt{0.70(1-0.70)/50} = (0.533, 0.867)$$

   (d) The margin of error is $E = z_{\alpha/2}\sqrt{\hat{p}(1-\hat{p})/n} = 2.575\sqrt{0.70(1-0.70)/50} = 0.1669$. The confidence interval in terms of point estimate ± margin of error is 0.70 ± 0.1669.

**11.29** (a) $\hat{p} = 16/20 = 0.8$
   (b) x = 16 and n - x = 4. n - x is less than 5, so the one-proportion z-interval procedure is not appropriate.

**11.31** $n = 0.25\left(\dfrac{z_{\alpha/2}}{E}\right)^2 = 0.25\left(\dfrac{1.96}{0.01}\right)^2 = 9604$

**11.33** $n = 0.25\left(\dfrac{z_{\alpha/2}}{E}\right)^2 = 0.25\left(\dfrac{1.645}{0.02}\right)^2 = 1691.27$  Round this up to n = 1692.

**11.35** $n = 0.25\left(\dfrac{z_{\alpha/2}}{E}\right)^2 = 0.25\left(\dfrac{2.575}{0.03}\right)^2 = 1841.84$  Round this up to n = 1842.

If you use $z_{\alpha/2} = 2.576$, you will get n = 1844.

**11.37** (a) $n = \hat{p}_g(1-\hat{p}_g)\left(\dfrac{z_{\alpha/2}}{E}\right)^2 = 0.3(1-0.3)\left(\dfrac{1.96}{0.01}\right)^2 = 8067.36$  Round this up to n = 8068.

   (b) This is smaller than the sample size in Exercise 11.31 because using an

**346** Chapter 11

educated guess will give us a more realistic sample, while using the constant 0.25 will give us a sample that will be generally larger than necessary.

**11.39** (a) $n = \hat{p}_g(1-\hat{p}_g)\left(\dfrac{z_{\alpha/2}}{E}\right)^2 = 0.1(1-0.1)\left(\dfrac{1.645}{0.02}\right)^2 = 608.86$ Round this up to n = 609.

(b) This is smaller than the sample size in Exercise 11.33 because using an educated guess will give us a more realistic sample, while using the constant 0.25 will give us a sample that will be generally larger than necessary.

**11.41** (a) $n = \hat{p}_g(1-\hat{p}_g)\left(\dfrac{z_{\alpha/2}}{E}\right)^2 = 0.5(1-0.5)\left(\dfrac{2.575}{0.03}\right)^2 = 1841.84$ Round this up to

n = 1842. If you use $z_{\alpha/2} = 2.576$, you will get n = 1844.

(b) This is the same as the sample size in Exercise 11.35 because an educated guess of 0.5 will give $\hat{p}_g(1-\hat{p}_g) = 0.5(1-0.5) = 0.25$ which is the same as using the constant 0.25.

**11.43** Use $\hat{p}_g = 0.4$ since it is the closest in the likely range to 0.5.

$$n = \hat{p}_g(1-\hat{p}_g)\left(\dfrac{z_{\alpha/2}}{E}\right)^2 = 0.4(1-0.4)\left(\dfrac{1.96}{0.01}\right)^2 = 9219.84$$

Round this up to n = 9220.

**11.45** Use $\hat{p}_g = 0.2$ since it is the closest in the likely range to 0.5.

$$n = \hat{p}_g(1-\hat{p}_g)\left(\dfrac{z_{\alpha/2}}{E}\right)^2 = 0.2(1-0.2)\left(\dfrac{1.645}{0.02}\right)^2 = 1082.41$$

Round this up to n = 1083.
Round this up to n = 5683.

**11.47** Use $\hat{p}_g = 0.5$ since it is the closest in the likely range to 0.5.

$$n = \hat{p}_g(1-\hat{p}_g)\left(\dfrac{z_{\alpha/2}}{E}\right)^2 = 0.5(1-0.5)\left(\dfrac{2.575}{0.03}\right)^2 = 1841.84$$

Round this up to n = 1842. If you use $z_{\alpha/2} = 2.576$, you will get n = 1844.

**11.49** $\hat{p} = 0.32$; The 95% confidence interval is

$$\hat{p} \pm z_{\alpha/2}\sqrt{\hat{p}(1-\hat{p})/n} = 0.32 \pm 1.96\sqrt{0.32(1-0.32)/1000} = (0.291, 0.349)$$

We are 95% confident that the proportion of all U.S. adults that never clothes-shop online is between 0.291 and 0.349.

**11.51** n = 500, x = 38, and n - x = 462. Both x and n - x are at least 5.

$\hat{p}$ = x/n = 38/500 = 0.076; $z_{\alpha/2}$ = $z_{0.025}$ = 1.96

$$0.076 - 1.96\sqrt{0.076(1-0.076)/500} \text{ to } 0.076 + 1.96\sqrt{0.076(1-0.076)/500}$$

$$0.0528 \text{ to } 0.0992$$

We can be 95% confident that the proportion of U.S. asthmatics who are allergic to sulfites is somewhere between 0.0528 and 0.0992.

**11.53** n = 1000, x = 800, and n - x = 200. Both x and n - x are at least 5.

$\hat{p}$ = x/n = 800/1000 = 0.80; $z_{\alpha/2}$ = $z_{0.005}$ = 2.575

$$0.80 - 2.575\sqrt{0.80(1-0.80)/1000} \text{ to } 0.80 + 2.575\sqrt{0.80(1-0.80)/1000}$$
$$0.767 \text{ to } 0.833$$

We can be 99% confident that the percentage of all registered voters who favor the creation of standards on CAFO pollution is somewhere between 76.7% and 83.3%.

**11.55** Here, n = 100, x = 96, and n - x = 4. To use Procedure 11.1, both x and n - x must be at least 5. Since n - x < 5, Procedure 11.1 should not have been used.

**11.57** $\hat{p}$ = 0.50, E = 0.03

$\hat{p}$ - E    to    $\hat{p}$ + E

0.50 - 0.03 to 0.50 + 0.03
0.47 to 0.53

We can be 95% confident that the percentage of all American adults who favor a plan to break up the 12 megabanks is between 47% and 53%.

**11.59** From Exercise 11.51, $\hat{p}$ = 0.076; $z_{\alpha/2}$ = 1.96; n = 500

(a) $E = z_{\alpha/2}\sqrt{\hat{p}(1-\hat{p})/n} = 1.96\sqrt{0.076(1-0.076)/500} = 0.0232$

(b) E = 0.01; $z_{\alpha/2}$ = $z_{0.025}$ = 1.96;

$$n = \hat{p}(1-\hat{p})\frac{z_{\alpha/2}^2}{E^2} = 0.5(1-0.5)\frac{1.96^2}{0.01^2} = 9604$$

(c) n = 9604; $\hat{p}$ = 0.071; $z_{\alpha/2}$ = $z_{0.025}$ = 1.96

$$0.071 - 1.96\sqrt{0.071(1-0.071)/9604} \text{ to } 0.3192 + 1.96\sqrt{0.071(1-0.071)/9604}$$
$$0.0659 \text{ to } 0.0761$$

(d) The margin of error for the estimate is 0.0051. As expected, this is less than the required margin of error of 0.01 specified in part (b).

(e) $\hat{p}_g$ = 0.10; $z_{\alpha/2}$ = $z_{0.025}$ = 1.96

part (b)    E = 0.01, $z_{0.025}$ = 1.96; sample size is:

$$n = \hat{p}(1-\hat{p})\frac{z_{\alpha/2}^2}{E^2} = 0.10(1-0.10)\frac{1.96^2}{0.01^2} = 3457.44 \to 3458$$

Thus the required sample size is n = 3458.
part (c)
$$0.071 - 1.96\sqrt{0.071(1-0.071)/3458} \text{ to } 0.3192 + 1.96\sqrt{0.071(1-0.071)/3458}$$
$$0.0624 \text{ to } 0.0796$$

part (d) The margin of error is 0.0086. As expected, this is less than what was specified in part (b).

(f) By employing the guess for $\hat{p}$ in part (d) we can reduce the required sample size by more than 6000 (from 9604 to 3458), saving considerable time and money. Moreover, the margin of error only rises from 0.0051 to 0.0086. The risk of using the guess 0.10 for $\hat{p}$ is that if the actual value of $\hat{p}$ turns out to be between 0.10 and 0.90, then the achieved margin of error will exceed the specified 0.01.

**11.61** (a) The confidence interval from Exercise 11.53 was (76.7%, 83.3%). To find the margin of error, take the width of the confidence interval and divide by 2. E = (83.3% − 76.7%)/2 = 3.3%.

(b) E = 0.015; $z_{\alpha/2} = z_{0.005} = 2.575$;

$$n = \hat{p}(1-\hat{p})\frac{z_{\alpha/2}^2}{E^2} = 0.5(1-0.5)\frac{2.575^2}{0.015^2} = 7367.36 \rightarrow 7368$$

(c) n = 7368; $\hat{p}$ = 0.822; $z_{\alpha/2} = z_{0.005} = 2.575$

$$0.822 - 2.575\sqrt{0.822(1-0.822)/7368} \text{ to } 0.822 + 2.575\sqrt{0.822(1-0.822)/7368}$$

$$0.811 \text{ to } 0.833$$

(d) The margin of error for the estimate is 0.011, which is less than the 0.015 required in part (b).

(e) $\hat{p}_g$ is between 0.75 and .85; $z_{\alpha/2} = z_{0.005} = 2.575$

part (b) E = 0.015, $z_{0.005} = 2.575$; sample size is:

$$n = \hat{p}(1-\hat{p})\frac{z_{\alpha/2}^2}{E^2} = 0.75(1-0.75)\frac{2.575^2}{0.015^2} = 5525.52 \rightarrow 5526$$

Thus the required sample size is n = 5526.
part (c)

$$0.822 - 2.575\sqrt{0.822(1-0.822)/5526} \text{ to } 0.822 + 2.575\sqrt{0.822(1-0.822)/5526}$$

$$0.809 \text{ to } 0.835$$

part (d)
The margin of error is 0.013, which is less than the 0.015 specified in part (b).

(f) By employing the guess for $\hat{p}$ in part (d) we can reduce the required sample size by 1842, from 7368 to 5526, saving considerable time and money. Moreover, the margin of error only rises from 0.011 to 0.013. The risk of using the guess 0.75 for $\hat{p}$ is that if the actual value of $\hat{p}$ turns out to be between 0.25 and 0.75, then the achieved margin of error will exceed the specified 0.015.

**11.63** (a) E = 0.01; $z_{\alpha/2} = z_{0.025} = 1.96$;

$$n = \hat{p}(1-\hat{p})\frac{z_{\alpha/2}^2}{E^2} = 0.5(1-0.5)\frac{1.96^2}{0.01^2} = 9604$$

(b) $\hat{p}_g$ is between 0.5% and 4.9%; $z_{\alpha/2} = z_{0.025} = 1.96$; E = 0.01

$$n = \hat{p}(1-\hat{p})\frac{z_{\alpha/2}^2}{E^2} = 0.049(1-0.049)\frac{1.96^2}{0.01^2} = 1790.15 \rightarrow 1791$$

Thus the required sample size is n = 1,791.

(c) The sample size has decreased from 9604 to 1791.

(d) If the actual value of $\hat{p}$ turns out to be between 0.049 and 0.951, then the achieved margin of error will exceed the specified error of 0.01.

**11.65** We will assume that both polls used a 95% confidence level. Then at the 95% confidence level, both polls are giving a range of believable values for the true population (Americans) proportion that felt that Obama was doing his job. The Gallup poll gave its range to be 0.39 to 0.45. The Quinnipiac University poll gave its range to be 0.36 to 0.40. We note that both ranges of values overlap and have common believable values from 0.39 to 0.40. Thus, it is possible that both of these polls were correct in their conclusions.

**11.67** The sample size is directly proportional to the quantity p(1 - p), which takes on its maximum value of 0.25 when p = .5. Sample sizes increase as p approaches 0.5 from either above or below. Thus to achieve a sample size adequate for any p in a given range, we should choose the largest sample that could result for any p in the range. This will always happen when we choose the value of p in the range that is closest to 0.5.

**11.69** (a) $\tilde{p} = (8+2)/(40+4) = 0.227$

The 95% one-proportion plus-four z-interval is

$$\tilde{p} \pm z_{\alpha/2}\sqrt{\tilde{p}(1-\tilde{p})/(n+4)} = 0.227 \pm 1.96\sqrt{0.227(1-0.227)/44} = (0.103, 0.351)$$

(b) This interval has the same level of precision as the interval found in Exercise 11.25. However, it is centered at 0.227 rather than 0.20.

**11.71** (a) $\tilde{p} = (35+2)/(50+4) = 0.685$

The 99% one-proportion plus-four z-interval is

$$\tilde{p} \pm z_{\alpha/2}\sqrt{\tilde{p}(1-\tilde{p})/(n+4)} = 0.685 \pm 2.575\sqrt{0.685(1-0.685)/54} = (0.522, 0.848)$$

(b) This interval is slightly more precise than the interval found in Exercise 11.27. Also, it is centered at 0.685 rather than 0.70.

**11.73** (a) $\tilde{p} = (16+2)/(20+4) = 0.75$

The 90% one-proportion plus-four z-interval is

$$\tilde{p} \pm z_{\alpha/2}\sqrt{\tilde{p}(1-\tilde{p})/(n+4)} = 0.75 \pm 1.645\sqrt{0.75(1-0.75)/24} = (0.605, 0.895)$$

(b) It was not appropriate to find the confidence interval in Exercise 11.29.

**11.75** $\tilde{p} = (707+2)/(1039+4) = 0.6798$

The 95% one-proportion plus-four z-interval is

$$\tilde{p} \pm z_{\alpha/2}\sqrt{\tilde{p}(1-\tilde{p})/(n+4)} = 0.6798 \pm 1.96\sqrt{0.6798(1-0.6798)/1043} = (0.651, 0.708)$$

We can be 95% confident that the proportion of all American adults who would continue working if they won 10 million dollars in the lottery is between 0.651 and 0.708.

**11.77** 77% of 434 is 334. Thus, x = 334 and n = 434.

$\tilde{p} = (334+2)/(434+4) = 0.767$

The 90% one-proportion plus-four z-interval is

$$\tilde{p} \pm z_{\alpha/2}\sqrt{\tilde{p}(1-\tilde{p})/(n+4)} = 0.767 \pm 1.645\sqrt{0.767(1-0.767)/438} = (0.734, 0.800)$$

We can be 90% confident that 73.4% to 80.0% of all new mothers breast-feed their infants at least briefly.

### Exercises 11.2

**11.79** (a) The sample proportion is $\hat{p} = x/n = 8/40 = 0.2$.

(b) $np_0 = 40(0.3) = 12$; $n(1 - p_0) = 40(1-0.30) = 28$
Since both are at least 5, we can employ Procedure 11.2.

(c) $z = \dfrac{0.2 - 0.3}{\sqrt{.03(1-0.3)/40}} = -1.38$; For $\alpha = 0.10$, the critical value is

$-z_\alpha = -1.28$. Since $-1.38 < -1.28$, reject $H_0$.

**11.81** (a) The sample proportion is $\hat{p} = x/n = 35/50 = 0.70$.

(b) $np_0 = 50(0.6) = 30$; $n(1 - p_0) = 50(1-0.60) = 20$
Since both are at least 5, we can employ Procedure 11.2.

(c) $z = \dfrac{0.70 - 0.60}{\sqrt{0.6(1-0.6)/50}} = 1.44$; For $\alpha = 0.05$, the critical value is $z_\alpha = 1.645$. Since $1.44 < 1.645$, do not reject $H_0$.

**11.83** (a) The sample proportion is $\hat{p} = x/n = 16/20 = 0.80$.

(b) $np_0 = 20(0.7) = 14$; $n(1 - p_0) = 20(1-0.70) = 6$
Since both are at least 5, we can employ Procedure 11.2.

(c) $z = \dfrac{0.80 - 0.70}{\sqrt{0.7(1-0.7)/20}} = 0.98$; For $\alpha = 0.05$, the critical values are $\pm z_{\alpha/2} = \pm 1.96$. Since $-1.96 < 0.98 < 1.96$, do not reject $H_0$

**11.85** (a) The sample proportion is $\hat{p} = x/n = 534/1008 = 0.5298$. 52.98% of those sampled disapprove of the government program.

(b) $\alpha = 0.05$, $p_0 = 0.50$
$np_0 = 1008(0.50) = 504$; $n(1 - p_0) = 1008(1-0.50) = 504$
Since both are at least 5, we can employ Procedure 11.2.

Step 1: $H_0$: $p = 0.50$, $H_a$: $p > 0.50$
Step 2: $\alpha = 0.05$
Step 3: $z = \dfrac{0.5298 - 0.50}{\sqrt{0.50(1-0.50)/1008}} = 1.89$
Step 4: Since $\alpha = 0.05$, the critical value is $z_\alpha = 1.645$

For the p-value approach, $P(z > 1.89) = 0.0294$
Step 5: Since $1.89 > 1.645$, reject $H_0$. P-value $< \alpha$. Thus, we reject $H_0$.
Step 6: The test results are statistically significant at the 5% level; that is, at the 5% significance level, the data do provide sufficient evidence to conclude that a majority of American adults disapprove of this government surveillance program.

**11.87** The sample proportion is $\hat{p} = 0.60$.

$\alpha = 0.10$, $p_0 = 0.58$
$np_0 = 1028(0.58) = 596.24$; $n(1 - p_0) = 1028(1 - 0.58) = 431.76$
Since both are at least 5, we can employ Procedure 11.2.
Step 1: $H_0$: $p = 0.58$, $H_a$: $p \neq 0.58$
Step 2: $\alpha = 0.10$
Step 3: $z = \dfrac{0.60 - 0.58}{\sqrt{0.58(1-0.58)/1028}} = 1.30$
Step 4: Since $\alpha = 0.10$, the critical values are $\pm z_{\alpha/2} = \pm 1.645$

For the p-value approach, $2P(Z > 1.30) = 2(0.0968) = 0.1936$
Step 5: Since $1.30 < 1.645$, reject $H_0$. P-value $> \alpha$. Thus, we do not reject $H_0$.
Step 6: The test results are not statistically significant at the 10% level; that is, at the 10% significance level, the data do not provide sufficient evidence to conclude that the percentage of all American adults who now think that a third major party is needed has changed from that in 2010.

**11.89** (a) The sample proportion is $\hat{p} = 0.650$.

$\alpha = 0.05$, $p_0 = 0.72$
$np_0 = 1003(0.72) = 722.2$; $n(1 - p_0) = 1003(1 - 0.72) = 280.8$
Since both are at least 5, we can employ Procedure 11.2.
Step 1: $H_0$: $p = 0.72$, $H_a$: $p < 0.72$
Step 2: $\alpha = 0.05$
Step 3: $z = \dfrac{0.65 - 0.72}{\sqrt{0.72(1-0.72)/1003}} = -4.94$

Step 4: Since $\alpha = 0.05$, the critical value is $-z_\alpha = -1.645$

For the p-value approach, $P(Z < -4.94) = 0.0000$

Step 5: Since $-4.94 < -1.645$, reject $H_0$. P-value $< \alpha$. Thus, we reject $H_0$.

Step 6: The test results are statistically significant at the 5% level; that is, at the 5% significance level, the data do provide sufficient evidence to conclude that the percentage of Americans who approve of labor unions has decreased since 1936.

(b) The sample proportion is $\hat{p} = 0.650$.

$\alpha = 0.05$, $p_0 = 2/3 = 0.667$
$np_0 = 1003(0.67) = 672$; $n(1 - p_0) = 1003(1 - 0.67) = 331$
Since both are at least 5, we can employ Procedure 11.2.
Step 1: $H_0$: $p = 0.67$, $H_a$: $p < 0.67$
Step 2: $\alpha = 0.05$
Step 3: $z = \dfrac{0.65 - 0.67}{\sqrt{0.67(1-0.67)/1003}} = -1.35$

Step 4: Since $\alpha = 0.05$, the critical value is $-z_\alpha = -1.645$

For the p-value approach, $P(Z < -1.35) = 0.0885$
Step 5: Since $-1.35 > -1.645$, do not reject $H_0$. P-value $> \alpha$. Thus, we do not reject $H_0$.
Step 6: The test results are not statistically significant at the 5% level; that is, at the 5% significance level, the data do not provide sufficient evidence to conclude that the percentage of Americans who approve of labor unions has decreased since 1963.

**11.91** (a) $n = 1053$, $p_0 = 0.5$, $x = 548$, $\hat{p} = 548/1053 = 0.5204$

$np_0 = 526.5$, $n(1 - p_0) = 526.5$
Since both are at least 5, we can employ Procedure 11.2.
Step 1: $H_0$: $p = 0.50$, $H_a$: $p > 0.50$
Step 2: $\alpha = 0.05$
Step 3: $z = \dfrac{0.5204 - 0.50}{\sqrt{0.50(1-0.50)/1053}} = 1.324$

Step 4: Since $\alpha = 0.05$, the critical value is $z_\alpha = 1.645$

For the p-value approach, $P(z > 1.32) = 0.0934$
Step 5: Since $1.32 < 1.645$, do not reject $H_0$. P-value $> \alpha$. Thus, do not reject $H_0$.

## 352   Chapter 11

Step 6: The test results are not statistically significant at the 5% level; that is, at the 5% significance level, the data do not provide sufficient evidence to conclude that a majority of U.S. adults favored passage of the legislation.

(b) The words "slim majority" still implies that there can be said a majority of U.S. adults favored passage. From part (a), we cannot conclude that a majority favor passage of the legislation.

(c) It could say that a slim majority of those polled favor passage of the legislation.

**11.93** (a) Using Minitab, choose **Stat, Basic Statistics,** then **1 Proportion.** Choose **Summarized data.** Enter 921 in the **Number of Events** textbox. Enter 1001 in the **Number of trials** textbox. Check **Perform Hypothesis test.** Enter 0.90 in the **Hypothesized proportion** textbox. Click **Options.** Enter 95.0 in the **Confidence level** textbox. Choose **greater than** in the **Alternative** pull-down menu. Check **Use test and interval based on normal distribution.** Click **OK** twice. The results are

Test of p = 0.9 vs p > 0.9

```
                              95% Lower
Sample    X     N  Sample p     Bound   Z-Value  P-Value
1        921  1001  0.920080   0.905982   2.12    0.017
```

Using the normal approximation.
The test-statistic is 2.12 and the P-value is 0.017. This P-value is less than a significance level of 5%. Therefore, we would reject the null hypothesis. At 5% significance, there is evidence that more than 9 out of 10 Americans always wash up after using the bathroom.

(b) Using the results in part (a) and changing the significance level to 1%, the test statistic and the P-value would still be 2.12 and 0.017 respectively. The P-value is greater than a significance level of 1%. Therefore, we would not reject the null hypothesis. At 1% significance, there is not significant evidence that more than 9 out of 10 Americans always wash up after using the bathroom.

**Exercises 11.3**

**11.95** We need to decide whether the difference between two sample proportions can reasonably be attributed to sampling error or are the population proportions really different.

**11.97** (a) Using sunscreen before going out in the sun
(b) Teen-age girls and teen-age boys
(c) The two proportions are sample proportions. The reference is specifically to those teen-age girls and boys who were surveyed

**11.99** (a) The parameters are $p_1$ and $p_2$. The rest are statistics.
(b) The fixed numbers are $p_1$ and $p_2$. The rest are variables.

**11.101** (a) $\hat{p}_1 = x_1/n_1 = 18/40 = 0.45$; $\hat{p}_2 = x_2/n_2 = 30/40 = 0.75$;
$\hat{p}_p = (x_1 + x_2)/(n_1 + n_2) = 48/80 = 0.6$

(b) $x_1 = 18$, $n_1 - x_1 = 22$, $x_2 = 30$, $n_2 - x_2 = 10$.
All are at least 5, so z-procedures are appropriate.

(c) $z = \dfrac{0.45 - 0.75}{\sqrt{0.6(1-0.6)[(1/40)+(1/40)]}} = -2.74$

α = 0.10, critical value is -1.28; since -2.74 < -1.28, reject $H_0$.

(d) The 80% confidence interval is

$$\hat{p}_1 - \hat{p}_2 \pm z_{\alpha/2}\sqrt{\hat{p}_1(1-\hat{p}_1)/n_1 + \hat{p}_2(1-\hat{p}_2)/n_2}$$

$$(0.45 - 0.75) \pm 1.28\sqrt{0.45(1-0.4)/40 + 0.75(1-0.75)/40} = (-0.43, -0.17)$$

**11.103** (a) $\hat{p}_1 = x_1/n_1 = 15/20 = 0.75$; $\hat{p}_2 = x_2/n_2 = 18/30 = 0.60$;
$\hat{p}_p = (x_1 + x_2)/(n_1 + n_2) = 33/50 = 0.66$

(b) $x_1 = 15$, $n_1 - x_1 = 5$, $x_2 = 18$, $n_2 - x_2 = 12$.
All are at least 5, so z-procedures are appropriate.

(c) $z = \dfrac{0.75 - 0.60}{\sqrt{0.66(1-0.66)[(1/20)+(1/30)]}} = 1.10$

$\alpha = 0.05$, critical value is 1.645; since 1.10 < 1.645, do not reject $H_0$.

(d) The 90% confidence interval is

$$\hat{p}_1 - \hat{p}_2 \pm z_{\alpha/2}\sqrt{\hat{p}_1(1-\hat{p}_1)/n_1 + \hat{p}_2(1-\hat{p}_2)/n_2}$$

$$(0.75 - 0.60) \pm 1.645\sqrt{0.75(1-0.75)/20 + 0.60(1-0.60)/30} = (-0.067, 0.367)$$

**11.105** (a) $\hat{p}_1 = x_1/n_1 = 30/80 = 0.375$; $\hat{p}_2 = x_2/n_2 = 15/20 = 0.750$;
$\hat{p}_p = (x_1 + x_2)/(n_1 + n_2) = 45/100 = 0.45$

(b) $x_1 = 30$, $n_1 - x_1 = 50$, $x_2 = 15$, $n_2 - x_2 = 5$.
All are at least 5, so z-procedures are appropriate.

(c) $z = \dfrac{0.375 - 0.750}{\sqrt{0.45(1-0.45)[(1/80)+(1/20)]}} = -3.02$

$\alpha = 0.05$, critical values are ±1.96; since -3.02 < -1.96, reject $H_0$.

(d) The 95% confidence interval is

$$\hat{p}_1 - \hat{p}_2 \pm z_{\alpha/2}\sqrt{\hat{p}_1(1-\hat{p}_1)/n_1 + \hat{p}_2(1-\hat{p}_2)/n_2}$$

$$(0.375 - 0.750) \pm 1.96\sqrt{0.375(1-0.375)/80 + 0.750(1-0.750)/20} =$$
$$(-0.592, -0.158)$$

**11.107** (a) Population 1: Women who took multivitamins containing folic acid
$\hat{p}_1 = 35/2701 = 0.01296$

Population 2: Women who received only trace elements
$\hat{p}_2 = 47/2052 = 0.02290$

$\hat{p}_p = (35 + 47)/(2{,}701 + 2{,}052) = 0.01725$

Step 1: $H_0$: $p_1 = p_2$, $H_a$: $p_1 < p_2$
Step 2: $\alpha = 0.01$

Step 3: $z = \dfrac{0.01296 - 0.0220}{\sqrt{0.01725(1-0.01725)[(1/2701)+(1/2052)]}} = -2.61$

Step 4: Since $\alpha = 0.01$, the critical value is $-z_\alpha = -2.33$

For the P-value approach, $P(z < -2.61) = 0.0045$

Step 5: Since -2.61 < -2.33, reject $H_0$. P-value < $\alpha$. Thus, we reject $H_0$.

**354** Chapter 11

Step 6: The test results are significant at the 1% level; that is, at the 1% significance level, the data do provide sufficient evidence to conclude that the women who take folic acid are at lesser risk of having children with major birth defects than those women who do not.

(b) This is a designed experiment. The researchers decided which women would take daily multivitamins.

(c) Yes. By using the basic principles of design (control, randomization, and replication) the doctors can conclude that the reduction in the rates of major birth defects in the folic acid group is likely caused by the folic acid.

**11.109** Population 1: Drivers of age 16-24, $\hat{p}_1 = 0.790 = 790/1000$

Population 2: Drivers of age 25-69, $\hat{p}_2 = 924/1100 = 0.840$

$\hat{p}_p = (790 + 924)/(1000 + 1100) = 0.8162$

Step 1: $H_0: p_1 = p_2$, $H_a: p_1 \neq p_2$
Step 2: $\alpha = 0.01$
Step 3: $z = \dfrac{0.790 - 0.840}{\sqrt{0.8162(1-0.8162)[(1/1000)+(1/1100)]}} = -2.95$

Step 4: Since $\alpha = 0.01$, the critical values is $\pm z_\alpha = \pm 2.575$

For the P-value approach, $2P(z < -2.95) = 0.0032$

Step 5: Since $-2.95 < -2.575$, reject $H_0$. P-value $< \alpha$. Thus, we reject $H_0$.

Step 6: The test results are significant at the 1% level; that is, at the 1% significance level, the data provide sufficient evidence to conclude that there is a difference in seat-belt usage between drivers 16-24 years old and those 25-69 years old.

**11.111** Population 1: Bachelors degree, $\hat{p}_1 = 386/750 = 0.5147$

Population 2: Graduate degree, $\hat{p}_2 = 237/500 = 0.4740$

$\hat{p}_p = (386 + 237)/(750 + 500) = 0.4984$

(a) The assumptions for using the two-sample z-test are simple random samples, independent samples, and $x_1$, $n_1 - x_1$, $x_2$, and $n_2 - x_2$ must all be greater than or equal to 5.

(b) Step 1: $H_0: p_1 = p_2$, $H_a: p_1 > p_2$
Step 2: $\alpha = 0.05$
Step 3: $z = \dfrac{0.5147 - 0.4740}{\sqrt{0.4984(1-0.4984)[(1/750)+(1/500)]}} = 1.41$

Step 4: Since $\alpha = 0.05$, the critical value is $z_\alpha = 1.645$

For the P-value approach, $P(z > 1.41) = 0.0793$

Step 5: Since $1.41 < 1.645$, do not reject $H_0$. P-value $> \alpha$. Thus, we do not reject $H_0$.

Step 6: The test results are not significant at the 5% level; that is, at the 5% significance level, the data do not provide sufficient evidence to conclude that a higher percentage of adults with Bachelors degrees are overweight than of adults with graduate degrees.

**11.113** From Exercise 11.107, the 99% confidence interval is

$$(0.012958 - 0.022904) \pm 2.33\sqrt{\frac{0.012958(1-0.012958)}{2701} + \frac{0.022904(1-0.022904)}{2052}}$$

$$-0.009946 \pm 0.009215 \text{ or } -0.019161 \text{ to } -0.000731$$

We can be 98% confident that the difference $p_1 - p_2$ between the rates of major birth defects for babies born to women who have taken folic acid and those born to women who have not taken folic acid is somewhere between -0.019161 and -0.000731.

**11.115** From Exercise 11.109, the 99% confidence interval is

$$\hat{p}_1 - \hat{p}_2 \pm z_{\alpha/2}\sqrt{\hat{p}_1(1-\hat{p}_1)/n_1 + \hat{p}_2(1-\hat{p}_2)/n_2}$$

$$(0.790 - 0.840) \pm 2.575\sqrt{0.790(1-0.790)/1000 + 0.840(1-0.840)/1100}$$

$$= (-0.0937, -0.0063)$$

We can be 90% confident that the difference between proportions of seat belt users for drivers in the age groups 16-24 years and 25-69 years is somewhere between -0.0937 and -0.0063.

**11.117** From Exercise 11.111, the 90% confidence interval is

$$\hat{p}_1 - \hat{p}_2 \pm z_{\alpha/2}\sqrt{\hat{p}_1(1-\hat{p}_1)/n_1 + \hat{p}_2(1-\hat{p}_2)/n_2}$$

$$(0.515 - 0.474) \pm 1.645\sqrt{0.515(1-0.515)/750 + 0.474(1-0.474)/500}$$

$$= (-0.007, 0.088)$$

We can be 90% confident that the difference between percentages of adults in the two degree categories who have an above healthy weight is somewhere between -0.007 and 0.088.

**11.119** (a) Using Minitab, choose **Stat ▶ Basic statistics ▶ 2 Proportions...**, select the **Summarized data** option button, click in the **Trials** text box for **First sample** and enter 300, click in the **Events** text box for **First sample** and type 215, click in the **Trials** text box for **Second sample** and type 250, click in the **Events** text box for **Second sample** and type 186, click the **Options...** button, click in the **Confidence level** text box and type 95, click in the **Test difference** text box and type 0, click the arrow button at the right of the **Alternative** drop-down list box and select **not equal**, select the **Use pooled estimate of p for test** check box, click **OK**, and click **OK**. The resulting output is

```
Sample    X    N    Sample p
1        215  300   0.716667
2        186  250   0.744000

Difference = p (1) - p (2)
Estimate for difference:  -0.0273333
95% CI for difference:  (-0.101675, 0.0470087)
Test for difference = 0 (vs not = 0):  Z = -0.72  P-Value = 0.473
```

Since the P-value = 0.473, which is greater than the significance level 0.05, do not reject the null hypothesis. The data do not provide sufficient evidence to conclude that there is a difference between the labor-force participation rates of U.S. and Canadian women.

(b) The confidence interval is part of the output for part (a). We can be 95% confident that the difference in labor-force participation rates between of U.S. and Canadian women is somewhere between -0.102 and 0.047.

**11.121** (a) Using Minitab, choose **Stat ▶ Basic statistics ▶ 2 Proportions...**, select the **Summarized data** option button, click in the **Trials** text box for **First sample** and enter 1020, click in the **Events** text box for **First sample** and enter 35, click in the **Trials** text box for **Second sample** and enter 1008, click in the **Events** text box for **Second sample** and enter 69, click the **Options...** button, click in the **Confidence level** text box and type 99, click in the **Test difference** text box and type 0, click the arrow button at the right of the **Alternative** drop-down list box and select **less than**, select the **Use pooled estimate of p for test** check box, click **OK**, and click **OK**. The resulting output is

Test and CI for Two Proportions

```
Sample    X      N    Sample p
1        35   1020    0.03431
2        69   1008    0.06845
```

Difference = p (1) - p (2)
Estimate for difference:  -0.03414
Test for difference = 0 (vs < 0):  Z = -3.48   P-Value = 0.0002

Since the P-value = 0.0002, which is less than the significance level 0.01, reject the null hypothesis. The data provide sufficient evidence to conclude that college graduates have a lower unemployment rate than high school graduates.

(b) Repeat part (a), selecting not equal in the Alternative drop down box and click in the **Confidence level** text book and type 98. The confidence interval is

98% CI for difference:  (-0.0569, -0.0114)

We can be 98% confident that the difference in the unemployment rates between college and high school graduates is somewhere between -0.0569 and -0.0114.

**11.123** (a) $\tilde{p}_1 = (18+1)/(40+2) = 0.452$ and $\tilde{p}_2 = (30+1)/(40+2) = 0.738$

The 80% two-proportions plus-four z-interval is

$$(\tilde{p}_1 - \tilde{p}_2) \pm z_{\alpha/2} \sqrt{\tilde{p}_1(1-\tilde{p}_1)/(n+2) + \tilde{p}_2(1-\tilde{p}_2)/(n+2)}$$

$$= (0.452 - 0.738) \pm 1.28\sqrt{0.452(1-0.452)/42 + 0.738(1-0.738)/42} = (-0.417, -0.155)$$

(b) This interval is close to the same level of precision as the interval found in Exercise 11.101. However, it is centered at -0.286 rather than -0.300.

**11.125** (a) $\tilde{p}_1 = (15+1)/(20+2) = 0.727$ and $\tilde{p}_2 = (18+1)/(30+2) = 0.594$

The 90% two-proportions plus-four z-interval is

$$(\tilde{p}_1 - \tilde{p}_2) \pm z_{\alpha/2} \sqrt{\tilde{p}_1(1-\tilde{p}_1)/(n+2) + \tilde{p}_2(1-\tilde{p}_2)/(n+2)}$$

$$= (0.727 - 0.594) \pm 1.645\sqrt{0.727(1-0.727)/22 + 0.594(1-0.594)/32} = (-0.079, 0.345)$$

(b) This interval is slightly more precise than the interval found in Exercise 11.103. In addition, it is centered at 0.133 rather than 0.150.

**11.127** (a) $\tilde{p}_1 = (30+1)/(80+2) = 0.378$ and $\tilde{p}_2 = (15+1)/(20+2) = 0.727$

The 95% two-proportions plus-four z-interval is

$$(\tilde{p}_1 - \tilde{p}_2) \pm z_{\alpha/2} \sqrt{\tilde{p}_1(1-\tilde{p}_1)/(n+2) + \tilde{p}_2(1-\tilde{p}_2)/(n+2)}$$

$$= (0.378 - 0.727) \pm 1.96\sqrt{0.378(1-0.378)/82 + 0.727(1-0.727)/22} = (-0.563, -0.135)$$

(b) This interval is more precise than the interval found in Exercise 11.105. In addition, it is centered at -0.349 rather than -0.375.

**11.129** Let population 1 be the civilian labor forces of Finland and population 2 be the civilian labor forces of Denmark.

$\tilde{p}_1 = (7+1)/(100+2) = 0.078$ and $\tilde{p}_2 = (3+1)/(75+2) = 0.052$

The 95% two-proportions plus-four z-interval is

$(\tilde{p}_1 - \tilde{p}_2) \pm z_{\alpha/2}\sqrt{\tilde{p}_1(1-\tilde{p}_1)/(n+2) + \tilde{p}_2(1-\tilde{p}_2)/(n+2)}$

$= (0.078 - 0.052) \pm 1.96\sqrt{0.078(1-0.078)/102 + 0.052(1-0.052)/77} = (-0.046, 0.098)$

We can be 95% confident that the difference between the unemployment rates in Finland and Denmark is between -4.6% and 9.8%. In other words, we can be 95% confident that the unemployment rate in Finland is between 4.6% smaller to 9.8% larger than the unemployment rate in Denmark.

**11.131** Let population 1 be elderly people taking calcium channel blockers and population 2 be elderly people taking beta blockers.

$\tilde{p}_1 = (27+1)/(202+2) = 0.137$ and $\tilde{p}_2 = (28+1)/(424+2) = 0.068$

The 95% two-proportions plus-four z-interval is

$(\tilde{p}_1 - \tilde{p}_2) \pm z_{\alpha/2}\sqrt{\tilde{p}_1(1-\tilde{p}_1)/(n+2) + \tilde{p}_2(1-\tilde{p}_2)/(n+2)}$

$= (0.137 - 0.068) \pm 1.645\sqrt{0.137(1-0.137)/204 + 0.068(1-0.068)/426}$

$= (0.025, 0.113)$

We can be 90% confident that the difference between the cancer rates of elderly people taking calcium channel blockers and those taking beta blockers is between 2.5% and 11.3%. In other words, we can be 90% confident that the cancer rate of elderly people taking calcium channel blockers is between 2.5% to 11.3% larger than the cancer rate of elderly people taking beta blockers.

**11.133** In the previous equation, replace the $n_1$ and $n_2$ with $n$.

$E = z_{\alpha/2}\sqrt{\hat{p}_1(1-\hat{p}_1)/n + \hat{p}_2(1-\hat{p}_2)/n}$. Divide both sides by $z_{\alpha/2}$, square both sides, and replace each of the $\hat{p}_1(1-\hat{p}_1)$ and $\hat{p}_2(1-\hat{p}_2)$ with 0.25 since that is the maximum value for those expressions. $\left(\dfrac{E}{z_{\alpha/2}}\right)^2 = \dfrac{0.25}{n} + \dfrac{0.25}{n} = \dfrac{0.50}{n}$.

Reciprocate this equation and you have $\dfrac{n}{0.50} = \left(\dfrac{z_{\alpha/2}}{E}\right)^2$. Multiplying by 0.50 results in $n = n_1 = n_2 = 0.50\left(\dfrac{z_{\alpha/2}}{E}\right)^2$.

**11.135** (a) The confidence interval found in Example 11.10 is (-0.129, -0.031). The length of this confidence interval is 0.098. Divide this in half to get the margin of error or 0.098/2 = 0.049. This is the amount of error that you add and subtract from the point estimate of the differences between the two sample proportions in order to form the confidence interval.

(b)

$E = z_{\alpha/2}\sqrt{\hat{p}_1(1-\hat{p}_1)/n_1 + \hat{p}_2(1-\hat{p}_2)/n_2}$

$= 1.645\sqrt{0.369(1-0.369)/747 + 0.449(1-0.449)/434} = 0.049$

**11.137** (a)

$$n = n_1 = n_2 = \left(\hat{p}_{1g}(1-\hat{p}_{1g}) + \hat{p}_{2g}(1-\hat{p}_{2g})\right)\left(\frac{z_{\alpha/2}}{E}\right)^2$$

$$= (0.41(1-0.41) + 0.49(1-0.49))\left(\frac{1.645}{0.01}\right)^2 = 13308.23 \to 13{,}309$$

(b)
$$(\hat{p}_1 - \hat{p}_2) \pm z_{\alpha/2}\sqrt{\hat{p}_1(1-\hat{p}_1)/n_1 + \hat{p}_2(1-\hat{p}_2)/n_2}$$

$$(0.383 - 0.437) \pm 1.645\sqrt{0.383(1-0.383)/13309 + 0.437(1-0.437)/13309}$$

$$(-0.0639, -0.0441)$$

(c) The margin of error in part (b) is (0.0639-0.0441)/2 = 0.00975. This is less than the required error specified in part (a). It is also only slightly less than the margin of error we obtained in part (c) of Exercise 11.136. This is due to the fact that the educated guess values of 0.41 and 0.49 are both pretty close to 0.50.

### Chapter 11 Review Problems

1. (a) Feeling that the overall medicinal benefits of marijuana outweigh the risks and potential harms.
   (b) All international physicians
   (c) Proportion of all international physicians who feel that the overall medicinal benefits of marijuana outweigh the risks and potential harms.
   (d) Proportion of international physicians in the sample who feel that the overall medicinal benefits of marijuana outweigh the risks and potential harms. Clearly the population of all international physicians is much larger than 1,446.
2. It is often impossible to take a census of an entire population. It is also expensive and time-consuming.
3. (a) "Number of successes" stands for the number of members of the sample that exhibit the specified attribute.
   (b) "Number of failures" stands for the number of members of the sample that do not exhibit the specified attribute.
4. (a) population proportion
   (b) normal
   (c) number of successes, number of failures, 5
5. The margin of error for the estimate of a population proportion tells us what the maximum difference between the sample proportion and the population proportion is likely to be. It is not an absolute maximum, and how likely it is depends on the confidence level used.
6. (a) Getting the "holiday blues"
   (b) Men and women
   (c) The proportion of men in the population who get the "holiday blues" and the proportion of women in the population who get the "holiday blues"
   (d) The proportion of men in the sample who get the "holiday blues" and the proportion of women in the sample who get the "holiday blues"
   (e) They are sample proportions since the information came from a poll, not a census. Also, it could not be a population proportion because I was not asked.
7. (a) The mean of all possible differences between the two sample proportions equals the difference of the population proportions.
   (b) For large samples, the possible differences between the two sample proportions have approximately a normal distribution.

8. The 95% confidence interval for the percentage of U.S. adults who would get a smallpox shot if it were available is 40% ± 3.0% or from 37% to 43%.

9. (a) n = 1218, $\hat{p}$ = 733/1218 = 0.6018, $z_{\alpha/2}$ = $z_{0.025}$ = 1.96

   $0.6018 - 1.96\sqrt{0.6018(1-0.6018)/1218}$ to $0.6018 + 1.96\sqrt{0.6018(1-0.6018)/1218}$

   0.5743 to 0.6293

   We can be 95% confident that the proportion, p, of students who expect difficulty finding a job is somewhere between 0.5743 and 0.6293.

   (b) The error is found by taking the width of the confidence interval and dividing by 2. So E = (0.6293 - 0.5743)/2 = 0.0275.

   (c) In terms of point estimate ± margin of error, the confidence interval is 0.6018 ± 0.0275.

10. (a) E = 0.02; $z_{\alpha/2}$ = $z_{0.025}$ = 1.96; $\hat{p}_g$ = 0.50

    $$n = 0.5^2 \frac{z_{\alpha/2}^2}{E^2} = 0.25 \cdot \frac{1.96^2}{0.02^2} = 2401$$

    (b) n = 2401; $\hat{p}$ = 0.587; $z_{\alpha/2}$ = $z_{0.025}$ = 1.96

    $0.587 - 1.96\sqrt{0.587(1-0.587)/2401}$ to $0.587 + 1.96\sqrt{0.587(1-0.587)/2401}$

    0.5673 to 0.6067

    (c) The margin of error for the estimate is 0.0197, slightly less than what is required in part (a).

    (d) $\hat{p}_g$ = 0.56; $z_{\alpha/2}$ = $z_{0.025}$ = 1.96

    part (a)    E = 0.02, $z_{0.025}$ = 1.96; sample size is:

    $$n = 0.56(1-0.56)\frac{z_{\alpha/2}^2}{E^2} = 0.2464 \cdot \frac{1.96^2}{0.02^2} = 2366.42 \to 2367$$

    Thus the required sample size is n = 2367.

    part (b)
    $0.587 - 1.96\sqrt{0.587(1-0.587)/2367}$ to $0.587 + 1.96\sqrt{0.587(1-0.587)/2367}$

    0.5672 to 0.6068

    part (c) The margin of error is 0.0198, which is the same as what is specified in part (a).

    (e) By employing the guess for $\hat{p}$ in part (d) we can reduce the required sample size by 34, from 2401 to 2367, saving a little time and money. Moreover, the margin of error stays the same. The risk of using the guess 0.56 for $\hat{p}$ is that if the actual value of $\hat{p}$ turns out to be between .44 and .56, then the achieved margin of error will exceed the specified 0.02.

11. n = 2512, x = 578, $\alpha$ = 0.05, $\hat{p}$ = 578/2512 = 0.2301

    $np_0$ = 2512(0.25) = 628; $n(1-p_0)$ = 2512(1-0.25) = 1884
    Since both are at least 5, we can employ Procedure 11.2.

    (a) Step 1: $H_0$: p = 0.25, $H_1$: p < 0.25
        Step 2: $\alpha$ = 0.05
        Step 3: $z = \dfrac{0.23 - 0.25}{\sqrt{0.25(1-0.25)/2512}} = -2.31$

        Step 4: Since $\alpha$ = 0.05, the critical value is $-z_\alpha$ = -1.645

For the P-value approach: P(z < -2.31) = 0.0104.
Step 5: Since -2.31 < -1.645, reject H₀. Since P-value < $\alpha$, reject H₀.
Step 6: The test results are statistically significant at the 5% level; that is, at the 5% significance level, the data do provide sufficient evidence to conclude that less than one in four Americans believe that juries "almost always" convict the guilty and free the innocent

12. (a) Observational study. The researchers had no control over any of the factors of the study.
    (b) Height may not be the only factor to be considered. Although there does appear to be an association, we don't know if it is a direct association.

13. Population 1: men, $\hat{p}_1$ = 607/1029 = 0.5899,

    Population 2: women, $\hat{p}_2$ = 648/1223 = 0.5298

    $\hat{p}_p$ = (607 + 648)/(1029 + 1223) = 0.5573

    Step 1: H₀: $p_1 = p_2$, Hₐ: $p_1 \neq p_2$
    Step 2: $\alpha$ = 0.01
    Step 3: $z = \dfrac{0.5899 - 0.5298}{\sqrt{0.5573(1-0.5573)((1/1029)+(1/1223))}} = 2.86$
    Step 4: Since $\alpha$ = 0.01, the critical values are $\pm z_\alpha = \pm 2.575$
    2*P(z > 2.86) = 0.0042
    Step 5: Since 2.86 > 2.575, reject H₀. P-value < $\alpha$. Thus, we reject H₀.
    Step 6: The test is significant at the 1% level; that is the data do provide evidence to conclude that a difference exists in the percentages of smartphone owners between men and women.

14. (a) From Problem 13
    $$(0.5899 - 0.5298) \pm 2.576 \sqrt{\dfrac{0.5899(1-0.5899)}{1029} + \dfrac{0.5298(1-0.5298)}{1223}}$$
    $0.0601 \pm 0.0540$ or $0.0061$ to $0.1141$

    (b) We can be 99% confident that the difference $p_1 - p_2$ between the proportions of smartphone owners between men and women is somewhere between 0.0061 to 0.1141. In other words, we can be 99% confident that the percentage of men who own a smartphone is somewhere between 0.61% to 11.41% greater than the percentage of women who own a smartphone.

15. (a) $\hat{p} = 4/103 = 0.039$
    The 95% one-proportion z-interval is
    $\hat{p} \pm z_{\alpha/2}\sqrt{\hat{p}(1-\hat{p})/n} = 0.039 \pm 1.96\sqrt{0.039(1-0.039)/103} = (0.002, 0.076)$
    (b) $\tilde{p} = (4+2)/(103+4) = 0.056$
    The 95% one-proportion plus-four z-interval is
    $\tilde{p} \pm z_{\alpha/2}\sqrt{\tilde{p}(1-\tilde{p})/(n+4)} = 0.056 \pm 1.96\sqrt{0.056(1-0.056)/107} = (0.012, 0.100)$
    (c) The discrepancy between the two methods is primarily due to using a z-interval when the conditions for using such an interval are not met. To get a good approximation to an exact interval, both x and n − x should be at least 5. In this case, x is only 4.
    (d) I would use the one-proportion plus-four z-interval because the conditions to use the one-proportion z-interval are not met.

# CHAPTER 12 SOLUTIONS

## Exercises 12.1

**12.1** A variable has a chi-square distribution if its distribution has the shape of a special type of right-skewed curve, called a chi-square curve.

**12.3** The $\chi^2$-curve with 20 degrees of freedom more closely resembles a normal curve. This follows from Property 4 of Key Fact 12.1, "As the number of degrees of freedom becomes larger, $\chi^2$-curves look increasingly like normal curves."

**12.5** (a) $\chi^2_{0.025} = 32.852$

(b) $\chi^2_{0.01} = 36.191$

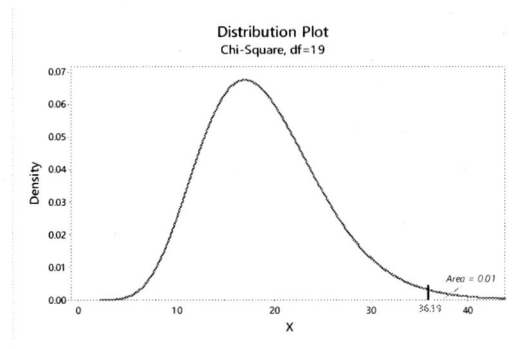

**12.7** (a) $\chi^2_{0.05} = 18.307$

(b) $\chi^2_{0.025} = 20.483$

## Exercises 12.2

**12.9** The term "goodness-of-fit" is used to describe the type of hypothesis test considered in this section because the test is carried out by determining how well the observed frequencies match or fit the expected frequencies.

**12.11** The assumptions are satisfied. The expected frequencies are np = 65, 30, and 5. All are 1 or more, and none are less than 5.

**12.13** The assumptions are satisfied. The expected frequencies are np = 10, 10, 12.5, 15, and 2.5. All are 1 or more, and exactly 20% are less than 5.

**12.15** The assumptions are not satisfied. The expected frequencies are np = 11, 11, 12.5, 15, and 0.5. One of them is not 1 or more.

**12.17** (a) The population consists of occupied housing units built after 2010. The variable under consideration is the primary heating fuel of each unit.

(b) The assumptions are not satisfied when n = 200. The expected frequencies are np = 101.6, 15.8, 67.8, 10.4, 3.8, 0.6. One of the frequencies is less than 1 and 33% are less than 5.
The assumptions are not satisfied when n = 300. The expected frequencies are np = 152.4, 23.7, 101.7, 15.6, 5.7, 0.9. One of the frequencies is less than 1 and only one (17%) is less than 5.
The assumptions are satisfied when n = 400. The expected frequencies are np = 203.2, 31.6, 135.6, 20.8, 7.6, 1.2. None of the frequencies are less than 1 and only one (17%) is less than 5.

(c) We need the count for the category "None" to be at least 1. Thus we must have n(0.003) > 1. This implies that n must be greater than 1/0.003 = 333.33. Thus the smallest possible value of n is 334.

**12.19** The procedure is summarized in the following table.

| Distribution | Observed Frequency O | Expected Frequency E | Difference O - E | Square of Difference $(O - E)^2$ | Chi-square subtotal $(O - E)^2/E$ |
|---|---|---|---|---|---|
| 0.2 | 85 | 100 | -15 | 225 | 2.250 |
| 0.4 | 215 | 200 | 15 | 225 | 1.125 |
| 0.3 | 130 | 150 | -20 | 400 | 2.667 |
| 0.1 | 70 | 50 | 20 | 400 | 8.000 |
|  | 500 |  |  |  | 14.042 |

df = 3; From Table VII, critical Value = 7.815. $\chi^2$ = 14.0422, which is greater than the critical value, so reject the null hypothesis. The data provide sufficient evidence that the variable differs from the given distribution.

**12.21** The procedure is summarized in the following table.

| Distribution | Observed Frequency O | Expected Frequency E | Difference O - E | Square of Difference $(O - E)^2$ | Chi-square subtotal $(O - E)^2/E$ |
|---|---|---|---|---|---|
| 0.2 | 9 | 10 | -1 | 1 | 0.100 |
| 0.1 | 7 | 5 | 2 | 4 | 0.800 |
| 0.1 | 1 | 5 | -4 | 16 | 3.200 |
| 0.3 | 12 | 15 | -3 | 9 | 0.600 |
| 0.3 | 21 | 15 | 6 | 36 | 2.400 |
|  | 50 |  |  |  | 7.100 |

df = 4; From Table VII, critical Value = 7.779. $\chi^2$ = 7.100, which is less than the critical value, so do not reject the null hypothesis. The data do not provide sufficient evidence that the variable differs from the given distribution.

**12.23** The procedure is summarized in the following table.

| Distribution | Observed Frequency O | Expected Frequency E | Difference O - E | Square of Difference $(O - E)^2$ | Chi-square subtotal $(O - E)^2/E$ |
|---|---|---|---|---|---|
| 0.5 | 147 | 175 | -28 | 784 | 4.480 |
| 0.3 | 115 | 105 | 10 | 100 | 0.952 |
| 0.2 | 88 | 70 | 18 | 324 | 4.629 |
|  | 350 |  |  |  | 10.061 |

df = 2; From Table VII, critical Value = 9.210. $\chi^2$ = 10.061, which is greater than the critical value, so reject the null hypothesis. The data provide sufficient evidence that the variable differs from the given distribution.

**12.25** (a) The population consists of this year's incoming college freshmen in the U.S. The variable under consideration is their political view.

(b)

| Political View | Distribution | Observed Frequency O | Expected Frequency E | Difference O - E | Square of Difference $(O - E)^2$ | Chi-square subtotal $(O - E)^2/E$ |
|---|---|---|---|---|---|---|
| Liberal | 0.277 | 147 | 138.500 | 8.500 | 72.250 | 0.522 |
| Moderate | 0.519 | 237 | 259.500 | -22.500 | 506.250 | 1.951 |
| Conservative | 0.204 | 116 | 102.000 | 14.000 | 196.000 | 1.922 |
|  |  | 500 |  |  |  | 4.395 |

Step 1: $H_0$: The distribution of political views for this year's incoming freshmen is the same as the 2000 distribution.
$H_a$: The distribution of political views for this year's incoming freshmen is different from the 2000 distribution.

Step 2: $\alpha = 0.05$

Step 3: $\chi^2 = 4.395$ (See column 7 of the table.)

Step 4: df = 2; From Table VII, critical value = 5.991

For the P-value approach, $P(\chi^2 > 4.395) > 0.10$

Step 5: Since 4.395 < 5.991, do not reject $H_0$. Since the P-value is greater than the significance level, do not reject $H_0$.

**364** Chapter 12

Step 6: The data do not provide sufficient evidence at the 5% level to conclude that the distribution of political views for this year's incoming freshmen is different from the 2000 distribution.

(c) Since the P-value is greater than 0.10, do not reject $H_0$ at the 10% significance level. The data do not provide sufficient evidence at the 10% level to conclude that the distribution of political views for this year's incoming freshmen is different from the 2000 distribution.

**12.27**

| Color | Distribution | Observed Frequency O | Expected Frequency E | Difference O - E | Square of Difference $(O-E)^2$ | Chi-square subtotal $(O-E)^2/E$ |
|---|---|---|---|---|---|---|
| Brown | 0.30 | 152 | 152.7 | -0.3 | 0.49 | 0.003 |
| Yellow | 0.20 | 114 | 101.8 | 12.2 | 148.84 | 1.462 |
| Red | 0.20 | 106 | 101.8 | 4.2 | 17.64 | 0.173 |
| Orange | 0.10 | 51 | 50.9 | 0.1 | 0.01 | 0.000 |
| Green | 0.10 | 43 | 50.9 | -7.9 | 62.41 | 1.226 |
| Blue | 0.10 | 43 | 50.9 | -7.9 | 62.41 | 1.226 |
|  |  | 509 |  |  |  | 4.091 |

Step 1: $H_0$: The color distribution of M&Ms is the same as that reported by M&M/Mars consumer affairs.
$H_a$: The color distribution of M&Ms is different from that reported by M&M/Mars consumer affairs.
Step 2: $\alpha = 0.05$
Step 3: $\chi^2 = 4.091$ (See column 7 of the table.)
Step 4: df = 5; From Table VII, the critical value = 11.070
For the P-value approach, $P(\chi^2 > 4.091) > 0.10$
Step 5: Since 4.091 < 11.070, do not reject $H_0$. Since the P-value is larger than the significance level, do not reject $H_0$.
Step 6: There is not sufficient evidence to conclude that the color distribution of M&Ms is different from that reported by M&M/Mars consumer affairs.

**12.29**

| Number | Distribution | Observed Frequency O | Expected Frequency E | Difference O - E | Square of Difference $(O-E)^2$ | Chi-square subtotal $(O-E)^2/E$ |
|---|---|---|---|---|---|---|
| 1 | 0.167 | 23 | 25 | -2 | 4 | 0.160 |
| 2 | 0.167 | 26 | 25 | 1 | 1 | 0.040 |
| 3 | 0.167 | 23 | 25 | -2 | 4 | 0.160 |
| 4 | 0.167 | 21 | 25 | -4 | 16 | 0.640 |
| 5 | 0.167 | 31 | 25 | 6 | 36 | 1.440 |
| 6 | 0.167 | 26 | 25 | 1 | 1 | 0.040 |
|  |  | 150 |  |  |  | 2.480 |

Step 1: $H_0$: The die is not loaded.
$H_a$: The die is loaded.
Step 2: $\alpha = 0.05$
Step 3: $\chi^2 = 2.480$ (See column 7 of the table.)
Step 4: df = 5; From Table VII, the critical value = 11.071
For the P-value approach, $P(\chi^2 > 2.480) > 0.10$.

Step 5: Since 2.480 < 11.071, do not reject $H_0$. Since the P-value is larger than the significance level, do not reject $H_0$.

Step 6: The data do not provide sufficient evidence to conclude that the die is loaded.

**12.31** (a) Using Minitab, choose **Stat/Tables/Chi-Square Goodness-of-Fit Test**, select the **Observed Counts** option button, enter O in the **Observed counts** text box, enter GAMES in the **Category names** text box, select the **Specific proportions** option button from the **Test** list, enter P in the **Specific proportions** text box, and click **OK**. The results are

Using category names in GAMES

| Category | Observed | Test Proportion | Expected | Contribution to Chi-Sq |
|---|---|---|---|---|
| 4 | 21 | 0.1250 | 13.1250 | 4.72500 |
| 5 | 24 | 0.2500 | 26.2500 | 0.19286 |
| 6 | 24 | 0.3125 | 32.8125 | 2.36679 |
| 7 | 36 | 0.3125 | 32.8125 | 0.30964 |

| N | DF | Chi-Sq | P-Value |
|---|---|---|---|
| 105 | 3 | 7.59429 | 0.055 |

The P-value = 0.055, which is larger than the significance level 0.05, so do not reject $H_0$. At the 5% significance, the data do not provide sufficient evidence to conclude that World Series teams are not evenly matched.

(b) If $\alpha = 0.10$, the P-value is less than the significance level. Now we reject $H_0$. At th 10% significance level, the data provide sufficient evidence to conclude that World Series teams are not evenly matched.

(c) The usual assumption is that the data comprise a random sample from some population about which we want to make some inference. In this case, the data include the entire population of World Series records. We can think of these records as a sample from the population of all World Series that have been played and have yet to be played, but they are not a random sample. Nevertheless, each World Series is independent of the others, and the procedure does allow us to gain some insight into whether the overall records of the number of games played in each series follow a theoretical model based on the teams being evenly matched.

**12.33** (a)

| Gender | Distribution | Observed Frequency O | Expected Frequency E | Difference O − E | Square of Difference $(O - E)^2$ | Chi-square subtotal $(O - E)^2/E$ |
|---|---|---|---|---|---|---|
| Two girls | 0.25 | 9523 | 10722 | −1199 | 1437601 | 134.080 |
| One girl and one boy | 0.50 | 22031 | 21444 | 587 | 344569 | 16.068 |
| Two boys | 0.25 | 11334 | 10722 | 612 | 374544 | 34.932 |
|  |  | 42888 |  |  |  | 185.080 |

Step 1: $H_0$: The distribution of genders follows the model
$H_a$: The distribution of genders does not follow the model

Step 2: $\alpha = 0.01$

Step 3: $\chi^2 = 185.080$ (See column 7 of the table.)

Step 4: df = 2; From Table VII, the critical value = 9.210

For the P-value approach, $P(\chi^2 > 185.080) < 0.005$.

Step 5: Since 185.080 > 9.210, reject $H_0$. Since the P-value is smaller than the significance level, reject $H_0$.

Step 6: At the 1% the data do provide sufficient evidence to conclude that the distribution of genders in two-children families differs from the distribution predicted by the model described.
(b) There seems to be strong evidence from part (a) that for child gender it is not true that a boy or a girl is equally likely to be born.

**12.35** (a) When c =2, the chi-square goodness-of-fit test is equivalent to the one-sample z-test for one proportion (Procedure 12.2) since when there are only two categories in the chi-square test, one of them can be thought of as a success and the other as a failure.

(b) The values of p in the chi-square test correspond to p and (1 - p) in the z-test. The two observed frequencies in the chi-square test correspond to x and n - x in the z-test. In other words, knowing one of the probabilities in the chi-square test when c = 2 allows us to know the other one. The same is true for the observed frequencies and for the expected frequencies. Finally, we compare the z and chi-square statistics when c = 2 and p = $p_0$.

$$\chi^2 = \sum \frac{(O-E)^2}{E} = \frac{(x-np)^2}{np} + \frac{[(n-x)-n(1-p)]^2}{n(1-p)}$$

$$= \frac{(x-np)^2}{np} + \frac{(np-x)^2}{n(1-p)} = (x-np)^2 \left[\frac{1}{np} + \frac{1}{n(1-p)}\right]$$

$$= (x-np)^2 \left[\frac{1}{np(1-p)}\right] = \frac{(x-np)^2}{np(1-p)}$$

Note that since $z = \dfrac{\frac{x}{n}-p}{\sqrt{\frac{p(1-p)}{n}}} = \dfrac{x-np}{\sqrt{np(1-p)}}$, $z^2 = \dfrac{(x-np)^2}{np(1-p)} = \chi^2$. Thus the computed value of the $\chi^2$ statistic is the same as the square of the z-test statistic. Since the $\chi^2$ statistic is always positive, either a positive or a negative z-value will lead to the same $\chi^2$ value, making the two tests equivalent when the alternative hypothesis for the z-test is two tailed.

### Exercises 12.3
**12.37** Cells
**12.39** To obtain the total number of observations of bivariate data in a contingency table, one can sum the individual cell frequencies, sum the row subtotals, or sum the column subtotals.
**12.41** Yes. If there were no association between the gender of the physician and specialty of the physician, then the same percentage of male and female physicians would choose internal medicine. Since different percentages of male and female physicians chose internal medicine, there is an association between the variables "gender" and "specialty." In other words, knowing the gender of a physician imparts information about the likelihood that the physician specialized in internal medicine.

**12.43** (a)

|     | M  | F  | Total |
|-----|----|----|-------|
| BUS | 2  | 7  | 9     |
| ENG | 10 | 2  | 12    |
| LIB | 3  | 1  | 4     |
| Total | 15 | 10 | 25  |

(b)

|       | BUS   | ENG   | LIB   | Total |
|-------|-------|-------|-------|-------|
| M     | 0.222 | 0.833 | 0.750 | 0.600 |
| F     | 0.778 | 0.167 | 0.250 | 0.400 |
| Total | 1.000 | 1.000 | 1.000 | 1.000 |

(c)

|       | M     | F     | Total |
|-------|-------|-------|-------|
| BUS   | 0.133 | 0.700 | 0.360 |
| ENG   | 0.667 | 0.200 | 0.480 |
| LIB   | 0.200 | 0.100 | 0.160 |
| Total | 1.000 | 1.000 | 1.000 |

(d) Yes. The conditional distributions of gender are different within each college.

**12.45** (a)

|       | Fresh | Soph | Jun | Sen | Total |
|-------|-------|------|-----|-----|-------|
| Dem   | 2     | 6    | 8   | 4   | 20    |
| Rep   | 3     | 9    | 12  | 6   | 30    |
| Other | 1     | 3    | 4   | 2   | 10    |
| Total | 6     | 18   | 24  | 12  | 60    |

(b)

|       | Fresh | Soph  | Jun   | Sen   |
|-------|-------|-------|-------|-------|
| Dem   | 0.333 | 0.333 | 0.333 | 0.333 |
| Rep   | 0.500 | 0.500 | 0.500 | 0.500 |
| Other | 0.167 | 0.167 | 0.167 | 0.167 |
| Total | 1.000 | 1.000 | 1.000 | 1.000 |

(c) There is no association between party affiliation and class level. All of the conditional distributions of political party within class level are identical.

(d) The marginal distribution of party affiliation will be identical to each of the conditional distributions, that is

| Party | Frequency |
|-------|-----------|
| Dem   | 0.333     |
| Rep   | 0.500     |
| Other | 0.167     |
| Total | 1.000     |

(e) True. Since party affiliation and class level are not associated, all of the conditional distributions of class level within political party will be identical and will be the same as the marginal distribution of class level.

**12.47** (a) 12

(b)

| Region\Race | White | Black | Other | Total |
|---|---|---|---|---|
| Northeast | 1,100 | 2,493 | 1,524 | 5,117 |
| Midwest | 1,137 | 1,580 | 504 | 3,221 |
| South | 2,761 | 7,848 | 2,258 | 12,867 |
| West | 1,700 | 764 | 1,766 | 4,230 |
| Total | 6,698 | 12,685 | 6,052 | 25,435 |

  (c) 25,435 (d) 6,698 (e) 12,867 (f) 764

**12.49** (a) 1,056,500 (b) 48,600 (c) 10,600
  (d) 88,800 + 226,500 − 10,600 = 304,700
  (e) 51,900 (f) 51,900 (g) 1,648,900 − 88,800 = 1,560,100

**12.51** (a) Complete the first column, then the second column, the third row, and then the remaining totals.

|  | Full Owner | Part Owner | Tenant | Total |
|---|---|---|---|---|
| Under 50 | **739** | 70 | 44 | **853** |
| 50 - 179 | 492 | 130 | 38 | 660 |
| 180 - 499 | 198 | **143** | **27** | 368 |
| 500 - 999 | 51 | 84 | 14 | 149 |
| 1000 & Over | 41 | 114 | 17 | 172 |
| Total | 1521 | 541 | **140** | **2202** |

  (b) 15 (c) 853,000 (d) 140,000 (e) 84,000 (f) 681,000
  (g) 58,000

**12.53**

|  | Northeast | Midwest | South | West | Total |
|---|---|---|---|---|---|
| White | 0.164 | 0.170 | 0.412 | 0.254 | 1.000 |
| Black | 0.197 | 0.125 | 0.619 | 0.060 | 1.000 |
| Other | 0.252 | 0.083 | 0.373 | 0.292 | 1.000 |
| Total | 0.201 | 0.127 | 0.506 | 0.166 | 1.000 |

(a) The conditional distributions of region within each race are given in rows 2, 3, and 4 of the above table. The conditional distributions of region by race indicate that in all populations, the largest percentage of AIDS cases is in the South, but there are differences in the percentages.
(b) The marginal distribution of region is given in the last row of the above table. The largest percentage of AIDS cases, regardless of race, is from the South.
(c) Yes. The conditional distributions for region are different for the three race categories.
(d) 50.6% of the AIDS cases were from the South.
(e) 41.2% of the AIDS cases among whites were from the South.
(f) True. Since there is an association between region and race category, the conditional distributions of race by region cannot be identical (If they were identical, there would be no association between the two variables).
(g)

|  | Northeast | Midwest | South | West | Total |
|---|---|---|---|---|---|
| White | 0.215 | 0.353 | 0.215 | 0.402 | 0.263 |
| Black | 0.487 | 0.491 | 0.610 | 0.181 | 0.499 |
| Other | 0.298 | 0.156 | 0.175 | 0.417 | 0.238 |
| Total | 1.000 | 1.000 | 1.000 | 1.000 | 1.000 |

The conditional distributions of race by region are given in columns 2, 3, 4, and 5 of the table. The marginal distribution of race is given in the last column of the table. The conditional distributions of race by region indicate that among all the regions, the largest percentage of cases is Black except for in the West where the largest race category is Other.

**12.55** (a)

|  | State | Federal | Local | Total |
|---|---|---|---|---|
| 8th grade or less | 0.142 | 0.119 | 0.131 | 0.137 |
| Some high school | 0.255 | 0.145 | 0.334 | 0.273 |
| GED | 0.285 | 0.226 | 0.141 | 0.238 |
| High school diploma | 0.205 | 0.270 | 0.259 | 0.225 |
| Post secondary | 0.090 | 0.158 | 0.103 | 0.098 |
| College grad or more | 0.024 | 0.081 | 0.032 | 0.029 |
| Total | 1.001 | 0.999 | 1.000 | 1.000 |

(b) Yes. The conditional distributions of educational attainment within the facility categories (columns 2, 3, and 4 of the table in part a) are not identical.

(c) The marginal distribution of educational attainment is given in the last column of the table in part (a).

(d)

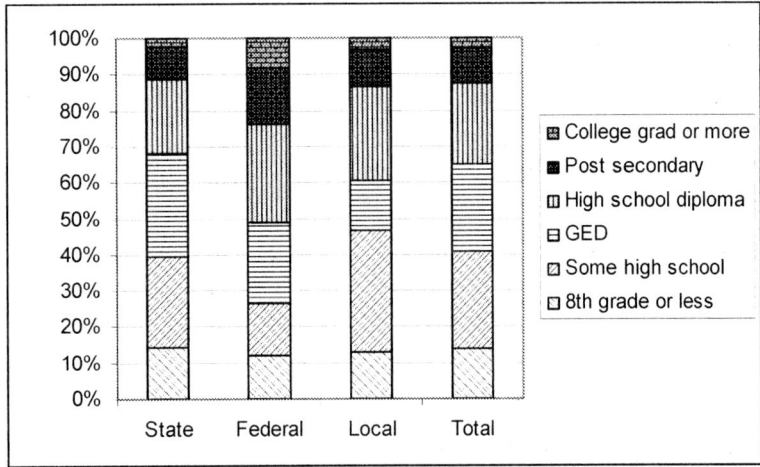

If the conditional distributions of educational attainment within the facility categories (first three bars) had been identical, each of the bars would have been segmented identically and would be identical to the fourth bar, the marginal distribution of educational attainment. The fact that they are not segmented identically means that there is an association between educational attainment and type of facility.

(e) False. Since educational attainment and type of facility are associated, the conditional distributions of facility type within educational attainment will be different.

(f) We interchanged the rows and columns of the table in Exercise 12.49 and then divided each cell entry by its associated column total to obtain the following table.

|  | 8th grade or less | Some high school | GED | High school diploma | Post secondary | College grad or more | Total |
|---|---|---|---|---|---|---|---|
| State | 0.662 | 0.598 | 0.768 | 0.584 | 0.590 | 0.521 | 0.641 |
| Federal | 0.047 | 0.029 | 0.051 | 0.065 | 0.087 | 0.148 | 0.054 |
| Local | 0.291 | 0.374 | 0.181 | 0.352 | 0.323 | 0.331 | 0.305 |
| Total | 1.000 | 1.001 | 1.000 | 1.001 | 1.000 | 1.000 | 1.000 |

The conditional distributions of facility type within educational attainment are given in columns 2, 3, 4, 5, 6, and 7. The marginal distribution of facility type is given by the last column.

(g) From the table in part (f), 5.4% of the prisoners are in federal facilities.

(h) From the table in part (f), 4.7% of the prisoners with at most an eighth grade education are in federal facilities.

(i) From the table in part (a), 11.9% of the prisoners in federal facilities have at most an 8th grade education.

**12.57** Using Minitab, choose **Stat ▶ Tables ▶ Cross Tabulation and Chi-Square**, enter REGION in the **Rows** text box, enter PARTY in the **Columns** text box, check the **Display** box for **Counts** and click **OK**. The following table is produced.

```
Rows: REGION    Columns: PARTY
           Democrat  Republican  All

Midwest         3          9       12
Northeast       6          3        9
South           6         10       16
West            6          7       13
All            21         29       50
```

(b) Repeat the procedure in part (a), but check the box for **Column percents** instead of **Counts**.

```
Rows: REGION    Columns: PARTY
           Democrat  Republican    All

Midwest      14.29      31.03    24.00
Northeast    28.57      10.34    18.00
South        28.57      34.48    32.00
West         28.57      24.14    26.00
All         100.00     100.00   100.00
```

(c) Repeat the procedure in part (a), but check the box for **Row percents** instead of **Counts**.

```
Rows: REGION    Columns: PARTY
           Democrat  Republican    All

Midwest      25.00      75.00   100.00
Northeast    66.67      33.33   100.00
South        37.50      62.50   100.00
West         46.15      53.85   100.00
All          42.00      58.00   100.00
```

(d) Yes. The conditional distributions of party within each region are not identical.

**12.59** Using Minitab, choose **Stat ▶ Tables ▶ Cross Tabulation and Chi-Square**, enter PARTY in the **Rows** text box, enter CLASS in the **Columns** text box, check the **Display** box for **Counts** and click **OK**. The following table is produced.

```
Rows: PARTY    Columns: CLASS

              I    II   III   All

Democrat     23    20    10    53
Independent   2     0     0     2
Republican    8    13    24    45
All          33    33    34   100
```

(b) Repeat the procedure in part (a), but check the box for **Column percents** instead of **Counts**.

```
Rows: PARTY    Columns: CLASS

                 I       II      III     All

Democrat      69.70   60.61   29.41    53.00
Independent    6.06    0.00    0.00     2.00
Republican    24.24   39.39   70.59    45.00
All          100.00  100.00  100.00   100.00
```

(c) Repeat the procedure in part (a), but check the box for **Row percents** instead of **Counts**.

```
Rows: PARTY    Columns: CLASS

                 I       II      III     All

Democrat      43.40   37.74   18.87   100.00
Independent  100.00    0.00    0.00   100.00
Republican    17.78   28.89   53.33   100.00
All           33.00   33.00   34.00   100.00
```

(d) Yes. The conditional distributions of party within each class are not identical.

**12.61** (a) If there were no association between age group and gender, the conditional distribution of age group within gender would be the same as the marginal distribution of age group. Therefore 7.3% of the resident population would be in the age group 20-24 years.

(b) If there were no association between age group and gender, the conditional distribution of age group within gender would be the same for each gender. Therefore 7.3% of the female residents would be in the age group 20-24 years.

(c) 7.3% of 158.0 million, i.e., 11,534,000, females would be in the age group 20-24 years.

(d) The number of female residents in the age group 20-24 years is not what we would expect IF there were NO association between age group and gender. Therefore there must be an association between age group and gender.

**Exercises 12.4**

**12.63** $H_0$: The two variables under consideration are statistically independent.
$H_a$: The two variables under consideration are statistically dependent.

**12.65** The degrees of freedom equal $(r - 1)(c - 1) = (6 - 1)(4 - 1) = 15$.

**12.67** (a) No. The existence of an association between two variables when both variables are observed does not imply a causal relationship. For example, there might exist another variable that causes the frequencies of both variables to rise and fall at the same time.

(b) Most people believe that the positive association between educational level has a causative effect on salary. Obviously, such a relationship is not perfect. There are those who never finished high school who are millionaires, and there are college graduates without jobs, but, in general, additional education enables people to qualify for better-paying jobs and to more easily adjust to changing job conditions.

# Chapter 12

**12.69** If there are four values for the first variable and five values for the second variable, there will be 20 expected frequencies. No more than 20% of these can be less than 5. 20% of 20 is 4. Therefore, a maximum number of 4 expected frequencies can be less than 5.

**12.71** If there are two values for the first variable and three values for the second variable, there will be 6 expected frequencies. No more than 20% of these can be less than 5. 20% of 6 is 1.2. Therefore, a maximum number of 1 expected frequency can be less than 5.

**12.73** If there are two values for the first variable and two values for the second variable, there will be 4 expected frequencies. No more than 20% of these can be less than 5. 20% of 4 is 0.8. Therefore, no frequencies can be less than 5.

**12.75** (a) Calculate the expected frequencies using the formula
$E=RC/n$ where R = row total, C = column total, and n = sample size. The following table on the left computes the row totals, column totals, and grand total. The following table on the right reports the expected frequencies.

|   | X: A | B | Total |
|---|---|---|---|
| Y: a | 10 | 20 | 30 |
| b | 30 | 40 | 70 |
| Total | 40 | 60 | 100 |

|   | X: A | B | Total |
|---|---|---|---|
| Y: a | 12 | 18 | 30 |
| b | 28 | 42 | 70 |
| Total | 40 | 60 | 100 |

(b) Compute the value of the test statistic, $\chi^2 = \sum \frac{(O-E)^2}{E}$ where O and E represent the observed and expected frequencies respectively. We show the contributions to this sum from each of the cells of the contingency table in the following table.

|   | A | B |
|---|---|---|
| a | 0.333 | 0.222 |
| b | 0.143 | 0.095 |

The total of the 4 table entries above is the value of the chi-square statistic, that is, $\chi^2 = 0.793$ Depending on rounding, your answer could differ slightly.

(c) The degrees of freedom are (2 - 1)(2 - 1) = 1, so the critical value from Table VII is $\chi^2_{0.05} = 3.841$.

For the P-value approach, P-value > 0.10

Since 0.793 < 3.841, we do not reject the null hypothesis. Also, since the P-value is greater than the significance level, we do not reject the null hypothesis. At the 5% significance level, the data do not provide sufficient evidence to conclude that the two variables are associated.

**12.77** (a) Calculate the expected frequencies using the formula
$E=RC/n$ where R = row total, C = column total, and n = sample size. The following table on the left computes the row totals, column totals, and grand total. The following table on the right reports the expected frequencies.

|   | X: A | B | C | Total |
|---|---|---|---|---|
| Y: a | 10 | 15 | 75 | 100 |
| b | 0 | 25 | 75 | 100 |
| Total | 10 | 40 | 150 | 200 |

|   | X: A | B | C | Total |
|---|---|---|---|---|
| Y: a | 5 | 20 | 75 | 100 |
| b | 5 | 20 | 75 | 100 |
| Total | 10 | 40 | 150 | 200 |

(b) Compute the value of the test statistic, $\chi^2 = \sum \frac{(O-E)^2}{E}$ where O and E represent the observed and expected frequencies respectively. We show the contributions to this sum from each of the cells of the contingency table in the following table.

|   | A | B | C |
|---|---|---|---|
| a | 5.000 | 1.250 | 0.000 |
| b | 5.000 | 1.250 | 0.000 |

The total of the 4 table entries above is the value of the chi-square statistic, that is, $\chi^2 = 12.500$ Depending on rounding, your answer could differ slightly.

(c) The degrees of freedom are (2 - 1)(3 - 1) = 2, so the critical value from Table VII is $\chi^2_{0.05} = 5.991$

For the P-value approach, P-value < 0.005

Since 12.500 > 5.991, we reject the null hypothesis. Also, since the P-value is less than the significance level, we reject the null hypothesis. At the 5% significance level, the data does provide sufficient evidence to conclude that the two variables are associated.

**12.79** Step 1: $H_0$: Siskel's ratings and Ebert's ratings of movies are not associated.

$H_a$: Siskel's ratings and Ebert's ratings of movies are associated.

Step 2: $\alpha = 0.01$

Step 3: Calculate the expected frequencies using the formula E=RC/n where R = row total, C = column total, and n = sample size. The results are shown in the following table.

Ebert's rating

|  |  | Thumbs down | Mixed | Thumbs up | Total |
|---|---|---|---|---|---|
| Siskel's Rating | Thumbs down | 11.8 | 8.4 | 24.8 | **45.0** |
|  | Mixed | 8.4 | 6.0 | 17.6 | **32.0** |
|  | Thumbs up | 21.8 | 15.6 | 45.6 | **83.0** |
|  | **Total** | **42.0** | **30.0** | **88.0** | **160.0** |

Compute the value of the test statistic, $\chi^2 = \sum \frac{(O-E)^2}{E}$ where O and E represent the observed and expected frequencies respectively. We show the contributions to this sum from each of the cells of the contingency table in the following table.

|  | Thumbs down | Mixed | Thumbs up |
|---|---|---|---|
| **Thumbs down** | 12.574 | 0.023 | 5.578 |
| **Mixed** | 0.019 | 8.167 | 2.475 |
| **Thumbs up** | 6.377 | 2.767 | 7.376 |

The total of the 9 table entries above is the value of the chi-square statistic, that is, $\chi^2 = 45.357$. Depending on rounding, your answer could differ slightly.

Step 4: The degrees of freedom are (3 - 1)(3 - 1) = 4, so the critical value from Table VII is $\chi^2_{0.01} = 13.277$.

For the P-value approach, P-value < 0.005

Step 5: Since 45.357 > 13.277, we reject the null hypothesis. Since the P-value is less than the significance level, we reject the null hypothesis

Step 6: We conclude that there is an association between Siskel's ratings and Ebert's ratings of movies.

**12.81** (a) The expected frequencies are shown below the observed frequencies in the following contingency table.

| Social Class | A few | Some | Lots | Total |
|---|---|---|---|---|
| Middle | 4<br>4.36 | 13<br>11.64 | 15<br>16.00 | 32 |
| Working | 5<br>4.64 | 11<br>12.36 | 18<br>17.00 | 34 |
| Total | 9 | 24 | 33 | 66 |

None of the expected values is less than one, but two of the six (33%) are less than 5. Thus Assumptions 1 and 2 are not both satisfied. The chi-square independence test should not be used.

(b)

| Social Class | Never | Sometimes | Often | Total |
|---|---|---|---|---|
| Middle | 2<br>6.303 | 8<br>8.727 | 22<br>16.970 | 32 |
| Working | 11<br>6.697 | 10<br>9.273 | 13<br>18.030 | 34 |
| Total | 13 | 18 | 35 | 66 |

None of the expected values is less than one, and none of the six are less than 5. Thus Assumptions 1 and 2 are both satisfied. The chi-square independence test may be used.

The following table gives the contributions from each cell to the chi-square statistic.

| Row, Column | O | E | $(O - E)^2/E$ |
|---|---|---|---|
| 1,1 | 2 | 6.303 | 2.9376 |
| 1,2 | 8 | 8.727 | 0.0606 |
| 1,3 | 22 | 16.970 | 1.4909 |
| 2,1 | 11 | 6.697 | 2.7648 |
| 2,2 | 10 | 9.273 | 0.0570 |
| 2,3 | 13 | 18.030 | 1.4033 |
|  | 66 |  | 8.7142 |

Step 1: $H_0$: The frequency with which parents play "I Spy" games with their children is independent of their economic class.

$H_a$: The frequency with which parents play "I Spy" games with their children is dependent of their economic class.

Step 2: $\alpha = 0.05$

Step 3: Observed and expected frequencies are presented in the frequency contingency table. Each expected frequency is placed below its corresponding observed frequency. Expected frequencies are calculated using the formula E = (R x C)/n.
The same information about the Os and Es is presented in the table below the contingency table.

$\chi^2$ = 8.7142  (See column 4 of the table below the contingency table.)

Step 4: Critical value = 5.991

For the P-value approach, 0.01 < P($\chi^2$ > 8.7142) < 0.025.

Step 5: Since 8.7142 > 5.991, reject $H_0$. Since the P-value is less than the significance level, reject $H_0$.

Step 6: The data do provide sufficient evidence to conclude that the frequency with which parents play "I Spy" with their children is dependent on economic class.

**12.83** The expected frequencies are shown below the observed frequencies in the following contingency table.

|  | Worry | | |
|---|---|---|---|
| Victim | Yes | No | Total |
| Yes | 18<br>7.323 | 21<br>31.676 | 39 |
| No | 22<br>32.676 | 152<br>141.324 | 174 |
| **Total** | **40** | **173** | **213** |

The following table gives the contributions from each cell to the chi-square statistic.

| Victim | Worry: Yes | Worry: NO |
|---|---|---|
| Yes | 15.562 | 3.598 |
| No | 3.488 | 0.807 |

Step 1: $H_0$: Worry about a gang attack and actually being a victim of a gang attack are not associated.
$H_a$: Worry about a gang attack and actually being a victim of a gang attack are associated.

Step 2: $\alpha$ = 0.01

Step 3: Observed and expected frequencies are presented in the frequency contingency table. Each expected frequency is placed below its corresponding observed frequency. Expected frequencies are calculated using the formula E = (R · C)/n.

$\chi^2$ = 23.455  (Find the sum of all the entries in the table above)
Note: Answers may vary slightly due to rounding.

Step 4: df = (2 - 1)(2 - 1) = 1; Critical value $\chi^2_{0.01}$ = 6.635

For the P-value approach, P($\chi^2$ > 23.455) < 0.005.

Step 5: Since 23.455 > 6.635, reject $H_0$. Since the P-value is smaller than the significance level, reject $H_0$.

## Chapter 12

Step 6: We conclude that an association exists between worry about a gang attack and actually being a victim of a gang attack are associated.

**12.85** (a) The expected frequencies are shown below the observed frequencies in the following contingency table.

Education

| Religiosity | Basic | Secondary | Advanced | Total |
|---|---|---|---|---|
| Religious | 77<br>68.086 | 149<br>145.729 | 78<br>90.185 | 304 |
| Not Religious | 23<br>25.756 | 56<br>55.128 | 36<br>34.116 | 115 |
| Atheist | 8<br>13.662 | 24<br>29.242 | 29<br>18.096 | 61 |
| Don't Know | 6<br>6.495 | 15<br>13.902 | 8<br>8.603 | 29 |
| Total | 114 | 244 | 151 | 509 |

The following table gives the contributions from each cell to the chi-square statistic.

Education

| Religiosity | Basic | Secondary | Advanced |
|---|---|---|---|
| Religious | 1.167 | 0.073 | 1.646 |
| Not Religious | 0.295 | 0.014 | 0.104 |
| Atheist | 2.347 | 0.940 | 6.570 |
| Don't Know | 0.038 | 0.087 | 0.042 |

Step 1: $H_0$: Religiosity and education are not associated.
$H_a$: Religiosity and education are associated.

Step 2: $\alpha = 0.05$

Step 3: Observed and expected frequencies are presented in the frequency contingency table. Each expected frequency is placed below its corresponding observed frequency. Expected frequencies are calculated using the formula $E = (R \cdot C)/n$.

$\chi^2 = 13.322$ (Find the sum of all the entries in the table above)

Note: Answers may vary slightly due to rounding.

Step 4: df = (4 − 1)(3 − 1) = 6; Critical value $\chi^2_{0.05} = 12.592$

For the P-value approach, $0.025 < P(\chi^2 > 13.322) < 0.05$.

Step 5: Since 13.322 > 12.592, reject $H_0$. Since the P-value is smaller than the significance level, reject $H_0$.

Step 6: At the 5% significance level, we conclude that an association exists between religiosity and education.

(b) The test statistic remains the same as part (a); $\chi^2 = 13.322$.

For the P-value approach, $0.025 < P(\chi^2 > 13.322) < 0.05$

Critical Value $\chi^2_{0.01} = 16.812$

Since 13.322 < 16.812 and since the P-value is greater than the significance level, do not reject $H_0$. At the 1% significance level, we do not conclude that an association exists between religiosity and education.

**12.87** Using Minitab, choose **Stat ▶ Tables ▶ Chi-Square Test (Table in Worksheet)**.

Choose **Summarized data in a two-way table**, and enter Male and Female in the **Columns containing the table** window. Click **OK**. The results are shown below using the P-value approach. Expected counts are printed below observed counts.

```
Chi-Square contributions are printed below expected counts
         Male   Female   Total
    1     77      77      154
         82.49   71.51
         0.366   0.422

    2    263     215      478
        256.05  221.95
         0.189   0.218

    3     28      27       55
         29.46   25.54
         0.072   0.084

Total    368     319      687
Chi-Sq = 1.350, DF = 2, P-Value = 0.509
```

Since the P-value = 0.509, which is greater than the significance level 0.05, do not reject $H_0$, the data do not provide evidence that gender and response are associated.

**12.89** Using Minitab, choose **Stat ▶ Tables ▶ Chi-Square Test (Table in Worksheet)**.

Choose **Summarized data in a two-way table**, and enter Urban, Suburban, and Rural in the **Columns containing the table** window. Click **OK**. Expected counts are printed below observed counts.

```
Chi-Square contributions are printed below expected counts

         Urban  Suburban  Rural   Total
    1     62       47       42     151
         60.44    53.80    36.75
         0.040    0.860    0.749

    2    197      171      109     477
        190.94   169.96   116.10
         0.192    0.006    0.435

    3     14       25       15      54
         21.62    19.24    13.14
         2.683    1.724    0.262
Total    273      243      166     682
Chi-Sq = 6.952, DF = 4, P-Value = 0.138
```

Since the P-value = 0.138, which is greater than the significance level 0.05, do not reject $H_0$, the data do not provide evidence that type of community and response are associated.

**12.91** Using Minitab, choose **Stat ▶ Tables ▶ Chi-Square Test (Table in Worksheet)**.

Choose **Summarized data in a two-way table**, and enter Under $20,000, $20,000-$29,000, $30,000-$49,999 and $50,000 and over in the **Columns containing the table** window. Click **OK**. Expected counts are printed below observed counts.

Chi-Square contributions are printed below expected counts

|   | Under $20,000 | $20,000-$29,999 | $30,000-$49,999 | $50,000 and over | Total |
|---|---|---|---|---|---|
| 1 | 40 | 32 | 41 | 34 | 147 |
|   | 29.18 | 34.52 | 40.09 | 43.21 |   |
|   | 4.014 | 0.184 | 0.021 | 1.963 |   |
| 2 | 80 | 116 | 131 | 138 | 465 |
|   | 92.30 | 109.20 | 126.82 | 136.68 |   |
|   | 1.638 | 0.423 | 0.138 | 0.013 |   |
| 3 | 11 | 7 | 8 | 22 | 48 |
|   | 9.53 | 11.27 | 13.09 | 14.11 |   |
|   | 0.228 | 1.620 | 1.980 | 4.413 |   |
| Total | 131 | 155 | 180 | 194 | 660 |

Chi-Sq = 16.634, DF = 6, P-Value = 0.011

Since the P-value = 0.011, which is less than the significance level 0.05, reject $H_0$, the data provide evidence that income and response are associated.

**Exercises 12.5**

**12.93** For the chi-square homogeneity test, the null hypothesis is that the distributions of the variable are the same for all the populations. This means there is one variable for two or more populations. For the chi-square independence test, the null hypothesis is that the two variables are independent. This means that there are two variables for one population.

**12.95** If populations are *homogeneous* with respect to a variable, it means that the distribution of the variable is similar among the different populations. If the populations are *nonhomogeneous,* it means the distribution of the variable is different among the different populations.

**12.97** If a variable has only two possible values, the chi-square homogeneity test provides a procedure for comparing several population proportions.

**12.99** If the variable has five possible values among six populations, the degrees of freedom for the chi-square statistic will be $(5-1)\cdot(6-1)=20$.

**12.101** (a) The expected frequencies are shown below the observed frequencies in the following contingency table.

Self Concept

|   | High | Moderate | Low | Total |
|---|---|---|---|---|
| Sighted | 13 | 73 | 14 | 100 |
|   | 10 | 70.625 | 19.375 |   |
| Blind | 3 | 40 | 17 | 60 |
|   | 6 | 42.375 | 11.625 |   |
| Total | 16 | 113 | 31 | 160 |

The table below gives the contributions from each cell to the chi-square statistic.

|   | High | Moderate | Low |
|---|---|---|---|
| Sighted | 0.900 | 0.080 | 1.491 |
| Blind | 1.500 | 0.133 | 2.485 |

Step 1: $H_0$: The distribution of self-concept is the same among the sightedness.
$H_a$: There is a difference in the distributions of self-concept among the sightedness.
Step 2: $\alpha = 0.05$
Step 3: Observed and expected frequencies are presented in the frequency contingency table. Each expected frequency is placed below its corresponding observed frequency. Expected frequencies are calculated using the formula $E = (R \cdot C)/n$.
$\chi^2 = 6.589$ (The sum of the cells in the table above.)
Note: Answers may vary slightly due to rounding.
Step 4: df = (2 - 1)(3 - 1) = 2; Critical value = $\chi^2_{0.05} = 5.991$
For the P-value approach, $0.025 < P(\chi^2 > 6.589) < 0.05$.
Step 5: Since 6.589 > 5.991, reject $H_0$. Since the P-value is smaller than the significance level, reject $H_0$.
Step 6: At the 5% significance level, we conclude that there is a difference in self-concept among those that are sighted and those that are blind.

(b) The test statistic remains the same as part (a); $\chi^2 = 6.589$.

For the P-value approach, $0.025 < P(\chi^2 > 6.589) < 0.05$

Critical Value $\chi^2_{0.01} = 9.210$

Since 6.589 < 9.210 and since the P-value is greater than the significance level, do not reject $H_0$. At the 1% significance level, we do not conclude that there is not a difference in self-concept among those that are sighted and those that are blind.

**12.103** The expected frequencies are shown below the observed frequencies in the following contingency table.

Race

| Region | White | Black | Other | Total |
|---|---|---|---|---|
| Northeast | 93 | 14 | 6 | 113 |
|  | 92.47 | 14.50 | 6.03 |  |
| Midwest | 118 | 14 | 4 | 136 |
|  | 111.29 | 17.45 | 7.25 |  |
| South | 167 | 42 | 7 | 216 |
|  | 176.76 | 27.72 | 11.52 |  |
| West | 113 | 7 | 15 | 135 |
|  | 110.48 | 17.33 | 7.20 |  |
| Total | 491 | 77 | 32 | 600 |

The table below gives the contributions from each cell to the chi-square statistic.

| Row, Column | O | E | $(O-E)^2/E$ |
|---|---|---|---|
| 1,1 | 93 | 92.47 | 0.003 |
| 1,2 | 14 | 14.50 | 0.017 |
| 1,3 | 6 | 6.03 | 0.000 |
| 2,1 | 118 | 111.29 | 0.405 |
| 2,2 | 14 | 17.45 | 0.682 |
| 2,3 | 4 | 7.25 | 1.457 |
| 3,1 | 167 | 176.76 | 0.539 |
| 3,2 | 42 | 27.72 | 7.356 |
| 3,3 | 7 | 11.52 | 1.773 |
| 4,1 | 113 | 110.48 | 0.057 |
| 4,2 | 7 | 17.33 | 6.157 |
| 4,3 | 15 | 7.20 | 8.450 |
| | 600 | 600.00 | 26.896 |

Step 1: $H_0$: The distribution of race is the same among the four U.S. regions.
$H_a$: There is a difference in the distributions of race among the four U.S. regions.
Step 2: $\alpha = 0.01$
Step 3: Observed and expected frequencies are presented in the frequency contingency table. Each expected frequency is placed below its corresponding observed frequency. Expected frequencies are calculated using the formula $E = (R \cdot C)/n$.
The same information about the Os and Es is presented in the table below the contingency table.

$\chi^2 = 26.896$ (See column 4 of the last table above.)
Note: Answers may vary slightly due to rounding.
Step 4: Critical value = 16.812

For the P-value approach, $P(\chi^2 > 26.896) < 0.005$.
Step 5: Since 26.896 > 16.812, reject $H_0$. Since the P-value is smaller than the significance level, reject $H_0$.
Step 6: At the 1% significance level, we conclude that there is a difference in race distributions among the four U.S. regions.

**12.105** The expected frequencies are shown below the observed frequencies in the following contingency table.

| Response | Region Urban | Suburban | Rural | Total |
|---|---|---|---|---|
| Support | 335<br>333.968 | 348<br>334.870 | 318<br>332.162 | 1001 |
| Oppose | 35<br>36.032 | 23<br>36.130 | 50<br>35.838 | 108 |
| Total | 370 | 371 | 369 | 1109 |

The table below gives the contributions from each cell to the chi-square statistic.

| Response | Urban | Suburban | Rural |
|---|---|---|---|
| Support | 0.003 | 0.515 | 0.604 |
| Oppose | 0.030 | 4.771 | 5.597 |

Step 1: $H_a$: There is no difference between the proportion of supporters among the three regions.
$H_a$: There is a difference between the proportion of supporters among the three regions.
Step 2: $\alpha = 0.01$

Step 3: Observed and expected frequencies are presented in the frequency contingency table. Each expected frequency is placed below its corresponding observed frequency. Expected frequencies are calculated using the formula E = (R · C)/n.

$\chi^2$ = 11.520 (The sum of the cells in the table above
Note: Answers may vary slightly due to rounding.

Step 4: df = (2 - 1)(3 - 1) = 2; Critical value = $\chi^2_{0.01}$ = 9.210

For the P-value approach, P($\chi^2$ > 11.520) < 0.005.

Step 5: Since 11.520 > 9.210, reject $H_0$. Since the P-value is less than the significance level, reject $H_0$.

Step 6: At the 1% significance level, we conclude that there is a difference between the proportion of supporters among the three regions.

**12.107** (a) Population 1: Men, $\hat{p}_1$ = 20/500 = 0.0400

Population 2: Women, $\hat{p}_2$ = 36/512 = 0.0703

$\hat{p}_P$ = (20 + 36)/(500 + 512) = 0.0553

Step 1: $H_0$: $p_1 = p_2$, $H_a$: $p_1 \neq p_2$
Step 2: $\alpha$ = 0.05

Step 3: $z = \dfrac{0.04 - 0.0703}{\sqrt{0.0553(1-0.0553)[(1/500)+(1/512)]}} = -2.108$

Step 4: Since $\alpha$ = 0.05, the critical values are $\pm z_{\alpha/2} = \pm 1.96$

For the P-value approach, 2*P(z < -2.11) = 0.0348

Step 5: Since -2.108 < -1.96, reject $H_0$. P-value < $\alpha$. Thus, reject $H_0$.

Step 6: The test results are significant at the 5% level; that is, at the 5% significance level, the data provide sufficient evidence to conclude that a difference exists in the proportion of male and female vegetarians.

(b) The expected frequencies are shown below the observed frequencies in the following contingency table.

Vegetarian

|       | Yes    | No      | Total |
|-------|--------|---------|-------|
| Men   | 20     | 480     | 500   |
|       | 27.668 | 472.332 |       |
| Women | 36     | 476     | 512   |
|       | 28.332 | 483.668 |       |
| Total | 56     | 956     | 1012  |

The table below gives the contributions from each cell to the chi-square statistic.

|       | Yes   | No    |
|-------|-------|-------|
| Men   | 2.125 | 0.124 |
| Women | 2.075 | 0.122 |

Step 1: $H_0$: The distribution of vegetarian preference is the same for men and women.
$H_a$: The distribution of vegetarian preference is not the same for men and women.

Step 2: $\alpha$ = 0.05

Step 3: Observed and expected frequencies are presented in the frequency contingency table. Each expected frequency is placed below its corresponding observed frequency. Expected frequencies are calculated using the formula E = (R · C)/n.

$\chi^2$ = 4.446 (sum of the cells in the table above)
Note: Answers may vary slightly due to rounding.

Step 4: df = (2-1)(2-1) = 1; Critical value $\chi^2_{0.05}$ = 3.841

For the P-value approach, 0.025 < P($\chi^2$ > 4.446) < 0.05.
From technology, the exact P-value is 0.03497

Step 5: Since 4.446 > 3.841, reject $H_0$. Since the P-value is smaller than the significance level, reject $H_0$.

Step 6: At the 5% significance level, we conclude that there is a difference in distribution of vegetarian preferences between the two genders.

(c) Both the homogeneity test and the two proportions z - test resulted in the same conclusion and had the exact same P-values.

(d) The homogeneity test is the same as a two-tailed two proportion z-test if there are only two populations.

**Review Problems For Chapter 12**

1. The distributions and curves are distinguished by their numbers of degrees of freedom.

2. (a) zero  (b) Right skewed  (c) Normal curve

3. (a) No. The degrees of freedom for the $\chi$-square goodness-of-fit test depends on the number of categories, not the number of observations.
   (b) No. The degrees of freedom for the $\chi$-square independence test is (r -1)(c-1) where r and c are the number of rows and columns, respectively, in the contingency table.
   (c) No. The degrees of freedom for the $\chi$-square homogeneity test is (r -1)(c-1) where r and c are the number of rows and columns, respectively, in the contingency table.

4. Values of the test statistic near zero arise when the observed and expected frequencies are in close agreement. It is only when these frequencies differ enough to produce large values of the test statistic that the null hypothesis is rejected. These values are in the right tail of the $\chi$-square distribution.

5. The value would be zero since O - E = 0 for pair of expected and observed frequencies.

6. (a) All expected frequencies are 1 or greater, and at most 20% of the expected frequencies are less than 5.
   (b) Very important. If either of these assumptions are not met, the test should not be carried out by these procedures.

7. (a) 18%
   (b) 18% of 37 million, or 6.66 million
   (c) Since the observed number of Americans of Irish ancestry who reside in the Northeast is not equal to the number expected if there were no association between ancestry and region, we conclude that there *is* an association between the ancestry and region of residence.

8. (a) Compare the conditional distributions of one of the variables within categories of the other variable. If all of the conditional distributions are identical, there is no association between the variables; if not, there is an association.
   (b) No. Since the data are for an entire population, we are not making an inference from a sample to the population. The association (or non-association) is a fact.

9. (a) Perform a chi-square test of independence. If the null hypothesis (of non-association) is rejected, we conclude that there is an association between the variables.

(b) Yes. It is possible (with probability α) that we could reject the null hypothesis when it is, in fact, true. It is also possible, that, even though there is actually an association between the variables, the evidence is not strong enough to draw that conclusion. Either one of these types of errors is due to randomness in selecting a sample which does not exactly reflect the characteristics of the population.

10. For df = 17:
    (a) $\chi^2_{0.10} = 24.769$  (b) $\chi^2_{0.01} = 33.409$  (c) $\chi^2_{0.05} = 27.587$

11.

| Highest level | O | p | E = np | (O-E)²/E |
|---|---|---|---|---|
| Not HS graduate | 45 | 0.129 | 64.5 | 5.895 |
| HS graduate | 163 | 0.312 | 156.0 | 0.314 |
| Some college | 76 | 0.168 | 84.0 | 0.762 |
| Associate's degree | 53 | 0.091 | 45.5 | 1.236 |
| Bachelor's degree | 102 | 0.195 | 97.5 | 0.208 |
| Advanced degree | 61 | 0.105 | 52.5 | 1.376 |
| Total | 500 | | 500.0 | 9.792 |

Step 1: $H_0$: The educational attainment distribution for adults 25 years old and over this year is the same as in 2010.
$H_a$: The educational attainment distribution for adults 25 years old and over this year differs from that of 2010.

Step 2: $\alpha$ = 0.05

Step 3: $\chi^2$ = 9.792  (See column 5 of the table.)

Step 4: df = 6 - 1 = 5; Critical value = 11.070
For the P-value approach, $0.05 < P(\chi^2 > 9.792) < 0.10$

Step 5: Since 9.792 < 11.070, do not reject $H_0$.
Since the P-value is larger than the significance level, do not reject $H_0$.

Step 6: At the 5% significance level, data does not provide evidence that the educational attainment distribution for adults 25 years old and over this year differs from that of 2010.

12. (a) The population consists of U.S. presidents.
    (b) The two variables are the presidents' region of birth and their political party.
    (c)

|  | D | DR | F | R | U | W | TOTAL |
|---|---|---|---|---|---|---|---|
| NE | 7 | 1 | 1 | 5 | 0 | 1 | 15 |
| MW | 1 | 0 | 0 | 10 | 0 | 0 | 11 |
| SO | 6 | 3 | 1 | 2 | 1 | 3 | 16 |
| WE | 1 | 0 | 0 | 1 | 0 | 0 | 2 |
| TOTAL | 15 | 4 | 2 | 18 | 1 | 4 | 44 |

13. (a) Dividing each cell entry by the row total in the table in part (a), we obtain the conditional distributions of party for each birth region and in the bottom row, the marginal distribution of party.

|  | D | DR | F | R | U | W | TOTAL |
|---|---|---|---|---|---|---|---|
| NE | 0.467 | 0.067 | 0.067 | 0.333 | 0.000 | 0.067 | 1.000 |
| MW | 0.091 | 0.000 | 0.000 | 0.909 | 0.000 | 0.000 | 1.000 |
| SO | 0.375 | 0.188 | 0.063 | 0.125 | 0.063 | 0.188 | 1.000 |
| WE | 0.500 | 0.000 | 0.000 | 0.500 | 0.000 | 0.000 | 1.000 |
| TOTAL | 0.341 | 0.093 | 0.046 | 0.409 | 0.023 | 0.091 | 1.000 |

(b) Yes. The conditional distributions of party for each birth region are not all the same.
(c) 40.9% of Presidents are Republicans.

(d) 40.9%
(e) 12.5% of Presidents born in the South are Republicans.

14. (a) Dividing each cell entry by the column total in the table in part (a), we obtain the conditional distributions of birth region for each party and in the right hand column, the marginal distribution of birth region.

|       | D     | DR    | F     | R     | U     | W     | TOTAL |
|-------|-------|-------|-------|-------|-------|-------|-------|
| NE    | 0.467 | 0.250 | 0.500 | 0.277 | 0.000 | 0.250 | 0.341 |
| MW    | 0.067 | 0.000 | 0.000 | 0.556 | 0.000 | 0.000 | 0.250 |
| SO    | 0.400 | 0.750 | 0.500 | 0.111 | 1.000 | 0.750 | 0.364 |
| WE    | 0.000 | 0.000 | 0.000 | 0.056 | 0.000 | 0.000 | 0.045 |
| TOTAL | 1.000 | 1.000 | 1.000 | 1.000 | 1.000 | 1.000 | 1.000 |

(b) Yes. The conditional distributions of birth region for each party are not all the same.
(c) 36.4% of presidents were born in the South.
(d) 36.4%
(e) 11.1% of Republican presidents were born in the South.

15. (a) 2046   (b) 737   (c) 266   (d) 3046   (e) 6580 - 1167 = 5413
    (f) 5403 + 1167 - 660 = 5910

16. (a)

|       | General | Psychiatric | Chronic | Tuberculosis | Other | Total |
|-------|---------|-------------|---------|--------------|-------|-------|
| GOV   | 0.314   | 0.361       | 0.808   | 0.750        | 0.144 | 0.311 |
| PROP  | 0.122   | 0.486       | 0.038   | 0.000        | 0.361 | 0.177 |
| NP    | 0.564   | 0.153       | 0.154   | 0.250        | 0.495 | 0.512 |
| Total | 1.000   | 1.000       | 1.000   | 1.000        | 1.000 | 1.000 |

The conditional distributions of control type with facility type are given in columns 2, 3, 4, 5, and 6 of the table.

(b) Yes. The conditional distributions of control type within facility type are not all identical.
(c) The marginal distribution of control type is given by the last column of the table above.
(d)

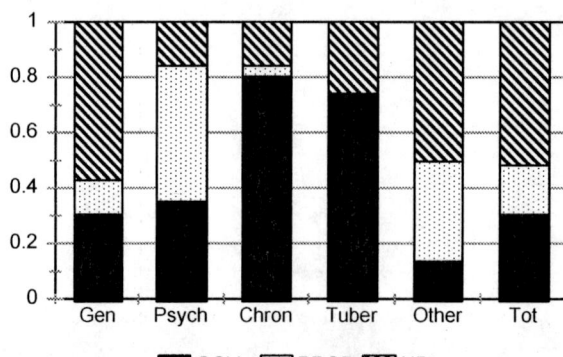

The conditional distributions of control type within facility type are shown be the first five bars of the graph. The marginal distribution of control type is given by the last bar. Since the bars are not identically shaded, control type and facility type are associated variables.

17. (a) False. Since we have established that facility type and control type are associated, the conditional distributions of facility type within control types will not be identical.

(b) After interchanging the rows and columns of the table given in Problem 15, we divided each cell entry by its associated column total to obtain the table below.

|  | GOV | PROP | NP | Total |
|---|---|---|---|---|
| General | 0.829 | 0.566 | 0.905 | 0.821 |
| Psychiatric | 0.130 | 0.307 | 0.034 | 0.112 |
| Chronic | 0.010 | 0.001 | 0.001 | 0.004 |
| Tuberculosis | 0.001 | 0.000 | 0.000 | 0.001 |
| Other | 0.029 | 0.127 | 0.060 | 0.062 |
| Total | 1.000 | 1.000 | 1.000 | 1.000 |

The conditional distributions of facility type within control type are given in columns 2, 3, and 4 of the table.

(c) The marginal distribution of facility type is given by the last column of the table.

18. (a) 17.7% of the hospitals are under proprietary control.
    (b) 48.6% of the psychiatric hospitals are under proprietary control.
    (c) 11.2% of hospitals are psychiatric hospitals.
    (d) 30.7% of the hospitals under proprietary control are psychiatric hospitals.

19. The expected frequencies are shown below the observed frequencies in the following contingency table.

| Usage | Age 18-24 | 25-64 | 65+ | Total |
|---|---|---|---|---|
| Never | 6<br>22.926 | 38<br>38.974 | 31<br>13.100 | 75 |
| Sometimes | 14<br>14.978 | 31<br>25.463 | 4<br>8.559 | 49 |
| Every day | 50<br>32.096 | 50<br>54.563 | 5<br>18.341 | 105 |
| Total | 70 | 119 | 40 | 229 |

The table below gives the contributions from each cell to the chi-square statistic.

| Usage | 18-24 | 25-64 | 65+ |
|---|---|---|---|
| Never | 12.496 | 0.024 | 24.457 |
| Sometimes | 0.064 | 1.204 | 2.428 |
| Every day | 9.987 | 0.382 | 9.704 |

Step 1: $H_a$: There is no association between age and internet usage.
$H_a$: There is an association between age and internet usage.

Step 2: $\alpha = 0.01$

Step 3: Observed and expected frequencies are presented in the frequency contingency table. Each expected frequency is placed below its corresponding observed frequency. Expected frequencies are calculated using the formula $E = (R \cdot C)/n$.

$\chi^2 = 60.746$ (The sum of the cells in the table above
Note: Answers may vary slightly due to rounding.

Step 4: df = (3 − 1)(2 − 1) = 4; Critical value = $\chi^2_{0.01} = 13.277$

For the P-value approach, $P(\chi^2 > 60.746) < 0.005$.

Step 5: Since 60.746 > 13.277, reject $H_0$. Since the P-value is less than the significance level, reject $H_0$.

Step 6: At the 1% significance level, we conclude that there is an association between age and internet usage.

20. (a) The populations are they types of residence of people. The can reside inside principal cities, outside principal cities but within metropolitan areas, or outside metropolitan areas.

(b) The variable is income level.
(c) The expected frequencies are shown below the observed frequencies in the following contingency table.

|  | IPC | OPC | OMA | Total |
|---|---|---|---|---|
| Under $15,000 | 75<br>70.560 | 106<br>119.110 | 46<br>37.330 | 227 |
| $15,000 - $34,999 | 106<br>101.954 | 161<br>172.107 | 61<br>53.939 | 328 |
| $35,000 - $74,999 | 98<br>103.509 | 183<br>174.730 | 52<br>54.761 | 333 |
| $75,000 & over | 48<br>50.977 | 102<br>86.053 | 14<br>26.970 | 164 |
| Total | 327 | 552 | 173 | 1052 |

The table below gives the contributions from each cell to the chi-square statistic.

|  | IPC | OPC | OMA |
|---|---|---|---|
| Under $15,000 | 0.279 | 1.443 | 2.014 |
| $15,000 - $34,999 | 0.161 | 0.717 | 0.924 |
| $35,000 - $74,999 | 0.293 | 0.391 | 0.139 |
| $75,000 & over | 0.174 | 2.955 | 6.237 |

Step 1: $H_0$: People residing in the three types of residence are homogeneous with respect to income level.
$H_a$: People residing in the three types of residence are nonhomogeneous with respect to income level.
Step 2: $\alpha = 0.05$
Step 3: Observed frequencies are presented in the frequency contingency table. Expected frequencies are calculated using the formula $E = (R \cdot C)/n$ and are included in the third column of the second table.
$\chi^2 = 15.728$ (sum up the cells in the table above)
Note: Answers may vary slightly due to rounding.
Step 4: df = (4 - 1)(3 - 1) = 6; Critical value = $\chi^2_{0.05} = 12.592$
For the P-value approach, $0.01 < P(\chi^2 > 15.728) < 0.025$.
Step 5: Since 15.728 > 12.592, reject $H_0$. Since the P-value is smaller than the significance level, reject $H_0$.
Step 6: At the 5% significance level, the data provide sufficient evidence to conclude that people residing in the three types of residence are nonhomogeneous with respect to income level.

21.

|  | Democrat | Republican | Independent | Total |
|---|---|---|---|---|
| Yes | 220 | 349 | 342 | 911 |
| No | 408 | 122 | 304 | 834 |
| Total | 628 | 471 | 646 | 1745 |

| Row, Column | O | E | $(O - E)^2/E$ |
|---|---|---|---|
| 1,1 | 220 | 327.86 | 35.482 |
| 1,2 | 349 | 245.89 | 43.236 |
| 1,3 | 342 | 337.25 | 0.067 |
| 2,1 | 408 | 300.14 | 38.761 |
| 2,2 | 122 | 225.11 | 47.228 |
| 2,3 | 304 | 308.75 | 0.073 |
|  | 1745 | 1745.00 | 164.847 |

Step 1: $H_0$: The percentages of Democrats, Republicans, and Independents who thought the U.S. economy was in a recession are not different
$H_a$: There is a difference in the percentages of Democrats, Republicans, and Independents who thought the U.S. economy was in a recession.

Step 2: $\alpha = 0.01$

Step 3: Observed frequencies are presented in the frequency contingency table. Expected frequencies are calculated using the formula E = (R · C)/n and are included in the third column of the second table.
$\chi^2 = 164.847$ (See column 4 of the table below the contingency table.)
Note: Answers may vary slightly due to rounding.

Step 4: Critical value = 9.210
For the P-value approach, $P(\chi^2 > 164.847) < 0.005$.

Step 5: Since 164.847 > 9.210, reject $H_0$. Since the P-value is smaller than the significance level, reject $H_0$.

Step 6: At the 1% significance level, the data provide sufficient evidence to conclude that there is a difference in the percentages of Democrats, Republicans, and Independents who thought the U.S. economy was in a recession.

22. (a) Using Minitab, **Stat ▶ Tables ▶ Cross Tabulation and Chi-Square,** enter LOT SIZE in the **For rows:** text box and HOUSE SIZE in the **For columns:** text box, click on the **Counts** box under **Display** to check it. Click **OK**. The results are

Rows: LOT SIZE   Columns: HOUSE SIZE

|  | H1 | H2 | H3 | All |
|---|---|---|---|---|
| L1 | 18 | 3 | 6 | 27 |
| L2 | 6 | 4 | 1 | 11 |
| L3 | 2 | 2 | 1 | 5 |
| L4 | 2 | 2 | 1 | 5 |
| All | 28 | 11 | 9 | 48 |

(b) Repeat the steps above, except click on the **Column Percents** box under **Display.** Click **OK**. The results are

Rows: LOT SIZE   Columns: HOUSE SIZE

|  | H1 | H2 | H3 | All |
|---|---|---|---|---|
| L1 | 64.29 | 27.27 | 66.67 | 56.25 |
| L2 | 21.43 | 36.36 | 11.11 | 22.92 |
| L3 | 7.14 | 18.18 | 11.11 | 10.42 |
| L4 | 7.14 | 18.18 | 11.11 | 10.42 |
| All | 100.00 | 100.00 | 100.00 | 100.00 |

The marginal distribution for lot size is given in the last column above.

(c) Repeat the steps above, except click on the **Row Percents** box under **Display.** Click **OK.** The results are

Rows: LOT SIZE   Columns: HOUSE SIZE

|     | H1    | H2    | H3    | All    |
|-----|-------|-------|-------|--------|
| L1  | 66.67 | 11.11 | 22.22 | 100.00 |
| L2  | 54.55 | 36.36 |  9.09 | 100.00 |
| L3  | 40.00 | 40.00 | 20.00 | 100.00 |
| L4  | 40.00 | 40.00 | 20.00 | 100.00 |
| All | 58.33 | 22.92 | 18.75 | 100.00 |

The marginal distribution for House Size is given in the last row above.

(d) By looking at the distributions, you could say that an association exists between the lot size and the house size, because the conditional distributions of house size by lot size are different. The assumptions are not met to conduct a chi-square independence test because too many of the expected counts are less than 5.

23. Using Minitab, **Stat ▶ Tables ▶ Cross Tabulation and Chi-Square,** enter REGION in the **For rows:** text box and AGE in the **For columns:** text box, click on the **Counts** box under **Display** to check it. Click on the Chi-square button, check the box for Chi-Square analysis and click **OK** twice. The results are

Rows: OPINION   Columns: EDUCATION

|            | College grad | Some college | HS grad | Not HS grad | All  |
|------------|--------------|--------------|---------|-------------|------|
| Favor      | 264          | 205          | 461     | 290         | 1220 |
| Oppose     | 17           | 26           | 81      | 81          | 205  |
| No opinion | 6            | 7            | 34      | 56          | 103  |
| All        | 287          | 238          | 576     | 427         | 1528 |

Cell Contents:     Count

Pearson Chi-Square = 77.837, DF = 6, P-Value = 0.000
Likelihood Ratio Chi-Square = 79.327, DF = 6, P-Value = 0.000

Since the Pearson Chi-Square P-value = 0.000, which is less than the significance level 0.01, reject the null hypothesis. The data provide sufficient evidence to conclude that opinion on this issue and educational level are associated.

# CHAPTER 13 SOLUTIONS

## Exercises 13.1

**13.1** We state the two numbers of degrees of freedom.

**13.3** $F_{0.05}$, $F_{0.025}$, $F_\alpha$

**13.5** (a) The first number in parentheses—12—is the number of degrees of freedom for the numerator.
(b) The second number in parentheses—7—is the number of degrees of freedom for the denominator.

**13.7** (a) $F_{0.05} = 1.89$  (b) $F_{0.01} = 2.47$  (c) $F_{0.025} = 2.14$

  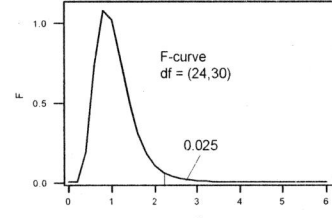

**13.9** (a) $F_{0.01} = 2.88$  (b) $F_{0.05} = 2.10$  (c) $F_{0.10} = 1.78$

  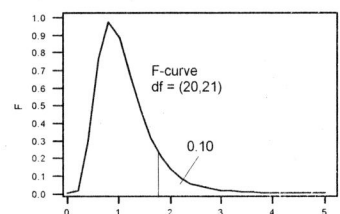

## Exercises 13.2

**13.11** The pooled-t procedure (Procedure 10.1) is a method for comparing the means of two populations. One-way ANOVA is a procedure for comparing the means of several populations. Thus, one-way ANOVA is a generalization of the pooled-t procedure.

**13.13** The reason for the word "variance" in the phrase "analysis of variance" is because the analysis-of-variance procedure for comparing means involves analyzing the *variation* in the sample data.

**13.15** A measure of variation among the sample means is called the <u>treatment mean square</u>. The mathematical abbreviation for it is <u>MSTR</u>.

**13.17** To compare the variation among the sample means to the variation within the samples, we use the ratio of *MSTR* to <u>MSE</u>. This ratio is called the <u>F-statistic</u>.

**13.19** The term *one-way* is used because the type of ANOVA compares the means of a variable for populations that result from a classification by *one* other variable, call the factor.

**13.21** No, because the variation among the sample means is not large relative to the variation within the samples.

**13.23** In a one-way ANOVA, the residual of an observation is the difference between an observation and the mean of the sample containing it.

**13.25**

$\bar{x}_1 = 18/3 = 6; \quad \bar{x}_2 = 6/3 = 2; \quad \bar{x}_3 = 16/4 = 4$

$\bar{x} = \sum x_i / n = (18 + 6 + 16)/(3 + 3 + 4) = 40/10 = 4$

$(n_1 - 1)s_1^2 = (8-6)^2 + (4-6)^2 + (6-6)^2 = 8$

$(n_2 - 1)s_2^2 = (2-2)^2 + (1-2)^2 + (3-2)^2 = 2$

$(n_3 - 1)s_3^2 = (4-4)^2 + (3-4)^2 + (6-4)^2 + (3-4)^2 = 6$

(a) $SSTR = n_1(\bar{x}_1 - \bar{x})^2 + n_2(\bar{x}_2 - \bar{x})^2 + n_3(\bar{x}_3 - \bar{x})^2 = 3(6-4)^2 + 3(2-4)^2 + 4(4-4)^2 = 24$

(b) $MSTR = SSTR/(k-1) = 24/(3-1) = 12$

(c–d) $SSE = (n_1 - 1)s_1^2 + (n_2 - 1)s_2^2 + (n_3 - 1)s_3^2 = 8 + 2 + 6 = 16$

$MSE = SSE/(n-k) = 16/(10-3) = 2.286$

(e) $F = MSTR/MSE = 12/2.286 = 5.25$

**13.27**

$\bar{x}_1 = 20/4 = 5; \quad \bar{x}_2 = 18/3 = 6; \quad \bar{x}_3 = 30/5 = 6; \quad \bar{x}_4 = 21/5 = 5; \quad \bar{x}_5 = 27/3 = 9$

$\bar{x} = \sum x_i / n = (20 + 18 + 30 + 21 + 27)/(4 + 3 + 5 + 5 + 3) = 120/20 = 6$

$(n_1 - 1)s_1^2 = (7-5)^2 + (4-5)^2 + (5-5)^2 + (4-5)^2 = 6$

$(n_2 - 1)s_2^2 = (5-6)^2 + (9-6)^2 + (4-6)^2 = 14$

$(n_3 - 1)s_3^2 = (6-6)^2 + (7-6)^2 + (5-6)^2 + (4-6)^2 + (8-6)^2 = 10$

$(n_4 - 1)s_4^2 = (3-5)^2 + (7-5)^2 + (7-5)^2 + (4-5)^2 + (4-5)^2 = 14$

$(n_5 - 1)s_5^2 = (7-9)^2 + (9-9)^2 + (11-9)^2 = 8$

(a) $SSTR = n_1(\bar{x}_1 - \bar{x})^2 + n_2(\bar{x}_2 - \bar{x})^2 + n_3(\bar{x}_3 - \bar{x})^2 + n_4(\bar{x}_4 - \bar{x})^2 + n_5(\bar{x}_5 - \bar{x})^2$
$= 4(5-6)^2 + 3(6-6)^2 + 5(6-6)^2 + 5(5-6)^2 + 3(9-6)^2 = 36$

(b) $MSTR = SSTR/(k-1) = 36/(5-1) = 9$

(c–d) $SSE = (n_1 - 1)s_1^2 + (n_2 - 1)s_2^2 + (n_3 - 1)s_3^2 + (n_4 - 1)s_4^2 + (n_5 - 1)s_5^2$
$= 6 + 14 + 10 + 14 + 8 = 52$

$MSE = SSE/(n-k) = 52/(20-5) = 3.467$

(e) $F = MSTR/MSE = 9/3.467 = 2.600$

**13.29**

$\bar{x}_1 = 24/3 = 8; \quad \bar{x}_2 = 15/3 = 5; \quad \bar{x}_3 = 36/3 = 12; \quad \bar{x}_4 = 9/3 = 3$

$\bar{x} = \sum x_i / n = (24 + 15 + 36 + 9)/(3 + 3 + 3 + 3) = 84/12 = 7$

$(n_1 - 1)s_1^2 = (11-8)^2 + (6-8)^2 + (7-8)^2 = 14$

$(n_2 - 1)s_2^2 = (9-5)^2 + (2-5)^2 + (4-5)^2 = 26$

$(n_3 - 1)s_3^2 = (16-12)^2 + (10-12)^2 + (10-12)^2 = 24$

$(n_4 - 1)s_4^2 = (5-3)^2 + (1-3)^2 + (3-3)^2 = 8$

(a) $SSTR = n_1(\bar{x}_1 - \bar{x})^2 + n_2(\bar{x}_2 - \bar{x})^2 + n_3(\bar{x}_3 - \bar{x})^2 + n_4(\bar{x}_4 - \bar{x})^2 + n_5(\bar{x}_5 - \bar{x})^2$
$= 3(8-7)^2 + 3(5-7)^2 + 3(12-7)^2 + 3(3-7)^2 = 138$

(b) $MSTR = SSTR/(k-1) = 138/(4-1) = 46$

(c-d)
$$SSE = (n_1-1)s_1^2 + (n_2-1)s_2^2 + (n_3-1)s_3^2 + (n_4-1)s_4^2 + (n_5-1)s_5^2$$
$$= 14 + 26 + 24 + 8 = 72$$
$$MSE = SSE/(n-k) = 72/(12-4) = 9$$

(e) $F = MSTR/MSE = 46/9 = 5.11$

**13.31** If the hypothesis test situation abides by the characteristics outlined in this exercise, we can use the pooled-t test or we can use the one-way ANOVA test discussed in this section using k = 2.

**Exercises 13.3**

**13.33** A small value of F results when MSTR is small compared to MSE, i.e., when the variation between sample means is small compared to the variation within samples. This describes what should happen when the null hypothesis is true; thus it does not comprise evidence that the null hypothesis is false. Only when the variation between sample means is large compared to the variation within samples, i.e., when F is large, do we have evidence that the null hypothesis is not true.

**13.35** SST = SSTR + SSE. This means that the total variation can be partitioned into a component representing variation among the sample means and another component representing variation within the samples.

**13.37** (a) One-way ANOVA
(b) Two-way ANOVA

**13.39** Since 0.708 = 2.124/df(Treatment), we have df(Treatment) = 3
It follows that df (Total) = 3 + 20 = 23.
Since 0.75 = 0.708/MSE, we have MSE = 0.708/0.75 = 0.944.
Then since 0.944 = SSE/20, we have SSE = 0.944(20) = 18.88.
Finally SST = SSTR + SSE = 2.124 + 18.880 = 21.004
Thus, the completed table is

| Source | df | SS | MS=SS/df | F-statistic |
|---|---|---|---|---|
| Treatment | 3 | 2.124 | 0.708 | 0.75 |
| Error | 20 | 18.880 | 0.944 | |
| Total | 23 | 21.004 | | |

**13.41** Since the Total df is 16, the df for Treatment is 14 - 12 = 2.
Since MSTR = SSTR/2, SSTR = 2(MSTR) = 2(1.4) = 2.8
Similarly, SSE = 12(MSE) = 12(0.9) = 10.8
Now SST = SSTR + SSE = 2.8 + 10.8 = 13.6
Finally, F = MSTR/MSE = 1.4/0.9 = 1.556
Thus, the completed table is

| Source | df | SS | MS=SS/df | F-statistic |
|---|---|---|---|---|
| Treatment | 2 | 2.8 | 1.4 | 1.556 |
| Error | 12 | 10.8 | 0.9 | |
| Total | 14 | 13.6 | | |

**13.43** We have the following:
k = 3
$n_1 = 3$  $n_2 = 3$,  $n_3 = 4$
$T_1 = 18$  $T_2 = 6$  $T_3 = 16$

$n = \sum n_j = 3 + 3 + 4 = 10$

$\sum x_i = \sum T_j = 18 + 6 + 16 = 40$

$\sum x_i^2 = (8)^2 + (4)^2 + (6)^2 + \ldots + (6)^2 + (3)^2 = 200$

(a) Consequently,

$$SST = \sum x_i^2 - (\sum x_i)^2 / n = 200 - 40^2/10 = 200 - 160 = 40$$

$$SSTR = \sum (T_j^2 / n_j) - (\sum x_i)^2 / n = (18)^2/3 + (6)^2/3 + (16)^2/4 - (40)^2/10 = 184 - 160 = 24$$

$$SSE = SST - SSTR = 40 - 24 = 16$$

(b) The results are the same as in Exercise 13.25.

(c)

| Source | df | SS | MS=SS/df | F-statistic |
|---|---|---|---|---|
| Treatment | 2 | 24 | 12.00 | 5.25 |
| Error | 7 | 16 | 2.29 | |
| Total | 9 | 40 | | |

(d) The df are (2,7). The critical value is 4.74. The P-value is $0.025 < P < 0.50$. Since 5.25 > 4.74 and since the p-value is less than the significance level, we reject the null hypothesis that the population means are equal.

**13.45** We have the following:

$k = 5$

| $n_1 = 4$ | $n_2 = 3$ | $n_3 = 5$ | $n_4 = 5$ | $n_5 = 3$ |
|---|---|---|---|---|
| $T_1 = 20$ | $T_2 = 18$ | $T_3 = 30$ | $T_4 = 25$ | $T_5 = 27$ |

$n = \sum n_j = 4 + 3 + 5 + 5 + 3 = 20$

$\sum x_i = \sum T_j = 20 + 18 + 30 + 25 + 27 = 120$

$\sum x_i^2 = (7)^2 + (4)^2 + (5)^2 + \ldots + (9)^2 + (11)^2 = 808$

(a) Consequently,

$$SST = \sum x_i^2 - (\sum x_i)^2 / n = 808 - 120^2/20 = 808 - 720 = 88$$

$$SSTR = \sum (T_j^2 / n_j) - (\sum x_i)^2 / n$$
$$= (20)^2/4 + (18)^2/3 + (30)^2/5 + (25)^2/5 + (27)^2/3 - (120)^2/20 = 756 - 720 = 36$$

$$SSE = SST - SSTR = 88 - 36 = 52$$

(b) The results are the same as in Exercise 13.27.

(c)

| Source | df | SS | MS=SS/df | F-statistic |
|---|---|---|---|---|
| Treatment | 4 | 36 | 9.00 | 2.60 |
| Error | 15 | 52 | 3.47 | |
| Total | 19 | 88 | | |

(d) The df are (4,15). The critical value is 3.06. The P-value is $0.05 < P < 0.10$. Since 2.60 > 3.06 and since the P-value is greater than the significance level, we do not reject the null hypothesis that the population means are equal.

**13.47** We have the following:
k = 4
$n_1 = 3$    $n_2 = 3$    $n_3 = 3$    $n_4 = 3$
$T_1 = 24$   $T_2 = 15$   $T_3 = 36$   $T_4 = 9$
$n = \sum n_j = 3 + 3 + 3 + 3 = 12$
$\sum x_i = \sum T_j = 24 + 15 + 36 + 9 = 84$
$\sum x_i^2 = (11)^2 + (6)^2 + (7)^2 + \ldots + (1)^2 + (3)^2 = 798$

(a) Consequently,
$$SST = \sum x_i^2 - (\sum x_i)^2 / n = 798 - 84^2/12 = 798 - 588 = 210$$
$$SSTR = \sum (T_j^2 / n_j) - (\sum x_i)^2 / n$$
$$= (24)^2/3 + (15)^2/3 + (36)^2/3 + (9)^2/3 - (84)^2/12 = 726 - 588 = 138$$
$$SSE = SST - SSTR = 210 - 138 = 72$$

(b) The results are the same as in Exercise 13.29.

(c)

| Source    | df | SS  | MS=SS/df | F-statistic |
|-----------|----|----|----------|-------------|
| Treatment | 3  | 138 | 46.0     | 5.11        |
| Error     | 8  | 72  | 9.0      |             |
| Total     | 11 | 210 |          |             |

(d) The df are (3,8). The critical value is 4.07. The P-value is 0.025 < P < 0.05. Since 5.11 > 4.07 and since the P-value is less than the significance level, we reject the null hypothesis that the population means are equal.

**13.49** The total number of populations being sampled is k = 3. Let the subscripts 1, 2, and 3 refer to diatoms, bacteria, and macroalgae, respectively. The total number of observations is n = 12. Also, $n_1 = 4$, $n_2 = 4$, and $n_3 = 4$.

Step 1: $H_0: \mu_1 = \mu_2 = \mu_3$ (population means are equal)
$H_a$: population means are different

Step 2: $\alpha = 0.05$

Step 3: Let $T_1$, $T_2$ and $T_3$ refer to the sum of the data values in each of the three samples, respectively. Thus:
$T_1 = 1828$    $T_2 = 1225$    $T_3 = 1175$.

Also, the sum of all the data values is $\sum x_i = 4228$, and their sum of squares is $\sum x_i^2 = 1,561,154$. Then,
$$SST = \sum x_i^2 - (\sum x_i)^2/n = 1561.154 - 4228^2/12 = 71488.67$$
$$SSTR = \frac{T_1^2}{n_1} + \frac{T_2^2}{n_2} + \frac{T_3^2}{n_3} - \frac{(\sum x_i)^2}{n} = \frac{1828^2}{4} + \frac{12255^2}{4} + \frac{1175^2}{4} - \frac{4228^2}{12} = 66043.17$$

and SSE = SST - SSTR = 71488.67 - 66043.17 = 5445.50
MSTR = SSTR/(k - 1) = 66043.17/(3 - 1) = 33021.585
MSE = SSE/(n - k) = 5445.50/(12 - 3) = 605.056
F = MSTR/MSE = 33021.585/605.056 = 54.98

The one-way ANOVA table is

| Source    | Df | SS       | MS = SS/df | F-statistic |
|-----------|----|---------|------------|-------------|
| Treatment | 2  | 66043.17 | 33021.585  | 54.58       |
| Error     | 9  | 5445.50  | 605.056    |             |
| Total     | 11 | 71488.67 |            |             |

Step 4: df = (k - 1, n - k) = (2, 9); critical value = 4.26
For the p-value approach, P(F > 54.58) < 0.005
Step 5: Since 54.58 > 4.26, we reject the null hypothesis.  Since P-value < 0.05, we reject the null hypothesis.
Step 6: At the 5% significance level, we reject the null hypothesis and conclude that there is a difference in the mean number of copepods among the three different diets.

**13.51** The number of populations being sampled is k = 5.  Let the subscripts 1-5 refer to strains A-E, respectively,  The total number observations is n = 25.  Also, $n_1 = n_2 = n_3 = n_4 = n_5 = 5$.

Step 1: $H_0$: $\mu_1 = \mu_2 = \mu_3 = \mu_4 = \mu_5$ (population means are equal)
$H_a$: Not all of the population means are equal
Step 2: $\alpha = 0.05$
Step 3: Let $T_1 - T_5$ refer to the sum of the data values in each of the five samples, respectively.  Thus,
$T_1 = 104$  $T_2 = 129$  $T_3 = 185$  $T_4 = 98$  $T_5 = 194$.

Also, the sum of all the data values is $\sum x_i = 710$, and their sum of squares is $\sum x_i^2 = 25424$.

Consequently, $SST = \sum x_i^2 - (\sum x_i)^2 / n = 25424 - 710^2 / 25 = 5260$

$$SSTR = \frac{T_1^2}{n_1} + \frac{T_2^2}{n_2} + \frac{T_3^2}{n_3} + \frac{T_4^2}{n_4} + \frac{T_5^2}{n_5} - \frac{(\sum x_i)^2}{n}$$

$$= \frac{104^2}{5} + \frac{129^2}{5} + \frac{185^2}{5} + \frac{98^2}{5} + \frac{194^2}{5} - \frac{710^2}{25} = 1620$$

and SSE = SST - SSTR = 5260 - 1620 = 3640.
MSTR = SSTR/(k - 1) = 1620/(5 - 1) = 405
MSE = SSE/(n - k) = 3640/(25 - 5) = 182
F = MSTR/MSE = 405/182 = 2.23

The one-way ANOVA table is:

| Source | df | SS | MS = SS/df | F-statistic |
|---|---|---|---|---|
| Treatment | 4 | 1620 | 405 | 2.23 |
| Error | 20 | 3640 | 182 | |
| Total | 24 | 5260 | | |

Step 4: df = (k - 1, n - k) = (4, 20); critical value = 2.87
For the P-value approach, P(F > 2.23) > 0.10.
Step 5: From Step 3, F = 2.23.  From Step 4, $F_{0.05}$ = 2.87.  Since 2.23 < 2.87, do not reject $H_0$.
Since the P-value is larger than the significance level, do not reject $H_0$.
Step 6: At the 5% significance level, the data do not provide sufficient evidence to conclude that the population means are different.

**13.53** The total number of populations being sampled is k = 4.  The total number of observations is n = 34.  Also, $n_1 = 8$, $n_2 = 8$, $n_3 = 10$, $n_4 = 8$.

Step 1: $H_0$: $\mu_1 = \mu_2 = \mu_3 = \mu_4$ (mean battery life among the four brands are equal)
$H_a$: Not all the means are equal.
Step 2: $\alpha = 0.05$
Step 3: Let $T_1$, $T_2$, $T_3$ and $T_4$, refer to the sum of the data values in each of

the four samples, respectively. Thus,
$T_1 = 61.50 \quad T_2 = 83.50 \quad T_3 = 85.50 \quad T_4 = 47.25$
Also, the sum of all the data values is $\sum x_i = 277.75$, and their sum of squares is $\sum x_i^2 = 2483.8125$.

Consequently, $SST = \sum x_i^2 - (\sum x_i)^2 / n = 2483.8125 - 277.75^2 / 34 = 214.840$

$$SSTR = \frac{T_1^2}{n_1} + \frac{T_2^2}{n_2} + \frac{T_3^2}{n_3} + \frac{T_4^2}{n_4} - \frac{(\sum x_i)^2}{n}$$

$$= \frac{61.5^2}{8} + \frac{83.5^2}{8} + \frac{85.5^2}{10} + \frac{47.25^2}{8} - \frac{277.75^2}{34} = 85.435$$

and SSE = SST - SSTR = 214.840 - 85.435 = 129.405
MSTR = SSTR/(k - 1) = 85.435/(4 - 1) = 28.478
MSE = SSE/(n - k) = 129.405/(34 - 4) = 4.314
F = MSTR/MSE = 28.478/4.314 = 6.602

The one-way ANOVA table is:

| Source | df | SS | MS = SS/df | F-statistic |
|---|---|---|---|---|
| Treatment | 3 | 85.435 | 28.478 | 6.602 |
| Error | 30 | 129.405 | 4.314 | |
| Total | 33 | 214.840 | | |

Step 4: df = (k - 1, n - k) = (3, 30); critical value = 2.92
For the p-value approach, $P(F > 6.602) < 0.005$.

Step 5: From Step 3, F = 6.602. From Step 4, $F_{0.05} = 2.92$. Since 6.602 > 2.92, reject $H_0$.
Since the p-value is smaller than the significance level, reject $H_0$.

Step 6: At the 5% significance level, the data provide sufficient evidence to conclude that a difference exists in mean battery life among the four brands of laptops.

**13.55** (a) To carry out the Analysis of variance and residual analysis, we will use the data in a single column, which is named RENT. In the next column, which is named REGION, are the names of each of the regions corresponding to each data value in RENT. The residual analysis consists of plotting the residuals against the means (or fits) and constructing a normal probability plot. This is done at the same time as the analysis of variance in Minitab. To do this, we choose **Stat ▶ Anova ▶One-way...**, select RENT for the **Response:** text box and REGION in the **Factor:** text box. Then click on the **Graphs** button and click to place check marks in the boxes for **Normal plot of residuals** and **Residuals versus fits**, and click **OK**. The printed results are

**One-way ANOVA: RENT versus REGION**

```
Source  DF     SS      MS      F     P
REGION   3  400513  133504  7.54  0.002
Error   16  283265   17704
Total   19  683778
S = 133.1   R-Sq = 58.57%   R-Sq(adj) = 50.81%
```

```
                                  Individual 99% CIs For Mean Based on
                                  Pooled StDev
Level       N     Mean    StDev   -+---------+---------+---------+--------
Midwest     6    743.0     92.1   (-------*-------)
Northeast   5   1055.2    150.2                         (--------*-------)
South       4    854.5    168.0          (---------*--------)
West        5   1064.4    128.4                         (-------*--------)
                                  -+---------+---------+---------+--------
                                  600       800      1000      1200

Pooled StDev = 133.1
```

(b) The F-value of 7.54 is significant at the 0.05 level (P-Value = 0.002). If the assumptions of normal populations and equal standard deviations are met, we reject the null hypothesis of equal population means and conclude that there is a difference in the mean monthly rents among newly completed apartments in the four U.S. regions.

(c) The graphs produced by the procedure in part (a) are shown following.

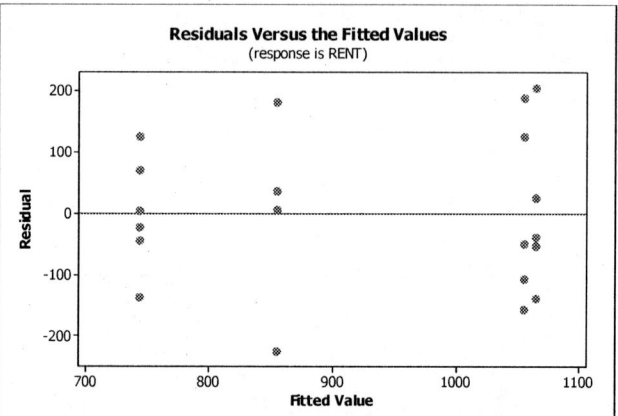

From the printed output, the ratio of the largest standard deviation to the smallest is 168.0/92.1 = 1.82. This is less than 2, indicating that the rule of 2 is not violated, leading us to conclude that the assumption of equal standard deviations is reasonable. This can also be seen in the right-hand graph above. There are no outliers in the normal probability plot and the points are linear, indicating that there are no problems with the normality assumption as well. [A Ryan-Joiner normality test of the residuals has a P-value of >0.10, leading us to conclude that the normality assumption is reasonable.]

**13.57** (a) To carry out the Analysis of variance and residual analysis, we will use the data in the single column named RATE. In the next column, which is named BREAST, are names of each of the three feather treatments corresponding to each data value in RATE. The residual analysis consists of plotting the residuals against the means (or fits) and constructing a normal probability plot. This is done at the same time as the analysis of variance in Minitab. To do this, we choose

**Stat ▶ Anova ▶ One-way...**, select RATE for the **Response:** text box and BREAST in the **Factor:** text box. Enter 99 in the **Confidence level** box. Then click on the **Graphs** button and click to place check marks in the boxes for **Normal plot of residuals** and **Residuals versus fits**, and click **OK**. The printed results are

```
One-way ANOVA: RATE versus BREAST
Source    DF       SS      MS      F      P
BREAST     2    960.3   480.1   6.09  0.008
Error     21   1655.3    78.8
Total     23   2615.6
```

```
S = 8.878   R-Sq = 36.71%   R-Sq(adj) = 30.69%

                                Individual 99% CIs For Mean Based on
                                Pooled StDev
Level      N    Mean    StDev   ----+---------+---------+---------+-----
Control    8   14.838   6.556              (--------*--------)
Enlarged   8   19.775  13.048                    (--------*--------)
Reduced    8    4.588   4.821   (--------*-------)
                                ----+---------+---------+---------+-----
                                    0        10        20        30

Pooled StDev = 8.878
```

(b) The F-value of 6.09 is significant at the 0.01 level (P-Value = 0.008). If the assumptions of normal populations and equal standard deviations are met, we reject the null hypothesis of equal population means. There is sufficient evidence to conclude that there is a difference in the mean singing rates among male rock sparrows exposed to the three types of breast treatments.

(c) The graphical results are

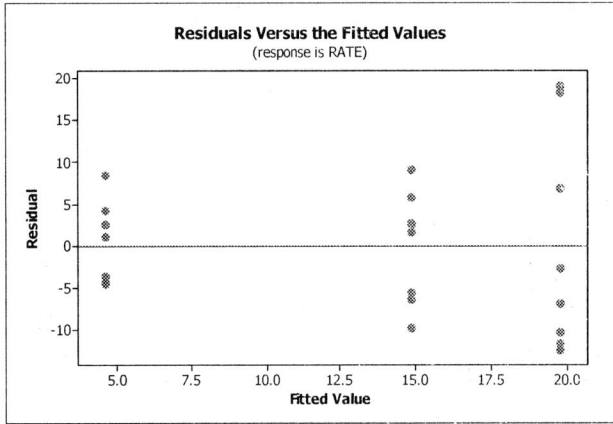

From the printed output, the ratio of the largest standard deviation to the smallest is 13.048/4.821 = 2.71. This is greater than 2, indicating that the rule of 2 is violated and leading us to conclude that the assumption of equal standard deviations may not be reasonable. The small sample sizes may be partly responsible for this occurrence, which can also be seen the right-hand graph above. The points in the probability plot are linear, indicating no problems with the normality assumption. [A Ryan-Joiner normality test of the residuals has a P-value of >0.10, leading us to conclude that the normality assumption is reasonable.]

**13.59** (a) To carry out the analysis of variance and residual analysis, we will use the data in the single column named VOLUME. In the next column, which is named MATERIAL, are names of each of the three feather treatments corresponding to each data value in VOLUME. The residual analysis consists of plotting the residuals against the means (or fits) and constructing a normal probability plot. This is done at the same time as the analysis of variance in Minitab. To do this, we choose

**Stat ▶ Anova ▶One-way...**, select VOLUME for the **Response:** text box and MATERIAL in the **Factor:** text box. Enter 95 in the **Confidence level** box. Then click on the **Graphs** button and click to place check marks in the boxes for **Normal plot of residuals** and **Residuals versus fits**, and click **OK**. The printed results are

## One-way ANOVA: HARDNESS versus MATERIAL

```
Source      DF      SS       MS        F       P
MATERIAL     2   805.31   402.66   114.71   0.000
Error       15    52.65     3.51
Total       17   857.97
```

S = 1.874    R-Sq = 93.86%    R-Sq(adj) = 93.04%

```
                                Individual 95% CIs For Mean Based on
                                Pooled StDev
Level       N    Mean   StDev   -----+---------+---------+---------+----
Duracross   6  39.633   3.120                                  (--*---)
Duradent    6  24.017   0.615   (--*--)
Endura      6  27.533   0.647         (--*--)
                                -----+---------+---------+---------+----
                                   25.0      30.0      35.0      40.0
```

Pooled StDev = 1.874

(b) The F-value of 114.71 is significant at the 0.05 level (P-Value = 0.000). If the assumptions of normal populations and equal standard deviations are met, we reject the null hypothesis of equal population means. There is sufficient evidence to conclude that there is a difference in the mean hardness among the three materials for making artificial teeth.

(c) The graphical results are

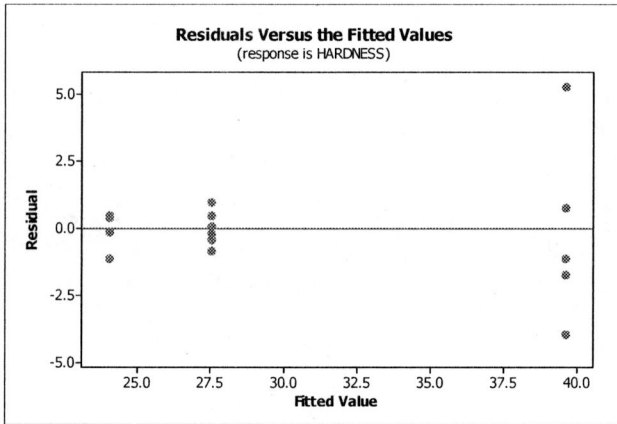

From the printed output, the ratio of the largest standard deviation to the smallest is 3.120/0.615 = 5.07. This is greater than 2. The rule of 2 is violated, leading us to conclude that the assumption of equal standard deviations may not be reasonable. The small sample sizes may be partly responsible for this occurrence, which can also be seen the right-hand graph above. The points in the probability plot are not linear, indicating that there may also be problems with the normality assumption. [A Ryan-Joiner normality test of the residuals has a P-value of < 0.010, so we conclude that the normality assumption is not reasonable.] Although the F value is so large that it is unlikely that either violation would have any serious effect on our conclusions, these data should probably be analyzed with a more robust method.

**13.61** The total number of populations being sampled is k = 5. The total number of observations is n = 236. Also, $n_1 = 46$, $n_2 = 54$, $n_3 = 47$, $n_4 = 42$, $n_5 = 47$.

Step 1: $H_0: \mu_1 = \mu_2 = \mu_3 = \mu_4 = \mu_5$ (mean BMIs are equal among the five treatment groups)
$H_a$: Not all the means are equal.
Step 2: $\alpha = 0.10$
Step 3: We can determine the mean of all the observations by the following

$$\bar{x} = \frac{n_1\bar{x}_1 + n_2\bar{x}_2 + n_3\bar{x}_3 + n_4\bar{x}_4 + n_5\bar{x}_5}{n_1 + n_2 + n_3 + n_4 + n_5}$$

$$= \frac{46(25.9) + 54(25.8) + 47(27.5) + 42(26.0) + 47(25.9)}{46 + 54 + 47 + 42 + 47} = 26.2136$$

Then, use the defining formulas to get the SSTR and SSE.

$$SSTR = \sum n_i(\bar{x}_i - \bar{x})^2 = 46(25.9 - 26.2136)^2 + 54(25.8 - 26.2136)^2 + 47(27.5 - 26.2136)^2$$
$$+ 42(26.0 - 26.2136)^2 + 47(25.9 - 26.2136)^2$$
$$= 98.0766$$

$$SSE = \sum (n_i - 1)s_i^2 = 45(4.3)^2 + 53(5.3)^2 + 46(5.8)^2 + 41(4.6)^2 + 46(4.3)^2$$
$$= 5586.36$$

and SST = SSE + SSTR = 5586.36 + 98.0766 = 5684.4366
MSTR = SSTR/(k - 1) = 98.0766/(5 - 1) = 24.5192
MSE = SSE/(n - k) = 5586.36/(236 - 5) = 24.1834
F = MSTR/MSE = 24.5192/24.1834 = 1.0139

The one-way ANOVA table is:

| Source | df | SS | MS = SS/df | F-statistic |
|---|---|---|---|---|
| Treatment | 4 | 98.0766 | 24.5192 | 1.0139 |
| Error | 231 | 5586.36 | 24.1834 | |
| Total | 235 | 5684.4366 | | |

Step 4: df = (k - 1, n - k) = (4, 231); From the table in the problem, critical value = 1.97
For the p-value approach, P(F > 1.0139) > 0.10.
Step 5: From Step 3, F = 1.0139. From Step 4, $F_{0.10} = 1.97$. Since 1.0139 < 1.97, do not reject $H_0$.
Since the p-value is larger than the significance level, do not reject $H_0$.
Step 6: At the 10% significance level, the data do not provide sufficient evidence to conclude that the mean BMI is different among the women in the five different treatment groups.

**13.63** The total number of populations being sampled is k = 6. The total number of observations is n = 160.

Step 1: $H_0: \mu_1 = \mu_2 = \mu_3 = \mu_4 = \mu_5 = \mu_6$ (mean starting salaries are the same)
$H_a$: Not all the means are equal.
Step 2: $\alpha = 0.01$
Step 3: We can determine the mean of all the observations by the following

$$\bar{x} = \frac{n_1\bar{x}_1 + n_2\bar{x}_2 + n_3\bar{x}_3 + n_4\bar{x}_4 + n_5\bar{x}_5 + n_5\bar{x}_5 + n_6\bar{x}_6}{n_1 + n_2 + n_3 + n_4 + n_5 + n_6}$$

$$= \frac{46(55.1) + 11(44.6) + 30(59.1) + 11(40.6) + 44(62.6) + 18(43.0)}{46 + 11 + 30 + 11 + 44 + 18} = 54.833$$

**400** Chapter 13

Then, use the defining formulas to get the SSTR and SSE.

$$SSTR = \sum n_i(\overline{x}_i - \overline{x})^2$$
$$= 46(55.1-54.833)^2 + 11(44.6-54.833)^2 + 30(59.1-54.833)^2$$
$$+ 11(40.6-54.833)^2 + 44(62.6-54.833)^2 + 18(43.0-54.833)^2$$
$$= 9104.431$$

$$SSE = \sum (n_i - 1)s_i^2 = 45(5.6)^2 + 10(4.7)^2 + 29(4.0)^2 + 10(5.0)^2 + 43(5.7)^2 + 17(4.8)^2$$
$$= 4134.850$$

and SST = SSE + SSTR = 4134.850 + 9104.431 = 13239.281
MSTR = SSTR/(k - 1) = 9104.431/(6 - 1) = 1820.886
MSE = SSE/(n - k) = 4134.850/(160 - 6) = 26.850
F = MSTR/MSE = 1820.886/26.850 = 67.817

The one-way ANOVA table is:

| Source | df | SS | MS = SS/df | F-statistic |
|---|---|---|---|---|
| Treatment | 5 | 9104.431 | 1820.886 | 67.817 |
| Error | 154 | 4134.850 | 26.850 | |
| Total | 159 | 13239.281 | | |

Step 4: df = (k - 1, n - k) = (5, 154); From the table in the problem, critical value = 3.14
For the p-value approach, P(F > 67.817) < 0.005.

Step 5: From Step 3, F = 67.817. From Step 4, $F_{0.01}$ = 3.14. Since 67.817 > 3.14, reject $H_0$.
Since the p-value is smaller than the significance level, reject $H_0$.

Step 6: At the 1% significance level, the data provide sufficient evidence to conclude that a difference exists in mean starting salaries among bachelor's degree candidates in the six fields.

**13.65** (a) We will obtain the plots and standard deviations using the normal probability plot in Minitab. Choose **Graph ▶ Probability plot...**, select the **Simple** version, enter WEIGHT_Lahna, WEIGHT_Siika, WEIGHT_Saerki, and WEIGHT_Parkki in the **Variables** text box, Click on the **Multiple graphs** button and select **In separate panels of the same graph**, and click **OK**. Repeat this process with LENGTH_Lahna, LENGTH_Siika, LENGTH_Saerki, and LENGTH_Parkki. The two sets of graphs are

The standard deviations shown at the right of each set of graphs are for the weights: 206.6, 309.6, 88.83, and 78.76. For the lengths they are 3.594, 5.581, 3.460, and 3.285

(b) To carry out the residual analysis (and possibly the ANOVA), we will use the data in the single columns named WEIGHT and LENGTH. In the next column, which is named SPECIES, are names of each of the four species corresponding to each data value in WEIGHT and LENGTH. The residual analysis consists of plotting the residuals against the means (or fits) and constructing a normal probability plot. This is done at the same time as the analysis of variance in Minitab. To do this, we choose **Stat ▶ Anova ▶One-way...**, select WEIGHT for the **Response:** text box and SPECIES in the **Factor:** text box. Enter 95 in the **Confidence level** box. Then click on the **Graphs** button and click to place check marks in the boxes for **Normal plot of residuals** and **Residuals versus fits**, and click **OK** twice. Repeat the process for LENGTH. The residual analysis results for WEIGHT and LENGTH are

(c) For the weights, the ratio of the largest to smallest standard deviation is 309.6/78.76 = 3.93. The rule of 2 is violated for the weights. Conducting a one-way ANOVA test on the weights data does not appear to be reasonable. Parts (d)-(e) are omitted for the weights. For the lengths, the ratio of the largest to smallest standard deviation is 5.581/3.285 = 1.70. The rule of 2 is not violated for the lengths. The normal probability plots of the individual samples and of the residuals are approximately linear in every plot. Conducting a one-way ANOVA test on these data does appear to be reasonable. Parts (d)-(e) will be performed for the lenths data.

(d) The printed portion of the output from the procedure in part (b) is

**One-way ANOVA: LENGTH versus SPECIES**

```
Source    DF      SS      MS      F       P
SPECIES    3  1845.9   615.3   44.98   0.000
Error     68   930.2    13.7
Total     71  2776.1

S = 3.699    R-Sq = 66.49%    R-Sq(adj) = 65.02%
```

```
                              Individual 95% CIs For Mean Based on
                              Pooled StDev
Level    N    Mean   StDev  ---------+---------+---------+---------+
Lahna   35  30.306   3.594                               (--*--)
Siika    6  28.800   5.581                             (-------*-------)
Saerki  20  20.645   3.460         (----*---)
Parkki  11  18.727   3.285  (-----*----)
                              ---------+---------+---------+---------+
                                    20.0      24.0      28.0      32.0
```

Pooled StDev = 3.699

Since the F-value of 44.98 has a P-value of 0.000, we reject the null hypothesis of equal population means.

(e) We conclude that at the 5% significance level, the data provide sufficient evidence that the mean lengths of the four species of fish caught in Lake Laengelmaevesi are different. Note: we determined in part (c) that it was not reasonable to perform the one-way ANOVA analysis on the weight data.

**13.67** (a) We will obtain the plots and standard deviations using the normal probability plot in Minitab. Choose **Graph ▶ Probability plot...**, select the **Simple** version, enter **Meadow Pipit**, **Tree Pipit**, and **Hedge Sparrow** in the **Variables** text box, Click on the **Multiple graphs** button and select **In separate panels of the same graph**, and click **OK**. Then repeat with Robin, Pied Wagtail, and Wren. The graphs are

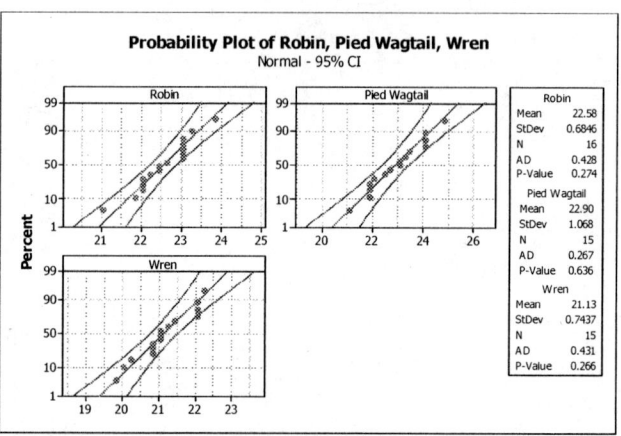

The standard deviations for the six birds are, respectively, 0.9206, 0.9014, 1.069, 0.6846, 1.068, and 0.7437.

(b) To carry out the residual analysis (and possibly the ANOVA), we will use the data in the single column named LENGTH. In the next column, which is named SPECIES, are the names of each of the six species corresponding to each data value in LENGTH. The residual analysis consists of plotting the residuals against the means (or fits) and constructing a normal probability plot. This is done at the same time as the analysis of variance in Minitab. To do this, we choose **Stat ▶ Anova ▶One-way...**, select LENGTH for the **Response:** text box and SPECIES in the **Factor:** text box. Enter 95 in the **Confidence level** box. Then click on the **Graphs** button and click to place check marks in the boxes for **Normal plot of residuals** and **Residuals versus fits**, and click **OK**. The residual analysis results are

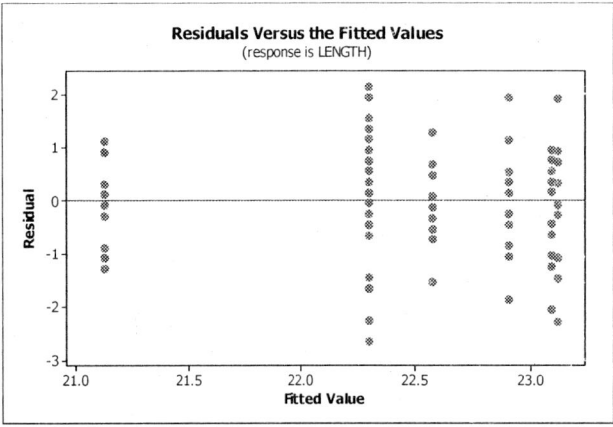

(c) The ratio of the largest to smallest standard deviation is 1.069/0.7437 = 1.44. The rule of 2 is not violated. This is also seen in the graph at the right above. The normal probability plots of the individual samples and of the residuals are all linear except for that of the Meadow Pipit, which appears to have two possible outliers with low values. Conducting a one-way ANOVA test on these data appears to be reasonable, although a more robust alternative procedure may be preferable.

(d) The printed portion of the output from the procedure in part (b) is

**One-way ANOVA: LENGTH versus SPECIES**

```
Source    DF      SS       MS      F      P
SPECIES    5   42.940   8.588   10.39  0.000
Error    114   94.248   0.827
Total    119  137.188

S = 0.9093    R-Sq = 31.30%    R-Sq(adj) = 28.29%
```

```
                             Individual 95% CIs For Mean Based on
                             Pooled StDev
Level            N    Mean   StDev
Hedge Sparrow   14  23.121   1.069
Meadow Pipit    45  22.299   0.921
Pied Wagtail    15  22.903   1.068
Robin           16  22.575   0.685
Tree Pipit      15  23.090   0.901
Wren            15  21.130   0.744
```

Pooled StDev = 0.909

Since the F-value of 10.39 has a P-value of 0.000, we reject the null hypothesis of equal population means.

(e) At the 5% significance level, we conclude that the data provide sufficient evidence that the mean egg lengths for the eggs laid by cuckoos in the nests of the six species are different.

**13.69** (a) We will obtain the plots and standard deviations using the normal probability plot in Minitab. Choose **Graph ▶ Probability plot...**, select the **Simple** version, enter Law, Science, Medicine, and Technology in the **Variables** text box, Click on the **Multiple graphs** button and select **In separate panels of the same graph**, and click **OK**. The graphs are

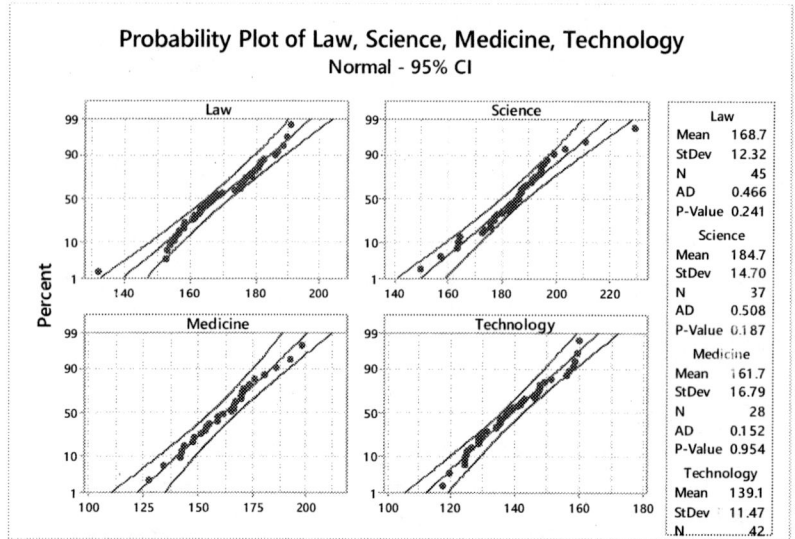

The standard deviations for the four subject areas are, respectively, 12.32, 14.70, 16.79, and 11.47.

(b) To carry out the residual analysis (and possibly the ANOVA), we will use the data in the single column named PRICE. In the next column, which is named SUBJECT, are the names of each of the four subject areas corresponding to each data value in PRICE. The residual analysis consists of plotting the residuals against the means (or fits) and constructing a normal probability plot. This is done at the same time as the analysis of variance in Minitab. To do this, we choose **Stat ▶ Anova ▶One-way...**, select PRICE for the **Response:** text box and SUBJECT in the **Factor:** text box. Enter 95 in the **Confidence level** box. Then click on the **Graphs** button and click to place check marks in the boxes for **Normal plot of residuals** and **Residuals versus fits**, and click **OK**. The residual analysis results are

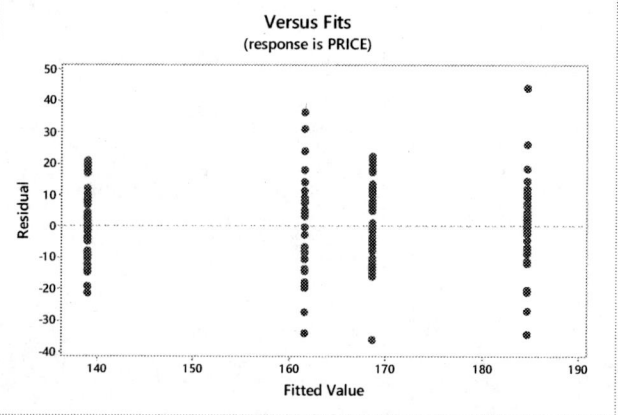

(c) The ratio of the largest to smallest standard deviation is 16.79/11.47 = 1.46. The rule of 2 is not violated. This is also seen in the graph at the right above. The normal probability plots of the individual samples and of the residuals are all linear. Conducting a one-way ANOVA test on these data appears to be reasonable.

(d) The printed portion of the output from the procedure in part (b) is

Method

  Null hypothesis         All means are equal
  Alternative hypothesis  At least one mean is different
  Significance level      α = 0.05

Equal variances were assumed for the analysis.

Factor Information

  Factor   Levels  Values
  SUBJECT     4    Law, Medicine, Science, Technology

Analysis of Variance

  Source    DF   Adj SS   Adj MS   F-Value   P-Value
  SUBJECT    3    42930   14310.0    77.13     0.000
  Error    148    27458     185.5
  Total    151    70388

Model Summary

       S   R-sq   R-sq(adj)   R-sq(pred)
  13.6208  60.99%    60.20%      58.74%

Since the F-value of 77.13 has a P-value of 0.000, we reject the null hypothesis of equal population means.

(e) At 5% significance, we conclude that the data provide sufficient evidence that the mean prices of hardcover books in the four subject areas are different.

**13.71** (a) We will obtain the plots and standard deviations using the normal probability plot in Minitab. Choose **Graph ▶ Probability plot...**, select the **Simple** version, enter HB SC, HB SS, and HB ST in the **Variables** text box, click on the **Multiple graphs** button and select **In separate panels of the same graph**, and click **OK**. The graphs are

The standard deviations for the three types of sickle cell disease are, respectively, 0.8445, 1.714, and 0.5379.

(b) To carry out the residual analysis (and possibly the ANOVA), we will use the data in the single column named HEMOLEVEL. In the next column, which is named TYPE, are the names of each of the three sickle cell types corresponding to each data value in HEMOLEVEL. The residual analysis consists of plotting the residuals against the means (or fits) and constructing a normal probability plot. This is done at the same time as the analysis of variance in Minitab. To do this, we choose

**Stat ▶ Anova ▶ One-way...**, select HEMOLEVEL for the **Response:** text box and TYPE in the **Factor:** text box. Enter 95 in the **Confidence level** box. Then click on the **Graphs** button and click to place check marks in the boxes for **Normal plot of residuals** and **Residuals versus fits**, and click **OK**. The residual analysis results are

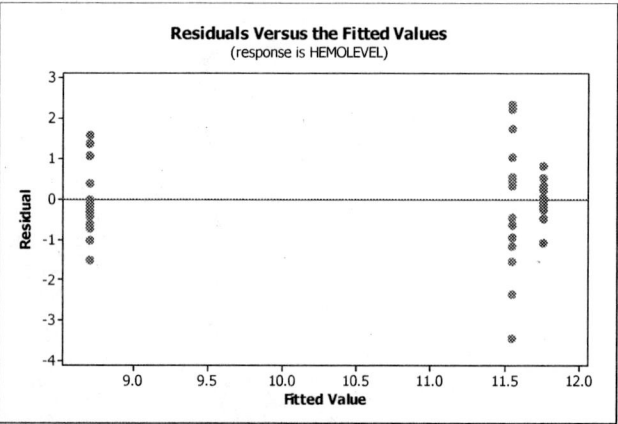

(c) The ratio of the largest to smallest standard deviation is 1.714/0.5379 = 3.19. The rule of 2 is violated. This is also seen in the graph at the right above. The normal probability plots of the individual samples and of the residuals are all linear. Conducting a one-way ANOVA test on these data may be questionable. We will proceed with the ANOVA and later we will compare the results with those of a more robust procedure.

(d) The printed portion of the output from the procedure in part (b) is

**One-way ANOVA: HEMOLEVEL versus TYPE**

```
Source   DF      SS      MS      F       P
TYPE      2   83.43   41.71   29.13   0.000
Error    38   54.42    1.43
Total    40  137.85
S = 1.197   R-Sq = 60.52%   R-Sq(adj) = 58.44%

                          Individual 95% CIs For Mean Based on
                          Pooled StDev
Level   N    Mean   StDev  --+---------+---------+---------+-------
HB SC  10  11.760   0.538                              (-----*-----)
HB SS  16   8.713   0.844  (----*----)
HB ST  15  11.547   1.714                       (----*----)
                           --+---------+---------+---------+-------
                           8.4       9.6      10.8      12.0
Pooled StDev = 1.197
```

Since the F-value of 29.13 has a P-value of 0.000, we reject the null hypothesis of equal population means, assuming that the assumptions for an ANOVA test are met.

(e) We conclude that the data provide sufficient evidence that the mean hemoglobin levels in the three types of sickle cell disease are different.

**13.73** (a) If you are given summary statistics and wish to conduct a one-way ANOVA, you can use the following formula to obtain the mean of all the observations.

$$\bar{x} = \frac{n_1\bar{x}_1 + n_2\bar{x}_2 + \cdots + n_k\bar{x}_k}{n_1 + n_2 + \cdots + n_k}$$

To verify this formula, recall that for the individual means for the samples from the separate populations, $\bar{x}_j = \frac{\sum x_{ji}}{n_j}$. Let the notation $\sum x_{1i}$ mean to sum up all of the individual observations in the sample from the first population. We can substitute this formula for the individual sample means into the formula for the overall sample mean.

$$\bar{x} = \frac{n_1 \frac{\sum x_{1i}}{n_1} + n_2 \frac{\sum x_{2i}}{n_2} + \cdots + n_k \frac{\sum x_{ki}}{n_k}}{n_1 + n_2 + \cdots + n_k}.$$

The sample sizes cancel out of the numerator and we are left with

$$\bar{x} = \frac{\sum x_{1i} + \sum x_{2i} + \cdots + \sum x_{ki}}{n_1 + n_2 + \cdots + n_k} = \frac{\sum x_i}{n}.$$

The numerator of this expression means that we will add up all of the observations from each of the individual samples. In other words, add up all of the observations. The denominator of the expression means that we will add up all of the individual sample sizes, which is the overall sample size. Therefore, we have verified the formula for obtaining the mean of all the observations from the summary statistics.

(b) If all of the sample sizes are the same then we can simplify the formula by the following:

$$\bar{x} = \frac{n_1\bar{x}_1 + n_2\bar{x}_2 + \cdots + n_k\bar{x}_k}{n_1 + n_2 + \cdots + n_k} = \frac{n\bar{x}_1 + n\bar{x}_2 + \cdots + n\bar{x}_k}{kn}.$$

We can factor an $n$ out of the numerator which will then cancel out with the $n$ in the denominator. We are left with the following:

$$\bar{x} = \frac{\bar{x}_1 + \bar{x}_2 + \cdots + \bar{x}_k}{k}.$$

This means that, if all the sample sizes are equal, the mean of all the observations is found by adding up all of the sample means and dividing by the number of populations. In other words, it is just the mean of the sample means.

(c) To get the value of the $F$-statistic from the summary statistics, we use the defining formulas for the SSTR and SSE calculations. First, calculate SSTR. $SSTR = \sum n_i(\bar{x}_i - \bar{x})^2$. For each of the sample means, find its deviation from the overall sample mean found using the formula in part (a), square this deviation, multiply it by the respective sample size, and find the sum of all these terms for each of the samples. For SSE, $SSE = \sum(n_i - 1)s_i^2$, square each sample standard deviation, multiply it by the sample size minus one, and find the sum of all these terms for each of the samples. Then, the MSTR value is found by dividing the SSTR term by the degrees of freedom for the treatment: $(k - 1)$. The MSE term is found by dividing the SSE term by the degrees of freedom for the error: $(n - k)$. Then, finally, the $F$-statistic is found by dividing MSTR by MSE.

**13.75** Suppose that we define the sample events A and B as follows:

A: the interval constructed around the difference $\mu_1 - \mu_2$

B: the interval constructed around the difference $\mu_1 - \mu_3$.

The 95% confidence interval for each event above is defined as:
P(A) = 0.95 and P(B) = 0.95.
The probability of both A and B occurring simultaneously is written P(A and B). Since A and B are realistically not independent, the general multiplication rule applies; i.e., P(A and B) = P(A)•P(B|A).
A difficulty arises in calculating a precise number for P(A and B) because we need additional information about P(B|A), which we do not have. However, P(B|A) is by no means equal to 1.00, which results in the product of P(A) and P(B|A) being less than 0.95 (because 0.95 times a probability less than 1.00 is clearly less than 0.95). Thus, the probability of both A and B occurring simultaneously is not 0.95.

### Review Problems For Chapter 13

1. One-way ANOVA is used to compare means of a variable for populations that result from a classification by one other variable called the factor.
2. (i) Simple random samples: Check by carefully studying the way that the sampling was done.
   (ii) Independent samples: Check by carefully studying the way that the sampling was done.
   (iii) Normal populations; Check with normal probability plots, histograms, dotplots.
   (iv) All populations have equal standard deviations; this is a reasonable assumption if the ratio of the largest standard deviation to the smallest one is less than 2.
3. F distribution
4. There are n = 17 observations and k = 3 samples. The degrees of freedom are (k - 1, n - k) = (2, 14).
5. (a) Variation among sample means is measured by the mean square for treatments, MSTR = SSTR/(k - 1).
   (b) The variation within samples is measured by the error mean square, MSE = SSE/(n - k).
6. (a) SST is the total sum of squares. It measures the total variation among all of the sample data. $SST = \sum (x - \bar{x})^2$

   SSTR is the treatment sum of squares. It measures the variation among the sample means. $SSTR = \sum n_i (\bar{x}_i - \bar{x})^2$

   SSE is the error sum of squares. It measures the variation within the samples. $SSE = \sum (x - \bar{x}_i)^2 = \sum (n_i - 1)s_i^2$

   (b) SST = SSTR + SSE. This means that the total variation in the sample can be broken down into two components, one representing the variation between the sample means and one representing the variation within the samples.
7. (a) One purpose of a one-way ANOVA table is to organize and summarize the quantities required for ANOVA.

(b)

| Source | df | SS | MS=SS/df | F-statistic |
|---|---|---|---|---|
| Treatment | k - 1 | SSTR | MSTR=SSTR/(k - 1) | F=MSTR/MSE |
| Error | n - k | SSE | MSE=SSE/(n - k) | |
| Total | n - 1 | SST | | |

8. 2
9. 14
10. 3.74
11. 6.51
12. 3.74
13. (a) The total number of populations being sampled is k = 3. Let the subscripts 1, 2, and 3 refer to A, B, and C, respectively. The total number of pieces of sample data is n = 12. Also, $n_1 = 3$, $n_2 = 5$, $n_3 = 4$. The following statistics for each sample are: $\bar{x}_1 = 3$, $\bar{x}_2 = 3$, and $\bar{x}_3 = 6$. Also, $s_1^2 = 4$, $s_2^2 = 6$, and $s_3^2 = 18$. Note, the mean of all the sample data is $\bar{x} = 4$. We use this information as follows:

(b) SST $= \sum (x - \bar{x})^2 = (1 - 4)^2 + \ldots + (3 - 4)^2 = 110.0$
SSTR $= n_1(\bar{x}_1 - \bar{x})^2 + n_2(\bar{x}_2 - \bar{x})^2 + n_3(\bar{x}_3 - \bar{x})^2$
$= 3(3 - 4)^2 + 5(3 - 4)^2 + 4(6 - 4)^2 = 24.0$
SSE $= (n_1 - 1)s_1^2 + (n_2 - 1)s_2^2 + (n_3 - 1)s_3^2$
$= (3 - 1)4 + (5 - 1)6 + (4 - 1)18 = 86.0$
SST = SSTR + SSE since 110.0 = 24.0 + 86.0

(c) Let $T_1$, $T_2$, and $T_3$ refer to the sum of the data values in each of the three samples, respectively. Thus: $T_1 = 9$   $T_2 = 15$   $T_3 = 24$. Also, the sum of all the data values is $\sum x = 48$, and their sum of squares is $\sum x^2 = 302$.

Consequently: $SST = \sum x^2 - \frac{(\sum x)^2}{n} = 302 - \frac{48^2}{12} = 110$

$SSTR = \frac{T_1^2}{n_1} + \frac{T_2^2}{n_2} + \frac{T_3^2}{n_3} - \frac{(\sum x)^2}{n} = \frac{9^2}{3} + \frac{15^2}{5} + \frac{24^2}{4} - \frac{48^2}{12} = 24$

and SSE = SST - SSTR = 110 - 24 = 86.

(d)

| Source | Df | SS | MS=SS/df | F-statistic |
|---|---|---|---|---|
| Treatment | 2 | 24 | 12.000 | 1.255 |
| Error | 9 | 86 | 9.556 | |
| Total | 11 | 110 | | |

14. (a) MSTR is a measure of the variation between the sample mean losses for highway robberies, gas station robberies, and convenience store robberies.

(b) MSE is a measure of the variation within the three samples.

(c) The four assumptions for one-way ANOVA, given in Key Fact 13.2: simple random samples, independent samples, normal populations, and equal standard deviations. Assumptions 1 and 2 on simple random independent samples are absolutely essential to the one-way ANOVA procedure. Assumption 3 on normality is not too critical as long as the populations are not too far from being normally distributed.

Assumption 4 on equal standard deviations is also not that important provided the sample sizes are roughly equal.

15. (a) The samples and their normal scores are shown in the table below along with the residuals which are $x_i - \bar{x}_j$.

| | Hwy | w | Residual | Gas | w | Residual | Conv. St. | w | Residual |
|---|---|---|---|---|---|---|---|---|---|
| | 952 | -1.18 | -27.8 | 1298 | -1.28 | 129.2 | 844 | -1.28 | 82.8 |
| | 996 | -0.50 | 16.2 | 1195 | -0.64 | 26.2 | 921 | -0.64 | 159.8 |
| | 839 | 0.00 | -140.8 | 1174 | -0.20 | 5.2 | 880 | -0.20 | 118.8 |
| | 1088 | 0.50 | 108.2 | 1113 | 0.20 | -55.8 | 706 | 0.20 | -55.2 |
| | 1024 | 1.18 | 44.2 | 953 | 0.64 | -215.8 | 602 | 0.64 | -159.2 |
| | | | | 1280 | 1.28 | 111.2 | 614 | 1.28 | -147.2 |
| Total | 4899 | | | 7013 | | | 4567 | | |
| Mean | 979.8 | | | 1168.8 | | | 761.2 | | |
| St.Dev. | 92.90 | | | 126.1 | | | 139.0 | | |

We will plot the normal scores w against the data for each type of store. The graphs have been produced using Minitab, but you can easily do them by hand. The standard deviations for each type of robbery are shown in the above table.

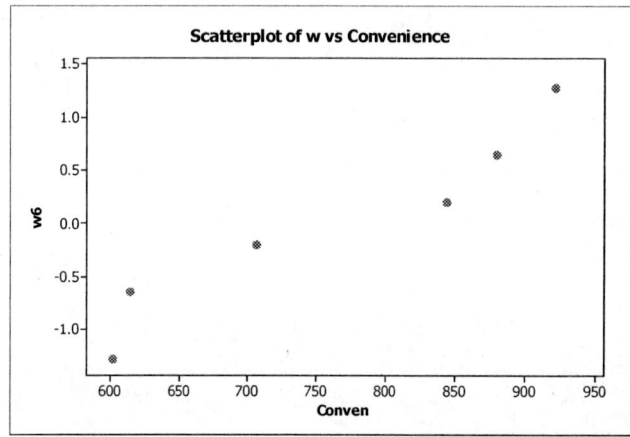

(b) Now order all of the residuals in the above table from smallest to largest and plot them against the normal scores in Table III for n = 17. Also plot the residuals against the means for each type of robbery. The results are

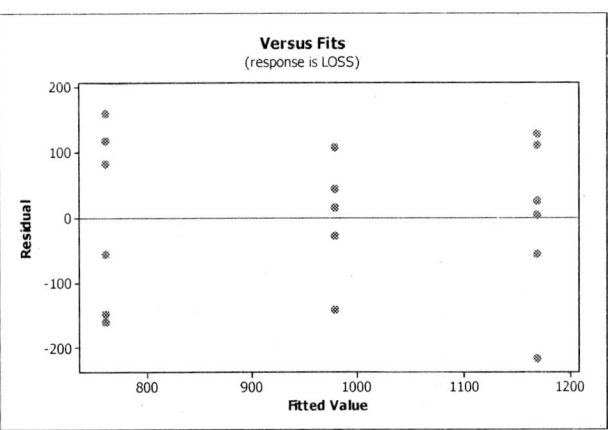

(c) All three of the normal probability plots are close to linear, and the ratio of the largest to smallest standard deviation is 1.50. The residual analysis shows that the normal probability plot of the residuals is close to linear and the last graph shows that the spreads of the residuals are about the same for each type of robbery. Thus the ANOVA assumptions of normality and equal standard deviations appear to be reasonable.

**16.** The total number of populations being sampled is k = 3. Let the subscripts 1, 2, and 3 refer to Highway, Gas station, and Convenience store, respectively. The total number of observations is n = 17. Also, $n_1$ = 5, $n_2$ = 6, and $n_3$ = 6.

Step 1: $H_0$: $\mu_1 = \mu_2 = \mu_3$ (population means are equal)
$H_a$: Not all the means are equal.
Step 2: $\alpha$ = 0.05
Step 3: The sums of squares are

$$SST = \sum x^2 - \frac{(\sum x)^2}{n} = 16,683,857 - \frac{16,479^2}{17} = 709,889.882$$

$$SSTR = \frac{T_1^2}{n_1} + \frac{T_2^2}{n_2} + \frac{T_3^2}{n_3} - \frac{(\sum x)^2}{n} = \frac{4899^2}{5} + \frac{7013^2}{6} + \frac{4567^2}{6} - \frac{16479^2}{17} = 499,349.416$$

and SSE = SST - SSTR = 709,889.882 - 499,349.416 = 210,540.466
The one-way ANOVA table is:

| Source | df | SS | MS = SS/df | F-statistic |
|---|---|---|---|---|
| Treatment | 2 | 499,349.416 | 249,674.708 | 16.602 |
| Error | 14 | 210,540.466 | 15,038.605 | |
| | 16 | 709,889.882 | | |

The F-statistic is defined and calculated as

$$F = \frac{MSTR}{MSE} = \frac{249,674.708}{15,038.605} = 16.602$$

Step 4: df = (k - 1, n - k) = (2, 14); critical value = 3.74
For the P-value approach, P(F > 16.602) < 0.005.
Step 5: From Step 3, F = 16.602. From Step 4, $F_{0.05}$ = 3.74. Since 16.602 > 3.74, reject $H_0$. Since the P-value is smaller than the significance level, reject $H_0$.
Step 6: At the 5% significance level, the data provide sufficient evidence to conclude that a difference in mean losses exists among the three types of robberies.

**17.** The total number of populations being sampled is k = 3. The total number of observations is n = 56. Also, $n_1$ = 29, $n_2$ = 11, $n_3$ = 16

Step 1: $H_0: \mu_1 = \mu_2 = \mu_3$ (mean angles of anterior-posterior foot pressure are equal among three groups)
$H_a$: Not all the means are equal.

Step 2: $\alpha = 0.01$

Step 3: We can determine the mean of all the observations by the following

$$\bar{x} = \frac{n_1\bar{x}_1 + n_2\bar{x}_2 + n_3\bar{x}_3}{n_1 + n_2 + n_3} = \frac{29(54.2) + 11(60.6) + 16(60.7)}{29 + 11 + 16} = 57.3143$$

Then, use the defining formulas to get the SSTR and SSE.

$$SSTR = \sum n_i(\bar{x}_i - \bar{x})^2 = 29(54.2 - 57.3143)^2 + 11(60.6 - 57.3143)^2 + 16(60.7 - 57.3143)^2$$
$$= 583.4286$$

$$SSE = \sum (n_i - 1)s_i^2 = 28(6)^2 + 10(8)^2 + 15(7)^2 = 2383$$

and SST = SSE + SSTR = 2383 + 583.4286 = 2966.4286
MSTR = SSTR/(k - 1) = 583.4286/(3 - 1) = 291.7143
MSE = SSE/(n - k) = 2383/(56 - 3) = 44.9623
F = MSTR/MSE = 291.7143/44.9623 = 6.4880

The one-way ANOVA table is:

| Source | df | SS | MS = SS/df | F-statistic |
|---|---|---|---|---|
| Treatment | 2 | 583.4286 | 291.7143 | 6.4880 |
| Error | 53 | 2383.0 | 44.9623 | |
| Total | 55 | 2966.4286 | | |

Step 4: df = (k - 1, n - k) = (2, 53); From the table in the problem, critical value = 5.03
For the p-value approach, P(F > 6.4880) < 0.005.

Step 5: From Step 3, F = 6.4880. From Step 4, $F_{0.01}$ = 5.03. Since 6.4880 > 5.03, reject $H_0$.
Since the p-value is smaller than the significance level, reject $H_0$.

Step 6: At the 1% significance level, the data provide sufficient evidence to conclude that there is a difference between the mean angle of anterior-posterior center of foot pressure for patients that have genu valgum, genu varum, or neither condition.

**18.** Using Minitab, choose **Graph ▶ Probability Plot**, select the **Simple** version and click **OK**. Enter Active, Passive, and Non in the **Graph Variables** text box, click on the **Multiple Graphs** button and select **In separate panels of the same graph**, and click **OK**. The result is

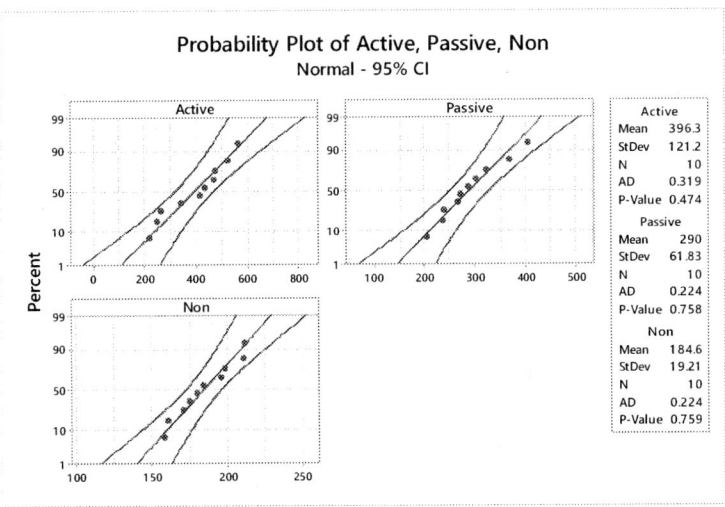

The standard deviations of the three groups are shown in the right hand panel of the graph above as 121.2, 61.83, and 19.21. The variables are approximately normal. However, the rule of 2 is violated since the ratio of the largest to smallest standard deviation is 121.2/19.21 = 6.31. Therefore, the assumption of equal population standard deviations is not met.

19. (a) Using Minitab, choose **Graph ▶ Probability Plot,** select the **Simple** version and click **OK**. Enter Loss, Stable, and Gain in the **Graph Variables** text box, click on the **Multiple Graphs** button and select **In separate panels of the same graph**, and click **OK**. The result is

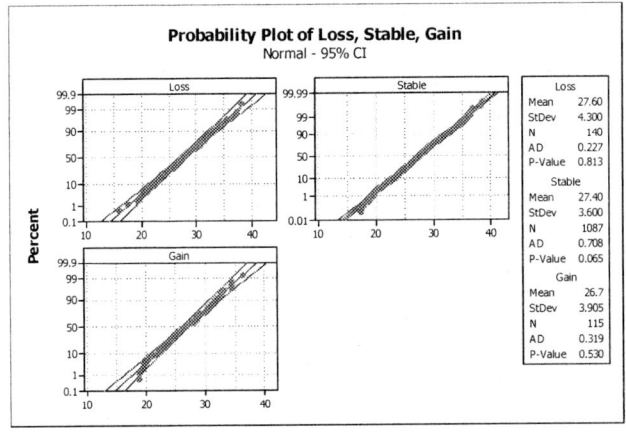

The standard deviations of the three groups are shown in the right hand panel as 4.300, 3.600, and 3.905.

(b) To carry out the Analysis of variance and residual analysis, we will use the data in a single column, which is named BMI. In the next column, which is named GROUP, are the names of each of the regions corresponding to each data value in BMI. The residual analysis consists of plotting the residuals against the means (or fits) and constructing a normal probability plot. This is done at the same time as the analysis of variance in Minitab. To do this, we choose **Stat ▶ Anova ▶ One-way...**, select BMI for the **Response:** text box and GROUP in the **Factor:** text box. Enter 95 in the **Confidence level** box. Then click on the **Graphs** button and click to place check marks in the boxes for **Normal plot of residuals** and **Residuals versus fits**, and click **OK**.

Click on the **Comparisons** button and select **Tukey's**, and enter 5 in the **family error rate** box, and click **OK**. The graphical results are

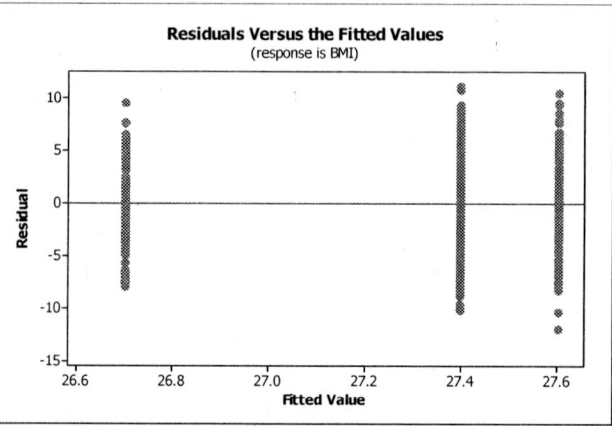

(c) The normal probability plots in part (a) are linear as is the normal probability plot of the residuals in part (b). The plot of the residuals against the fitted values shows approximately equal spreads. The rule of 2 is not violated since the ratio of the largest to smallest standard deviation is 4.3/3.6 = 1.19. A one-way ANOVA is reasonable.

(d) The one-way ANOVA results from the procedure in part (b) are

**One-way ANOVA: BMI versus GROUP**

```
Source    DF      SS      MS     F      P
GROUP      2    60.0    30.0  2.18  0.113
Error   1339 18379.7    13.7
Total   1341 18439.6

S = 3.705   R-Sq = 0.33%   R-Sq(adj) = 0.18%
```

```
                          Individual 95% CIs For Mean Based on
                          Pooled StDev
Level     N    Mean  StDev  ------+---------+---------+---------+---
Gain    115  26.700  3.905  (----------*----------)
Loss    140  27.603  4.300                    (---------*---------)
Stable 1087  27.399  3.600                       (---*--)
                           ------+---------+---------+---------+---
                             26.40     27.00     27.60     28.20
```

(e) The P-value of the test is 0.113, which is larger than the 0.05 significance level. Do not reject the null hypothesis of equal means. The data do not provide sufficient evidence to conclude that there is a difference in body mass index among the three groups of men with different weight losses.

20. (a) Using Minitab, choose **Graph ▶ Probability Plot,** select the **Simple** version and click **OK**. Enter Loss, Stable, and Gain in the **Graph Variables** text box, click on the **Multiple Graphs** button and select **In separate panels of the same graph**, and click **OK**. The result is

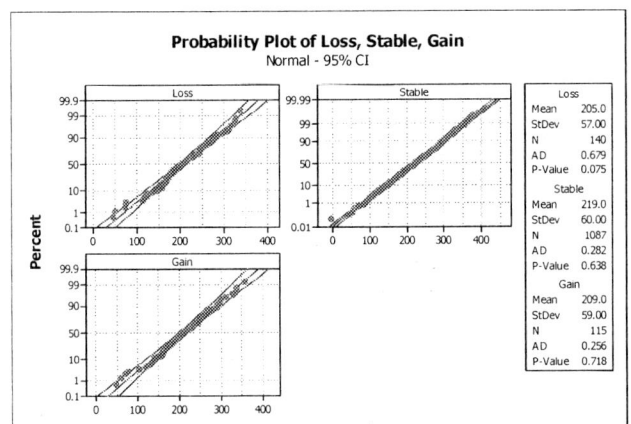

The standard deviations of the three groups are shown in the right hand panel as 57, 60, and 59.

(b) To carry out the Analysis of variance and residual analysis, we will use the data in a single column, which is named POWER. In the next column, which is named GROUP, are the names of each of the regions corresponding to each data value in POWER. The residual analysis consists of plotting the residuals against the means (or fits) and constructing a normal probability plot. This is done at the same time as the analysis of variance in Minitab. To do this, we choose **Stat ▶ Anova ▶ One-way...**, select POWER for the **Response:** text box and GROUP in the **Factor:** text box. Enter 95 in the **Confidence level** box. Then click on the **Graphs** button and click to place check marks in the boxes for **Normal plot of residuals** and **Residuals versus fits**, and click **OK**. Click on the **Comparisons** button and select **Tukey's**, and enter 5 in the **family error rate** box, and click **OK**. The graphical results are

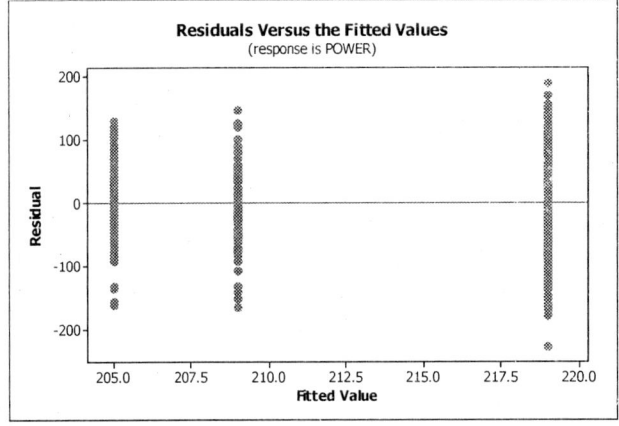

(c) The normal probability plots in part (a) are linear as is the normal probability plot of the residuals in part (b). The plot of the residuals against the fitted values shows approximately equal spreads. The rule of 2 is not violated since the ratio of the largest to smallest standard deviation is 60/57 = 1.05. A one-way ANOVA is reasonable.

(d) The one-way ANOVA results from the procedure in part (b) are

One-way ANOVA: POWER versus GROUP

```
Source   DF       SS      MS     F      P
GROUP     2    31742   15871  4.47  0.012
Error  1339  4758204    3554
Total  1341  4789946
```

```
                S = 59.61    R-Sq = 0.66%    R-Sq(adj) = 0.51%

                                        Individual 95% CIs For Mean Based on Pooled
                                        StDev
        Level      N     Mean    StDev  -+---------+---------+---------+--------
        Gain     115   209.00    59.00           (---------------*--------------)
        Loss     140   205.00    57.00   (--------------*-------------)
        Stable  1087   219.00    60.00                                   (----*----)
                                        -+---------+---------+---------+--------
                                        196.0     203.0     210.0     217.0
```

Pooled StDev = 59.61

(e) The P-value of the test is 0.012, which is smaller than the 0.05 significance level. Reject the null hypothesis of equal means. The data provide sufficient evidence to conclude that there is a difference in leg power among the three groups of men with different weight losses.

**21.** (a) Using Minitab, choose **Graph ▶ Probability Plot,** select the **Simple** version and click **OK**. Enter 25-34, 35-44, 45-54 and 55-64 in the **Graph Variables** text box, click on the **Multiple Graphs** button and select **In separate panels of the same graph**, and click **OK**. The result is

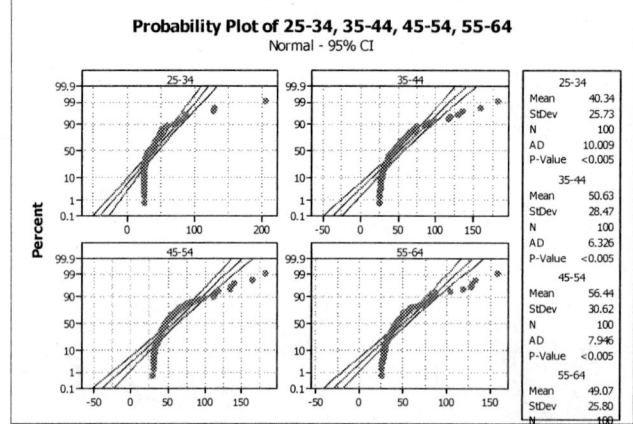

The standard deviations of the incomes in the four age groups are 25.73, 28.47, 30.62, and 25.80.

(b) To carry out the Analysis of variance and residual analysis, we will use the data in a single column, which is named INCOME. In the next column, which is named AGE, are the categories for ages corresponding to each data value in INCOME. The residual analysis consists of plotting the residuals against the means (or fits) and constructing a normal probability plot. This is done at the same time as the analysis of variance in Minitab. To do this, we choose **Stat ▶ Anova ▶One-way...,** select INCOME for the **Response:** text box and AGE in the **Factor:** text box. Enter 95 in the **Confidence level** box. Then click on the **Graphs** button and click to place check marks in the boxes for **Normal plot of residuals** and **Residuals versus fits**, and click **OK**. Click on the **Comparisons** button and select **Tukey's**, and enter 5 in the **family error rate** box, and click **OK**. The graphical results are

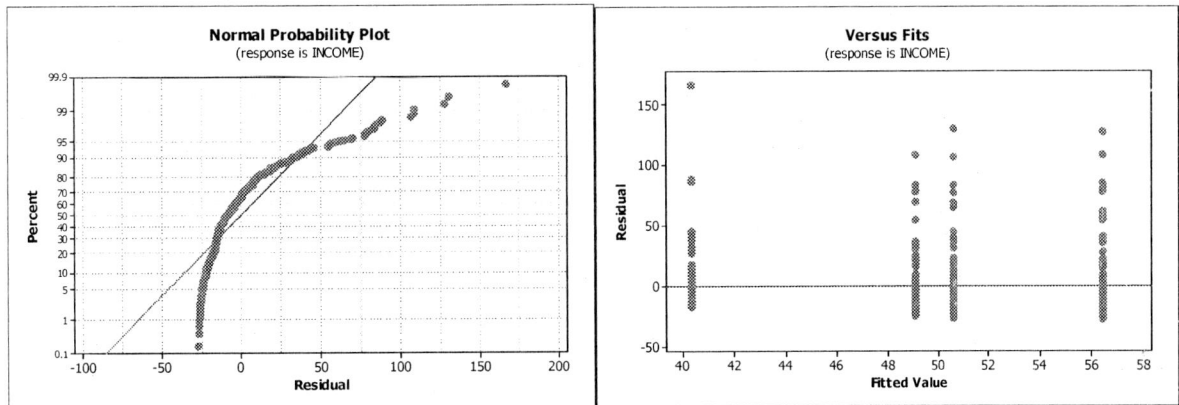

(c) The normal probability plots for each age group are non-linear, as is the residual normal probability plot in part (b). The rule of 2 is not violated as the ratio of the largest to smallest standard deviation is 30.62/25.73 = 1.19. The normality assumption is not reasonable, so parts (d)-(e) are omitted.

## CHAPTER 14 SOLUTIONS

### Exercises 14.1

**14.1** Conditional distribution, conditional mean, and conditional standard deviation

**14.3** (a) population regression line  (b) $\sigma$
(c) normal, $\beta_0 + 6\beta_1$, $\sigma$

**14.5** The sample regression line is the best estimate of the population regression line.

**14.7** Residual

**14.9** The plot of the residuals against the values of the predictor variable provides the same information as a scatter diagram of the data points. However, it has the advantage of making it easier to spot patterns such as curvature and non-constant standard deviation.

**14.11** (a) The assumption of linearity (Assumption 1) may be violated since the band is not horizontal, as may be the assumption of equal standard deviations (Assumption 2) since there is more variation in the residuals for small x than for large x.
(b) It appears that the standard deviation does not remain constant; thus, Assumption 2 is violated.
(c) The graph does not suggest violation of one or more of the assumptions for regression inferences.
(d) The normal probability plot appears to be more curved than linear; thus, the assumption of normality is violated.

**14.13** (a) We first create the table below.

| x | y | $\hat{y}$ | $e = y - \hat{y}$ | $e^2$ |
|---|---|---|---|---|
| 3 | -4 | -5 | 1 | 1 |
| 1 | 0 | -1 | 1 | 1 |
| 2 | -5 | -3 | -2 | 4 |
|   |   |   |   | 6 |

$$s_e = \sqrt{\frac{SSE}{n-2}} = \sqrt{\frac{6}{1}} = 2.449$$

(b-c) For the residual plot, we plot e against x in the graph below at the left. We can use Excel for the normal probability plot. With the values of e in a column named 'e', highlight the data with the mouse, choose **Charts and Plots**, click on the **Function Type** drop down box, and click on **Normal Probability Plot**. Drag the 'e' from the **Names and Columns** box into the **Quantitative Variable** box and click **OK**. The graph is shown below at the right.

**14.15** (a) We first create the table below.

| x | y | $\hat{y}$ | $e = y - \hat{y}$ | $e^2$ |
|---|---|---|---|---|
| 3 | 4 | 3 | 1 | 1 |
| 4 | 5 | 5 | 0 | 0 |
| 1 | 0 | -1 | 1 | 1 |
| 2 | -1 | 1 | -2 | 4 |
|   |   |   |   | 6 |

$$s_e = \sqrt{\frac{SSE}{n-2}} = \sqrt{\frac{6}{2}} = 1.732$$

(b-c) For the residual plot, we plot e against x in the graph following at the left. We can use Excel for the normal probability plot. With the values of e in a column named 'e', highlight the data with the mouse, choose **Charts and Plots**, click on the **Function Type** drop down box, and click on **Normal Probability Plot**. Drag the 'e' from the **Names and Columns** box into the **Quantitative Variable** box and click **OK**. The graph is shown following at the right.

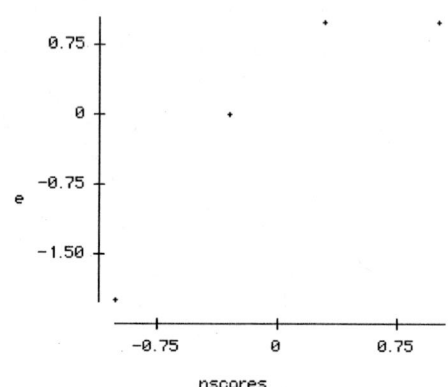

**14.17** (a) We first create the table below.

| x | y | $\hat{y}$ | $e = y - \hat{y}$ | $e^2$ |
|---|---|---|---|---|
| 2 | 3 | 3 | 0 | 0 |
| 2 | 4 | 3 | 1 | 1 |
| 3 | 0 | 2 | -2 | 4 |
| 4 | 2 | 1 | 1 | 1 |
| 4 | 1 | 1 | 0 | 0 |
|   |   |   |   | 6 |

$$s_e = \sqrt{\frac{SSE}{n-2}} = \sqrt{\frac{6}{3}} = 1.414$$

(b-c) For the residual plot, we plot e against x in the graph following at the left. We can use Minitab for the normal probability plot. With the values of e in a column named 'e', choose **Graph, Probability Plot**. Click **Simple** and click **OK**. Enter e into the Graph variables text box. Click **OK**. The graph is shown following at the right.

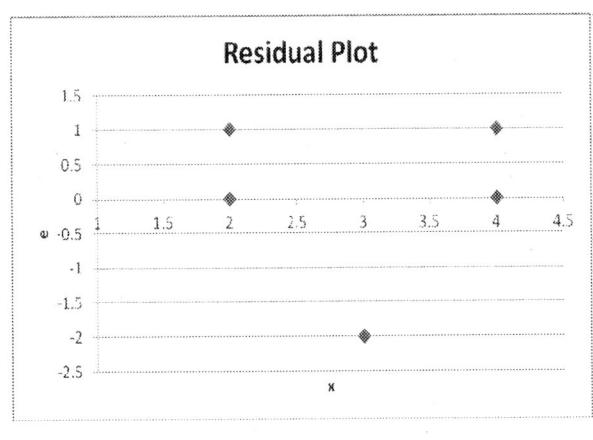

**14.19** (a) We first create the table below.

| X | Y | $\hat{y}$ | $e = y - \hat{y}$ | $e^2$ |
|---|---|---|---|---|
| 1 | 4 | 3 | 1 | 1 |
| 2 | 3 | 5 | -2 | 4 |
| 3 | 8 | 7 | 1 | 1 |
|   |   |   |   | 6 |

$$s_e = \sqrt{\frac{SSE}{n-2}} = \sqrt{\frac{6}{1}} = 2.449$$

(b-c) For the residual plot, we plot e against x in the graph following at the left. We can use Minitab for the normal probability plot. With the values of e in a column named 'e', choose **Graph, Probability Plot.** Click **Simple** and click **OK**. Enter e into the Graph variables text box. Click **OK**. The graph is shown following at the right.

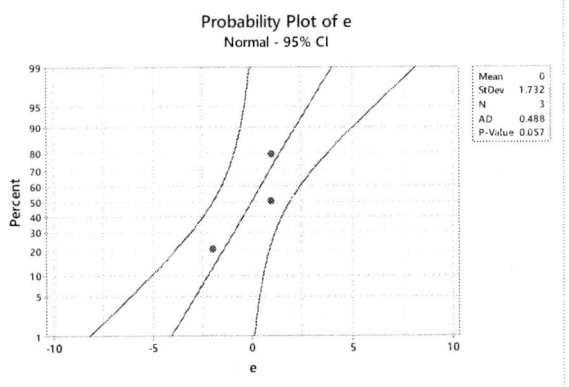

**14.21** (a) We first create the table below.

| x | y | $\hat{y}$ | $e = y - \hat{y}$ | $e^2$ |
|---|---|---|---|---|
| 0 | 4 | 2.875 | 1.125 | 1.265625 |
| 2 | 2 | 1.625 | 0.375 | 0.140625 |
| 2 | 0 | 1.625 | -1.625 | 2.640625 |
| 5 | -2 | -0.250 | -1.750 | 3.062500 |
| 6 | 1 | -0.875 | 1.875 | 3.515625 |
|   |   |   |   | 10.625000 |

$$s_e = \sqrt{\frac{SSE}{n-2}} = \sqrt{\frac{10.625}{3}} = 1.882$$

**422 Chapter 14**

(b-c) For the residual plot, we plot e against x in the graph following at the left. We can use Excel for the normal probability plot. With the values of e in a column named 'e', highlight the data with the mouse, choose **Charts and Plots**, click on the **Function Type** drop down box, and click on **Normal Probability Plot**. Drag the 'e' from the **Names and Columns** box into the **Quantitative Variable** box and click **OK**. The graph is shown following at the right.

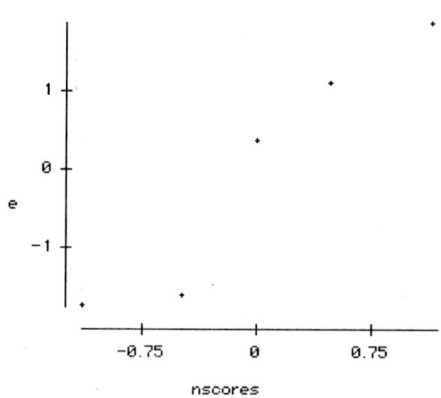

**14.23** If the assumptions for regression inferences are satisfied for a model relating a Corvette's age to its price, this means that there are constants $\beta_0$, $\beta_1$, and $\sigma$ such that, for each age x, the prices for Corvettes of that age are normally distributed with mean $\beta_0 + \beta_1 x$ and standard deviation $\sigma$.

**14.25** If the assumptions for regression inferences are satisfied for a model relating the volume of plant emissions of volatile compounds to the weight of the plant, this means that there are constants $\beta_0$, $\beta_1$, and $\sigma$ such that, for each weight x, the volumes of emissions y are normally distributed with mean $\beta_0 + \beta_1 x$ and standard deviation $\sigma$.

**14.27** If the assumptions for regression inferences are satisfied for a model relating the test scores of calculus students to their study-times, this means that there are constants $\beta_0$, $\beta_1$, and $\sigma$ such that, for each study-time x, the test scores y are normally distributed with mean $\beta_0 + \beta_1 x$ and standard deviation $\sigma$.

**14.29** To compute the standard error of the estimate, first retrieve the computation for the error sum of squares (SSE) in part (a) of Exercise 4.99 and then apply the formula for $s_e$ in Definition 14.2 of the text.

(a) In part (a) of Exercise 4.99, n = 10 and SSE was computed as 1,623.7. Applying Definition 14.2, the standard error of the estimate is

$$s_e = \sqrt{\frac{SSE}{n-2}} = \sqrt{\frac{1623.7}{10-2}} = 14.2465$$

Roughly speaking, the predicted prices for the Corvettes differ, on average, from the observed prices by $1424.65.

(b) Presuming that the variables age (x) and price (y) for Corvettes satisfy Assumptions (1) - (3) for regression inferences, the standard error of the estimate $s_e$ = 14.2465 provides an estimate for the common population standard deviation $\sigma$ of prices for all Corvettes of any given age.

Copyright © 2016 Pearson Education, Inc.

(c) Obtain the necessary normal scores from Table III in the text.

| Age x | Residual e |
|---|---|
| 6 | 0.82 |
| 6 | -9.18 |
| 6 | 5.82 |
| 2 | 24.20 |
| 2 | -16.80 |
| 5 | -2.09 |
| 4 | 10.01 |
| 5 | 10.91 |
| 1 | -3.70 |
| 4 | -19.99 |

| Residual e | Normal score n |
|---|---|
| -19.99 | -1.55 |
| -16.80 | -1.00 |
| -9.18 | -0.65 |
| -3.70 | -0.37 |
| -2.09 | -0.12 |
| 0.82 | 0.12 |
| 5.82 | 0.37 |
| 10.01 | 0.65 |
| 10.91 | 1.00 |
| 24.20 | 1.55 |

(d) Taking into account the small sample size, we can say that the residuals fall roughly in a horizontal band centered and symmetric about the x-axis. We can also say that the normal probability plot for residuals is very roughly linear. Therefore, it appears reasonable to consider the assumptions for regression inferences for the variables age and price of Corvettes to be met.

**14.31** To compute the standard error of the estimate, first retrieve the computation for the error sum of squares (SSE) in part (a) of Exercise 4.101 and then apply the formula for $s_e$ in Definition 14.2 of the text.
(a) In part (a) of Exercise 4.101, n = 10 and SSE was computed as 88.5. Applying Definition 14.2, the standard error of the estimate is

$$s_e = \sqrt{\frac{SSE}{n-2}} = \sqrt{\frac{264.16}{11-2}} = 5.4177 \text{ hundred nanograms}$$

Roughly speaking, the predicted volumes of volatile emissions from the plants differ, on average, from the observed volumes by 541.77 nanograms.

(b) Presuming that the weight (x) of the potato plants *Solanum tubersom* and volume of volatile emissions (y) from the plants satisfy Assumptions (1) - (3) for regression inferences, the standard error of the estimate $s_e$ = 5.4177 hundred nanograms (541.77 nanograms) provides an estimate for the common population standard deviation $\sigma$ of emission volumes for all plants of any particular weight.

(c)

Residuals Versus X
(response is Y)

| Weight x | Residual e |
|---|---|
| 57 | -4.81 |
| 85 | 4.63 |
| 57 | -2.31 |
| 65 | 8.39 |
| 52 | 0.01 |
| 67 | -2.93 |
| 62 | -6.12 |
| 80 | -3.55 |
| 77 | 0.44 |
| 53 | 8.85 |
| 68 | -2.60 |

Normal Probability Plot of the Residuals
(response is EMISSION)

| Residual e | Normal score n |
|---|---|
| -6.12 | -1.59 |
| -4.81 | -1.06 |
| -3.55 | -0.73 |
| -2.93 | -0.46 |
| -2.60 | -0.22 |
| -2.31 | 0.00 |
| 0.01 | 0.22 |
| 0.44 | 0.46 |
| 4.63 | 0.73 |
| 8.39 | 1.06 |
| 8.85 | 1.59 |

(d) Taking into account the small sample size, we can say that the residuals fall roughly in a horizontal band centered and symmetric about the x-axis. We can also say that the normal probability plot for residuals is approximately linear.
Therefore, based on the sample data, there are no obvious violations of the assumptions for regression inferences for the variables plant weight and volume of volatile emissions.

**14.33** To compute the standard error of the estimate, first retrieve the computation for the error sum of squares (SSE) in part (a) of Exercise 4.103 and then apply the formula for $s_e$ in Definition 14.2 of the text.

(a) In part (a) of Exercise 4.103, n = 8 and SSE was computed as 75.11. Applying Definition 14.2, the standard error of the estimate is

$$s_e = \sqrt{\frac{SSE}{n-2}} = \sqrt{\frac{75.11}{8-2}} = 3.5382$$

Roughly speaking, the predicted test scores differ, on average, from the observed scores by 3.5382 points.

(b) Presuming that the study-time ($x$) of the calculus students and their test scores ($y$) satisfy Assumptions (1) - (3) for regression inferences, the standard error of the estimate $s_e$ = 3.5382 provides an estimate for the common population standard deviation $\sigma$ of test scores for all calculus students with any particular study-time.

(c) Obtain the necessary normal scores from Table III in the text.

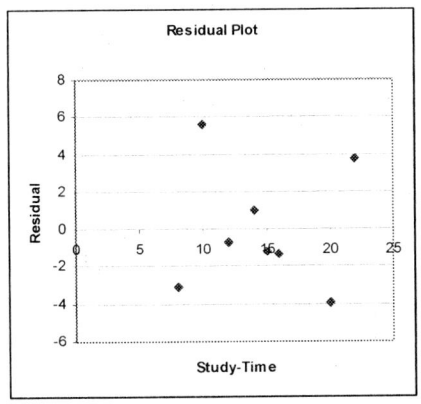

| Study Time x | Residual e |
|---|---|
| 10 | 5.59 |
| 15 | -1.18 |
| 12 | -0.72 |
| 20 | -3.95 |
| 8 | -3.10 |
| 16 | -1.34 |
| 14 | 0.97 |
| 22 | 3.74 |

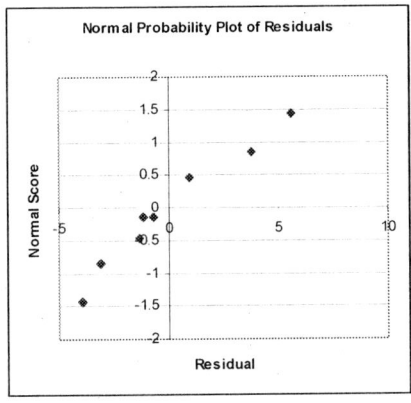

| Residual e | Normal score n |
|---|---|
| -3.95 | -1.43 |
| -3.10 | -0.85 |
| -1.34 | -0.47 |
| -1.18 | -0.15 |
| -0.72 | 0.15 |
| 0.97 | 0.47 |
| 3.74 | 0.85 |
| 5.59 | 1.43 |

(d) Taking into account the small sample size, we can say that the residuals fall roughly in a horizontal band centered and symmetric about the x-axis. We can also say that the normal probability plot for residuals is approximately linear. Therefore, based on the sample data, there are no obvious violations of the assumptions for regression inferences for the variables study-time and test scores of calculus students.

**14.35** Using Minitab, with the data in columns named INAUGURATION and DEATH, we choose **Stat ▶ Regression ▶ Regression...**, select <u>DEATH</u> in the **Response** text box, and select <u>INAUGURATION</u> in the **Predictors** text box. Click the **Graphs** button, select the **Regular** option button from the **Residuals for Plots** list, select the **Normal plot of residuals** check box from the **Residual Plots** list, click in the **Residuals versus the variables** text box and specify <u>INAUGURATION</u>, click **OK**, and click **OK**. The results are

Analysis of Variance

| Source | DF | Adj SS | Adj MS | F-Value | P-Value |
|---|---|---|---|---|---|
| Regression | 1 | 2021 | 2020.57 | 21.45 | 0.000 |
|   INAUGURATION | 1 | 2021 | 2020.57 | 21.45 | 0.000 |
| Error | 37 | 3485 | 94.19 | | |
|   Lack-of-Fit | 19 | 2179 | 114.68 | 1.58 | 0.168 |
|   Pure Error | 18 | 1306 | 72.56 | | |
| Total | 38 | 5506 | | | |

**426** Chapter 14

```
Model Summary

     S    R-sq   R-sq(adj)  R-sq(pred)
9.70514  36.70%    34.99%      28.71%

Coefficients

Term           Coef   SE Coef  T-Value  P-Value   VIF
Constant        5.7      14.1     0.40    0.690
INAUGURATION  1.179     0.255     4.63    0.000  1.00
Regression Equation

DEATH = 5.7 + 1.179 INAUGURATION
```

(a) The standard error of the estimate is 9.70514. Roughly speaking, the predicted death ages differ, on average, from the observed death ages by 9.70514 years.

(b)

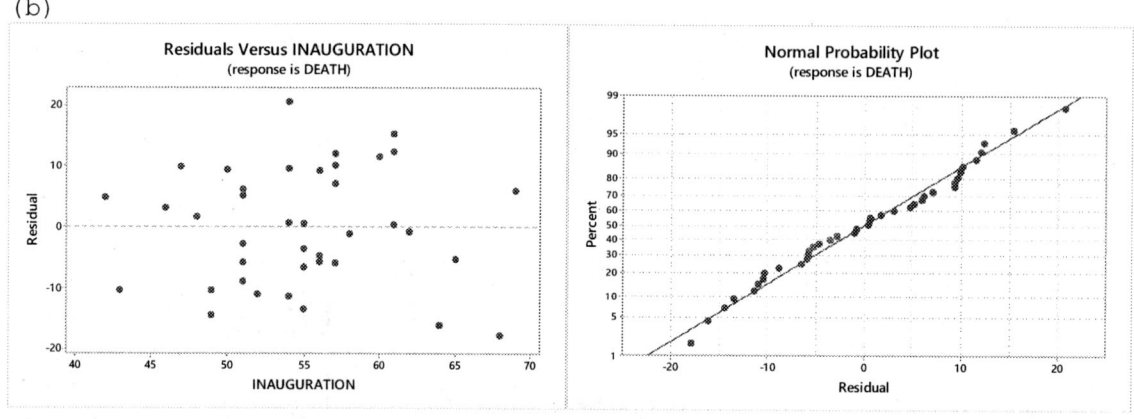

(c) The residuals fall roughly in a horizontal band centered and symmetric about the x-axis. We can also say that the normal probability plot for residuals is approximately linear. Therefore, based on the sample data, there are no obvious violations of the assumptions for regression inferences for the variables inauguration age and death age of U.S. Presidents.

**14.37** Using Minitab, with the data in columns named LOT SIZE and VALUE, we choose

**Stat ▶ Regression ▶ Regression...**, select <u>VALUE</u> in the **Response** text box, and select '<u>LOT SIZE</u>' in the **Predictors** text box. Click the **Graphs** button, select the **Regular** option button from the **Residuals for Plots** list, select the **Normal plot of residuals** check box from the **Residual Plots** list, click in the **Residuals versus the variables** text box and specify '<u>LOT SIZE</u>', click **OK**, and click **OK**. The results are

```
The regression equation is
VALUE = 292 + 67.1 LOT SIZE

Predictor    Coef   SE Coef      T      P
Constant    292.0     140.3   2.08  0.043
LOT SIZE    67.11     59.87   1.12  0.269

S = 139.012    R-Sq = 2.9%    R-Sq(adj) = 0.6%

Analysis of Variance
Source            DF       SS      MS     F      P
Regression         1    24286   24286  1.26  0.269
Residual Error    42   811619   19324
Total             43   835905
```

(a) The standard error of the estimate is 139.012 thousand dollars. Roughly speaking, the predicted values differ, on average, from the observed values by 139.012 thousand dollars.

(b)

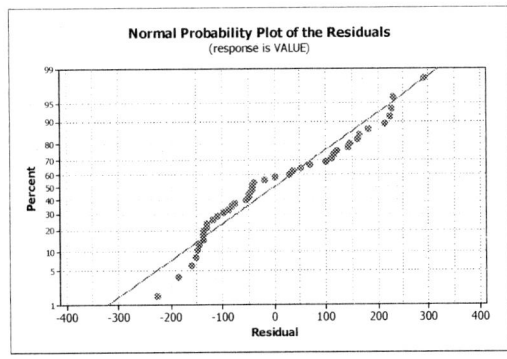

(c) The residuals appear to fall roughly in a band centered and symmetric about the x-axis. We can also say that the normal probability plot for residuals deviates from a linear pattern, but it is difficult to say whether the deviation is enough to rule out normality. Therefore, for the time being, based on the sample data, the assumptions for regression inferences seem to be reasonable for the variables lot size and value. More on these data later.

**14.39** Using Minitab, with the data in columns named HIGH and LOW, we choose **Stat** ▶ **Regression** ▶ **Regression...**, select LOW in the **Response** text box, and select HIGH in the **Predictors** text box. Click the **Graphs** button, select the **Regular** option button from the **Residuals for Plots** list, select the **Normal plot of residuals** check box from the **Residual Plots** list, click in the **Residuals versus the variables** text box and specify HIGH, click **OK**, and click **OK**. The results are

```
The regression equation is
LOW = - 7.57 + 0.917 HIGH

Predictor       Coef    SE Coef       T      P
Constant      -7.572      1.786   -4.24  0.000
HIGH         0.91685    0.02967   30.91  0.000

S = 4.15111    R-Sq = 95.2%    R-Sq(adj) = 95.1%

Analysis of Variance

Source           DF       SS      MS       F      P
Regression        1    16459   16459  955.17  0.000
Residual Error   48      827      17
Total            49    17286
```

(a) The standard error of the estimate is 4.15111 degrees. Roughly speaking, the predicted low temperatures differ, on average, from the observed low temperatures by 4.15111 degrees.

(b)

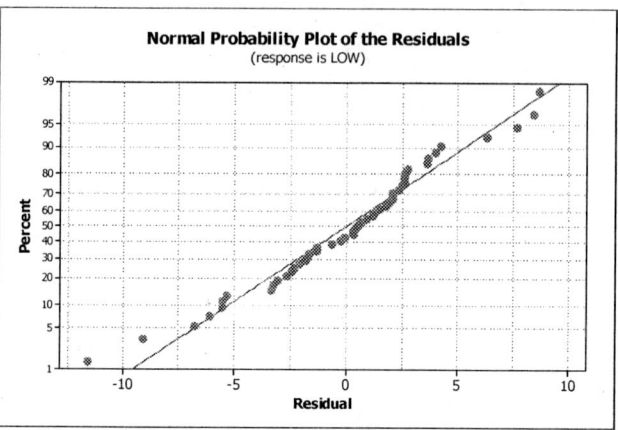

(c) The residuals appear to fall roughly in a band centered and symmetric about the x-axis. We can also say that the normal probability plot for residuals deviates only slightly from a linear pattern. Therefore, based on the sample data, the assumptions for regression inferences seem to be reasonable for the variables average high January temperature and average low January temperature.

**14.41** Using Minitab, with the data in columns named DISP and MPG, we choose **Stat** ▶ **Regression** ▶ **Regression...**, select DISP in the **Response** text box, select MPG in the **Predictors** text box, select **Residuals** from the **Storage** check-box list. Click the **Graphs...** button, select the **Regular** option button from the **Residuals for Plots** list, select the **Normal plot of residuals** check box from the **Residual Plots** list, click in the **Residuals versus the variables** text box and specify DISP, click **OK**, and click **OK**. The results are

(a)    The regression equation is
       MPG = 32.3 - 3.58 DISP

```
Predictor      Coef      StDev         T         P
Constant    32.2732     0.6704     48.14     0.000
DISP        -3.5762     0.2099    -17.03     0.000

S = 2.248      R-Sq = 70.9%     R-Sq(adj) = 70.7%

Analysis of Variance

Source            DF        SS        MS         F         P
Regression         1    1465.8    1465.8    290.18     0.000
Residual Error   119     601.1       5.1
Total            120    2067.0
```

The standard error of the estimate is the first entry in the sixth line of the computer output. It is reported as s = 2.248. Presuming that the variables DISP and MPG satisfy the assumptions for regression inferences, the standard error of the estimate, 2.248, provides an estimate for the common population standard deviation, $\sigma$, of miles per gallon for all cars with a particular engine displacement.

(b)   The two graphs that result follow.

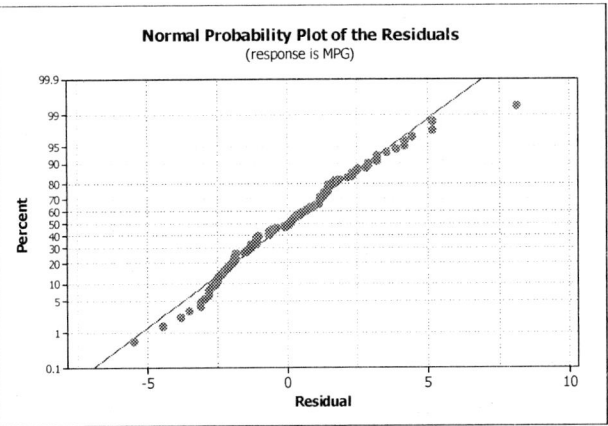

(c)   The residuals plotted against engine displacement display a concave upward curve, indicating that the linear model is not appropriate for the data. Thus, Assumption 1 is violated and the statement made in part (a) interpreting the meaning of the standard error does not hold.

**14.43** Using Minitab, with the data in columns named DIAMETER and VOLUME, we choose **Stat ▶ Regression ▶ Regression...**, select VOLUME in the **Response** text box, select DIAMETER in the **Predictors** text box, select **Residuals** from the **Storage** check-box list. Click the **Graphs...** button, select the **Regular** option button from the **Residuals for Plots** list, select the **Normal plot of residuals** check box from the **Residual Plots** list, click in the **Residuals versus the variables** text box and specify DIAMETER, click **OK**, and click **OK**. The results are

(a)   The regression equation is
VOLUME = - 41.6 + 6.84 DIAMETER

```
Predictor      Coef       StDev         T         P
Constant     -41.568      3.427     -12.13     0.000
DIAMETER      6.8367      0.2877     23.77     0.000

S = 9.875      R-Sq = 89.3%    R-Sq(adj) = 89.1%

Analysis of Variance

Source           DF         SS         MS          F         P
Regression        1      55083      55083     564.88     0.000
Residual Error   68       6631         98
Total            69      61714
```

The standard error of the estimate is the first entry in the sixth line of the computer output. It is reported as s = 9.875. Presuming that the variables DIAMETER and VOLUME satisfy the assumptions for regression inferences, the standard error of the estimate, 9.875, provides an estimate for the common population standard deviation, $\sigma$, of volumes for all trees having a particular diameter at breast height.

(b) The two graphs that result follow.

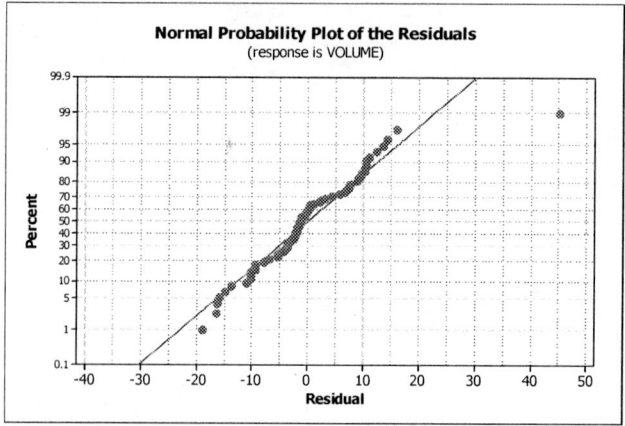

(c) The residuals plotted against diameter are displayed in a curved concave upward pattern, indicating that the linear model is not appropriate. This is a violation of Assumption 1. In addition, both plots indicate the presence of an outlier. Finally, it appears that there is more variability in the center of the first graph than at the left side, so Assumption 2 is also violated. Removal of the outlier will not change the curved pattern of the other residuals or the unequal variation in the residuals, so this data should not be analyzed using a linear model.

### Exercises 14.2

**14.45** normal distribution with mean $\beta_1 = -3.5$

**14.47** We can also use the coefficient of determination, $r^2$, and the linear correlation coefficient, $r$, as a basis for a test to decide whether a regression equation is useful for prediction.

**14.49** From Exercise 14.13, we have $s_e = 2.44949$.

Also, $\sum x_i = 6$; $\sum x_i^2 = 14$; n=3; $S_{xx} = \sum x_i^2 - \left(\sum x_i\right)^2 / n = 14 - 6^2/3 = 2$

(a) $t = \dfrac{b_1}{s_e / \sqrt{S_{xx}}} = \dfrac{-2}{2.44949/\sqrt{2}} = -1.155$; the critical values with 1 df are $\pm 6.314$. Since $-6.314 < t < 6.314$, we cannot reject $\beta_1 = 0$ and conclude that x is not useful for predicting y.

(b) The 90% confidence interval for the slope of the regression line is

$b_1 \pm t_{\alpha/2} \dfrac{s_e}{\sqrt{S_{xx}}} = -2 \pm 6.314 \dfrac{2.44949}{\sqrt{2}} = (-12.936, 8.936)$

**14.51** From Exercise 14.15, we have $s_e = 1.732$.

Also, $\sum x_i = 10$; $\sum x_i^2 = 30$; n=4; $S_{xx} = \sum x_i^2 - \left(\sum x_i\right)^2 / n = 30 - 10^2/4 = 5$

(a) $t = \dfrac{b_1}{s_e / \sqrt{S_{xx}}} = \dfrac{2}{1.732/\sqrt{5}} = 2.582$; the critical values with 2 df are $\pm 2.920$.

Since $-2.920 < t < 2.920$, we cannot reject $\beta_1 = 0$ and conclude that x is not useful for predicting y.

(b) The 90% confidence interval for the slope of the regression line is

$$b_1 \pm t_{\alpha/2} \frac{s_e}{\sqrt{S_{xx}}} = 2 \pm 2.920 \frac{1.732}{\sqrt{5}} = (-0.260, 4.262)$$

**14.53** From Exercise 14.17, we have $s_e = 1.414$.

Also, $\sum x_i = 15; \sum x_i^2 = 49;$ n=5; $S_{xx} = \sum x_i^2 - (\sum x_i)^2/n = 49 - 15^2/5 = 4$

(a) $t = \dfrac{b_1}{s_e/\sqrt{S_{xx}}} = \dfrac{-1}{1.414/\sqrt{4}} = -1.414$; the critical values with 3 df are ±2.353.

Since $-2.353 < t < 2.353$, we cannot reject $\beta_1 = 0$ and conclude that x is not useful for predicting y.

(b) The 90% confidence interval for the slope of the regression line is

$$b_1 \pm t_{\alpha/2} \frac{s_e}{\sqrt{S_{xx}}} = -1 \pm 2.353 \frac{1.414}{\sqrt{4}} = (-2.664, 0.664)$$

**14.55** From Exercise 14.19, we have $s_e = 2.449$.

Also, $\sum x_i = 6; \sum x_i^2 = 14;$ n=3; $S_{xx} = \sum x_i^2 - (\sum x_i)^2/n = 14 - 6^2/3 = 2$

(a) $t = \dfrac{b_1}{s_e/\sqrt{S_{xx}}} = \dfrac{2}{2.449/\sqrt{2}} = 1.154$; the critical values with 1 df are ±6.314.

Since $-6.314 < t < 6.314$, we cannot reject $\beta_1 = 0$ and conclude that x is not useful for predicting y.

(b) The 90% confidence interval for the slope of the regression line is

$$b_1 \pm t_{\alpha/2} \frac{s_e}{\sqrt{S_{xx}}} = 2 \pm 6.314 \frac{2.449}{\sqrt{2}} = (-8.938, 12.938)$$

**14.57** From Exercise 14.21, we have $s_e = 1.88193$.

Also, $\sum x_i = 15; \sum x_i^2 = 69;$ n=5; $S_{xx} = \sum x_i^2 - (\sum x_i)^2/n = 69 - 15^2/5 = 24$

(a) $t = \dfrac{b_1}{s_e/\sqrt{S_{xx}}} = \dfrac{-0.625}{1.88193/\sqrt{24}} = -1.627$; the critical values with 3 df are ±2.353. Since $-2.353 < t < 2.353$, we cannot reject $\beta_1 = 0$ and conclude that x is not useful for predicting y.

(b) The 90% confidence interval for the slope of the regression line is

$$b_1 \pm t_{\alpha/2} \frac{s_e}{\sqrt{S_{xx}}} = -0.625 \pm 2.353 \frac{1.88193}{\sqrt{24}} = (-1.512, 0.282)$$

**14.59** From Exercise 4.59, $\sum x = 41$, $\sum x^2 = 199$, and $b_1 = -27.9029$. From Exercise 14.29, $s_e = 14.2464$.

Step 1: $H_0: \beta_1 = 0$, $H_a: \beta_1 \neq 0$

Step 2: $\alpha = 0.10$

Step 3: $t = \dfrac{b_1}{s_e/\sqrt{\sum x^2 - (\sum x)^2/n}} = \dfrac{-27.9029}{14.2465/\sqrt{199 - 41^2/10}} = -10.887$

Step 4: df = n - 2 = 8; critical values = ±1.860
For the p-value approach, p < 0.01.

**432** Chapter 14

Step 5: Since −10.887 < −1.860, reject $H_0$.
Because the p-value is less than the significance level of 0.10, we can reject $H_0$.

Step 6: At 10% significance level, the data provide sufficient evidence to conclude that the slope of the population regression line is not zero and, hence, that age is useful as a predictor of price for Corvettes.

**14.61** From Exercise 4.61, $\sum x = 723$, $\sum x^2 = 48747$, and $b_1 = 0.16285$. From Exercise 14.31, $s_e = 5.418$.

Step 1: $H_0$: $\beta_1 = 0$, $H_a$: $\beta_1 \neq 0$

Step 2: $\alpha = 0.05$

Step 3: $t = \dfrac{b_1}{s_e / \sqrt{\sum x^2 - (\sum x)^2 / n}} = \dfrac{0.16285}{5.418 / \sqrt{48747 - 723^2 / 11}} = 1.053$

Step 4: df = n − 2 = 9; critical values = ±2.262
For the p-value approach, p > 0.20.

Step 5: Since 1.053 < 2.262, do not reject $H_0$.
Because the p-value is greater than the significance level of 0.05, we cannot reject $H_0$.

Step 6: At the 5% significance level, we conclude that plant weight is not useful as a predictor of volume of volatile emissions.

**14.63** From Exercise 4.63, $\sum x = 117$, $\sum x^2 = 1869$, and $b_1 = -0.846$. From Exercise 14.33, $s_e = 3.538$.

Step 1: $H_0$: $\beta_1 = 0$, $H_a$: $\beta_1 \neq 0$

Step 2: $\alpha = 0.01$

Step 3: $t = \dfrac{b_1}{s_e / \sqrt{\sum x^2 - (\sum x)^2 / n}} = \dfrac{-0.846}{3.538 / \sqrt{1869 - 117^2 / 8}} = -3.004$

Step 4: df = n − 2 = 6; critical values = ±3.707
For the p-value approach, 0.02 < p < 0.05.

Step 5: Since −3.707 < −3.004 < 3.707, do not reject $H_0$.
Because the p-value is less than the significance level of 0.01, we cannot reject $H_0$.

Step 6: At the 1% significance level, we conclude that study-time is not useful as a predictor of test scores of calculus students.

**14.65** From Exercise 4.59, $\sum x = 41$, $\sum x^2 = 199$, and $b_1 = -27.9029$. From Exercise 14.29, $s_e = 14.2465$.

Step 1: For a 90% confidence interval, $\alpha = 0.10$. With df = n − 2 = 8, $t_{\alpha/2} = t_{0.05} = 1.860$.

Step 2: The endpoints of the confidence interval for $\beta_1$ are

$$b_1 \pm t_{\alpha/2} \cdot s_e / \sqrt{\sum x^2 - (\sum x)^2 / n}$$

$-27.9029 \pm 1.860 \cdot 14.2465 / \sqrt{199 - 41^2 / 10}$

$-27.9029 \pm 4.7670$

$-32.6699$ to $-23.1359$

We can be 90% confident that the yearly decrease in mean price for Corvettes is somewhere between $2314 and $3267.

**14.67** From Exercise 4.61, $\sum x = 723$, $\sum x^2 = 48747$, and $b_1 = 0.16285$. From Exercise 14.31, $s_e = 5.4177$.

Step 1:  For a 95% confidence interval, $\alpha = 0.05$. With df = n − 2 = 9, $t_{\alpha/2}$ = $t_{0.025}$ = 2.262.

Step 2:  The endpoints of the confidence interval for $\beta_1$ are

$$b_1 \pm t_{\alpha/2} \cdot s_e / \sqrt{\sum x^2 - (\sum x)^2 / n}$$

$$0.16285 \pm 2.262 \cdot 5.4177 / \sqrt{48747 - 723^2/11}$$

$$0.16285 \pm 0.34997$$

$$-0.187 \text{ to } 0.513$$

We can be 95% confident that the increase in mean volatile plant emissions per one gram increase in weight is somewhere between −0.187 and 0.513 hundred nanograms.

**14.69** From Exercise 4.63, $\sum x = 117$, $\sum x^2 = 1869$, and $b_1 = -0.846$. From Exercise 14.33, $s_e = 3.53816$.

Step 1:  For a 99% confidence interval, $\alpha = 0.01$. With df = n − 2 = 6, $t_{\alpha/2}$ = $t_{0.005}$ = 3.707.

Step 2:  The endpoints of the confidence interval for $\beta_1$ are

$$b_1 \pm t_{\alpha/2} \cdot s_e / \sqrt{\sum x^2 - (\sum x)^2 / n}$$

$$-0.846 \pm 3.707 \cdot 3.53816 / \sqrt{1869 - 117^2/8}$$

$$-0.846 \pm 1.044$$

$$-1.890 \text{ to } 0.198$$

We can be 99% confident that the change in test score per one hour increase in study-time for calculus students is somewhere between −1.890 and 0.198.

**14.71** (a) In Exercise 14.35, part (c), we determined that it was reasonable to consider Assumptions 1-3 for regression inferences met by these variables.

(b) Part of the Minitab output from Exercise 14.35 is reproduced below.

```
Term              Coef    SE Coef   T-Value   P-Value   VIF
Constant           5.7      14.1      0.40     0.690
INAUGURATION      1.179    0.255      4.63     0.000    1.00
Regression Equation
```

The t value and the P-value for the regression t-test are 4.63 and 0.000, respectively. Since 0.000 < 0.05, we conclude that the inauguration ages are useful for predicting the presidents' death ages.

**14.73** (a) In Exercise 14.37, part (c), we determined that it was temporarily reasonable to consider Assumptions 1-3 for regression inferences met by these variables, although Assumption 3 was questionable. Note: In Exercise 4.73, we concluded that the observations did not follow a linear pattern. This exercise will help to resolve the conflict.

(b) Part of the Minitab output from Exercise 14.37 is reproduced below.

```
Predictor    Coef    SE Coef     T       P
Constant    292.0     140.3    2.08    0.043
LOT SIZE     67.11     59.87   1.12    0.269
```

The t value and the P-value for the regression t-test are 1.12 and 0.269, respectively. Since 0.269 > 0.05, we conclude that the lot size is not useful for predicting the home values.

**14.75** (a) In Exercise 14.39, part (c), we determined that it was reasonable to consider Assumptions 1-3 for regression inferences met by these variables.

## 434  Chapter 14

(b) Part of the Minitab output from Exercise 14.39 is reproduced below.

```
Predictor      Coef   SE Coef      T       P
Constant     -7.572    1.786    -4.24   0.000
HIGH         0.91685  0.02967   30.91   0.000
```

The t value and the P-value for the regression t-test are 30.91 and 0.000, respectively. Since 0.000 < 0.05, we conclude that average high temperatures are useful for predicting average low temperatures in January.

**14.77** (a) In Exercise 14.41, part (c), we determined that it was not reasonable to consider Assumption 1 for regression inferences met by these variables. The residual plot was convex upward. Therefore, parts (b) and (c) are omitted.

**14.79** (a) In Exercise 14.43, part (c), we determined that it was not reasonable to consider Assumptions 1 and 2 for regression inferences met by these variables. Therefore, parts (b) and (c) are omitted.

### Exercises 14.3

**14.81** $11,443

**14.83** From Exercise 14.13, $\hat{y} = 1 - 2x$. From Exercise 14.49, $s_e = 2.44949$ and $S_{xx} = 2$.

(a) The point estimate for the conditional mean of y at x = 2 is
$\hat{y}_p = 1 - 2x = 1 - 2(2) = -3$.

(b) The 95% confidence interval for the conditional mean at x = 2 is

$$\hat{y}_p \pm t_{\alpha/2} s_e \sqrt{\frac{1}{n} + \frac{(x_p - \sum x_i / n)^2}{S_{xx}}} = -3 \pm 12.706(2.44949)\sqrt{\frac{1}{3} + \frac{(2 - 6/3)^2}{2}} = -3 \pm 17.97$$

$$= (-20.97, 14.97)$$

(c) The predicted value of y at x = 2 is $\hat{y}_p = 1 - 2x = 1 - 2(2) = -3$.

(d) The 95% prediction interval for y at x = 2 is

$$\hat{y}_p \pm t_{\alpha/2} s_e \sqrt{1 + \frac{1}{n} + \frac{(x_p - \sum x_i / n)^2}{S_{xx}}} = -3 \pm 12.706(2.44949)\sqrt{1 + \frac{1}{3} + \frac{(2 - 6/3)^2}{2}}$$

$$= -3 \pm 35.94 = (-38.94, 32.94)$$

**14.85** From Exercise 14.15, $\hat{y} = -3 + 2x$. From Exercise 14.51, $s_e = 1.73205$ and $S_{xx} = 24$.

(a) The point estimate for the conditional mean of y at x = 4 is
$\hat{y}_p = -3 + 2x = -3 + 2(4) = 5$.

(b) The 95% confidence interval for the conditional mean at x = 4 is

$$\hat{y}_p \pm t_{\alpha/2} s_e \sqrt{\frac{1}{n} + \frac{(x_p - \sum x_i / n)^2}{S_{xx}}} = 5 \pm 4.303(1.73205)\sqrt{\frac{1}{4} + \frac{(4 - 10/4)^2}{5}} = 5 \pm 6.24$$

$$= (-1.24, 11.24)$$

(c) The predicted value of y at x = 4 is $\hat{y}_p = -3 + 2x = -3 + 2(4) = 5$.

(d) The 95% prediction interval for y at x = 4 is

$$\hat{y}_p \pm t_{\alpha/2} s_e \sqrt{1 + \frac{1}{n} + \frac{(x_p - \sum x_i / n)^2}{S_{xx}}} = 5 \pm 4.303(1.73205)\sqrt{1 + \frac{1}{4} + \frac{(4 - 10/4)^2}{5}}$$

$$= 5 \pm 9.72 = (-4.72, 14.72)$$

**14.87** From Exercise 14.17, $\hat{y}=5-x$. From Exercise 14.53, $s_e=1.414$ and $S_{xx}=4$.

(a) The point estimate for the conditional mean of y at x = 3 is
$\hat{y}_p = 5-x = 5-3 = 2$.

(b) The 95% confidence interval for the conditional mean at x = 3 is
$$\hat{y}_p \pm t_{\alpha/2} s_e \sqrt{\frac{1}{n} + \frac{(x_p - \sum x_i/n)^2}{S_{xx}}} = 2 \pm 3.182(1.414)\sqrt{\frac{1}{5} + \frac{(3-15/5)^2}{4}} = 2 \pm 2.01$$
$$= (-0.01, 4.01)$$

(c) The predicted value of y at x = 3 is $\hat{y}_p = 5-x = 5-3 = 2$.

(d) The 95% prediction interval for y at x = 3 is
$$\hat{y}_p \pm t_{\alpha/2} s_e \sqrt{1+\frac{1}{n} + \frac{(x_p - \sum x_i/n)^2}{S_{xx}}} = 2 \pm 3.182(1.414)\sqrt{1+\frac{1}{5} + \frac{(3-15/5)^2}{4}}$$
$$= 2 \pm 4.93 = (-2.93, 6.93)$$

**14.89** From Exercise 14.19, $\hat{y}=1+2x$. From Exercise 14.55, $s_e=2.449$ and $S_{xx}=2$.

(a) The point estimate for the conditional mean of y at x = 2 is
$\hat{y}_p = 1+2x = 1+2(2) = 5$.

(b) The 95% confidence interval for the conditional mean at x = 2 is
$$\hat{y}_p \pm t_{\alpha/2} s_e \sqrt{\frac{1}{n} + \frac{(x_p - \sum x_i/n)^2}{S_{xx}}} = 5 \pm 12.706(2.449)\sqrt{\frac{1}{3} + \frac{(2-6/3)^2}{2}} = 5 \pm 17.97$$
$$= (-12.97, 22.97)$$

(c) The predicted value of y at x = 2 is $\hat{y}_p = 1+2x = 1+2(2) = 5$.

(d) The 95% prediction interval for y at x = 2 is
$$\hat{y}_p \pm t_{\alpha/2} s_e \sqrt{1+\frac{1}{n} + \frac{(x_p - \sum x_i/n)^2}{S_{xx}}} = 5 \pm 12.706(2.449)\sqrt{1+\frac{1}{3} + \frac{(2-6/3)^2}{2}}$$
$$= 5 \pm 35.95 = (-30.95, 40.95)$$

**14.91** From Exercise 14.21, $\hat{y}=2.875+0.625x$. From Exercise 14.57, $s_e=1.88193$ and $S_{xx}=24$.

(a) The point estimate for the conditional mean of y at x = 3 is
$\hat{y}_p = 2.875+0.625x = 2.875+0.625(3) = 4.75$.

(b) The 95% confidence interval for the conditional mean at x = 3 is
$$\hat{y}_p \pm t_{\alpha/2} s_e \sqrt{\frac{1}{n} + \frac{(x_p - \sum x_i/n)^2}{S_{xx}}} = 4.75 \pm 3.182(1.88193)\sqrt{\frac{1}{5} + \frac{(3-15/5)^2}{24}} = 4.75 \pm 2.68$$
$$= (2.07, 7.43)$$

(c) The predicted value of y at x = 3 is
$\hat{y}_p = 2.875+0.625x = 2.875+0.625(3) = 4.75$.

(d) The 95% prediction interval for y at x = 3 is
$$\hat{y}_p \pm t_{\alpha/2} s_e \sqrt{1+\frac{1}{n} + \frac{(x_p - \sum x_i/n)^2}{S_{xx}}} = 4.75 \pm 3.182(1.88193)\sqrt{1+\frac{1}{5} + \frac{(3-15/5)^2}{24}}$$
$$= 4.75 \pm 6.56 = (-1.81, 11.31)$$

**14.93** From Exercise 4.59, $\hat{y} = 456.602 - 27.9029x$, $\sum x = 41$, and $\sum x^2 = 199$. From Exercise 14.29, $s_e = 14.2465$.

(a) $\hat{y}_p = 456.602 - 27.9029(4) = 344.9904 = \$34,499$.

(b) Step 1: For a 90% confidence interval, $\alpha = 0.10$.

With df = n - 2 = 8, $t_{\alpha/2} = t_{0.05} = 1.860$.

Step 2: $\hat{y}_p = 344.9904$

Step 3: The endpoints of the confidence interval are

$$\hat{y}_p \pm t_{\alpha/2} \cdot s_e \cdot \sqrt{\frac{1}{n} + \frac{(x_p - \sum x/n)^2}{\sum x^2 - (\sum x)^2/n}}$$

$$= 344.9904 \pm 1.860 \cdot 14.2465 \sqrt{\frac{1}{10} + \frac{(4 - 41/10)^2}{199 - 41^2/10}}$$

$$= 344.9904 \pm 8.3931 = (336.60, 353.38)$$

The interpretation of this interval is as follows: We can be 90% confident that the mean price of all four-year-old Corvettes is somewhere between \$33,660 and \$35,338.

(c) This is the same as the answer in part (a): $\hat{y}_p = 344.9904$.

(d) For a 90% prediction interval, Steps 1 and 2 are the same as Steps 1 and 2, respectively, in part (b). Thus, they are not repeated here. Only Step 3 is presented.

Step 3: The endpoints of the prediction interval are

$$\hat{y}_p \pm t_{\alpha/2} \cdot s_e \cdot \sqrt{1 + \frac{1}{n} + \frac{(x_p - \sum x/n)^2}{\sum x^2 - (\sum x)^2/n}}$$

$$= 344.9904 \pm 1.860 \cdot 14.265 \sqrt{1 + \frac{1}{10} + \frac{(4 - 41/10)^2}{199 - 41^2/10}}$$

$$= 344.9904 \pm 27.7959 = (317.20, 372.78)$$

The interpretation of this interval is as follows: We can be 90% certain that the price of a randomly selected four-year-old Corvette will be somewhere between \$31,720 and \$37,278.

(e)

90% confidence interval for mean Corvette price: 33660 to 35338

90% prediction interval for Corvette prices: 31720 to 37278

(f) The error in the estimate of the mean price of four-year-old Corvettes is due only to the fact that the population regression line is being estimated by a sample regression line; whereas, the error in the prediction of the price of a randomly selected four-year-old Corvette is due to that fact plus the variation in prices for four-year-old Corvettes.

**14.95** From Exercise 4.61, $\hat{y} = 3.523688 + 0.162848x$, $\sum x = 726$, and $\sum x^2 = 48747$.
From Exercise 14.31, $s_e = 5.4177$.

(a) $\hat{y}_p = 3.523688 + 0.162848(60) = 13.2946$

(b) Step 1: For a 95% confidence interval, $\alpha = 0.05$.
With df $= n - 2 = 9$, $t_{\alpha/2} = t_{0.025} = 2.262$.

Step 2: $\hat{y}_p = 13.2946$

Step 3: The endpoints of the confidence interval are

$$\hat{y}_p \pm t_{\alpha/2} \cdot s_e \cdot \sqrt{\frac{1}{n} + \frac{(x_p - \sum x/n)^2}{\sum x^2 - (\sum x)^2/n}}$$

$$= 13.2946 \pm 2.262 \cdot 5.4177 \cdot \sqrt{\frac{1}{11} + \frac{(60 - 723/11)^2}{48747 - 723^2/11}}$$

$$= 813.2946 \pm 4.2036 = (9.0910, 17.4962)$$

The interpretation of this interval is as follows: We can be 95% confident that the mean quantity of volatile emissions of all plants that weigh 60 grams is somewhere between 9.0910 and 17.4982 hundred nanograms.

(c) This is the same as the answer in part (a): $\hat{y}_p = 13.2946$.

(d) For a 95% prediction interval, Steps 1 and 2 are the same as Steps 1 and 2, respectively, in part (b). Thus, they are not repeated here. Only Step 3 is presented.

Step 3: The endpoints of the prediction interval are

$$\hat{y}_p \pm t_{\alpha/2} \cdot s_e \cdot \sqrt{1 + \frac{1}{n} + \frac{(x_p - \sum x/n)^2}{\sum x^2 - (\sum x)^2/n}}$$

$$= 13.2946 \pm 2.262 \cdot 5.4177 \cdot \sqrt{1 + \frac{1}{11} + \frac{(60 - 723/11)^2}{48747 - 723^2/11}}$$

$$= 813.2946 \pm 12.9557 = (0.3389, 26.2503)$$

The interpretation of this interval is as follows: We can be 95% certain that the quantity of volatile emissions of a randomly selected plant that weighs 60 grams is somewhere between 0.3389 and 26.2503 hundred anograms.

**14.97** From Exercise 4.63, $\hat{y} = 94.86698 - 0.84561x$, $\sum x = 117$, and $\sum x^2 = 1869$.
From Exercise 14.33, $s_e = 3.5382$.

(a) $\hat{y}_p = 94.86698 - 0.84561(15) = 82.1828$

(b) Step 1: For a 99% confidence interval, $\alpha = 0.01$.
With df $= n - 2 = 6$, $t_{\alpha/2} = t_{0.005} = 3.707$.

Step 2: $\hat{y}_p = 82.1828$

Step 3:   The endpoints of the confidence interval are

$$\hat{y}_p \pm t_{\alpha/2} \cdot s_e \cdot \sqrt{\frac{1}{n} + \frac{(x_p - \sum x/n)^2}{\sum x^2 - (\sum x)^2/n}}$$

$$= 82.1828 \pm 3.707 \cdot (3.5382) \cdot \sqrt{\frac{1}{8} + \frac{(15 - 117/8)^2}{1869 - 117^2/8}}$$

$$= 82.1828 \pm 4.6537 = (77.5291, 86.8365)$$

The interpretation of this interval is as follows: We can be 99% confident that the mean calculus test score of students who study 15 hours is somewhere between 77.53 and 86.84 mm.

(c) This is the same as the answer in part (a): $\hat{y}_p$ = 82.1828.

(d) For a 99% prediction interval, Steps 1 and 2 are the same as Steps 1 and 2, respectively, in part (b). Thus, they are not repeated here. Only Step 3 is presented.

Step 3:   The endpoints of the prediction interval are

$$\hat{y}_p \pm t_{\alpha/2} \cdot s_e \cdot \sqrt{1 + \frac{1}{n} + \frac{(x_p - \sum x/n)^2}{\sum x^2 - (\sum x)^2/n}}$$

$$= 82.1828 \pm 3.707 \cdot (3.5382) \cdot \sqrt{1 + \frac{1}{8} + \frac{(15 - 117/8)^2}{1869 - 117^2/8}}$$

$$= 82.1828 \pm 13.9172 = (68.2656, 96.0100)$$

The interpretation of this interval is as follows: We can be 99% certain that a randomly selected calculus student who studies 125 hours will have a test score somewhere between 68.27 and 96.01 mm.

**14.99** (a) In Exercise 14.35, part (c), we determined that it was reasonable to consider Assumptions 1-3 for regression inferences met by these variables.

(b) Using Minitab, with the data in columns named INAUGURATION and DEATH, we choose **Stat ▶ Regression ▶ Regression...**, select DEATH in the **Response** text box, and select INAUGURATION in the **Predictors** text box. Click the **Options** button, select the **Regular** option button from the **Residuals for Plots** list, enter 53 in the **Prediction intervals for new observations** box and 95 in the **Confidence level** box, click **OK**, and click **OK**. Since much of the output was shown in Exercise 14.35, we show below only that part of the output pertinent to this exercise.

```
Variable       Setting
INAUGURATION      53

   Fit    SE Fit      95% CI              95% PI
68.1679  1.62938  (64.8665, 71.4694)  (48.2282, 88.1076)
```

The point estimate for the conditional mean of DEATH for inauguration age 53 is found under the word 'Fit' in the output as 68.17. This is the same as the predicted score for inauguration at age 53.

(c) The 95% confidence interval for the conditional mean death age for presidents inaugurated at age 53 is (64.87, 71.47). We can be 95% confident that the mean death age of U.S. Presidents inaugurated at age 53 lies somewhere between 64.87 and 71.47 years.

(d) The predicted death age for Presidents inaugurated at age 53 is found under the word 'Fit' in the output as 68.17.

(e) The 95% prediction interval for the death age of a President inaugurated at age 53 is (48.23, 88.11). We can be 95% certain that the death age of a president inaugurated at age 53 will fall somewhere between 48.23 and 88.11. Note: The lower end of this prediction interval can be safely increased to 53 since a president cannot die prior to being inaugurated.

(f) The confidence interval and the prediction interval are both centered at the predicted value for an inauguration age of 53. The prediction interval is longer because there is variation associated with the additional observation that is not present in the confidence interval for the mean.

**14.101** (a) In Exercise 14.37, part (c), we determined that it was reasonable to consider Assumptions 1-3 for regression inferences met by these variables. Note: In Exercise 4.73, we concluded that the pattern of the observations was NOT linear, meaning that Assumption 1 for regressions inferences was not met by these data. In Exercise 14.73, we found that the evidence for a linear pattern was weak or none. Thus we will not continue with parts (b)-(f).

**14.103** (a) In Exercise 14.39, part (c), we determined that it was reasonable to consider Assumptions 1-3 for regression inferences met by these variables.

(b) Using Minitab, with the data in columns named HIGH and LOW, we choose

**Stat ▶ Regression ▶ Regression...**, select LOW in the **Response** text box, and select HIGH in the **Predictors** text box. Click the **Options** button, select the **Regular** option button from the **Residuals for Plots** list, enter 55 in the **Prediction intervals for new observations** box and 95 in the **Confidence level** box, click **OK**, and click **OK**. Since much of the output was shown in Exercise 14.39, we show below only that part of the output pertinent to this exercise.

Predicted Values for New Observations

```
New
Obs    Fit    SE Fit    95% CI              95% PI
  1   42.855  0.590   (41.669, 44.040)   (34.425, 51.285)
```

The point estimate for the conditional mean January low temperature for an average high temperature of 55 degrees is found under the word 'Fit' in the output as 42.855 degrees. This is the same as the predicted value for an average high temperature of 55 degrees.

(c) The 95% confidence interval for the conditional mean low January temperature for cities with mean high temperature of 55 degrees is (41.669, 44.040). We can be 95% confident that the mean low January temperature for cities with a mean January high temperature of 55 degrees lies somewhere between 41.669 and 44.040 degrees.

(d) The predicted mean low January temperature for a city with a mean high January temperature of 55 degrees is found under the word 'Fit' in the output as 42.855 degrees.

(e) The 95% prediction interval for the mean January low temperature for a city with a mean January high temperature is (34.425, 51.285). We can be 95% certain that the mean low January temperature for a city with a mean January high temperature of 55 degrees will lie somewhere between 34.425 and 51.285 degrees.

(f) The confidence interval and the prediction interval are both centered at the predicted value for a city with a mean January high temperature of 55 degrees. The prediction interval is longer because there is

**440** Chapter 14

variation associated with the additional observation that is not present in the confidence interval for the mean.

**14.105**(a) In Exercise 14.41, part (c), we determined that it was not reasonable to consider Assumption 1 for regression inferences met by these variables. The residual plot was convex upward. Therefore, parts (b)-(f) are omitted.

**14.107**(a) In Exercise 14.43, part (c), we determined that it was not reasonable to consider Assumptions 1 and 2 for regression inferences met by these variables. Therefore, parts (b)-(f) are omitted.

**14.109**(a)-(b) In Example 14.9 of the text, the 95% confidence interval for the mean price of all Orions of a given age is given by

$$134.69 \pm 2.262 \cdot 12.58 \cdot \sqrt{\frac{1}{11} + \frac{(x_p - 58/11)^2}{326 - 58^2/11}}$$

Substituting the values 2 through 7 into this formula for $x_p$, we obtain the following confidence intervals and a plot showing the lower and upper confidence limits for the six different values of $x_p$. The margin of error (half of the difference between the upper and lower confidence limits) is shown in the last column of the table and is plotted against $x_p$ in the second graph.

| $x_p$ | Lower Confidence Limit | Upper Confidence Limit | Margin of Error |
|---|---|---|---|
| 2 | 112.25 | 157.13 | 22.44 |
| 3 | 117.93 | 151.45 | 16.76 |
| 4 | 122.92 | 146.46 | 11.77 |
| 5 | 125.94 | 143.44 | 8.75 |
| 6 | 124.95 | 144.43 | 9.74 |
| 7 | 120.79 | 148.59 | 13.90 |

We see from the table and the first plot that the confidence intervals get wider as the predictor variable gets farther from the mean of the predictor variable. This is also seen in the second plot, which shows that the margin of error first decreases as $x_p$ increases to the mean

5.27 of the predictor variable and then increases again as $x_p$ moves away from 5.27.

(c) In Example 14.10 of the text, the 95% prediction interval for an Orion of a given age is given by

$$= 134.69 \pm 2.262 \cdot 12.58 \cdot \sqrt{1 + \frac{1}{11} + \frac{(x_p - 58/11)^2}{326 - 58^2/11}}$$

Substituting the values 2 through 7 into this formula for $x_p$, we obtain the following confidence intervals and a plot showing the lower and upper confidence limits for the six different values of $x_p$. The margin of error (half of the difference between the upper and lower confidence limits) is shown in the last column of the table and is plotted against $x_p$ in the second graph.

| $x_p$ | Lower Prediction Limit | Upper Prediction Limit | Margin of Error |
|---|---|---|---|
| 2 | 98.45 | 170.93 | 36.24 |
| 3 | 101.67 | 167.71 | 33.02 |
| 4 | 103.89 | 165.49 | 30.80 |
| 5 | 104.92 | 164.46 | 29.77 |
| 6 | 104.61 | 164.77 | 30.08 |
| 7 | 103.02 | 166.36 | 31.67 |

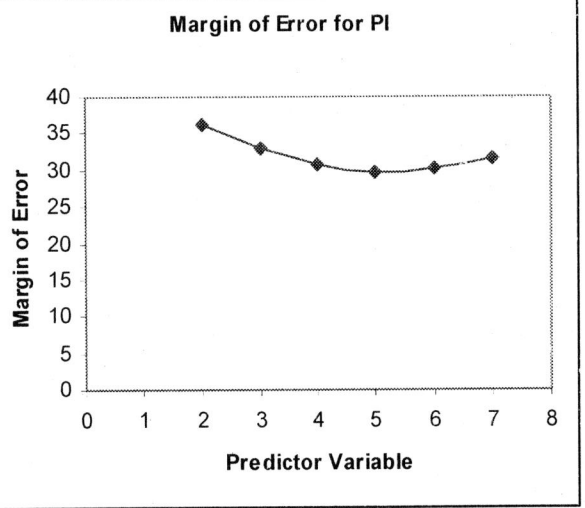

We see from the table and the first plot that the prediction intervals get wider as the predictor variable gets farther from the mean of the predictor variable. This is also seen in the second plot which shows that the margin of error first decreases as $x_p$ increases to the mean 5.27 of the predictor variable and then increases again as $x_p$ moves away from 5.27.

### Exercises 14.4

**14.111** r, the sample linear correlation coefficient

**14.113** No. The variables could be negatively correlated, but not necessarily in a linear pattern.

**14.115** uncorrelated

**14.117** negatively

**14.119** Step 1 $H_0: \rho = 0$, $H_a: \rho < 0$
Step 2 $\alpha = 0.10$
Step 3 From Exercise 4.135, $r = -0.76$, $n = 3$
$$t = \frac{r}{\sqrt{\frac{1-r^2}{n-2}}} = \frac{-0.76}{\sqrt{\frac{1-(-0.76)^2}{1}}} = -1.169$$
Step 4 The critical value is $-3.0784$
For the P-value approach, $P > 0.10$.
Step 5 Since $-1.169 > -3.078$, do not reject $H_0$.
Since the P-value is greater than $0.10$, do not reject $H_0$.

**14.121** Step 1 $H_0: \rho = 0$, $H_a: \rho > 0$
Step 2 $\alpha = 0.10$
Step 3 From Exercise 4.137, $r = 0.877$, $n = 4$
$$t = \frac{r}{\sqrt{\frac{1-r^2}{n-2}}} = \frac{0.877}{\sqrt{\frac{1-0.877^2}{2}}} = 2.581$$
Step 4 The critical value is $1.886$
For the P-value approach, $0.05 < P < 0.10$
Step 5 Since $2.581 > 1.886$, reject $H_0$.
Since the P-value is less than $0.10$, reject $H_0$.

**14.123** Step 1 $H_0: \rho = 0$, $H_a: \rho \neq 0$
Step 2 $\alpha = 0.10$
Step 3 From Exercise 4.139, $r = -0.633$, $n = 5$
$$t = \frac{r}{\sqrt{\frac{1-r^2}{n-2}}} = \frac{-0.633}{\sqrt{\frac{1-(-0.633)^2}{3}}} = -1.41$$
Step 4 The critical values are $\pm 2.353$
For the P-value approach, $P > 0.20$.
Step 5 Since $-1.41 > -2.353$, do not reject $H_0$.
Since the P-value is greater than $0.10$, do not reject $H_0$.

**14.125** Step 1 $H_0: \rho = 0$, $H_a: \rho > 0$
Step 2 $\alpha = 0.10$
Step 3 From Exercise 4.141, $r = 0.756$, $n = 3$
$$t = \frac{r}{\sqrt{\frac{1-r^2}{n-2}}} = \frac{0.756}{\sqrt{\frac{1-(0.756)^2}{1}}} = 1.15$$
Step 4 The critical value is $3.078$
For the P-value approach, $P > 0.10$
Step 5 Since $1.15 < 3.078$, do not reject $H_0$.
Since the P-value is greater than $0.10$, do not reject $H_0$.

**14.127** Step 1 $H_0: \rho = 0$, $H_a: \rho \neq 0$
Step 2 $\alpha = 0.10$
Step 3 From Exercise 4.143, $r = -0.685$, $n = 5$
$$t = \frac{r}{\sqrt{\frac{1-r^2}{n-2}}} = \frac{-0.685}{\sqrt{\frac{1-(-0.685)^2}{3}}} = -1.629$$
Step 4 The critical values are $\pm 2.353$
For the P-value approach, $P > 0.20$.

Step 5  Since −2.353 < −1.629 < 2.353, do not reject $H_0$.
Since the P-value is greater than 0.10, do not reject $H_0$.

**14.129** From Exercise 4.145, r = −0.96787 and n = 10.
Step 1: $H_0$: $\rho = 0$   $H_a$: $\rho < 0$
Step 2: $\alpha = 0.05$
Step 3: $t = \dfrac{r}{\sqrt{\dfrac{1-r^2}{n-2}}} = \dfrac{-0.96787}{\sqrt{\dfrac{1-(-0.96787)^2}{10-2}}} = -10.8870$

Step 4: The critical value for n − 2 = 8 df is −1.860.
For the p-value approach, p < 0.005.
Step 5: Since −10.8870 < −1.860, reject $H_0$ and conclude that $\rho < 0$.
Since p < 0.05, reject $H_0$ and conclude that $\rho < 0$.

**14.131** From Exercise 4.147, r = 0.33107 and n = 11.
Step 1: $H_0$: $\rho = 0$   $H_a$: $\rho \neq 0$
Step 2: $\alpha = 0.05$
Step 3: $t = \dfrac{r}{\sqrt{\dfrac{1-r^2}{n-2}}} = \dfrac{0.33107}{\sqrt{\dfrac{1-(0.33107)^2}{11-2}}} = 1.0526$

Step 4: The critical values for n − 2 = 9 df are ±2.262.
For the p-value approach, p > 0.200.
Step 5: Since −2.262 < 1.0526 < 2.262, do not reject $H_0$ and conclude that $\rho = 0$ is reasonable.
Since p > 0.05, do not reject $H_0$ and conclude that $\rho = 0$ is reasonable.

**14.133** From Exercise 4.149, r = −0.7749 and n = 8.
Step 1: $H_0$: $\rho = 0$   $H_a$: $\rho < 0$
Step 2: $\alpha = 0.01$
Step 3: $t = \dfrac{r}{\sqrt{\dfrac{1-r^2}{n-2}}} = \dfrac{-0.7749}{\sqrt{\dfrac{1-(-0.7749)^2}{8-2}}} = -3.003$

Step 4: The critical value for n − 2 = 6 df is −3.143.
For the p-value approach, 0.010 < p < 0.025.
Step 5: Since −3.003 > 3.143, do not reject $H_0$. There is not sufficient evidence to conclude that $\rho \neq 0$.
Since p > 0.01, do not reject $H_0$.

**14.135** In Exercise 14.35, part (c), we determined that it was reasonable to consider Assumptions 1-3 for regression inferences met by these variables.

Using Minitab, choose **Stat ▶ Basic Statistics ▶ Correlation...**, select INAUGURATION and DEATH in the **Variables** text box, and click **OK**. The result is
    Pearson correlation of INAUGURATION and DEATH = 0.606
    P-Value = 0.000
Note: The p-value shown is for a two-tailed test. The p-value for the right-tail test would be even smaller. Since the p-value < 0.05, reject $H_0$ and conclude that the inauguration ages and death ages of U.S. Presidents are positively linearly correlated.

**14.137** In Exercise 14.73, part (a), we determined that it was not reasonable to consider Assumptions 1 for regression inferences met by these

variables. The correlation t-test should not be used. Over several exercises, there was some controversy regarding this decision. If you choose to continue with the correlation t-test, the results from Minitab are
   Pearson correlation of LOT SIZE and VALUE = 0.170
   P-Value = 0.269

The high P-value confirms our decision not to reject ρ = 0.

**14.139** In Exercise 14.35, part (c), we determined that it was reasonable to consider Assumptions 1-3 for regression inferences met by these variables.

Using Minitab, choose **Stat ▶ Basic Statistics ▶ Correlation...**, select HIGH and LOW in the **Variables** text box, and click **OK**. The result is
   Pearson correlation of HIGH and LOW = 0.976
   P-Value = 0.000

Since the P-value < 0.05, reject ρ = o.

**14.141** In Exercise 14.37, part (c), we determined that it was not reasonable to consider Assumption 1 for regression inferences met by these variables. The correlation t-test should not be used.

**14.143** In Exercise 14.39, part (c), we determined that it was not reasonable to consider Assumptions 1 and 2 for regression inferences met by these variables. The correlation t-test should not be used.

### Review Problems for Chapter 14

1. (a) conditional
   (b) The four assumptions for regression inferences are:
      (1) *Population regression line:* There is a line $y = \beta_0 + \beta_1 x$ such that, for each x-value, the mean of the corresponding population of y-values lies on that line.
      (2) *Equal standard deviations:* The standard deviation σ of the population of y-values corresponding to a particular x-value is the same, regardless of the x-value.
      (3) *Normality:* For each x-value, the corresponding population of y-values is normally distributed.
      (4) *Independence:* The observations of the response variable are independent of one another.

2. (a) The slope of the sample regression line, $b_1$
   (b) The y-intercept of the sample regression line, $b_0$
   (c) $s_e$

3. We used a plot of the residuals against the values of the predictor variable x and a normal probability plot of the residuals. In the first plot, the residuals should lie in a horizontal band centered on and symmetric about the x-axis. In the second plot, the points should lie roughly in a line.

4. (a) A residual plot showing curvature indicates that the first assumption, that of linearity, is probably not valid.
   (b) This type of plot indicates that the second assumption, that of constant standard deviation, is probably not valid.
   (c) A normal probability plot with extreme curvature indicates that the third assumption, that the conditional distribution of the response variable is normally distributed, is not valid.
   (d) A normal probability plot that is roughly linear, but shows outliers may be indicating that: the linear model is appropriate for most values of x, but not for all; or, that the standard deviation is not constant for all values of x, allowing for a few y values to lie far from the regression line; or, that a few data values are 'faulty', that is, the

5. If we reject the null hypothesis, we are claiming that $\beta_1$ is not zero. This means that different values of x will lead to different values of y. Hence, the regression equation is useful for making predictions.
6. t (for $b_1$), r, $r^2$
7. No. The best estimate of conditional mean of the response variable and the best prediction of a single future value of y are the same.
8. A confidence interval estimates the value of a parameter; a prediction interval is used to predict a future value of a random variable.
9. $\rho$
10. (a) If $\rho > 0$, the variables are positively correlated, that is, there is a tendency for one variable to increase linearly as the other one increases.
    (b) If we know only that $\rho \neq 0$, the two variables are linearly correlated, that is, there is a linear relationship between them.
    (c) If $\rho < 0$, the variables are negatively correlated, that is, there is a tendency for one variable to decrease linearly as the other one increases.
11. If the regression assumptions are satisfied, then there is a linear relationship between the student/faculty ratio and the graduation rate, the conditional standard deviation of the graduation rate is the same for all values of the student/faculty ratio, the conditional distribution of the graduation rate is a normal distribution for all values of the student/faculty ratio, and the values of the graduation rate are independent of each other.
12. (a) To determine $b_0$ and $b_1$ for the regression line, construct a table of values for x, y, $x^2$, xy and their sums. A column for $y^2$ is also presented for later use.

| x | y | $x^2$ | xy | $y^2$ |
|---|---|---|---|---|
| 16 | 45 | 256 | 720 | 2025 |
| 20 | 55 | 400 | 1100 | 3025 |
| 17 | 70 | 289 | 1190 | 4900 |
| 19 | 50 | 361 | 950 | 2500 |
| 22 | 47 | 484 | 1034 | 2209 |
| 17 | 46 | 289 | 782 | 2116 |
| 17 | 50 | 289 | 850 | 2500 |
| 17 | 66 | 289 | 1122 | 4356 |
| 10 | 26 | 100 | 260 | 676 |
| 18 | 60 | 324 | 1080 | 3600 |
| 173 | 515 | 3081 | 9088 | 27907 |

$S_{xx} = 3081 - 173^2/10 = 88.10$
$S_{xy} = 9088 - (173)(515)/10 = 178.50$
$b_1 = 178.50/88.10 = 2.0261$
$b_0 = 515/10 - 2.0261(173/10) = 16.4484$
$\hat{y}_p = 16.448 + 2.0261x$ is the equation of the regression line.

(b) $S_{yy} = 27907 - 515^2/10 = 1384.50$
$SSE = S_{yy} - S^2_{xy}/S_{xx} = 1384.50 - 178.50^2/88.10 = 1022.8400$

$$s_e = \sqrt{\frac{SSE}{n-2}} = \sqrt{\frac{1922.8400}{10-2}} = 11.3073$$

This value of $s_e$ indicates that, roughly speaking, the predicted values of the graduation rate differ from the observed values of the graduation rate by about 11.31.

(c) Presuming that the variables Student/Faculty ratio and Graduation rate satisfy the assumptions for regression inferences, the standard error of the estimate, $s_e = 11.31$, provides an estimate for the common population standard deviation, $\sigma$, of graduation rates of entering freshmen at universities with any particular student/faculty ratio.

13. For each value of x, we compute $e = y - \hat{y}$ in the following table and plot e against x.

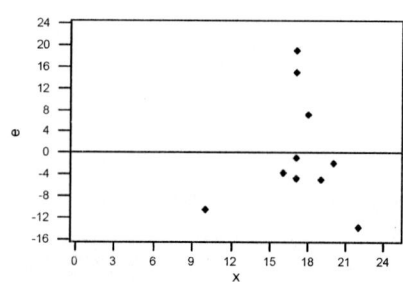

| x | e |
|---|---|
| 16 | -3.87 |
| 20 | -1.97 |
| 17 | 19.11 |
| 19 | -4.94 |
| 22 | -14.02 |
| 17 | -4.89 |
| 17 | -0.89 |
| 17 | 15.11 |
| 10 | -10.71 |
| 18 | 7.08 |

| Residual | Normal score |
|---|---|
| -14.02 | -1.55 |
| -10.71 | -1.00 |
| -4.94 | -0.65 |
| -4.89 | -0.37 |
| -3.87 | -0.12 |
| -1.97 | 0.12 |
| -0.89 | 0.37 |
| 7.08 | 0.65 |
| 15.11 | 1.00 |
| 19.11 | 1.55 |

The small sample size makes it difficult to evaluate the plot of e against x. While there are no obvious patterns in the first plot, there are two points (in the lower left and right corners of the plot) that are cause for concern. The normal probability plot is not far from linear.

14. (a) From Review Problem 11, $\sum x = 173$, $\sum x^2 = 3,081$, and $b_1 = 2.0261$. From Review Problem 12, $s_e = 11.3073$.

Step 1: $H_0: \beta_1 = 0$
$H_a: \beta_1 \neq 0$
Step 2: $\alpha = 0.05$
Step 3: $t = \dfrac{b_1}{s_e/\sqrt{\sum x^2 - (\sum x)^2/n}} = \dfrac{2.0261}{11.3073/\sqrt{3081 - 173^2/10}} = 1.68$

Step 4: df = n - 2 = 8; critical values = ±2.306
For the P-value approach, note that
0.10 < p < 0.20.

Step 5: Since -2.306 < 1.68 < 2.306, do not reject the null hypothesis at the 5% significance level.
Since p > $\alpha$, do not reject the null hypothesis.

Step 6: The data do not provide sufficient evidence to conclude that the student-to-faculty ratio is useful as a predictor of graduation rate.

(b) Step 1: For a 95% confidence interval, $\alpha$ = 0.05. With df = n - 2 = 8, $t_{\alpha/2}$ = $t_{0.025}$ = 2.306.

Step 2: The endpoints of the confidence interval for $\beta_1$ are

$$b_1 \pm t_{\alpha/2} \cdot s_e / \sqrt{\sum x^2 - (\sum x)^2 / n}$$

$2.0261 \pm 2.306 \cdot 11.3073 / \sqrt{3081 - 173^2/10}$

$2.0261 \pm 2.7780$

$-0.7519$ to $4.8041$

We can be 95% confident that the change in graduation rate for universities per 1 unit increase in the student/faculty ratio is somewhere between -0.7519 and 4.8041 percent.

15. From Review Problem 12, $\hat{y}$ = 16.4484 + 2.0261x, $\sum x$ = 173, and $\sum x^2$ = 3,081.
From Review Problem 12, $s_e$ = 11.3073.

(a) $\hat{y}_p$ = 16.4484 + 2.0261(17) = 50.89

(b) Step 1: For a 95% confidence interval, $\alpha$ = 0.05.
With df = n - 2 = 8, $t_{\alpha/2}$ = $t_{0.025}$ = 2.306.

Step 2: $\hat{y}_p$ = 50.89
Step 3: The endpoints of the confidence interval are:

$$\hat{y}_p \pm t_{\alpha/2} \cdot s_e \cdot \sqrt{\frac{1}{n} + \frac{(x_p - \sum x/n)^2}{\sum x^2 - (\sum x)^2/n}}$$

$= 50.89 \pm 2.306 \cdot 11.3073 \cdot \sqrt{\frac{1}{10} + \frac{(17 - 173/10)^2}{3081 - 173^2/10}}$

$= 50.89 \pm 8.29 = (42.60, 59.18)$

The interpretation of this interval is as follows: We can be 95% confident that the graduation rate of universities with a student/faculty ratio of 17 is somewhere between 42.60% and 59.18%.

16. (a) This is the same as the answer in Problem 15 part (a): $\hat{y}_p$ = 50.89.
(b) For a 95% prediction interval, Steps 1 and 2 are the same as Steps 1 and 2, respectively, in Problem 15 part (b). Thus, they are not repeated here. Only Step 3 is presented.
Step 3: The endpoints of the prediction interval are:

**448** Chapter 14

$$\hat{y}_p \pm t_{\alpha/2} \cdot s_e \cdot \sqrt{1 + \frac{1}{n} + \frac{(x_p - \sum x/n)^2}{\sum x^2 - (\sum x)^2 n}}$$

$$= 50.89 \pm 2.306 \cdot 11.3073 \cdot \sqrt{1 + \frac{1}{10} + \frac{(17 - 173/10)^2}{3081 - 173^2/10}}$$

$$= 50.89 \pm 27.36 = (23.53, 78.25)$$

The interpretation of this interval is as follows: We can be 95% certain that the graduation rate of a randomly selected university with a student/faculty ratio of 17 will be somewhere between 23.53% and 78.25%.

(c) The error in the estimate of the mean graduation rate for universities with a student/faculty ratio of 17 is due only to the fact that the population regression line is being estimated by a sample regression line. The error in the prediction of the graduation rate of a randomly chosen university with a student/faculty ratio is due to the estimation error mentioned above plus the variation in graduation rates of universities with a student/faculty ratio of 17.

17. From Review Problem 12,
$S_{xx}$ = 88.10
$S_{xy}$ = 178.50
$S_{yy}$ = 1384.50

$$r = \frac{S_{xy}}{\sqrt{S_{xx}S_{yy}}} = \frac{178.50}{\sqrt{(88.10)(1384.50)}} = 0.5111$$

Step 1: $H_0$: $\rho$ = 0, $H_a$: $\rho$ > 0
Step 2: $\alpha$ = 0.025
Step 3:

$$t = \frac{r}{\sqrt{\frac{1-r^2}{n-2}}} = \frac{0.5111}{\sqrt{\frac{1-(0.5111)^2}{10-2}}} = 1.682$$

Step 4: df = n - 2 = 8; critical value = 2.306
For the P-value approach, 0.05 < p < 0.10.
Step 5: Since 1.682 < 2.306, do not reject $H_0$.
Because the p-value is larger than the significance level of 0.025, we cannot reject $H_0$.
Step 6: The data do not provide sufficient evidence to conclude that graduation rate and student/faculty ratio are positively linearly correlated.

18. Using Minitab, with the data in columns named IMR and LE, we choose **Stat ▶ Regression ▶ Regression...**, select 'LE' in the **Response:** text box, select IMR in the **Predictors:** text box, click on the **Graphs** button, click in the **Residuals versus the variables** text box and select IMR, and click **OK**. Looking ahead to parts (b-h) and (j), click on the **Options...** button, enter 30 in the **Prediction intervals for new observations:** text box, enter 95 in the **Confidence level:** text box, and click **OK**. Click on the **Storage...** button, check the **Residuals** box, click **OK**, and click **OK**. Then choose **Stat ▶ Basic statistics ▶ Normality test...**, select RESI1 in the **Variable** text box, select the **Ryan-Joiner** option button from the **Tests for Normality**

field, and click **OK**. We will first show all of the written output and then answer the questions. The graphical output will be shown in part (k).

```
The regression equation is
LE = 79.4 - 0.354 IMR

Predictor      Coef    SE Coef       T       P
Constant     79.396      1.009   78.72   0.000
IMR         -0.35374    0.02601  -13.60   0.000

S = 5.17598    R-Sq = 76.1%    R-Sq(adj) = 75.7%

Analysis of Variance

Source           DF      SS       MS       F       P
Regression        1   4956.2   4956.2  185.00   0.000
Residual Error   58   1553.9     26.8
Total            59   6510.1

Unusual Observations

Obs   IMR    LE      Fit    SE Fit  Residual  St Resid
 11    46  49.890  63.251   0.795   -13.361    -2.61R
 25   100  47.430  43.894   1.971     3.536     0.74 X
 34    34  44.280  67.418   0.680   -23.138    -4.51R
 48    44  50.160  63.828   0.773   -13.668    -2.67R
 52    91  43.450  47.365   1.734    -3.915    -0.80 X

R denotes an observation with a large standardized residual.
X denotes an observation whose X value gives it large leverage.

Predicted Values for New Observations

New Obs   Fit    SE Fit       95% CI              95% PI
   1    68.784   0.669   (67.445, 70.122)   (58.337, 79.231)

Values of Predictors for New Observations

New Obs   IMR
  1      30.0
```

(a) LE = 79.4 - 0.354 IMR
(b) The standard error of the estimate is 5.17598. Roughly speaking, this indicates how much, on average, the predicted life expectancies differ from the observed values of life expectancies.
(c) The coefficient of determination is 0.761, indicating that 76.1% of the variation in life expectancy is explained by a linear relationship with the infant mortality rate. The t-test for the slope results in t = -13.60 with a P-value of 0.000. Thus, we would reject the null hypothesis that $\beta_1 = 0$. This is a fairly strong relationship and indicates that the IMR may be useful for predicting life expectancy if the assumptions for making such inferences are valid.
(d) The point estimate for the conditional mean of life expectancy for IMR = 30 is 68.784 years. This is the point on the sample regression line that best estimates the corresponding point on the population regression line.
(e) The 95% confidence interval for the conditional mean of life expectancy for IMR = 30 is 67.445 to 70.122 years. We are 95% confident that the mean life expectancy is somewhere between these limits when the IMR is 30.
(f) The predicted value of life expectancy for IMR = 30 is 68.784 years. This is our best estimate of life expectancy for a country with an IMR equal to 30.

(g) The 95% prediction interval for the life expectancy of a country with IMR = 30 is 58.337 to 79.231 years. We are 95% certain that the life expectancy is somewhere between these limits when the IMR is 30.

(h) The error in the estimate of the mean life expectancy of countries with IMR = 30 is due only to the fact that the population regression line is being estimated by a sample regression line; whereas, the error in the prediction of the life expectancy of a single country with an IMR = 30 is due to that fact plus the variation in life expectancies for countries with IMR = 30.

(i) Choose **Stat ▸ Basic Statistics ▸ Correlation,** enter IMR and "LE' in the **Variables** text box, check the box to **Display P-values**, and click **OK**. The result is

Pearson correlation of IMR and LE = −0.873
P-Value = 0.000

The P-value for a left-tail test is half of that shown (still 0.000). Since the P-value is less than 0.05, we reject the null hypothesis that ρ = 0 and since r is negative, the data provide evidence that there is a negative correlation between IMR and life expectancy assuming that the assumptions for making regression inferences are reasonable.

(j) For the residual analysis, we look at the graph produced by the procedure at the beginning of this solution and also the results of a Ryan-Joiner normality test on the residuals, which were stored earlier.

The first graph shows several residuals somewhat farther from the horizontal axis than the rest of the points and there are two residuals at the far right that indicate potential influential observations. This graph does not show a band of residuals centered on and symmetric about the horizontal axis. The normal probability plot shows several points well off the line and the Ryan-Joiner test has a P-value that is less than 0.01. Thus, Assumption 3 for making regression inferences appears to be violated. In addition, the concern about several outliers and potential influential observations indicates that the tests, confidence interval and prediction interval produced earlier may not be valid.

19. Using Minitab, with the data in columns named HIGH and PRECIPITATION, we choose **Stat ▸ Regression ▸ Regression...,** select PRECIPITATION in the **Response:** text box, select HIGH in the **Predictors:** text box, click on the **Graphs** button, click in the **Residuals versus the variables** text box and select HIGH, and click **OK**. Looking ahead to parts (b–h) and (j), click on the **Options...** button, enter 83 in the **Prediction intervals for new observations:** text box, enter 95 in the **Confidence level:** text box, and click **OK**. Click on the **Storage...** button, check the **Residuals** box, click

OK, and click **OK**. Then choose **Stat** ▸ **Basic statistics** ▸ **Normality test...**, select RESI1 in the **Variable** text box, select the **Ryan-Joiner** option button from the **Tests for Normality** field, and click **OK**. We will first show all of the written output and then answer the questions. The graphical output will be shown in part (k).

```
The regression equation is
PRECIPITATION = 2.05 + 0.0067 HIGH


Predictor      Coef    SE Coef      T      P
Constant      2.048      2.170   0.94  0.350
HIGH        0.00667    0.02742   0.24  0.809


S = 2.41336    R-Sq = 0.1%    R-Sq(adj) = 0.0%

Analysis of Variance

Source          DF       SS      MS      F      P
Regression       1    0.344   0.344   0.06  0.809
Residual Error  46  267.919   5.824
Total           47  268.263

Unusual Observations

Obs  HIGH  PRECIPITATION    Fit   SE Fit  Residual  St Resid
  6    86          8.100  2.621    0.410     5.479     2.30R
 10    87          9.600  2.628    0.425     6.972     2.93R
 22    45          0.500  2.348    0.972    -1.848    -0.84 X
 45    93          7.900  2.668    0.537     5.232     2.22R

R denotes an observation with a large standardized residual.
X denotes an observation whose X value gives it large influence.

Predicted Values for New Observations

New
Obs    Fit   SE Fit      95% CI           95% PI
  1  2.601    0.373  (1.850, 3.353)  (-2.314, 7.517)
```

(a) PRECIPITATION = 2.05 + 0.0067 HIGH

(b) The standard error of the estimate is 2.41336. Roughly speaking, this indicates how much, on average, the predicted average July precipitations differ from the observed values of precipitations.

(c) The coefficient of determination is 0.001, indicating that 0.1% of the variation in average July precipitation is explained by a linear relationship with the average July high temperature. The t-test for the slope results in t = 0.24 with a P-value of 0.809. Thus, we would not reject the null hypothesis that $\beta_1 = 0$. This is a very weak relationship and indicates that the average July high temperature is useless for predicting average July precipitation if the assumptions for making such inferences are valid.

(d) The point estimate for the conditional mean of average July precipitation for an average July high temperature of 83 degrees is 0.373 inches. This is the point on the sample regression line that best estimates the corresponding point on the population regression line.

(e) The 95% confidence interval for the conditional mean of average July precipitation for an average July high temperature of i3 degrees is 1.850 to 3.353 inches. We are 95% confident that the mean average July precipitation is somewhere between these limits when the average July high temperature is 83 degrees.

**452** Chapter 14

(f) The predicted value of average July precipitation for a city with an average July high temperature of 83 degrees is 0.373 inches. This is our best estimate of average July precipitation for a city with an average July high temperature of 83 degrees.

(g) The 95% prediction interval for the average July precipitation of a city with an average high July temperature of 83 degrees is −2.314 to 7.517 inches. The lower prediction limit can be changed to 0.00 since negative precipitation amounts are not possible. We are 95% certain that the average July precipitation is somewhere between these limits for a city with an average July high temperature of 83 degrees.

(h) The error in the estimate of the mean average July precipitation of cities with average July high temperature of 83 degrees is due only to the fact that the population regression line is being estimated by a sample regression line; whereas, the error in the prediction of the average July precipitation of a single city with an average July high temperature of 83 degrees is due to that fact plus the variation in average July precipitation for cities with average high July temperatures of 83 degrees.

(i) Choose **Stat ▶ Basic Statistics ▶ Correlation,** enter HIGH and PRECIPITATION in the **Variables** text box, check the box to **Display P-values**, and click **OK**. The result is
   Pearson correlation of HIGH and PRECIPITATION = 0.036
   P-Value = 0.809
Since the P-value is greater than 0.05, we do not reject the null hypothesis that $\rho = 0$.

(j) For the residual analysis, we look at the graph produced by the procedure at the beginning of this solution and also the results of a Ryan-Joiner normality test on the residuals, which were stored earlier.

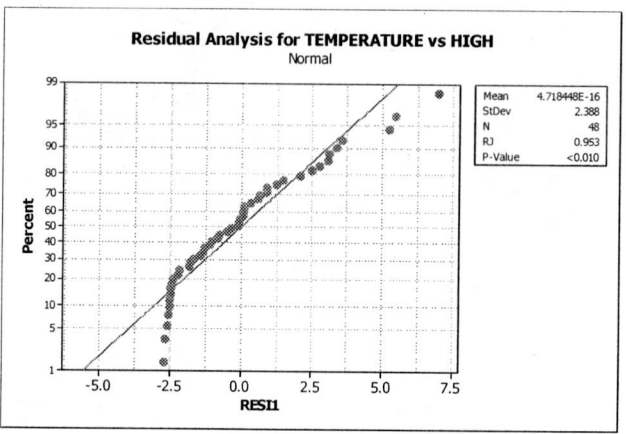

The first graph does not show a band of residuals centered on and symmetric about the horizontal axis. They may even be somewhat U-shaped. The normal probability plot shows several points well off the line and the Ryan-Joiner test has a P-value that is less than 0.01. Thus, Assumption 3 for making regression inferences appears to be violated. In addition, since a linear model accounts for only 0.1% of the variation in average July precipitation, it appears that Assumption 1 (of a linear model) is also violated. The regression inferences above are very unlikely to be valid and are pretty useless.

20. Using Minitab, with the data in columns named FIRST and SECOND, we choose **Stat ▶ Regression ▶ Regression...**, select SECOND in the **Response:** text box, select FIRST in the **Predictors:** text box, click on the **Graphs** button, click

in the **Residuals versus the variables** text box and select FIRST, and click **OK**. Looking ahead to parts (b-h) and (j), click on the **Options...** button, enter 72 in the **Prediction intervals for new observations:** text box, enter 95 in the **Confidence level:** text box, and click **OK**. Click on the **Storage...** button, check the **Residuals** box, click **OK**, and click **OK**. Then choose **Stat**

▶ **Basic statistics** ▶ **Normality test...**, select RESI1 in the **Variable** text box, select the **Ryan-Joiner** option button from the **Tests for Normality** field, and click **OK**. We will first show all of the written output and then answer the questions. The graphical output will be shown in part (k).

Analysis of Variance

| Source | DF | Adj SS | Adj MS | F-Value | P-Value |
|---|---|---|---|---|---|
| Regression | 1 | 95.13 | 95.132 | 8.29 | 0.005 |
|   FIRST | 1 | 95.13 | 95.132 | 8.29 | 0.005 |
| Error | 150 | 1721.07 | 11.474 | | |
|   Lack-of-Fit | 15 | 113.54 | 7.569 | 0.64 | 0.841 |
|   Pure Error | 135 | 1607.53 | 11.908 | | |
| Total | 151 | 1816.20 | | | |

Model Summary

| S | R-sq | R-sq(adj) | R-sq(pred) |
|---|---|---|---|
| 3.38730 | 5.24% | 4.61% | 2.66% |

Coefficients

| Term | Coef | SE Coef | T-Value | P-Value | VIF |
|---|---|---|---|---|---|
| Constant | 56.09 | 6.60 | 8.50 | 0.000 | |
| FIRST | 0.2557 | 0.0888 | 2.88 | 0.005 | 1.00 |

Regression Equation

SECOND = 56.09 + 0.2557 FIRST

Fits and Diagnostics for Unusual Observations

| Obs | SECOND | Fit | Resid | Std Resid | |
|---|---|---|---|---|---|
| 2 | 72.000 | 73.215 | -1.215 | -0.37 | X |
| 5 | 67.000 | 74.494 | -7.494 | -2.22 | R |
| 21 | 68.000 | 75.005 | -7.005 | -2.07 | R |
| 26 | 69.000 | 76.028 | -7.028 | -2.09 | R |
| 103 | 70.000 | 76.795 | -6.795 | -2.04 | R |
| 121 | 83.000 | 73.982 | 9.018 | 2.69 | R |
| 127 | 83.000 | 74.238 | 8.762 | 2.60 | R |
| 136 | 82.000 | 75.005 | 6.995 | 2.07 | R |
| 142 | 76.000 | 77.050 | -1.050 | -0.32 | X |
| 149 | 83.000 | 76.028 | 6.972 | 2.08 | R |
| 150 | 84.000 | 76.028 | 7.972 | 2.37 | R |
| 152 | 81.000 | 77.306 | 3.694 | 1.12 | X |

R  Large residual
X  Unusual X

| Variable | Setting |
|---|---|
| FIRST | 72 |

| Fit | SE Fit | 95% CI | 95% PI |
|---|---|---|---|
| 74.4938 | 0.340386 | (73.8212, 75.1663) | (67.7671, 81.2205) |

(a) The regression equation is SECOND = 56.09 + 0.2557 FIRST
(b) The standard error of the estimate is 3.38730. Roughly speaking, this

indicates how much, on average, the predicted values of the second round differ from the observed values of the second round.

(c) The coefficient of determination is 0.0524, indicating that 5.24% of the variation in second round scores is explained by a linear relationship with the first round scores. The t-test for the slope results in t = 2.88 with a P-value of 0.005. Thus, we would reject the null hypothesis that $\beta_1$ = 0. Even though the P-value is small, the low coefficient of determination indicates that this is a fairly weak relationship and that first round score is not very useful in making predictions of the second round score if the assumptions for making such inferences are valid.

(d) The point estimate for the conditional mean second round score for a player with a first round score of 72 is 74.4938. This is the point on the sample regression line that best estimates the corresponding point on the population regression line.

(e) The 95% confidence interval for the conditional mean second round score for players with a first round score of 72 is 73.8212 to 75.1663. We are 95% confident that the mean second round score is somewhere between these limits for players with a first round score of 72.

(f) The predicted second round score for a player with a first round score of 72 is 74.4938. This is our best estimate of the second round score for a player with a first round score of 72.

(g) The 95% prediction interval for the second round score for a player with a first round score of 72 is 67.7671 to 81.2205. We are 95% certain that the second round score is somewhere between these limits for a player with a first round score of 72.

(h) The error in the estimate of the mean second round score for players with a first round score of 72 is due only to the fact that the population regression line is being estimated by a sample regression line; whereas, the error in the prediction of the second round score of a player with a first round score of 72 is due to that fact plus the variation in second round scores for players with a first round score of 72.

(i) Choose **Stat ▶ Basic Statistics ▶ Correlation,** enter FIRST and SECOND in the **Variables** text box, check the box to **Display P-values**, and click **OK**. The result is

   Pearson correlation of FIRST ROUND and SECOND ROUND = 0.229
   P-Value = 0.005

The P-value for a right-tailed test is half of that shown (0.005/2 = 0.0025) if r is positive. Since the P-value is less than 0.05, we reject the null hypothesis that $\rho$ = 0.

(j) For the residual analysis, we look at the graph produced by the procedure at the beginning of this solution and also the results of a Ryan-Joiner normality test on the residuals, which were stored earlier.

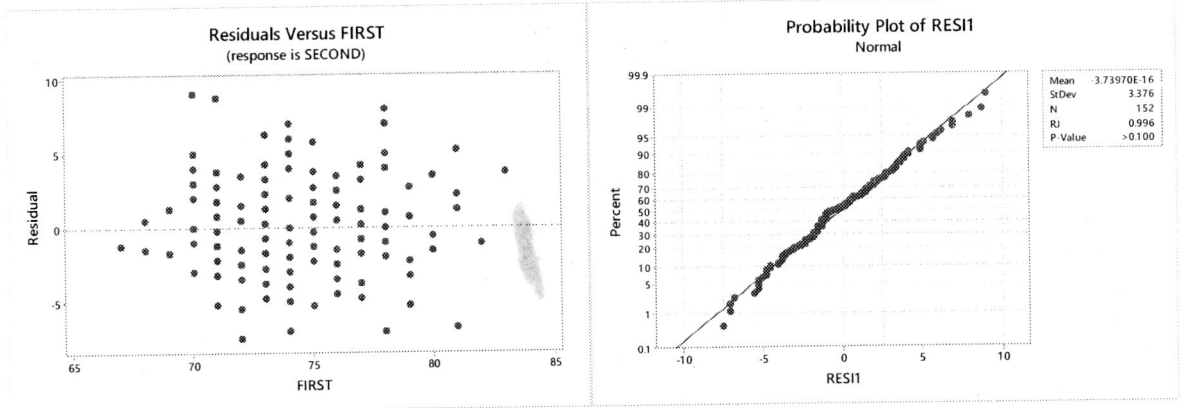

In the left-hand graph, the residuals are located in a horizontal band centered on and somewhat symmetrical about the horizontal axis. The P-value for the Ryan-Joiner test is > 0.100 indicating that normality of the residuals is a reasonable assumption. Since the linear model also appears to be appropriate, although not very useful, the regression inferences made earlier in this problem are likely to be reasonable.